Ultrasonic Nondestructive Evaluation Systems
Models and Measurements

Ultrasonic Nondestructive Evaluation Systems
Models and Measurements

Lester W. Schmerr Jr.
Iowa State University
Center for Nondestructive Evaluation

Sung-Jin Song
Sungkyunkwan University
School of Mechanical Engineering

 Springer

Lester W. Schmerr Jr.
Iowa State University
Center for Nondestructive Evaluation
Ames, IA 50011-3041
USA

Sung-Jin Song
Sungkyunkwan University
School of Mechanical Engineering
Suwon, Kyonggi-do 440-746
Korea

Library of Congress Control Number: 2007923062

ISBN 0-387-49061-2   e-ISBN 0-387-49063-9
ISBN 978-0-387-49061-8   e-ISBN 978-0-387-49063-2

Printed on acid-free paper.

springer.com

Dedications

To my parents, Ida Clara (Streithorst) Schmerr and Lester William Schmerr Sr., and to my wife Mary Jo (Freiburger) Schmerr

Lester W. Schmerr Jr.

To my mother, Byung-Nam Kim, and to my wife, Hea Young Jung

Sung-Jin Song

Preface

This book deals with ultrasonic nondestructive evaluation (NDE) inspections where high frequency waves are used to locate and characterize dangerous flaws (such as cracks) in materials. Ultrasonic NDE flaw inspections involve a very complex combination of electrical, electromechanical, and acoustic/elastic components so that it is important to understand the behavior of those components and their interactions in order to make quantitative flaw measurements. It will be shown that through the use of models and measurements it is now possible to characterize all the elements of an ultrasonic NDE flaw inspection system. Those elements include the pulser/receiver, the cabling, the transducers, and the wave propagation and scattering processes present in an ultrasonic NDE flaw measurement. It will also be demonstrated how to combine models and measurements of those elements to form ultrasonic measurement models which can simulate the flaw signals seen in ultrasonic NDE tests. This comprehensive modeling and measurement capability is described for the first time in this book.

There are important engineering applications of this new technology. For example, these ultrasonic models and measurements can be used to design new ultrasonic inspections as well as optimize existing ones. This technology can also help one to extract information on the nature of the flaw present from the measured ultrasonic flaw signals that can then be used to evaluate the safety and reliability of the material being inspected.

The topics covered in this book include Fourier analysis, linear system theory, and wave propagation and scattering theory for fluids and solids. A series of Appendices provide some background materials for all these topics. Additional background information in these areas can be found in *Fundamentals of Ultrasonic Nondestructive Evaluation – A Modeling Approach* by L. W. Schmerr Jr. This book will also provide many details of the fundamentals of the ultrasonic measurement process but the primary purpose here is to show how the elements of an ultrasonic measurement system combine to generate a measured signal received from a flaw in a material and to give models and measurements that make it practical to simulate those measured flaw signals. In addition to giving the equations and models that govern the behavior of an ultrasonic system we also develop some simple but powerful MATLAB functions and scripts. Those functions/scripts can be used by the reader to conduct simulated inspections and to quickly learn how to implement this modeling technology. The validity of the models discussed is also demonstrated by comparing them to experiments.

There are two parts of this book that warrant special notice. First, a recently developed pulse-echo method for measuring the sensitivity of an ultrasonic transducer is given in Chapter 6. This method makes the experimental characterization of transducers much easier than previous methods. Since transducer characterization is an important part of the series of measurements needed to characterize completely all the components an ultrasonic measurement system, having this simple method for calculating sensitivity also makes that entire chain of measurements more practical. Second, in Chapter 9 we give a complete description of Gaussian beam theory and its use for simulating the wave fields generated by ultrasonic transducers in the form of a multi-Gaussian beam model. Although there are other methods for calculating these wave fields, multi-Gaussian beam models are generally the most effective ultrasonic beam models available. Gaussian beams have been described in other application areas such as Laser science and Geophysics, but the underlying theory as it relates to NDE problems has not been previously given in a complete and unified manner. Chapter 9, therefore, provides a detailed discussion of Gaussian beams as used for modeling sound beams in fluids and isotropic, homogeneous elastic solids. Because the general treatment in Chapter 9 necessarily leads to a lengthy and detailed description of Gaussian beam theory, Appendix F describes the propagation and transmission/reflection of circularly symmetric Gaussian beams along a single direction, a simple case where the properties of these beams can be more clearly illustrated and explained.

This book is an outgrowth of over thirty years of ultrasonic NDE modeling research by the two authors, their colleagues from around the world, and many students. It is designed to communicate that research in an organized fashion and to serve as the foundation for solving many important ultrasonic NDE problems. However, it is also our vision that this modeling technology is not just for the "modelers". We believe that modeling can affect the NDE community at all levels. Thus, the book was developed as part of a workshop series sponsored by the World Federation of NDE Centers (www.wfndec.org). One purpose of that series is to "teach the teacher", that is to provide materials to those with a responsibility for supervising and educating others in the NDE field so that they in turn could communicate the materials and resulting knowledge to others. This book is written at an advanced undergraduate or graduate education level, but by combining the concepts presented here with the simulation capabilities that the MATLAB functions provide one can use or deliver this material at a number of levels. We hope that the reader will enjoy learning about how ultrasonic NDE systems work as much as we have and will pass that learning on to others. We have placed exercises at the end of some of the Chapters and Appendices (most of them MATLAB-based) to help in that learning process.

We would especially like to thank Prof. Alexander Sedov and Drs. Hak-Joon Kim, Ana Lopez-Sanchez, Ruiju Huang, and Changjiu Dang for both their contributions to the research that has helped make this book possible and for their assistance in its preparation.

L.W. Schmerr
S.J. Song

Contents

1 Introduction

1.1 Prologue

In the following Chapters we will describe in detail models that can be used to characterize all the elements of an ultrasonic nondestructive evaluation (NDE) flaw measurement system. We will also discuss the measurements needed to obtain the system parameters that appear in the models. These models can be used to optimize existing inspections, design new inspections, and analyze inspection results. This technology can also be a major cost-saving tool for industry if the models are used to replace expensive tests and sample fabrications. For this to occur, it must be clear that the models are accurate and reliable. We hope to provide sufficient information on current ultrasonic NDE modeling efforts so that the reader can better judge for himself/herself the maturity of this field.

Many aspects of modeling ultrasonic NDE systems require a background in linear system theory and wave propagation and scattering theory. We will provide some of that background in the Appendices and later Chapters but in many cases we will state results without proof and point the reader to other sources. One source in particular that will be referred to frequently is the book *Fundamentals of Ultrasonic Nondestructive Evaluation – A Modeling Approach* by L.W. Schmerr Jr. which is listed as a reference at the end of this Chapter. In subsequent discussions that source will be referred to as the reference [Fundamentals].

In this Chapter we will provide an overview of the models and methods that will be discussed in later Chapters, using the flaw measurement setup of Fig. 1.1 as an example. We will highlight the major results that allow us to model all the components of Fig.1.1 and ultimately obtain an explicit model of the entire measurement system. Although most of our discussions will refer to the immersion system of Fig. 1.1, the models are also applicable to other NDE setups that involve angle beam and contact transducers. Some angle beam inspection applications of the models, for example, are described in Chapter 13.

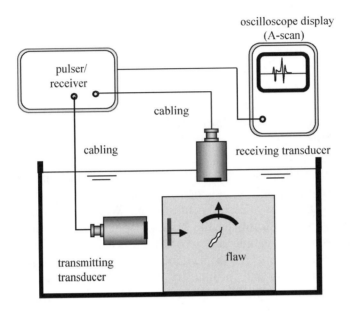

Fig. 1.1. The components of an ultrasonic flaw measurement system.

Throughout this book we will only model inspection systems that use bulk waves. Appendix E gives a brief introduction to the properties of other types of waves such as surface (Rayleigh) waves and guided waves but models of inspections with those wave types require transducer models and wave propagation and scattering models that are not treated here.

1.2 Ultrasonic System Modeling – An Overview

An ultrasonic measurement system involves the generation, propagation, and reception of short transient signals. In the electrical elements of the system shown in Fig. 1.1 such as the pulser/receiver and cabling, these signals are electrical pulses. In the acoustic/elastic parts of the system, the signals are short time duration acoustical pulses traveling in either fluids or solids. The ultrasonic transducers are "mixed" devices that transform electrical pulses into acoustic pulses, and vice-versa. In modeling ultrasonic systems it is convenient not to deal with these transient signals directly but to work instead with their spectral (frequency domain) components.

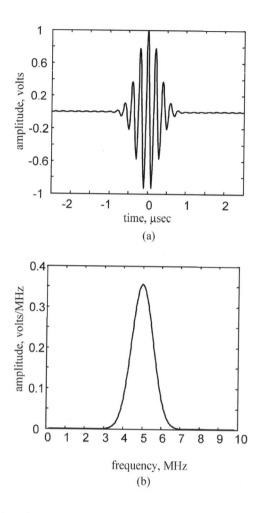

Fig. 1.2. (a) A voltage versus time trace and **(b)** the magnitude of its frequency domain spectrum (for positive frequencies).

Thus, Fourier analysis becomes an essential part of any discussion of ultrasonic system modeling. Figure 1.2, for example, shows a simulated transient voltage versus time signal that might be measured in an ultrasonic NDE system and its corresponding spectral amplitude. It can be seen that the pulse in Fig. 1.2 is very short (typically on a microsecond scale) and the corresponding frequencies in the pulse spectrum are in the 10^6 Hz (MHz) range. These values are similar to the pulses and spectra one often finds in NDE tests. Appendix A gives a brief introduction to Fourier

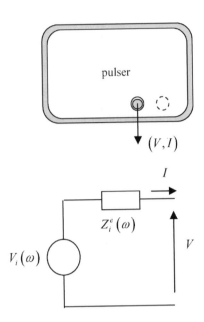

Fig. 1.3. An ultrasonic pulser and an equivalent circuit model as a voltage source and electrical impedance.

transforms, Fast Fourier transforms and related concepts that form some of the fundamental foundations for transforming time signals into frequency domain signals and vice-versa.

      Chapter 2 discusses the modeling of the pulser section of a pulser/receiver and the basic characteristics of the signals generated by the pulser. The pulser is an active electrical network, i.e. it contains a driving energy source as well as complex circuits that shape the output electrical pulse. If the pulser acts as a linear device, then it can be replaced by a very simple equivalent model (in the frequency domain) consisting of a voltage source, $V_i(\omega)$, and impedance, $Z_i^e(\omega)$, both of which are complex functions of the circular frequency, ω, as shown in Fig. 1.3. This representation is possible because of a fundamental theorem of electrical circuits called Thévenin's Theorem. Appendix B gives a brief proof of Thévenin's theorem and discusses the concept of impedance. It is demonstrated in Chapter 2 that one can experimentally determine the voltage source and impedance terms shown in Fig. 1.3 by performing a set of electrical voltage measurements on the pulser under different loading conditions.

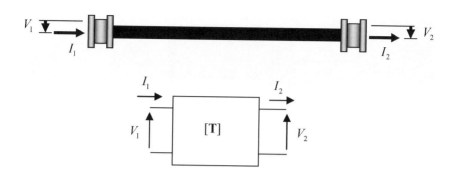

Fig. 1.4 A cable and its model as a two port system characterized by a 2x2 transfer matrix, [**T**].

Note that these measurements are done for a particular set of pulser settings (such as energy and damping). When the pulser settings are changed, the equivalent source and impedance also change.

The cabling in a measurement system is discussed in Chapter 3. The cable is modeled as a two port system of the type shown in Fig. 1.4, where an input voltage and current at one end of the cable is transformed into an output voltage and current at the other end. The relationship between these inputs and outputs can be expressed in terms of a 2x2 transfer matrix, $[\mathbf{T}]$, where:

$$\begin{Bmatrix} V_1 \\ I_1 \end{Bmatrix} = \begin{bmatrix} T_{11} & T_{12} \\ T_{21} & T_{22} \end{bmatrix} \begin{Bmatrix} V_2 \\ I_2 \end{Bmatrix}. \tag{1.1}$$

Two port systems and related concepts such as linear time-shift invariant (LTI) systems are also important fundamental foundations for analyzing linear systems. These concepts are discussed in Appendix C. It is shown in Chapter 3 that the transfer matrix components of the cabling can also be obtained by performing a set of electrical measurements at the ends of the cable under different driving/termination conditions. From those measurements it can be seen that at the MHz frequencies found in ultrasonic systems unless the cabling is very short (typically much less than a meter in length) the cables do not act as pure "pass-through" devices that simply transfer the signal unchanged from one end of the cable to the other. Thus, cabling has an effect on the measured signals and this part of the ultrasonic system needs to be characterized as part of any system modeling effort.

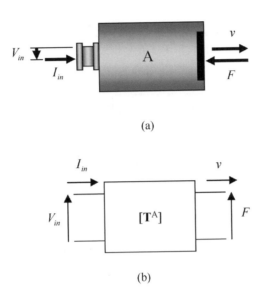

(a)

(b)

Fig. 1.5. (a) A transmitting ultrasonic transducer A as a transformer of voltage, V_{in}, and current, I_{in}, at its electrical port into a compressive force, F, and average velocity, v, at its acoustic port, and **(b)** a model of the transducer characterized by a 2x2 transfer matrix, $\left[\mathbf{T}^A \right]$.

Chapter 4 discusses a transducer when it is used as a generator of sound in an ultrasonic system. Like the cabling, the sending transducer can be modeled as a two port system where the voltage and current at the input electrical port are converted into a compressive force and average velocity on the output side (Fig. 1.5). To characterize the transducer's transfer matrix in the same manner as done for the cabling, one would have to perform a series of both electrical and acoustic measurements at the input/output ports under different driving/termination conditions. This is possible in principle but in general it is not practical since it is difficult (and expensive) to make the precise acoustic measurements this type of characterization would require. It is shown in Chapter 4, however, that it is not necessary to know all the elements of the transducer transfer matrix directly since when the transducer A of Fig. 1.5 is used in practice it is always radiating waves into a known medium. For radiation into a fluid or a linear elastic solid the output compressive force, $F_t(\omega)$, and the average

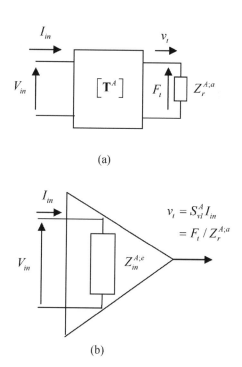

(a)

(b)

Fig. 1.6. (a) A transmitting transducer radiating into a medium characterized by a terminated two-port system, and **(b)** a simpler equivalent model of the transducer as an electrical impedance, $Z_{in}^{A;e}(\omega)$, and a sensitivity, $S_{vI}^{A}(\omega)$.

output velocity, $v_t(\omega)$, are proportional to each other through the relationship $F_t = Z_r^{A;a} v_t$, where $Z_r^{A;a}(\omega)$ is the acoustic radiation impedance of transducer A as shown in Fig. 1.6 (a). This relationship results in the two port transfer matrix model of the transducer being terminated at its acoustic port with the acoustic impedance $Z_r^{A;a}(\omega)$. Under these conditions it is shown in Chapter 4 that one can replace the terminated transducer transfer matrix by an equivalent reduced transducer model consisting only of an electrical impedance, $Z_{in}^{A;e}$, and a transducer sensitivity, S_{vI}^{A}, as shown in Fig. 1.6 (b), where the sensitivity is modeled as an ideal "converter" that transforms input current to output velocity or force. The transducer impedance, $Z_{in}^{A;e}$, is by definition the input voltage divided by the input current, while the sensitivity, S_{vI}^{A}, is defined as the

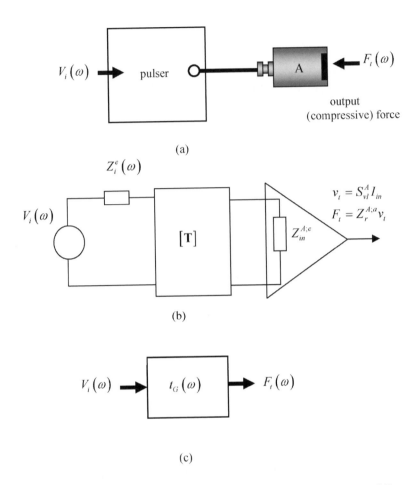

$$Z_i^e(\omega)$$

$$v_t = S_{vI}^A I_{in}$$
$$F_t = Z_r^{A;a} v_t$$

$$[\mathbf{T}]$$

$$Z_{in}^{A;e}$$

(b)

$$V_i(\omega) \rightarrow \boxed{t_G(\omega)} \rightarrow F_t(\omega)$$

(c)

Fig. 1.7. (a) The sound generation process consisting of the pulser, cabling, and sending transducer. **(b)** The detailed models of each of those components. **(c)** A single input-output LTI system model characterized by the sound generation transfer function, $t_G(\omega)$.

average output velocity divided by the input current. The advantage of using this reduced model is that both the transducer impedance and sensitivity can be obtained by purely electrical measurements, making it possible to readily characterize a transducer in terms of these parameters. Chapter 6 outlines a new pulse-echo method for determining transducer electrical impedance and sensitivity that makes it easy to obtain these quantities in a simple calibration setup. Chapter 6 also describes the measurement of "effective" transducer parameters such as effective radius and focal length. Those effective values are needed to accurately model the wave field of a transducer.

In Chapter 4 it is demonstrated that at high frequencies the acoustic radiation impedance is a known constant if the transmitting transducer acts as a piston source (i.e. if the velocity distribution is uniform over the transducer surface at the acoustic port). For an immersion piston transducer, for example, $Z_r^{A;a} = \rho_f c_f S_A$, where ρ_f is the density of the fluid, c_f is the compressional wave speed of the fluid, and S_A is the area of the transducer face at the acoustic port. Thus, with measurements of $Z_{in}^{A;e}, S_{vl}^{A}$ and the transducer effective parameters and with $Z_r^{A;a}$ easily found, it is possible to completely characterize the transmitting transducer's role in the ultrasonic measurement system.

Since the model parameters of the pulser, cabling and transducer shown in Figs. 1.3-1.6 can all be obtained with a series of measurements, it is also possible to combine these models together into a single linear time-shift invariant (LTI) system that characterizes the entire sound generation process, as shown in Fig. 1.7. From the concepts discussed in Appendix C, the LTI system for the sound generation process can be represented in terms of a transfer function, $t_G(\omega)$, that relates the voltage source, $V_i(\omega)$, of the pulser to the output force, $F_t(\omega)$, of the transducer. In Chapter 4, this sound generation transfer function is given explicitly in terms of the pulser, cabling, and transducer parameters as:

$$t_G(\omega) = \frac{F_t(\omega)}{V_i(\omega)} = \frac{Z_r^{A;a} S_{vl}^{A}}{\left(Z_{in}^{A;e} T_{11} + T_{12}\right) + \left(Z_{in}^{A;e} T_{21} + T_{22}\right) Z_i^{e}}. \tag{1.2}$$

Since the pulser voltage source, $V_i(\omega)$, and all the quantities appearing in this sound generation transfer function can be measured the output force on the transducer can be found as

$$F_t(\omega) = t_G(\omega) V_i(\omega). \tag{1.3}$$

In Chapter 5 LTI system concepts are used again to relate the output force on the transmitting transducer, $F_t(\omega)$, to the blocked force, $F_B(\omega)$ acting on the receiving transducer through an acoustic/elastic transfer function, $t_A(\omega)$, that describes all the three-dimensional wave propagation and scattering processes occurring between the sending and receiving transducers. This relationship is given by:

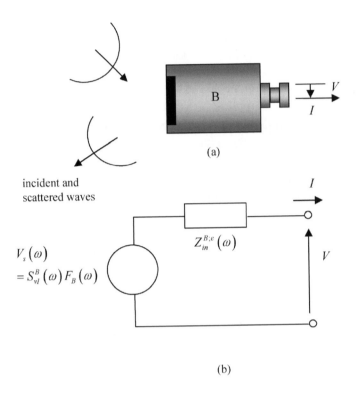

incident and
scattered waves

$V_s(\omega)$
$= S_{vI}^{B}(\omega) F_B(\omega)$

$Z_{in}^{B;e}(\omega)$

(b)

Fig. 1.8. (a) A receiving transducer transforming the incident and scattered waves at the transducer face into output voltage, V, and current, I , and (b) a model of the receiving transducer and acoustic sources as a voltage source and   electrical impedance.

$$F_B(\omega) = t_A(\omega) F_I(\omega). \tag{1.4}$$

The blocked force is defined in Chapter 5 as the compressive force exerted on the receiving transducer by the incident waves when its face is held rigidly fixed. As shown in Chapter 5, it is this particular force that arises naturally in the reception process and it is also shown that a receiving transducer B and the acoustic sources that drive it can be modeled as a voltage source $V_s = F_B S_{vI}^{B}$ in series with an electrical impe-dance, $Z_{in}^{B;e}$, where S_{vI}^{B} and $Z_{in}^{B;e}$ are the sensitivity and impedance of transducer B (see Fig. 1.8). This result shows that the same transducer

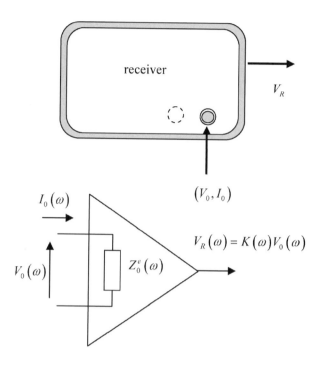

Fig. 1.9. The receiver modeled as an electrical impedance, $Z_0^e(\omega)$, and a gain factor, $K(\omega)$.

impedance and sensitivity that are used to characterize a transmitting transducer are also the terms needed to model the transducer when it is acting as a receiver. The receiving transducer is also connected to the receiver through cabling that can again be modeled as a 2x2 transfer matrix. The components of this matrix can be found by the same electrical measurements discussed in Chapter 3. The receiver, like the pulser, is an electrical network that needs to be characterized. In many ultrasonic pulser/receivers the receiver section performs both amplification and filtering functions. We will not model the filters present in ultrasonic receivers because in quantitative NDE measurements where one wants as wide a frequency response as possible these filtering functions are typically disabled. However, filtering can always be easily added to our receiver model when necessary. Thus, in Chapter 5 the receiver is modeled only as an electrical impedance, $Z_0^e(\omega)$, and a gain factor, $K(\omega)$, (see Fig. 1.9)

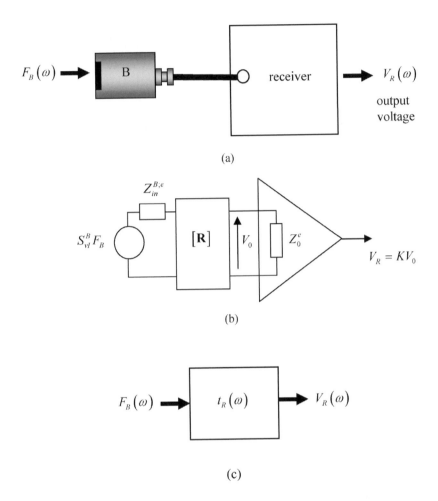

(a)

(b)

(c)

Fig. 1.10. (a) The sound reception process consisting of the receiving transducer, cabling, and receiver and **(b)** the detailed models of each of those components, which can be combined into **(c)** a single input-output LTI system model characterized by the sound reception transfer function, $t_R(\omega)$.

both of which can also be found by a series of electrical measurements. Thus, the entire sound reception process can also be described by a reception transfer function, $t_R(\omega)$, that relates the frequency components of the output voltage of the receiver, $V_R(\omega)$, to the blocked force, $F_B(\omega)$,

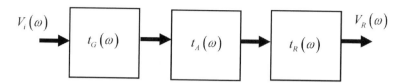

Fig. 1.11. An ultrasonic measurement system modeled as a series of LTI systems, each characterized by their transfer functions.

(see Fig. 1.10). This transfer function is given by

$$t_R(\omega) = \frac{V_R(\omega)}{F_B(\omega)} = \frac{K Z_o^e S_{vI}^B}{\left(Z_{in}^{B;e} R_{11} + R_{12}\right) + \left(Z_{in}^{B;e} R_{21} + R_{22}\right) Z_o^e}, \tag{1.5}$$

where the R_{ij} $(i,j=1,2)$ terms are the components of the transfer matrix of the receiving cable shown in Fig. 1.10 (b). Since all the terms appearing in Eq. (1.5) can also be measured, this receiving transfer function can be obtained explicitly and we can write the output voltage, $V_R(\omega)$, as

$$V_R(\omega) = t_G(\omega) t_R(\omega) t_A(\omega) V_i(\omega). \tag{1.6}$$

Equation (1.6) gives a model of the entire measurement process as simply a product of transfer functions multiplied by the pulser source voltage, $V_i(\omega)$ (see Fig. 1.11). Equations (1.2) and (1.5) show that the generation and reception transfer functions can be determined by making electrical measurements of all the electrical and electromechanical components that make up those functions. Similarly, the pulser source voltage, $V_i(\omega)$ can be measured. Thus, the only remaining unknown in Eq. (1.6) is the acoustic/elastic transfer function, $t_A(\omega)$, where, from Eq. (1.4)

$$t_A(\omega) = \frac{F_B(\omega)}{F_t(\omega)} \tag{1.7}$$

(see Fig. 1.12). It is not possible to directly measure this transfer function, since it is determined by inaccessible quantities such as the displacements in the sound beam generated in the solid surrounding a flaw and the

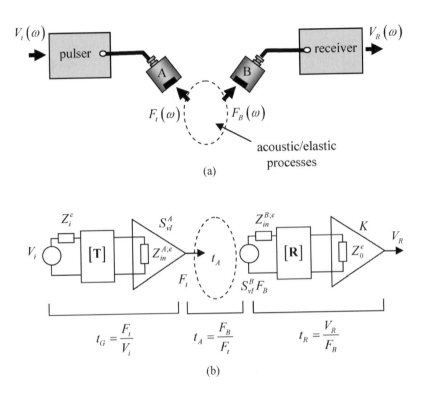

(a)

(b)

Fig. 1.12. (a) The components of an ultrasonic measurement system, showing the sound generation process that transforms the voltage source, $V_i(\omega)$, into the transmitted (compressive) force, $F_t(\omega)$, and the sound reception process which transforms the blocked force, $F_B(\omega)$, into the frequency components of the measured output voltage, $V_R(\omega)$. The acoustic/elastic transfer function, $t_A(\omega)$, describes all the wave propagation and scattering processes that occur between the transmitting and receiving transducers. **(b)** The corresponding model of all the components of the measurement system showing the system elements and the corresponding transfer functions that define the system.

resulting displacements of the waves scattered from the flaw. However, it is possible to model those quantities if one has sufficiently general ultrasonic beam and flaw scattering models. Appendix D and Chapter 8 both provide some basic background into wave propagation theory and the properties of sound beams in fluids and solids that is needed for beam models and flaw scattering models. In Chapter 9 an ultrasonic beam model

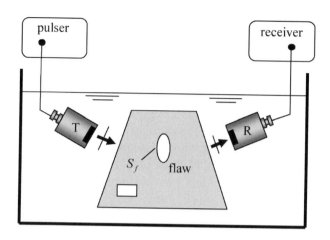

Fig. 1.13. An ultrasonic immersion system for inspecting a flawed component.

that uses the paraxial approximation and a superposition of Gaussian beams is developed for fluids and isotropic elastic solids. This multi-Gaussian beam model allows one to simulate the ultrasonic sound beams found in many ultrasonic testing geometries and is computationally very efficient. Appendix F provides some background in Gaussian beam theory needed for a more thorough understanding of the models discussed in Chapter 9. Chapter 10 describes models of the waves scattered from flaws, using the Kirchhoff and Born approximations – two approximate methods that have been found to be very useful in modeling NDE problems. Chapter 10 also describes some exact scattering models for simple flaw shapes that can be used to validate those more approximate models. In Chapter 11, a general expression for the acoustic/elastic transfer function is derived for an ultrasonic immersion flaw inspection system of the type shown in Fig. 1.13. This transfer function is shown to be given by

$$t_A(\omega) = \frac{1}{Z_r^{T;a} v_T^{(1)} v_R^{(2)}} \int_{S_f} \left(\mathbf{t}^{(1)} \cdot \mathbf{v}^{(2)} - \mathbf{t}^{(2)} \cdot \mathbf{v}^{(1)} \right) dS , \tag{1.8}$$

where $Z_r^{T;a}$ is the acoustic radiation impedance of the transmitting transducer, T. The quantities $\left(\mathbf{t}^{(1)}, \mathbf{v}^{(1)} \right)$ are the traction vector and velocity vector, respectively, on the surface, S_f, of the flaw when the transmitting transducer, T, is firing and the flaw is present (labeled as state (1)), while

$\left(\mathbf{t}^{(2)}, \mathbf{v}^{(2)}\right)$ are the traction vector and velocity vector, respectively, if the receiving transducer, R , was acting as a transmitter and the flaw was absent (labeled as state (2)). The quantities $v_T^{(1)}(\omega), v_R^{(2)}(\omega)$ are just the average velocities on the faces of the transmitting transducer in state (1) and the receiving transducer in state (2), respectively.

Equation (1.8) is a very general result as it relies primarily on the assumptions of linearity and reciprocity. It is also a very useful result since it shows that if one has beam models and flaw scattering models that can predict the fields on the surface of the flaw in states (1) and (2), then those fields can be inserted into Eq. (1.8) to obtain the acoustic/elastic transfer function that is needed to predict the measured output voltage, $V_R(\omega)$, in Eq. (1.6).

By making some additional assumptions, Eq. (1.8) can be reduced to a very modular model. For example, it is shown in Chapter 11 that if the flaw is small enough so the beam variations across the flaw surface can be neglected and if the incident beam can be expressed as a quasi-plane wave acting on the flaw, then the transfer function of Eq. (1.8) can be written in the form

$$t_A(\omega) = \hat{V}_0^{(1)}(\omega)\hat{V}_0^{(2)}(\omega)A(\omega)\left[\frac{4\pi\rho_2 C_{\alpha 2}}{-ik_{\alpha 2}Z_r^{T;a}}\right], \tag{1.9}$$

where $\hat{V}_0^{(m)}$ $(m=1,2)$ are the velocity fields incident on the flaw in states (1) and (2) (normalized by the average velocities on the face of the transmitting transducer), $A(\omega)$ is a particular component of the vector plane wave far-field scattering amplitude of the flaw, and the remaining term in Eq. (1.9) is a combination of known material and geometrical parameters that are defined explicitly in Chapter 11. Equation (1.9) is in a very useful form since the velocity field terms, $\hat{V}_0^{(m)}, (m=1,2)$, which involve ultrasonic beam model calculations, and the flaw response, which is contained entirely in the $A(\omega)$ term, are separated. This modularity allows one to easily perform engineering parametric studies and to isolate the contribution of the flaw from the overall measured response. The latter capability is particularly important since ultimately one must extract information on the flaw itself for sizing and classification purposes, and that information is contained only in $A(\omega)$.

Equations (1.8) or (1.9) complete the overall measurement model defined by Eq. (1.6) and Fig. 1.11, since it is possible to measure

$t_G(\omega), t_R(\omega), V_i(\omega)$ and with beam and flaw scattering models we can obtain $t_A(\omega)$, leading to a prediction of the output voltage frequency components, $V_R(\omega)$. This voltage can then be transformed into the time-domain to obtain the A-scan flaw signal that would be seen on an oscilloscope screen of the system shown in Fig. 1.1. Figure 1.12 (b) shows all the measurement system components and the transfer functions that combine those components into the model of Eq. (1.6). Chapter 7 gives some examples where A-scan signals determined experimentally in a pitch-catch measurement calibration setup are compared to the signals synthesized by measuring/modeling all the system components of Fig. 1.12 (b) and combining them to predict the output response.

There are, of course, many electrical measurements that underlie the determination of the transfer functions $t_G(\omega), t_R(\omega)$ and the voltage source term, $V_i(\omega)$. Obtaining these individual terms is essential if one wants to quantify how a particular component, such as a transducer or a cable, affects the measured result. However, in many cases one is only interested in the net combined contribution of all of the electrical and electromechanical components to the measured response. In that case all of these terms can be combined into a single system function, $s(\omega)$, where

$$s(\omega) = t_G(\omega) t_R(\omega) V_i(\omega). \tag{1.10}$$

It is shown in Chapter 7 that if one measures the output voltage in a reference experiment, $V_R^{ref}(\omega)$, where the acoustic/elastic transfer function, $t_A^{ref}(\omega)$, is known explicitly, then one can obtain this system function directly by deconvolution, i.e.

$$s(\omega) = \frac{V_R^{ref}(\omega)}{t_A^{ref}(\omega)}. \tag{1.11}$$

In practice, this deconvolution is carried out with a Wiener filter to desensitize the result to noise, but the basic process is still primarily the simple complex division of Eq. (1.11). If a subsequent flaw measurement is then made with the same electrical and electromechanical components (pulser/receiver, cabling, transducers) and at the same system settings as in the reference experiment, then the system function, $s(\omega)$, obtained from Eq. (1.11) is the same in both the reference experiment and the flaw measurement. Thus the measured flaw response $V_R^f(\omega) = s(\omega) t_A^f(\omega)$ is known once the acoustic/elastic transfer function for the flaw measurement setup, $t_A^f(\omega)$, is obtained, using either Eq. (1.8) or Eq. (1.9) and the appropriate beam models and flaw scattering models. Obtaining the system function experimentally in this fashion makes it very practical to develop measurement models that can predict the measured output signals of very complex inspection problems. In Chapter 12, we demonstrate the versatility of this approach by combining a system function with the acoustic/elastic transfer function of Eq. (1.8) to produce an overall ultrasonic measurement model of the form

$$V_R(\omega) = \frac{s(\omega)}{Z_r^{T;a} v_T^{(1)} v_R^{(2)}} \int_{S_f} \left(\mathbf{t}^{(1)} \cdot \mathbf{v}^{(2)} - \mathbf{t}^{(2)} \cdot \mathbf{v}^{(1)} \right) dS . \tag{1.12}$$

If the transfer function of Eq. (1.9) is used instead we obtain the Thompson-Gray measurement model

$$V_R(\omega) = s(\omega) \hat{V}_0^{(1)}(\omega) \hat{V}_0^{(2)}(\omega) A(\omega) \left[\frac{4\pi \rho_2 c_{\alpha 2}}{-i k_{\alpha 2} Z_r^{T;a}} \right] . \tag{1.13}$$

In Chapter 12, MATLAB codes are developed that implement these measurement models as well as a measurement model suitable for cylindrical-shaped scatterers such as a side-drilled hole, which is a commonly used reference reflector in ultrasonic testing. These measurement models are combined with measurements of $s(\omega)$, the multi-Gaussian beam model of Chapter 9 and flaw scattering models of Chapter 10 to predict the output signals for spherical pores, flat-bottom holes, and side-drilled holes. It is shown that these measurement model predictions agree well with the responses measured experimentally for these reflectors.

Finally, in Chapter 13, we discuss some of the ways in which ultrasonic measurement models can be used as tools in NDE applications. For example, the use of the models to determine flaw scattering amplitudes experimentally is demonstrated. This is an important capability since if we can extract the flaw response from the total measured response, this flaw response can be directly used in quantitative flaw classification and sizing algorithms [Fundamentals]. We also discuss in Chapter 13 how models can predict distance amplitude correction (DAC) transfer curves. DAC curves are commonly used for calibration purposes but in current practice their determination requires the construction of sets of reference specimens for every different testing situation. Model-based DAC transfer curves allow one to perform calibrations on a simple specimen and then transform those calibrations to other more complex testing configurations, thus avoiding the considerable expense of fabrication of many different test specimens. Chapter 13 also applies our ultrasonic measurement model to angle beam shear wave tests, demonstrating that the concepts presented for immersion systems also can be applied to other setups as well. These angle beam inspection models are then used in model-assisted flaw identification and sizing applications. All the examples shown in Chapter 13, however, only illustrate a very small fraction of the areas where these models are useful. Model-based applications are still in their infancy, so there is considerable work that can be done with these models (and others) to help solve fundamental NDE problems.

1.3 Some Remarks on Notation

In some of the following Chapters it will be necessary to occasionally use Einstein summation notation to avoid overly complex expressions. In that notation a repeated subscript is understood to imply a summation over the values (1, 2, 3) of the indices. For example, in calculating the scalar (dot) product of two vectors we can write $\mathbf{u} \cdot \mathbf{v} = u_i v_i = u_1 v_1 + u_2 v_2 + u_3 v_3$. In contrast an unrepeated (free) subscript takes on any of the values (1, 2, 3). For example: the expression $u_j = \partial \phi / \partial x_j$ implies the three equations: $u_1 = \partial \phi / \partial x_1$, $u_2 = \partial \phi / \partial x_2$, $u_3 = \partial \phi / \partial x_3$. For more details see the reference [Fundamentals].

1.4 Organization of the Book

Models that characterize the individual electrical and electromechanical components (pulser/receiver, transducer(s), cabling) of an ultrasonic measurement system are discussed in Chapters 2-6. Appendices A, B, and C provide some of the necessary background material for those Chapters. As discussed previously, all those components can be lumped together into a single system function, $s(\omega)$, that can be determined experimentally in a calibration experiment. Thus, if the reader wants to concentrate primarily on the wave processes present in an ultrasonic system, he/she can begin with the discussion of $s(\omega)$ in Chapter 7 and then cover Chapters 8-13 that discuss beam models, flaw scattering models, and ultrasonic measurement models in detail and describe a number of applications.

Appendices D and E provide background material on waves needed for the Chapters on wave modeling and Appendix F gives some of the fundamental properties of Gaussian beams.

A number of short exercises are given throughout the book. In most cases those exercises involve the use of MATLAB and MATLAB-based functions. MATLAB functions and scripts are also developed and described at a number of places in the book. A complete set of all the MATLAB resources used in this book can be found on the Web at www.springer.com/978-0-387-49061-8. Appendix G also gives listings of the MATLAB functions and scripts used to develop a complete ultrasonic NDE flaw measurement model.

References to all the topics discussed in this Chapter can be found at the ends of each of the following Chapters. For more information on ultrasonic nondestructive evaluation methods and applications we have listed a few suggested reading references below.

1.5 Reference

Schmerr LW (1998) Fundamentals of ultrasonic nondestructive evaluation – a modeling approach. Plenum Press, New York (referred to as [Fundamentals] in this book)

1.6 Suggested Reading

Blitz J and Simpson G (1996) Ultrasonic methods of non-destructive testing. Chapman & Hall, London, UK

Harker AH (1988) Elastic waves in solids. Adam Hilger, Philadelphia

Krautkramer J, Krautkramer H (1990) Ultrasonic testing of materials, 4th ed. Springer-Verlag, Berlin, Germany

Lempriere BM (2002) Ultrasound and elastic waves – frequently asked questions. Academic Press, San Diego, CA

Rose JL (1999) Ultrasonic waves in solid media. Cambridge University Press, Cambridge, UK

2 The Pulser

A pulser/receiver is a complex electrical network that generates the energy that drives the transmitting transducer in an ultrasonic measurement system. The pulser/receiver also amplifies and/or filters the electrical response arriving from the receiving transducer. In this Chapter we will examine only the pulser section of a pulser/receiver and describe some of the important overall characteristics of its output signals and how those signals are affected by instrument setting changes. Simple models that can describe the pulser output are also discussed.

2.1 Characteristics of a Pulser

Figure 2.1 shows a sketch of the front panel of a typical laboratory "spike" pulser/receiver while Fig. 2.2 shows a highly idealized circuit schematic of this same instrument. The pulser side of this instrument has three controls. One control is the "energy" setting. The energy setting basically controls the amount of energy stored in the capacitor, C_0, of Fig. 2.2. This energy is periodically discharged into the sending transducer by closing the switch shown in that figure. The "rep rate" controls the frequency at which this switch is closed, which typically may be varied from several hundred closings/sec to several thousand closings/sec. Generally this rate is set to ensure that the waves traveling in a component have had time to decay in amplitude to very small values before the next discharge occurs. In this case there is no overlapping of the received responses from one closing to the next which, if it occurred, could cause triggering problems when the received signals are displayed on an oscilloscope screen since the oscilloscope is triggered by a signal generated in synchronization with the pulser discharges. The "damping" control on the pulser changes the value of a damping resistance, R_d, in the pulser/receiver.

In addition to a spike-like pulser, which uses a capacitive discharge to drive a transducer, there are also square wave pulser/receivers like the UTEX 340 shown in Fig. 2.3 which drive a transducer with circuits that produce a rectangular-shaped voltage pulse. This particular pulser has

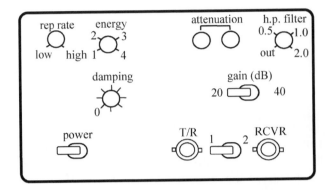

Fig. 2.1. The front panel controls of a typical laboratory "spike" pulser/receiver.

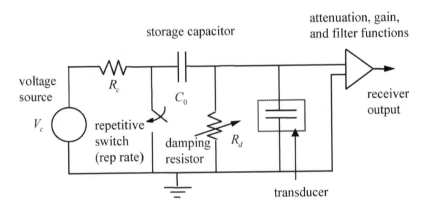

Fig. 2.2. A highly simplified circuit representation of a "spike" pulser/receiver.

most of its controls also available under computer control. An image of the UTEX 340 computer control panel is shown in Fig. 2.4. It can be seen from that figure that on the pulser side of this instrument there are primarily three settings- the pulse repetition rate, the pulse voltage amplitude (in volts), and the pulse width (in nanoseconds). The energy/ damping settings of the spike pulser and the voltage/pulse width settings of the square wave pulser control the amplitude and shape of the voltage and current at the output port of the pulser. In the next section we will show how the output behavior of these pulsers can be described in terms of

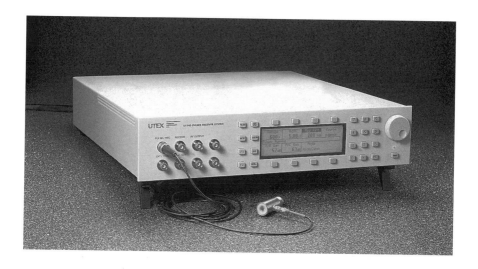

Fig. 2.3. A UTEX 340 square wave pulser/receiver. Photo courtesy of UTEX Scientific Instruments, Inc. , Mississauga, Ontario, Canada.

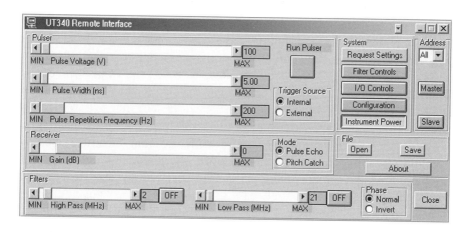

Fig. 2.4. The control panel of the UTEX 340 pulser/receiver. Photo courtesy of UTEX Scientific Instruments, Inc., Mississauga, Ontario, Canada.

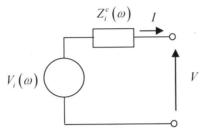

Fig. 2.5. The Thévenin equivalent voltage source and impedance for a pulser.

a simple equivalent circuit whose parameters can be obtained with several electrical measurements. The properties of that equivalent circuit, however, are dependent on these pulser settings so that if they are changed, the equivalent parameters will change.

2.2 Measurement of the Circuit Parameters of a Pulser

As shown in Appendix B, Thévenin's theorem allows us to replace the pulser, which is a circuit network with sources, with the equivalent voltage source and equivalent impedance of Fig. 2.5 if one assumes that the pulser is a linear device. Several authors have used either the simple model of Fig. 2.5 or other similar equivalent circuits to model both the pulser and receiver circuits [2.1-2.4]. As pointed out in these studies, because of the internal diode protection circuits and other elements present in pulser/receivers, strictly speaking those devices may not act in a linear fashion. However, if the measurement of $V_i(\omega)$ and $Z_i^e(\omega)$ are made for a specific set of pulser settings at the same external electrical loading conditions (cabling, transducer) found in the measurement system, then the simple equivalent circuit of Fig.2.5 can be successfully used to model a given pulser [2.4]. It is relatively easy to measure the Thévenin equivalent voltage source for the pulser, $V_i(\omega)$, by measuring the open-circuit voltage, $V_0(t)$, at the output terminals of the pulser and then Fourier-transforming this measured voltage to obtain $V_0(\omega)$. Since there is no current flowing from the pulser under open-circuit conditions we have $V_i(\omega) = V_0(\omega)$.

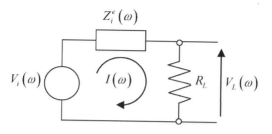

Fig. 2.6. The Thévenin equivalent circuit for a pulser attached to a known external resistance, R_L, for measuring the impedance, $Z_i^e(\omega)$.

To find the electrical impedance of the pulser we can place a known load resistance, R_L, at the output terminals of the pulser and measure the voltage, $V_L(t)$, across this load. Fourier transforming this voltage then gives $V_L(\omega)$. But from the Thévenin equivalent circuit of the pulser shown in Fig. 2.6, we see that

$$V_i - V_L = Z_i^e I$$
$$V_L = R_L I. \tag{2.1}$$

So eliminating the current, I, we find

$$Z_i^e(\omega) = R_L \left(\frac{V_i(\omega)}{V_L(\omega)} - 1 \right). \tag{2.2}$$

Since the values of the Thévenin equivalent parameters $\left(V_i, Z_i^e\right)$ depend on the instrument settings of the pulser we have shown these parameters at several different settings. Figure 2.7, for example, shows the magnitude of the Thévenin equivalent voltage measured for a Panametrics 5052 PR pulser/receiver (spike pulser) at combinations of two different energy settings and two damping settings. In the same fashion Fig. 2.8 shows the magnitude of the Thévenin equivalent voltage obtained for a UTEX 320 pulser/receiver (square wave pulser) at combinations of two different voltage settings and two pulse width settings. Figures 2.9 and 2.10 show the corresponding dependency of the equivalent impedance of

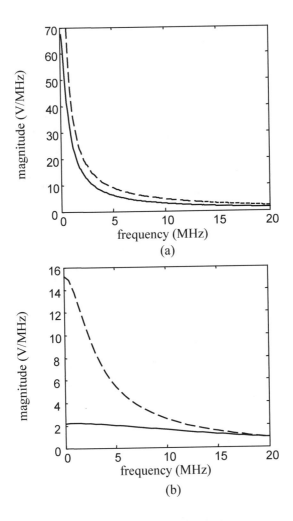

Fig. 2.7. Magnitude of the Thévenin equivalent voltage source versus frequency obtained for a Panametrics 5052PR pulser/receiver for **(a)** damping setting = 0 and energy setting =1 (solid line) or energy setting = 4 (dashed line), and **(b)** damping setting = 7 and energy setting = 1 (solid line) or energy setting = 4 (dashed line).

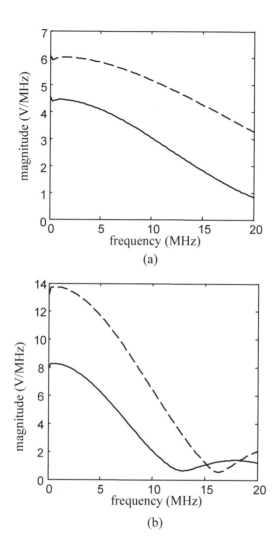

Fig. 2.8. Magnitude of the Thévenin equivalent voltage source versus frequency obtained for a UTEX 320 pulser/receiver at: **(a)** pulse width = 10 and voltage = 100V (solid line) or voltage = 200V (dashed line), and **(b)** pulse width = 50 and voltage = 100 V (solid line) or voltage = 200 V (dashed line).

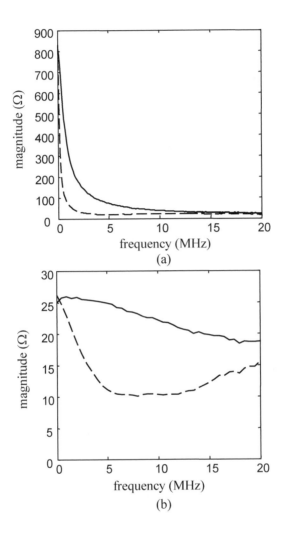

Fig. 2.9. Magnitude of the Thévenin equivalent pulser impedance versus frequency obtained for a Panametrics 5052PR pulser/receiver for **(a)** damping setting = 0, energy setting =1 (solid line), energy setting = 4 (dashed line), and **(b)** damping setting = 7, energy setting = 1 (solid line), energy setting = 4 (dashed line).

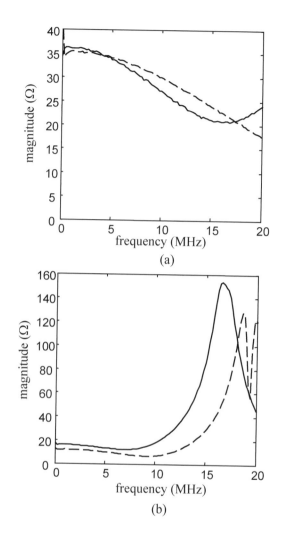

Fig. 2.10. Magnitude of the Thévenin equivalent impedance versus frequency for a UTEX 320 pulser/receiver at: **(a)** pulse width = 10, voltage = 100V (solid line), voltage = 200V (dashed line), and **(b)** pulse width = 50, voltage = 100 V (solid line), voltage = 200 V (dashed line).

the pulser at the same pulser settings for these two pulser/receivers. It can be seen that the energy and voltage settings do increase the magnitude of the Thévenin equivalent voltage source for these pulsers, as expected, but that there are also changes in the shape of the voltage source and impedance with frequency so that the overall behavior of such pulsers is a rather complex function of the pulser settings Although the resistance, R_L, appears in Eq. (2.2) the impedance. $Z_i^e(\omega)$ should not depend on that resistance, as discussed in Appendix B. Pulser impedance measurements made in this fashion with spike and square wave pulsers, however, do show some variations with the load used, possibly due the non-linear elements present in those instruments, as discussed previously. Figure 2.11 shows, for example, the magnitude of the equivalent impedance of the Panametrics 5052 PR pulser obtained when a 50 Ω resistor was used at the pulser output versus the impedance obtained when a transducer and cable were attached to the output port instead. In the latter case the voltage and current were both measured at the output port of the pulser in order to calculate the impedance of the loading induced by the cabling and transducer. The R_L in Eq. (2.2) was then replaced by that load impedance to calculate the pulser impedance. It can be seen from Fig 2.11 that there are indeed differences in the calculated impedance of the pulser under these different external loads. Similar changes have been observed when calculating the equivalent impedance of square wave pulsers. In general our experience has been that it is best to make these measurements of the pulser impedance under the actual loading conditions that will be found when using the pulser in ultrasonic flaw measurements, but we have also

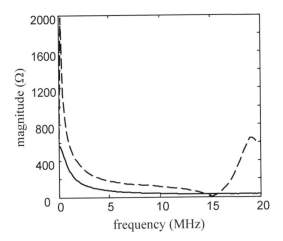

Fig. 2.11. The magnitude of the Thévenin equivalent impedance of a Panametrics 5052 PR pulser/receiver versus frequency found using a 50 ohm resistor loading (solid line) and a loading consisting of a cable and transducer (dashed line).

been successful in using the pulser impedance values measured with Eq. (2.2) and purely resistive loads to simulate the pulser effects in an overall ultrasonic system measurement model of the type discussed in Chapter 7. Thus, while the loading at the pulser output port does change the measured values of the equivalent impedance of the pulser it appears that these loading effects do not significantly affect the measured output voltage in an ultrasonic measurement system, where other parameters, such as transducer sensitivity, play a more important role.

2.3 Pulser Models

It is possible to set up a simple model of the open-circuit output voltage of a typical spike or square wave pulser by directly specifying this voltage in the form of a four parameter model given by

$$V_i(t) = \begin{cases} 0 & t \leq 0 \\ -V_\infty \left[1 - \exp(-\alpha_1 t)\right] & 0 \leq t \leq t_0 \\ -V_0 \exp\left[-\alpha_2 (t - t_0)\right] & t \geq t_0 \end{cases} \tag{2.3}$$

where $V_\infty = V_0 / \left(1 - e^{-\alpha_1 t_0}\right)$ and the four parameters $(t_0, \alpha_1, \alpha_2, V_0)$ control the amplitude and rise and fall characteristics of the pulse. Figure 2.12 (a) shows a plot of this modeled voltage which is very similar in form to a measured Thévenin equivalent open-circuit voltage from the Panametrics 5052PR pulser/receiver, as shown in Fig. 2.12 (b). This same model, with the appropriate choice of parameters, can also be used to model a square wave pulse output (see Fig. 2.13 (a)). The actual open-circuit output voltage of a UTEX 320 square wave pulser/receiver is shown in Fig. 2.13 (b). The spectrum generated by this simple source model can be obtained from Eq. (2.3) by numerically evaluating the FFT of this time domain response or one can use the explicit Fourier transform of the $V_i(t)$ of Eq. (2.3), which is given by:

$$V_i(\omega) = \frac{V_\infty \left\{1 - \exp\left[-(\alpha_1 - i\omega)t_0\right]\right\}}{\alpha_1 - i\omega} + \frac{V_\infty \left\{1 - \exp\left[i\omega t_0\right]\right\}}{i\omega}$$

$$-\frac{V_0 \exp\left[i\omega t_0\right]}{\alpha_2 - i\omega}. \tag{2.4}$$

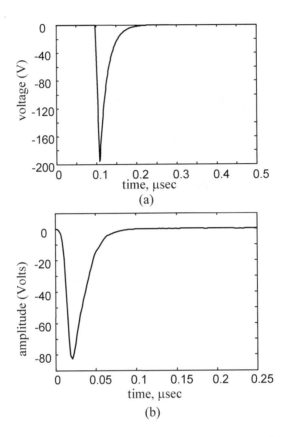

Fig. 2.12. (a) Voltage pulse (volts) versus time (μsec) obtained from Eq. (2.3) with $t_0 = 0.01$, $\alpha_1 = 0.2$, $\alpha_2 = 50$, $V_0 = 200$ (shifted for better visualization). **(b)** Measured open-circuit voltage versus time for a Panametrics 5052PR pulser/receiver at energy setting 1, damping setting 5.

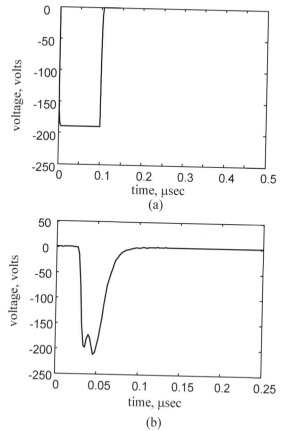

Fig. 2.13. **(a)** Voltage pulse (volts) versus time (μsec) obtained from Eq. (2.3) with $t_0 = 0.1$, $\alpha_1 = 1000$, $\alpha_2 = 1000$, $V_0 = 190$. **(b)** Measured open-circuit output voltage versus time for a UTEX 320 pulser/receiver.

It is not as easy to obtain an explicit parametric model of the impedance of a pulser since this impedance changes significantly in both amplitude and shape with the pulser settings and as a function of frequency, as shown in Figs. 2.9 and 2.10. However, one could try to model the pulser impedance by an equivalent RLC circuit whose parameters are adjusted to match the measured impedance values (as a function of frequency) at various damping settings, as done by Brown [2.1]. Brown found that the equivalent RLC parameters obtained for a Panametrics 5052PR did change significantly, particularly at the higher damping settings.

2.4 References

2.1 Brown LF (2000) Design considerations for piezoelectric polymer ultrasound transducers. IEEE Trans. Ultrasonics, Ferroelectrics, and Frequency Control 47: 1377-1396

2.2 Dang CJ, Schmerr LW, Sedov A (2002) Modeling and measuring all the elements of an ultrasonic nondestructive evaluation system. I: Modeling foundations. Research in Nondestructive Evaluation 14: 141-176

2.3 Dang CJ, Schmerr LW, Sedov A (2002) Modeling and measuring all the elements of an ultrasonic nondestructive evaluation system. II: Model-based measurements. Research in Nondestructive Evaluation 14: 177-201

2.4 Ramos A, San Emterio JL, Sanz PT (2000) Dependence of pulser driving responses on electrical and motional characteristics of NDE ultrasonic probes. Ultrasonics 38: 553-558

2.5 Exercises

1. The MATLAB function model_pulser takes as inputs an energy setting (energy = 1, 2, 3, 4), a damping setting (damping = 0, 5, 10), a resistance loading , RL, (in ohms) across the output terminals of the pulser and returns the sampled voltage, vt, across RL (in volts) and the sampled time values, t, (in μsec). The form of the calling sequence of this function is:

```
>> [t , vt] =model_pulser( energy, damping, RL);
```

Use this model pulser at energy = 2, damping = 5 settings for both open circuit conditions (RL = inf) and a given load (RL = 250 ohms) to determine the Thévenin equivalent source voltage (in volts) and impedance (in ohms) of the pulser at these settings as functions of frequency over the range of frequencies from 0-20 MHz and plot the magnitude and phase of these functions over the same frequency range. Use the MATLAB unwrap function to eliminate any artificial jumps of 2π in the phase plots. Example:

```
>> plot(f, unwrap(angle(Vf)))
```

Show and explain all the steps you used to obtain your answers.

3 The Cabling

At the MHz frequencies involved in NDE tests, the electrical cables that transfer the electrical pulses from the pulser to the sending transducer and from the receiving transducer to the receiver do not just pass those signals unchanged. Thus, significant cabling effects may be present in some ultrasonic testing setups. Here we will discuss models and measurements that can help us to quantitatively determine the effects of the cables. These models and measurements will enable us to predict how the voltage and current change from one end of the cable to the other (Fig. 3.1).

3.1 Cable Modeling

At the most fundamental level we can model a cable as a set of coaxial conductors transferring electrical and magnetic fields ($\mathbf{E}, \mathbf{H}$) from one end of the cable to the other, as shown in Fig. 3.2. It is shown in many texts on electromagnetism [3.1-3.7] that the fields at each end of the cable are related by the reciprocity relationship

$$\int_{S_1} \left(\mathbf{E}_1^{(2)} \times \mathbf{H}_1^{(1)} - \mathbf{E}_1^{(1)} \times \mathbf{H}_1^{(2)} \right) \cdot \mathbf{n}_1 \, dS$$
$$= \int_{S_2} \left(\mathbf{E}_2^{(2)} \times \mathbf{H}_2^{(1)} - \mathbf{E}_2^{(1)} \times \mathbf{H}_2^{(2)} \right) \cdot \mathbf{n}_2 \, dS, \tag{3.1}$$

where ($\mathbf{E}_1, \mathbf{H}_1$) are fields at the left end of the cable acting over an area S_1 whose unit normal (pointing out from the cable) is $\mathbf{n}_1$, and ($\mathbf{E}_2, \mathbf{H}_2$) are the corresponding fields at the other end, S_2, whose outward normal is $\mathbf{n}_2$ as shown in Fig. 3.2. The superscripts (1) and (2) on the field variables in Eq. (3.1) designate these fields when the cable is under two different driving/termination conditions at its ends. These two driving/termination conditions are labeled as states (1) and (2). If the fields are carried in the cable as a fundamental propagating electromagnetic wave mode called a *TEM mode*, then it can be shown that the electric field, $\mathbf{E}$, can be expressed in terms of a potential (voltage), V, across the two conductors in the cable

Fig. 3.1. A cable and the voltages and currents at its end connectors.

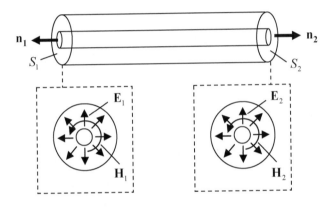

Fig. 3.2. The electrical and magnetic fields at the ends of a coaxial cable.

and the magnetic field, **H**, can be related to the current, I, flowing through the central conductor [3.4]. These relations are

$$\mathbf{E} = -\nabla V$$
$$I = \int_c \mathbf{H} \cdot d\mathbf{l}, \tag{3.2}$$

where c is a closed path taken around the central conductor of the cable and $d\mathbf{l}$ is a vector differential element along that path.

For such a propagating TEM mode it can also be shown that the reciprocity relationship of Eq. (3.1) reduces to a similar reciprocity relationship between the voltages and the currents in states (1) and (2) given by

Fig. 3.3. A cable modeled in terms of the voltages and currents at its two ends (ports).

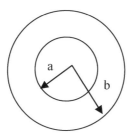

Fig. 3.4. Cross-section of an ideal circular coaxial cable where the radius of the inner conductor is a and the radius of the outer conductor is b.

$$V_1^{(1)}I_1^{(2)} - V_1^{(2)}I_1^{(1)} = V_2^{(1)}I_2^{(2)} - V_2^{(2)}I_2^{(1)} \qquad (3.3)$$

so that we can then consider our cable as modeled in terms of these voltages and currents where I_1 is the current flowing into the cable at the left end and I_2 is the current flowing out of the cable at the other end (see Fig. 3.3). If the reciprocity relationship of Eq. (3.3) is satisfied for any set of driving/termination conditions, then it can also be shown that the voltage and current at one end (port) of the cable are linearly related to the voltage and current at the other end (port) and we can model the cable as a reciprocal two port system (see Appendix C) where one has

$$\begin{Bmatrix} V_1 \\ I_1 \end{Bmatrix} = \begin{bmatrix} T_{11} & T_{12} \\ T_{21} & T_{22} \end{bmatrix} \begin{Bmatrix} V_2 \\ I_2 \end{Bmatrix} \qquad (3.4)$$

and $\det[\mathbf{T}] = T_{11}T_{22} - T_{12}T_{21} = 1$.

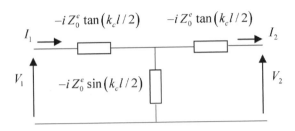

Fig. 3.5. An equivalent circuit model of a cable.

As developed in many electrical engineering texts, one can use a simple transmission line model of the cable and obtain an explicit expression for this transfer matrix $[\mathbf{T}]$ in the form [3.5]

$$\begin{Bmatrix} V_1 \\ I_1 \end{Bmatrix} = \begin{bmatrix} \cos(k_c l) & -iZ_0^e \sin(k_c l) \\ -i\sin(k_c l)/Z_0^e & \cos(k_c l) \end{bmatrix} \begin{Bmatrix} V_2 \\ I_2 \end{Bmatrix}, \tag{3.5}$$

where l is the length of the cable, Z_0^e is the characteristic impedance of the cable (in ohms), and $k_c = \omega/c$ is the wave number and c is the wave speed of signals in the cable. For an ideal circular coaxial cable as shown in Fig. 3.4 where the inner conductor is of radius a and the outer conductor is of radius b the characteristic impedance of the cable is given by [3.5]

$$Z_0^e = \frac{1}{2\pi} \sqrt{\frac{\mu}{\varepsilon}} \ln\left(\frac{b}{a}\right), \tag{3.6}$$

where μ is the permeability and ε the permittivity of the material in the cable between the inner and outer conductors.

In Appendix C we showed how a simple RC circuit could be expressed in transfer matrix form as a two port system. Thus it is not surprising that conversely a two port system can also be expressed as an equivalent circuit. There are actually many different equivalent circuits that yield the same results as the transfer matrix. Figure 3.5 shows one commonly used circuit [3.1] that uses three impedances arranged in a T-shape to model the cable.

If our cable model is terminated with a impedance, Z_2^e, as shown in Fig. 3.6 (a) then the cable and its termination can be represented as a single equivalent impedance, Z_1^e, as shown in Fig. 3.6 (b). The behavior of

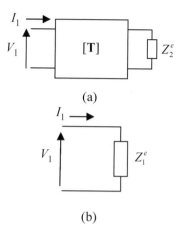

(a)

(b)

Fig. 3.6. (a) A cable terminated with an impedance, Z_2^e, and **(b)** the equivalent impedance, Z_1^e, of this terminated cable.

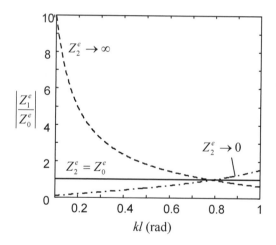

Fig. 3.7. The effect of different termination conditions on the equivalent impedance of a cable.

this equivalent impedance versus the non-dimension frequency $k_c l$ is shown in Fig. 3.7 for open-circuit ($Z_2^e \to \infty$) termination, short-circuit ($Z_2^e = 0$) termination, and termination at the characteristic impedance of

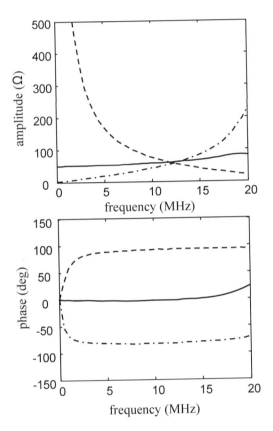

Fig. 3.8. Measured values of the magnitude and phase of a 50 ohm cable under open-circuit (dashed line), short-circuit (dashed-dotted line), and 50 ohm (solid line) termination conditions.

the cable ($Z_2^e = Z_0^e$). It can be seen that the open- and short-circuit cases generate frequency dependent equivalent impedances while in the matched termination case the equivalent impedance is frequency independent. This same behavior is seen when the equivalent impedance of a 50 ohm cable is measured experimentally, as shown in Fig. 3.8. The cables used in an ultrasonic test for sound generation and reception are terminated/driven by ultrasonic transducers which in general are not matched in impedance to the cable so that inherently we can expect some frequency dependent effects due to the cabling in NDE tests.

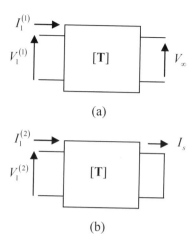

Fig. 3.9. (a) A cable, modeled as a two port system, under open-circuit conditions, and **(b)** under short-circuit conditions. Measurements of the voltages and currents shown can be used to determine the transfer matrix of the cable.

3.2 Measurement of the Cabling Transfer Matrix

As can be seen from Figs. 3.7 and 3.8 a simple two port model can accurately represent the behavior of an ordinary coaxial cable. However, we do not ordinarily know all the detailed parameters that are needed to obtain the transfer matrix components in Eq. (3.5). Furthermore, in immersion NDE testing, such cabling is connected to fixtures that support the transducer in an immersion tank and the details of the cabling within the fixtures are in general also not known. This is not a problem since it is possible to directly measure the transfer matrix components of the combined cabling and fixtures in situ by attaching the cable/fixture to a driving source, such as the ultrasonic pulser, and making a series of voltage and current measurements under different cable/fixture termination conditions. Figure 3.9 (a) shows a two port model of a cable under open-circuit conditions at its output port and driving voltage $V_1^{(1)}$ and current $I_1^{(1)}$ at its input port while Fig. 3.9 (b) shows the same model under short-circuit conditions at the output port with driving voltage $V_1^{(2)}$ and current $I_1^{(2)}$ at the input port. From Eq. (3.4) it is easy to see that:

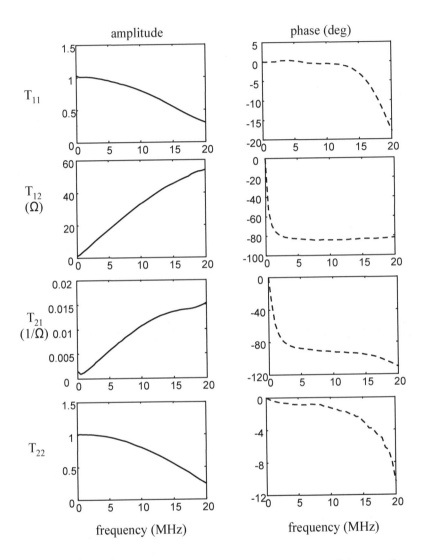

Fig. 3.10. Measured values of the magnitudes and phases of the transfer matrix components versus frequency for a cable.

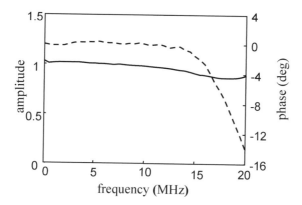

Fig. 3.11. A check on the satisfaction of reciprocity ($\det[\mathbf{T}] = 1$) for the measured transfer matrix components of Fig. 3.10. The amplitude (solid line) and phase (dashed line) of the determinant are shown.

$$T_{11}(\omega) = V_1^{(1)}(\omega) / V_\infty(\omega)$$
$$T_{21}(\omega) = I_1^{(1)}(\omega) / V_\infty(\omega)$$
$$T_{12}(\omega) = V_1^{(2)}(\omega) / I_s(\omega)$$
$$T_{22}(\omega) = I_1^{(2)}(\omega) / I_s(\omega)$$

(3.7)

so that all the transfer matrix elements can be obtained by making measurements of the voltages and currents in these two states: $v_1^{(m)}(t), i_1^{(m)}(t), v_\infty(t)$, $i_s(t)$ $(m = 1,2)$ and Fourier transforming them to obtain $V_1^{(m)}(\omega), I_1^{(m)}(\omega)$, $V_\infty(\omega), I_s(\omega)$ $(m = 1,2)$. The consistency of these measured transfer matrix elements can be checked by the reciprocity relationship $T_{11}T_{22} - T_{12}T_{21} = 1$.

Figure 3.10 shows the transfer matrix components found in this manner as a function of frequency for a cable (both amplitude and phase are plotted). It can be seen that the measured magnitudes of these components do exhibit the cosine and sine function behavior of Eq. (3.5) and the measured phase terms also generally follow that simple model behavior. As a reciprocity check on these measurements we can compute the determinant of the measured transfer matrix. Figure 3.11 shows that det[**T**] = 1 is well satisfied over a wide range of frequencies.

3.3 References

3.1 Pozar DM (1998) Microwave engineering, 2nd ed. John Wiley and Sons, New York, NY

3.2 Magnusson PC, Alexander GC, Tripathi V (1992) Transmission lines and wave propagation, 3rd ed. CRC Press, Boca Raton, FL

3.3 Seshadre SR (1971) Fundamentals of transmission lines and electromagnetic fields. Addison-Wesley Publishing Company, Reading, MA

3.4 Balanis CA (1989) Advanced engineering electromagnetics. John Wiley and Sons, New York, NY

3.5 Staelin DH, Morgenthaler AW, Kong JA (1994) Electromagnetic waves. Prentice Hall, Englewood Cliffs, NJ

3.6 Bladel JV(1985) Electromagnetic fields. Hemisphere Publishing Co., New York, NY

3.7 Karmel PR, Colef GD, Camisa RL (1998) Introduction to electromagnetic and microwave engineering. John Wiley and Sons, New York, NY

3.4 Exercises

1. Consider a 1 meter long, 50 ohm cable, where the wave speed in the cable is one half the wave speed of light, c_0 ,in a vacuum ($c_0 = 2.998 \times 10^8$ m/sec). Determine the transfer matrix components of the cable at 10 kHz, 100 kHz, 1 MHz, 20 MHz.

2. Consider a cable for which we wish to measure the transfer matrix components (as a function of frequency). We can do this in MATLAB for a function cable_X which has the calling sequence:

```
>> [ v1, i1, vt, it] = cable_X( V, dt, R,  L, 'term');
```

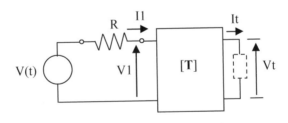

Fig. 3.12. A measurement setup for obtaining the transfer matrix components of a cable.

The input arguments of cable_X are as follows. V is a sampled voltage source versus time, where the sampling interval is dt. R is an external resistance (in ohms). This source and resistance are connected in series to one end of the cable, which is of length L (in m) as shown in Fig. 3.12. The other end of cable can be either open-circuited or short-circuited. The string 'term' specifies the termination conditions. It can be either 'oc' for open-circuit or 'sc' for short-circuit. Cable_X then returns the "measured" sampled voltages and currents versus time (v1, i1, vt, it) where (v1, i1) are on the input side of the cable and (vt, it) are at the terminated end (Note: for open-circuit conditions it = 0 and for short-circuit conditions vt = 0). As a voltage source to supply the V input to cable_X use the MATLAB function pulserVT. For a set of sampled times this function returns a sampled voltage output that is typical of a "spike" pulser. Make a vector, t, of 512 sampled times ranging from 0 to 5 μsec with the MATLAB call:

```
>> t = s_space(0, 5, 512);
```

(see the discussion of the s_space function in Appendix A; a code listing of the function is given in Appendix G) and call the pulserVT function as follows:

```
>> V = pulserVT(200, 0.05, 0.2, 12, t);
```

For the resistance, take R = 200 ohms, and specify the length of the cable as L = 2 m.

Using Eq. (3.7), determine the four cable transfer matrix components and plot their magnitudes and phases from 0 to 30 MHz. Note that the outputs of cable_X are all time domain signals but the quantities in Eq. (3.7) are all in the frequency domain so you will need to define a set of 512 sampled frequency values, f, through:

```
>> dt = t(2) –t(1);
>> f =s_space(0, 1/dt, 512);
```

What is the range of frequencies contained in f here?

4 Transmitting Transducer and the Sound Generation Process

In this Chapter we will discuss models of the ultrasonic transducer as a transmitting device that converts electrical energy into acoustic energy. We will also combine the models of the pulser and cabling from Chapters 2 and 3 with the transducer model of this Chapter to describe a model of the entire sound generation process.

4.1 Transducer Modeling

An ultrasonic transducer is normally based on a piezoelectric material that has the ability to convert electrical energy at its electrical port into acoustic energy (motion) at its acoustic port and, conversely, to also convert acoustic energy back into electrical energy. Thus a piezoelectric ultrasonic transducer can act as both a transmitter and receiver of sound. In this Chapter we will examine the transducer in its role as a transmitter. By treating the coupled electromagnetic and elastic fields contained in the transducer as those of a piezoelectric medium and considering the fields at the two transducer ports as purely electrical fields and acoustic fields that arise from those internal piezoelectric interactions, one can define a reciprocity relationship between the fields at the two ports in the form [4.1-4.3]

$$\int_{S_c} \left(\mathbf{E}^{(2)} \times \mathbf{H}^{(1)} - \mathbf{E}^{(1)} \times \mathbf{H}^{(2)} \right) \cdot \mathbf{n} \, dS = \int_{S} \left(p^{(1)} \mathbf{v}^{(2)} - p^{(2)} \mathbf{v}^{(1)} \right) \cdot \mathbf{n} \, dS, \tag{4.1}$$

where $(\mathbf{E}, \mathbf{H})$ are the electrical and magnetic fields at the transducer's electrical port (over area S_c) and $(p, \mathbf{v})$ are the pressure and velocity fields at the acoustic port (over area S), and $\mathbf{n}$ is the unit normal pointing outwards from each port (see Fig. 4.1). Only the pressure appears on the right side of Eq. (4.1) since for an immersion transducer this is the only component of the stress tensor that can exist for a fluid. Even for a contact transducer, however, there is normally a thin fluid couplant layer between the transducer

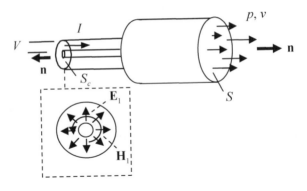

Fig. 4.1. The electrical and magnetic fields at a transducer's electrical port and the corresponding voltage and current flowing into that port. At the acoustic port distributed pressure and velocity fields are generated, as shown.

and the solid component so that in contact testing again only a pressure exists at the transducer face. The superscripts (1) and (2) indicate these fields for two different states (i.e. under two different sets of driving and termination conditions). If we assume that the electrical and magnetic fields at the electrical port are in the form of TEM waves, as done for the cable in the previous Chapter, then we have [4.3]

$$V^{(1)}I^{(2)} - V^{(2)}I^{(1)} = -\int_{S}\left(p^{(2)}\mathbf{v}^{(1)} - p^{(1)}\mathbf{v}^{(2)}\right)\cdot\mathbf{n}\,dS, \qquad (4.2)$$

where V and I are the voltage and current flowing into the electrical port, as shown in Fig. 4.1. At the acoustic port, we will assume the transducer acts as a piston transducer, i.e. the velocity is constant over the area S. This is an assumption frequently used to model ultrasonic transducers and is one we will adopt here. In that case, the right side of Eq.(4.2) can be expressed in terms of the two quantities, F and v, where

$$F(\omega) = \int_{P} p(\mathbf{x},\omega)\,dS(\mathbf{x})$$

$$v(\omega) = \mathbf{v}(\omega)\cdot\mathbf{n} \qquad (4.3)$$

so that F is the compressive force acting at the transducer face and v is the uniform outward normal velocity on this face. In this case Eq. (4.2) becomes

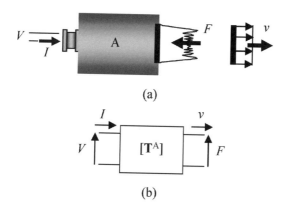

Fig. 4.2. (a) An ultrasonic transducer represented as a device that converts voltage and current into force and velocity and **(b)** its corresponding two port system representation. The pressure distribution over the acoustic port that generates the force F is generally non-uniform, as shown. However, we assume the velocity distribution at the acoustic port is uniform as shown, i.e. the transducer acts as a piston.

$$V^{(1)}I^{(2)} - V^{(2)}I^{(1)} = F^{(1)}v^{(2)} - F^{(2)}v^{(1)}, \tag{4.4}$$

which is the reciprocity relation in terms of "lumped" parameters. Even if the transducer does not act as a piston, it is possible to use Eq. (4.4). The details can be found in [4.3] but we will not discuss that generalization here. In terms of these parameters, therefore, we can consider a transducer as a two port device that converts voltage and current into force and velocity, as shown in Fig. 4.2. If the reciprocity relation Eq. (4.4) is satisfied for all states then this two port system can be written in terms of a reciprocal transfer matrix $\left[\mathbf{T}^A \right]$, where

$$\begin{Bmatrix} V \\ I \end{Bmatrix} = \begin{bmatrix} T_{11}^A & T_{12}^A \\ T_{21}^A & T_{22}^A \end{bmatrix} \begin{Bmatrix} F \\ v \end{Bmatrix} \quad , \quad \det\left[\mathbf{T}^A \right] = 1. \tag{4.5}$$

By modeling the fields in the transducer as 1-D fields, Sittig [4.4], [4.5] developed an explicit expression for the transfer matrix components that describe a compressional wave transducer. In the Sittig model, the transfer matrix of a transducer A, $\left[\mathbf{T}^A\right]$, can be written as a product of two 2x2 transfer matrices, $\left[\mathbf{T}_e^A\right],\left[\mathbf{T}_a^A\right]$, as $\left[\mathbf{T}^A\right]=\left[\mathbf{T}_e^A\right]\left[\mathbf{T}_a^A\right]$, where

$$\left[\mathbf{T}_e^A\right]=\begin{bmatrix} 1/n & n/i\omega C_o \\ -i\omega C_o & 0 \end{bmatrix}$$

$$\left[\mathbf{T}_a^A\right]=\frac{1}{Z_b^a - iZ_0^a \tan(kd/2)}$$

$$\cdot\begin{bmatrix} Z_b^a + iZ_0^a \cot(kd) & \left(Z_0^a\right)^2 + iZ_0^a Z_b^a \cot(kd) \\ 1 & Z_b^a - 2iZ_0^a \tan(kd/2) \end{bmatrix}. \tag{4.6}$$

The multiple parameters appearing in this model are as follows. The parameter k is the wave number for the piezoelectric plate, $k = \omega/v_0$, where v_0 is the wave speed of compressional waves in the piezoelectric plate given by $v_0 = \sqrt{c_{33}^D/\rho_P}$ in terms of the elastic constant of the plate, c_{33}^D, at constant electric flux density, and ρ_P, the density of the plate. The constant $n = h_{33}C_0$ is given in terms of h_{33}, a piezoelectric stiffness constant for the plate, and C_0, the clamped capacitance of the plate, which is given by $C_0 = S/\beta_{33}^S d$, where S is the area of the piezoelectric plate, β_{33}^S is the dielectric impermeability of the plate at constant strain, and d is the plate thickness. The quantity $Z_0^a = \rho_P v_0 S$ is the plane wave acoustic impedance of the piezoelectric plate, while $Z_b^a(\omega)$ is the corresponding acoustic impedance of the backing (which is a function of frequency since the backing normally consists of one or more layers and is highly attenuating).

It can be seen from Eq. (4.6) that in order to use the Sittig model one must know in considerable detail the internal material and geometry parameters of the transducer. When designing and manufacturing transducers, such details are known explicitly but it is not possible to obtain such detailed knowledge of transducers that are purchased commercially. Thus, one must rely instead on experimental means to determine the transfer matrix of the transducer. Unfortunately, at present a practical experimental method does not exist that can determine the complete transfer matrix of an ultrasonic transducer. The problem lies in the fact that it is difficult to enforce different known termination conditions at the acoustic port (as was done in the cable case for one of the electrical

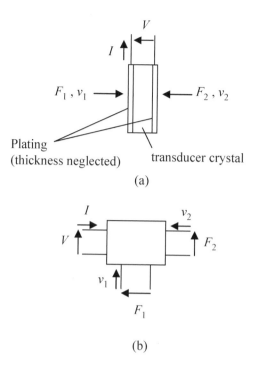

Fig. 4.3. (a) A 1-D model of the electrical and acoustic parameters for a plated piezoelectric crystal and **(b)** its representation as a three port system.

ports of the cable). Also, while it is easy to measure the voltage and current at the electrical port of the transducer it is more difficult to measure the force and velocity parameters at the acoustic port without investing in expensive equipment. Fortunately, as we will show later, we can characterize the role of the transducer in an ultrasonic measurement setup in terms of only two parameters that are related to the transducer's transfer matrix. These two parameters are the transducer's sensitivity and its equivalent electrical impedance. We will also show that it is possible to determine the sensitivity and impedance with purely electrical measurements at the transducer's electrical port. Thus, we can bypass the need to have the full set of transfer matrix components for characterizing the transducer.

In designing ultrasonic transducers, many designers find it convenient to use a three port model instead of the two port Sittig model. The Mason model and the KLM model are two models of this type that are commonly used in practice [4.6],[4.7]. Like the Sittig model, both models

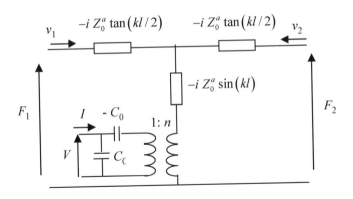

Fig. 4.4. The Mason equivalent circuit model of the three port system defined by Eq. (4.7).

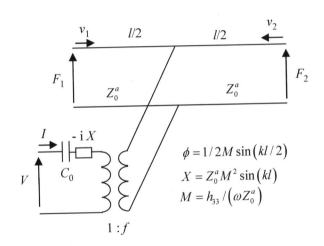

Fig. 4.5. The KLM equivalent circuit model for the three port system defined by Eq. (4.7).

treat the transducer as a plated piezoelectric element where 1-D electrical and mechanical fields are present, as shown in Fig. 4.3. The electrical port is where electrical connections are made to the plated faces of the piezoelectric plate while the two acoustic ports are the two faces of the plate (Fig. 4.3). The electrical and mechanical lumped parameters for this three port model can be shown to satisfy the relations [4.8]

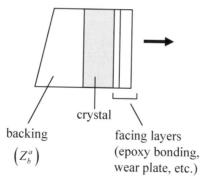

crystal

backing
(Z_b^a)

facing layers
(epoxy bonding,
wear plate, etc.)

Fig. 4.6. Construction of a typical commercial transducer showing the crystal backing and one or more facing acoustic layers at the transducer acoustic output port.

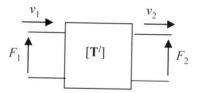

Fig. 4.7. The acoustic two port system model of an acoustic layer.

$$\begin{Bmatrix} F_1 \\ F_2 \\ V \end{Bmatrix} = i \begin{bmatrix} Z_0^a \cot(kl) & Z_0^a/\sin(kl) & h_{33}/\omega \\ Z_0^a/\sin(kl) & Z_0^a \cot(kl) & h_{33}/\omega \\ h_{33}/\omega & h_{33}/\omega & 1/\omega C_0 \end{bmatrix} \begin{Bmatrix} v_1 \\ v_2 \\ I \end{Bmatrix}, \qquad (4.7)$$

which can be seen to be given in the form of a 3x3 impedance matrix. Note that in Eq. (4.7) the velocities are assumed to be flowing into the transducer at the acoustic ports. This convention is opposite to what is assumed (at the acoustic output port) of a transfer matrix model (see Fig. 4.2 (b)). If the material backing on the piezoelectric element is specified as a given acoustic impedance, $Z_b^a(\omega)$, as done for the Sittig model, then this three port model reduces to a two port model. The Sittig model is just a transfer matrix representation of the resulting two port system. In contrast, the Mason and KLM models are just equivalent circuit representations of the three port system described by Eq. (4.7) where the acoustic impedance of the backing of the piezoelectric element is not specified. Figure 4.4 shows a schematic of the Mason equivalent circuit model and Fig. 4.5 shows the KLM equivalent circuit.

The Sittig model is a particularly useful model to use to consider additional acoustic layers in the transducer model at the transducer output port. Such layers are normally present in the form of wear plates to protect the piezoelectric element or impedance matching plates (Fig. 4.6) and can be represented as acoustic two port systems (Fig. 4.7). The transfer matrix $\left[\mathbf{T}^l\right]$ for an acoustic layer containing 1-D propagating compressional waves is given by

$$
\begin{Bmatrix} F_1 \\ v_1 \end{Bmatrix} = \begin{bmatrix} \cos\left(k_a l_a\right) & -iZ_0^a \sin\left(k_a l_a\right) \\ -i\sin\left(k_a l_a\right)/Z_0^a & \cos\left(k_a l_a\right) \end{bmatrix} \begin{Bmatrix} F_2 \\ v_2 \end{Bmatrix}, \tag{4.8}
$$

where $k_a = \omega/c$ is the wave number for waves traveling in the layer with compressional wave speed, c, l_a is the layer thickness, and $Z_0^a = \rho c S$ is the acoustic impedance of the layer, with ρ the density of the layer and S is the cross-sectional area. Note that this transfer matrix has exactly the same form as the matrix obtained for a cable, so this matrix is the acoustic analog of that electrical model. A transducer containing such an acoustic layer can be joined with the Sittig model by simply multiplying that model by an additional acoustic transfer matrix so that the entire transfer matrix for the transducer, $[\mathbf{T}^A]$, is given by

$$
\left[\mathbf{T}^A\right] = \left[\mathbf{T}_e^A\right]\left[\mathbf{T}_a^A\right]\left[\mathbf{T}^l\right] \tag{4.9}
$$

and more layers can be handled in exactly the same fashion.

4.2 Transducer Acoustic Radiation Impedance

When an ultrasonic transducer is used in an ultrasonic measurement system its acoustic port is always terminated, i.e. the output force and velocity are related to one another. For an immersion transducer radiating into a fluid, for example, we will show in Chapter 8 that for a planar piston transducer the pressure field, $p(\mathbf{x}, \omega)$, on the face of the acoustic output port of the transducer is given in terms of the uniform normal velocity, $v_t(\omega)$, at that port by the Rayleigh-Sommerfeld integral:

Fig. 4.8. An ultrasonic immersion transducer radiating into a fluid.

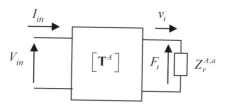

Fig. 4.9. A transducer A, whose acoustic radiation impedance is $Z_r^{A;a}$, radiating into a material and modeled as a acoustically terminated two port system.

$$p(\mathbf{x},\omega) = \frac{-i\omega\rho v_t(\omega)}{2\pi} \int_S \frac{\exp(ikr)}{r} dS(\mathbf{y}),$$ (4.10)

where $\mathbf{x}$ and $\mathbf{y}$ are two points on the surface, S, of the transducer face, ρ is the density of the fluid, $k = \omega/c$ is the wave number for waves propagating in the fluid whose compressional wave speed is c, and $r = |\mathbf{x} - \mathbf{y}|$ is the distance between $\mathbf{x}$ and $\mathbf{y}$. Since the compressional force, F_t, at the transducer's output port is just the integral of this pressure, we have

$$F_t(\omega) = \left[\frac{-i\omega\rho}{2\pi} \int_S \int_S \frac{\exp(ikr)}{r} dS(\mathbf{y}) dS(\mathbf{x}) \right] v_t(\omega)$$

$$= Z_r^a(\omega) v_t(\omega),$$ (4.11)

where the term in brackets in Eq. (4.11), Z_r^a, is called the *transducer radiation impedance*. The radiating transducer A of Fig. 4.8, therefore, can be represented as a terminated two port system as shown in Fig. 4.9. Greenspan [4.9] has shown that the two integrals in Eq. (4.11) can be performed for

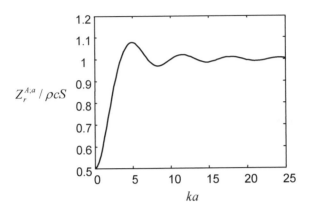

Fig. 4.10. The normalized acoustic radiation impedance of a circular, planar, piston transducer A of radius a versus the non-dimensional frequency, ka.

a circular planar piston transducer of radius a, to obtain an explicit expression for the radiation impedance given by

$$Z_r^{A;a} / \rho c S_A = 1 - \left[J_1(ka) - i S_1(ka) \right] / ka, \tag{4.12}$$

where J_1, S_1 are first order Bessel and Struve functions, respectively, and $S_A = \pi a^2$ is the area of the "active" face of the transducer at its acoustic port. Figure 4.10 shows a plot of this normalized radiation impedance versus ka, which is a non-dimensional frequency.

It can be seen from Fig. 4.10 that for approximately $ka > 10$ we can take $Z_r^a = \rho c S_A$ which is just the value of the acoustic impedance of a traveling plane wave. For most ultrasonic transducers, the ka value at the MHz frequencies used in testing is large. For example, at 5 MHz a 6.35 mm radius piston transducer radiating into water has a ka value of approximately 135. This same transducer radiating into steel would have a ka value of approximately 34. Thus, even though such ultrasonic transducers generate sound beams that are not just plane waves, their acoustic radiation impedances can generally be taken as simply as the constant value, $\rho c S$, of a plane wave. This is true for any shaped piston transducer, not just the circular case considered by Greenspan. To see this consider Eq. (4.11) again and with $\mathbf{x}$ fixed let $dS(\mathbf{y}) = r dr d\phi'$ (see Fig. 4.11 (a)). Then the radial integration can be performed to yield

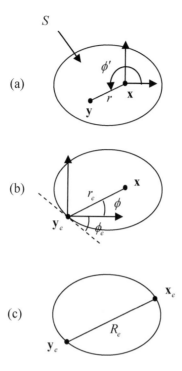

Fig. 4.11. (a) Integration over points **y** on the transducer face, and **(b)** averaging over points **x** on the transducer face, leading to **(c)** remaining integrations in terms of the distance, R_e , between points on the transducer edge.

$$F_t(\omega) = \frac{-i\omega\rho v_t(\omega)}{2\pi} \int_S \left[\int_0^{2\pi} \int_0^{r_e} \exp(ikr) \, dr \, d\phi' \right] dS(\mathbf{x})$$

$$= \frac{\rho c v_t(\omega)}{2\pi} \int_S \left[2\pi - \int_0^{2\pi} \exp(ikr_e) \, d\phi' \right] dS(\mathbf{x}),$$

(4.13)

where $r_e = r_e(\mathbf{x}, \mathbf{y}_e(\phi'))$ is the radius from point **x** to a general point on the edge of the transducer surface, S (see Fig. 4.11 (b)). With $\mathbf{y}_e(\phi')$ fixed, we can let $dS(\mathbf{x}) = r_e \, dr_e \, d\phi$ and Eq. (4.13) becomes

$$F_t(\omega) = \rho c S v_t(\omega) - \frac{\rho c v_t(\omega)}{2\pi} \int_{-\phi_e}^{\pi-\phi_e} \int_0^{2\pi} \left[\int_0^{R_e} \exp(ikr_e) r_e dr_e \right] d\phi d\phi', \qquad (4.14)$$

where R_e is shown in Fig. 4.11 (c). Performing the integral on r_e by parts, it follows that

$$\int_0^{R_e} \exp(ikr_e) r_e dr_e = \frac{1}{ikR_e} \left[R_e^2 \exp(ikR_e) + \frac{R_e^2}{ikR_e} (1 - \exp(ikR_e)) \right]$$
$$= O(1/kR_e) \qquad (4.15)$$

so that at high frequencies the integral in Eq. (4.14) can be neglected and we have

$$F_t(\omega) = \rho c S \, v_t(\omega) \qquad (4.16)$$

4.3 Transducer Impedance and Sensitivity

Since to date there is not a practical method available to determine experimentally all the transfer matrix components of a radiating transducer, it is necessary to re-examine the terminated model of Fig. 4.9 and express it in terms of quantities that can be easily measured. In this case we can write the transfer matrix relations for a transmitting transducer A either in terms of the transmitted output force, F_t, or the transmitted output velocity, v_t, since

$$\begin{Bmatrix} V_{in} \\ I_{in} \end{Bmatrix} = \begin{bmatrix} T_{11}^A & T_{12}^A \\ T_{21}^A & T_{22}^A \end{bmatrix} \begin{Bmatrix} Z_r^{A;a} v_t \\ v_t \end{Bmatrix}$$
$$= \begin{bmatrix} T_{11}^A & T_{12}^A \\ T_{21}^A & T_{22}^A \end{bmatrix} \begin{Bmatrix} F_t \\ F_t / Z_r^{A;a} \end{Bmatrix}. \qquad (4.17)$$

The effects of this transducer on the other electrical components connected to it through its electrical port are determined by the transducer's electrical impedance, $Z_{in}^{A;e}(\omega)$, which is given by

$$Z_{in}^{A;e} = \frac{V_{in}}{I_{in}} = \frac{Z_r^{A;a} T_{11}^A + T_{12}^A}{Z_r^{A;a} T_{21}^A + T_{22}^A} \qquad (4.18)$$

However, this quantity can obviously be obtained by measuring V_{in} and I_{in}, the driving voltage and current at the transducer's electrical port, respectively, when it is radiating into a material and it is not necessary to know the underlying transfer matrix components in Eq. (4.18) [4.10]. If the transducer's electrical impedance $Z_{in}^{A;e}(\omega)$ were found in this fashion by electrical measurements and if we also had characterized the pulser and cabling by the methods discussed in Chapters 2 and 3 for a given ultrasonic setup, we could then find explicitly both the voltage V_{in} and the current I_{in} at the transducer's electrical port for this setup. Thus, $Z_{in}^{A;e}(\omega)$ is all that is needed to characterize the electrical properties of the transducer in an ultrasonic measurement system. In addition, if we knew the transducer's radiation impedance, $Z_r^{A;a}$ and also obtained a measure of a quantity such as v_t / I_{in} or F_t / V_{in}, we could determine both the output force and velocity of the transducer and we would have characterized the transducer completely, i.e. both electrically and acoustically. Such quantities, which are just ratios of a transducer output to a transducer input, are called transducer transmitting sensitivities, S_{OI}, where O is an output quantity such as force or velocity, and I is an input quantity such as voltage or current, and $S_{OI} = O / I$. There are, obviously, a number of different sensitivities one could define. For example we have

$$S_{vI}^A = \frac{v_t}{I_{in}}$$

$$S_{FI}^A = \frac{F_t}{I_{in}} = Z_r^{A;a} S_{vI}^A$$

$$\tag{4.19}$$

$$S_{vV}^A = \frac{v_t}{V_{in}} = S_{vI}^A / Z_{in}^{A;e}$$

$$S_{FV}^A = \frac{F_t}{V_{in}} = Z_r^{A;a} S_{vI}^A / Z_{in}^{A;e}$$

We will choose to describe the transducer A in terms of its sensitivity S_{vI}^A. As Eq. (4.19) shows, if we also know the transducer's electrical impedance, $Z_{in}^{A;e}(\omega)$, and its acoustic radiation impedance, $Z_r^{A;a}$, we could then also obtain any of the other sensitivities listed in Eq. (4.19). From Eq. (4.17) it follows that:

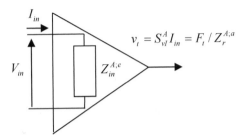

Fig. 4.12. A model of a transmitting ultrasonic transducer as an electrical impedance and an ideal "converter" that transforms the input electrical signals into the acoustic output signals.

$$S_{vI}^A \equiv \frac{v_t}{I_{in}} = \frac{1}{Z_r^{A;a} T_{21}^A + T_{22}^A}. \tag{4.20}$$

It will be shown in Chapter 7 that it is possible to obtain this sensitivity by direct electrical measurements of the voltage and current at the transducer's electrical port, so that there is a practical way to determine all the transducer parameters, $Z_{in}^{A;e}, S_{vI}^A, Z_r^{A;a}$ .Thus, we can replace the two port transfer matrix model of the transmitting transducer by the simpler model shown in Fig. 4.12, where we have represented the transducer as an electrical impedance and an ideal "converter" that transforms the input current to output velocity (or force).

4.4 The Sound Generation Process

We can combine our pulser, cabling and transducer models into a complete model of the entire sound generation process in an ultrasonic measurement system [4.10]. This generation process model is shown schematically in Fig. 4.13. We can treat this whole process as a single input, single output LTI system that is characterized by a transfer function, $t_G(\omega)$, as shown in Fig. 4.14. We will choose to write this transfer function in terms of the output force rather than the output velocity as $t_G(\omega) = F_t(\omega)/V_i(\omega)$. Since we have defined all of the elements contained in the sound generation process, we can obtain an explicit expression for this transfer function. From Fig. 4.13 we have

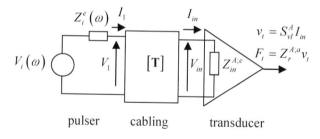

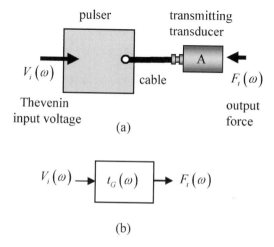

pulser   cabling   transducer

Fig. 4.13. A model of the entire sound generation process in an ultrasonic system.

pulser   transmitting
                         transducer

$V_i(\omega)$ $F_t(\omega)$

Thevenin   output
input voltage   force
            (a)

$V_i(\omega) \rightarrow$ $t_G(\omega)$ $\rightarrow F_t(\omega)$

(b)

Fig. 4.14. (a) The elements in the sound generation process – the pulser, the cabling, and the transmitting transducer and **(b)** an LTI system model of the sound generation process whose transfer matrix is $t_G(\omega)$.

$$V_i - V_1 = Z_i^e I_1,$$ (4.21)

$$\begin{Bmatrix} V_1 \\ I_1 \end{Bmatrix} = \begin{bmatrix} T_{11} & T_{12} \\ T_{21} & T_{22} \end{bmatrix} \begin{Bmatrix} V_{in} \\ I_{in} \end{Bmatrix},$$ (4.22)

$$F_t = Z_r^{A,a} S_{vl}^A I_{in},$$ (4.23)

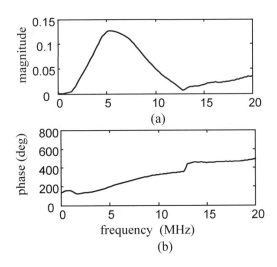

Fig. 4.15. A sound generation transfer function obtained experimentally. **(a)** Magnitude of the transfer function versus frequency and **(b)** its phase versus frequency.

$$V_{in} = Z_{in}^{A;e} I_{in}. \qquad (4.24)$$

Using Eqs. (4.21 - 4.24) it is easy to show that [4.10]

$$t_G(\omega) = \frac{F_t(\omega)}{V_i(\omega)} = \frac{Z_r^{A;a} S_{vI}^A}{\left(Z_{in}^{A;e} T_{11} + T_{12}\right) + \left(Z_{in}^{A;e} T_{21} + T_{22}\right) Z_i^e}, \qquad (4.25)$$

where $(T_{11}, T_{12}, T_{21}, T_{22})$ are the components of the transfer matrix, $[\mathbf{T}]$, for the cabling between the pulser and the transmitting transducer, $Z_{in}^{A;e}$ is the electrical impedance of the transmitting transducer A and S_{vI}^A is its sensitivity, and $Z_r^{A;a}$ is the acoustic radiation impedance of the transducer. With this transfer function we can model completely the effect of the pulser, the cabling and the transducer and predict the output force, $F_t(\omega)$. Figure 4.15 shows an example where the magnitude and phase of a sound generation transfer function, $t_G(\omega)$, was experimentally determined by characterizing all the components contained in Eq. (4.21). In this case the pulser was the pulser section of a Panametrics 5052 PR pulser/receiver (measured at a set of specific energy and damping settings). The cabling consisted of 1.83 m of flexible 50 ohm coaxial cable connected to a 0.61 m fixture rod. The rod also contained internal cabling and was terminated by a right-angle adapter to which the transducer was connected. The transducer was a relatively broadband 6.35 mm diameter 5 MHz immersion transducer. The sensitivity and impedance of the transducer were obtained by the methods which will be discussed in Chapter 6.

4.5 References

4.1 Foldy LL., Primakoff H (1945) A general theory of passive linear electro-acoustic transducers and the electroacoustic reciprocal theorem. J. Acoust. Soc. Am. 17: 109-120

4.2 Primakoff H, Foldy LL (1947) A general theory of passive linear electro-acoustic transducers and the electroacoustic reciprocal theorem, II. J. Acoust. Soc. Am. 19: 50-58

4.3 Dang, CJ, Schmerr LW, Sedov A (2002) Modeling and measuring all the elements of an ultrasonic nondestructive evaluation system I: Modeling foundations. Research in Nondestructive Evaluation 14: 141-176

4.4 Sittig EK, (1967) Transmission parameters of thickness-driven piezoelectric transducers arranged in multilayer configurations. IEEE Trans. Sonics and Ultrasonics SU-14: 167-174

4.5 Sittig EK (1969) Effects of bonding and electrode layers on the transmission parameters of piezoelectric transducers used in ultrasonic digital delay lines. IEEE Trans. Sonics and Ultrasonics SU-16: 2-10

4.6 Mason WP (Ed.) (1964) Physical Acoustics, Vol. 1- Part A. Academic Press, New York, NY

4.7 Krimholtz R, Leedom DA, Matthaei GL (1970) New equivalent circuits for elementary piezoelectric transducers. Electronics Letters 16: 398-399

4.8 Kino GS (1987) Acoustic Waves: Devices, Imaging, and Analog Signal Processing. Prentice Hall, Englewood Cliffs, NJ

4.9 Greenspan M (1979) Piston radiator: some extensions of the theory. J. Acoust. Soc. Am. 65: 608-621

4.10 Dang CJ, Schmerr LW, Sedov A (2002) Modeling and measuring all the elements of an ultrasonic nondestructive evaluation system. II : Model-based measurements. Research in Nondestructive Evaluation 14: 177-201

4.6 Exercises

1. Equation (4.8) gave the transfer matrix for a layer in terms of the force and the velocity on both sides of an elastic layer. Consider the case where a plane compressional (P-) wave is incident on a layer, generating both

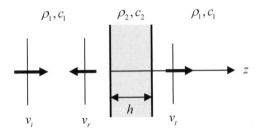

Fig. 4.16. A plane wave incident on a layer.

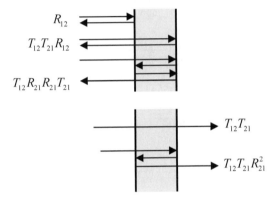

Fig. 4.17. Waves incident on a layer, showing the first few reflected and transmitted waves.

transmitted and reflected waves, as shown in Fig. 4.16. Let the velocities of these waves in their directions of propagation be given by

$$v_i = V_i \exp\left[ik_1 x\right]$$
$$v_r = V_r \exp\left[-ik_1 x\right]$$
$$v_t = V_t \exp\left[ik_1 (x-h)\right],$$

where we have written the transmitted wave in terms of the coordinate $x_2 = x - h$ since that wave only exists for $x_2 \geq 0$. Then the corresponding forces in these waves are

$$F_i = (\rho_1 c_1 S) v_i = Z_1 v_i$$
$$F_r = (\rho_1 c_1 S) v_r = Z_1 v_r$$
$$F_t = (\rho_1 c_1 S) v_t = Z_1 v_t.$$

On the sides of the layer we have $F_1 = F_i + F_r, F_2 = F_t, v_1 = V_i + V_r, v_2 = V_t$. Using Eq. (4.8) for the layer then we can obtain the reflection and transmission coefficients of the layer in the forms

$$R = \frac{F_r}{F_i} = R_{12} + \frac{R_{21} T_{12} T_{21} \exp(2ik_2 h)}{1 - R_{21}^2 \exp(2ik_2 h)}$$

$$T = \frac{F_t}{F_i} = \frac{T_{12} T_{21} \exp(ik_2 h)}{1 - R_{21}^2 \exp(2ik_2 h)},$$

where R_{ij}, T_{ij} are the plane wave reflection and transmission coefficients for a single interface going from medium i to medium j given by (see Appendix D):

$$R_{12} = -R_{21} = \frac{\rho_2 c_2 - \rho_1 c_1}{\rho_1 c_1 + \rho_2 c_2}$$

$$T_{12} = \frac{2 \rho_2 c_2}{\rho_1 c_1 + \rho_2 c_2}$$

$$T_{21} = \frac{2 \rho_1 c_1}{\rho_1 c_1 + \rho_2 c_2}.$$

The layer reflection and transmission coefficients (R, T) are functions of frequency because they contain all the waves that bounce back and forth in the layer and emerge into the adjacent media. To examine this behavior in frequency use MATLAB to plot the magnitude of these coefficients for 500 frequency values ranging from zero to 20 MHz for a thin (1 mm thick) aluminum plate in water. Can you explain the frequency dependent behavior of this plot?

To see the individual reflected and transmitted waves, we can expand the denominators of the (R, T) expressions and obtain

$$R = R_{12} + R_{21} T_{12} T_{21} \exp(2ik_2 h)\{1 + R_{21}^2 \exp(2ik_2 h) + ...\}$$
$$T = T_{12} T_{21} \exp(ik_2 h)\{1 + R_{21}^2 \exp(2ik_2 h) + ...\},$$

which are the first few reflected and transmitted waves as shown in Fig. 4.17. Use MATLAB to calculate the magnitude of (R, T) for just these first few terms. How do your results here compare to your previous results?

5 The Acoustic/Elastic Transfer Function and the Sound Reception Process

5.1 Wave Processes and Sound Reception

The last Chapter showed how to characterize the relationship between the Thévenin equivalent driving voltage of the pulser and the output force, $F_t(\omega)$, at the face of the transmitting transducer. That output force will launch waves from the transducer, waves that will propagate and interact with the component being inspected as well as with whatever flaws may be present. A portion of these waves will be captured by a receiving transducer as shown in Fig. 5.1. The waves incident on the receiving transducer will generate a force on that transducer, labeled $F_B(\omega)$ in Fig. 5.1. All the acoustic/elastic wave propagation and scattering interactions that occur between the transmitting transducer and the receiving transducer are complex 3-D wave phenomena.

Later Chapters will describe in detail how models can describe these waves. Here, we are interested in characterizing the role that the acoustic/elastic interactions play in the overall ultrasonic measurement system and we will give some simple examples of those interactions. We will also describe models for characterizing the entire reception process (see Fig. 5.2) where the force, $F_B(\omega)$, is converted into electrical energy at the receiving transducer, transmitted by a cable to the receiver, and then amplified to generate a final system output voltage, $V_R(\omega)$. Like the process of sound generation both the acoustic/elastic process and the reception process can be modeled as transfer functions. The acoustic/elastic transfer function is defined as:

$$t_A(\omega) = \frac{F_B(\omega)}{F_t(\omega)} \tag{5.1}$$

and the reception process transfer function is defined as:

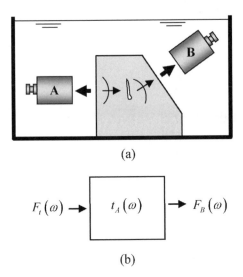

(a)

$$F_t(\omega) \rightarrow \boxed{t_A(\omega)} \rightarrow F_B(\omega)$$

(b)

Fig. 5.1. (a) An ultrasonic pitch-catch immersion inspection, showing the acoustic/elastic waves present between the sending transducer and the receiving transducer, and **(b)** an LTI system model of those acoustic/elastic processes whose transfer function is $t_A(\omega)$.

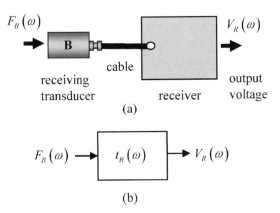

Fig. 5.2. (a) The elements of the reception process – the receiving transducer, the cabling, and the receiver portion of a pulser/receiver, and **(b)** an LTI system model of the reception process whose transfer function is $t_R(\omega)$.

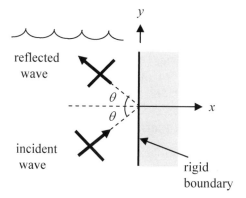

Fig. 5.3. Modeling the interaction of the waves incident on a "blocked" receiving transducer where the waves are treated as plane waves and the transducer surface is treated as a infinite, planar and rigid (immobile) boundary.

$$t_R(\omega) = \frac{V_R(\omega)}{F_B(\omega)} \qquad (5.2)$$

5.2 The Blocked Force

The force, $F_B(\omega)$, appearing in both Eqs. (5.1) and (5.2) is a particular force acting on the receiving transducer called the *blocked force*. This blocked force is defined as the force that would be exerted on the receiving transducer if its face was held rigidly fixed (immobile). We will see shortly why this specific force arises naturally when we discuss the reception process. However, we can use a simple model to gain some additional understanding of this force. Consider, for example, the waves incident on a receiving transducer in an immersion setup. Let θ be the angle that these incident waves make with the normal to the transducer and assume that these incident waves behave like harmonic plane waves, as shown in Fig. 5.3. If we neglect any wave diffraction effects at the edges of the receiving transducer, we can model the face of that transducer, when its face is held rigidly fixed, as an infinite plane rigid surface, as shown in Fig. 5.3. The pressure of the incident plane wave can be given as

$$p_{inc} = P_i \exp\left[ik\left(x\cos\theta + y\sin\theta \right) - i\omega t \right] \qquad (5.3)$$

and the pressure in the plane reflected wave given by

$$p_{reflt} = P_r \exp\left[ik\left(-x\cos\theta + y\sin\theta\right) - i\omega t\right] \tag{5.4}$$

since it reflects from the surface with the same angle as the incident wave as shown in Appendix D. At the transducer face, $x = 0$, which is held rigidly fixed, the total displacement and velocity normal to the transducer (in the x-direction) must be zero. Thus, from the equation of motion (see Appendix D) we have at the transducer face

$$v_x\left(x,y,t\right)\Big|_{x=0} = \frac{1}{i\omega\rho} \frac{\partial p\left(x,y,t\right)}{\partial x}\Big|_{x=0} = 0 \tag{5.5}$$

where $p = p_{inc} + p_{reflt}$ is the total pressure. Placing Eqs. (5.3) and (5.4) into Eq. (5.5) we find

$$\frac{ik\cos\theta}{i\omega\rho}\left(P_i - P_r\right)\exp\left(iky\sin\theta - i\omega t\right) = 0 \tag{5.6}$$

so that $P_i = P_r$ and the total pressure, p_B, at the blocked transducer face is just $p_B = 2p_{inc}$. If we let S be the area of the face of the transducer then we see that the blocked force acting on the face of the transducer, $F_B\left(\omega\right) = \iint_S p_B dS$, is just twice that of the force $F_{inc} = \iint_S p_{inc} dS$, exerted by the incident wave over the same area, i.e.

$$F_B\left(\omega\right) = 2F_{inc}\left(\omega\right) \tag{5.7}$$

To summarize: If we assume plane wave interactions at the receiving transducer, the blocked force, $F_B\left(\omega\right)$, is just twice the force, $F_{inc}\left(\omega\right)$, exerted by the waves incident on the area of the receiver. The force, $F_{inc}\left(\omega\right)$, acting on S is computed from the incident waves *as if the transducer were absent*.

Many authors use Eq. (5.7) without further discussion since the plane wave interaction assumption on which it is based is likely a good assumption in most cases. We will also find it useful to use Eq. (5.7) when obtaining the acoustic/elastic transfer function since then we can model the pressure wave field of only the incident waves at the receiving transducer and use Eq. (5.7) to obtain the blocked force, without having to consider explicitly any more complex interactions of the incident waves with the receiving transducer.

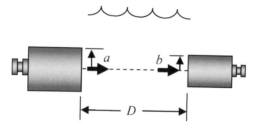

Fig. 5.4. An ultrasonic pitch-catch calibration setup where the waves generated by a circular planar piston transducer are received by a second circular planar transducer and where the transducer axes are aligned.

5.3 The Acoustic/Elastic Transfer Function

To obtain the acoustic/elastic transfer function, $t_A(\omega)$, in a general ultrasonic NDE measurement system requires a knowledge of the waves propagating in the component being inspected as well as the waves generated by any flaws present. We will develop models needed to describe those waves in Chapters 9 and 10. Here, however, we will discuss some simple setups where there are explicit analytical expressions for the acoustic/elastic transfer function. One setup that is commonly used for calibrating pitch-catch setups is shown in Fig. 5.4 where a circular planar piston transducer, of radius a, radiates waves into a fluid which are captured by a circular planar piston receiving transducer of radius b, where the two central axes of the transducers are aligned and the transducer faces are parallel to one another. In this case an explicit model has been developed for $F_{inc}(\omega)$, the force of the waves incident on the area of the receiver (in the absence of that receiver). This force is given by [5.1]

$$F_{inc}(\omega) = \rho c_p v_0(\omega)\Big\{\Theta\exp\big(ik_p D\big)$$

$$-16a^2 b^2 \int_0^{\pi/2} \frac{\sin^2 u \cos^2 u}{(a-b)^2 + 4ab\cos^2 u}$$

$$\cdot\exp\Big[ik_p \sqrt{D^2 + (a-b)^2 + 4ab\cos^2 u}\,\Big]du\Big\}, \qquad (5.8)$$

where

$$\Theta = \begin{cases} \pi b^2 & a \geq b \\ \pi a^2 & b \geq a \end{cases} \tag{5.9}$$

and ρ, c_p are the density and compressional wave speed of the fluid, respectively, $k_p = \omega / c_p$, $v_0(\omega)$ is the velocity on the face of the transmitting transducer, and D is the distance between the transducers. If we take the acoustic radiation impedance of the transmitter as $Z_r^a = \pi a^2 \rho c_p$ and the blocked force at the receiver as $F_B = 2F_{inc}$, then we have for the transfer function

$$t_A(\omega) = \frac{2}{\pi a^2} \Big\{ \Theta \exp(ik_p D)$$

$$-16a^2 b^2 \int_0^{\pi/2} \frac{\sin^2 u \cos^2 u}{(a-b)^2 + 4ab \cos^2 u} \tag{5.10}$$

$$\cdot \exp\left[ik_p \sqrt{D^2 + (a-b)^2 + 4ab \cos^2 u} \,\right] du \Big\}.$$

In the special case when the transducers are both of the same size ($b = a$), Eq. (5.10) reduces to

$$t_A(\omega) = 2 \Big\{ \exp(ik_p D) - \frac{4}{\pi}$$

$$\cdot \int_0^{\pi/2} \sin^2 u \exp\left[ik_p \sqrt{D^2 + 4a^2 \cos^2 u} \,\right] du \Big\}. \tag{5.11}$$

At high frequencies the integral in Eq. (5.11) can be evaluated analytically, yielding [Fundamentals]

$$t_A(\omega) = 2 \exp(ikD) \Big[1 - \exp(ik_p a^2 / D)$$

$$\cdot \{ J_0(k_p a^2 / D) - iJ_1(k_p a^2 / D) \} \Big], \tag{5.12}$$

where J_0, J_1 are Bessel functions of order zero and one, respectively.

Although Eq. (5.12) is only an approximation of Eq. (5.11) it has been found to give accurate results when $k_p a \gg 1$ which is well satisfied for the size of transducers and frequencies used in NDE testing. Thus, Eq. (5.12) can be regularly used in place of Eq. (5.11). This eliminates the need to numerically evaluate any integrals.

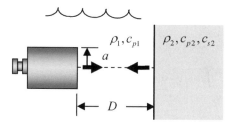

Fig. 5.5. An ultrasonic pulse-echo calibration setup where the waves generated by a circular, planar, piston transducer are reflected from a plane fluid-solid interface at normal incidence and the reflected waves are received by the same transducer.

Some more explicit results can also be obtained from Eq. (5.10) for other cases as well. For example, if we assume $a \gg b$ Eq. (5.10) reduces to

$$t_A(\omega) = 2\frac{b^2}{a^2}\left\{\exp\left[ik_p D\right] - \exp\left[ik_p\sqrt{D^2 + a^2}\right]\right\}. \qquad (5.13)$$

This is just the case where the receiver is small enough so that it acts as a point source and the transfer function is just proportional to the on-axis pressure of the transmitting transducer (see Chapter 8). Similarly, if we assume $b \gg a$ then Eq. (5.10) becomes

$$t_A(\omega) = 2\left\{\exp\left[ik_p D\right] - \exp\left[ik_p\sqrt{D^2 + a^2}\right]\right\}, \qquad (5.14)$$

which again is proportional to the on-axis pressure. For the case where the transducers are separated by a large distance D, where $D \gg a, b$, Eq. (5.10) becomes

$$t_A(\omega) = -ik_p a^2 \frac{\exp\left(ik_p D\right)}{D}, \qquad (5.15)$$

which has the behavior of a spherically spreading wave, a behavior that is characteristic of point sources and the transducer far-field (again, see Chapter 8).

A similar immersion calibration setup that is useful for pulse-echo testing is shown in Fig. 5.5 where a circular planar piston transducer of radius a is oriented at normal incidence to the planar surface of a solid block. In this case, the force in the waves incident on the receiver from the front face of the solid, $F_{inc}(\omega)$, can be obtained as [Fundamentals]:

$$F_{inc}(\omega) = \pi a^2 \rho_1 c_{p1} v_0(\omega) R_{12} \exp(2ik_{p1}D) \Big[1 - \exp(ik_{p1}a^2/2D)$$
$$\cdot \Big\{ J_0(k_{p1}a^2/2D) - iJ_1(k_{p1}a^2/2D) \Big\} \Big], \tag{5.16}$$

where ρ_1, c_{p1} are the density and compressional wave speed of the fluid, $k_{p1} = \omega/c_{p1}$ is the wave number, $v_0(\omega)$ is the velocity on the face of the transmitting transducer when it is firing, and D is the distance from the transducer to the fluid-solid interface (Fig. 5.5). The quantity R_{12} is the plane wave reflection coefficient for the interface, based on the ratio of the reflected pressure to that of the incident pressure (see appendix D) given by

$$R_{12} = \frac{\rho_2 c_{p2} - \rho_1 c_{p1}}{\rho_2 c_{p2} + \rho_1 c_{p1}}, \tag{5.17}$$

where ρ_2, c_{p2} are the density and compressional wave speed of the solid, respectively. If we again take the radiation impedance as $Z_r^a = \pi a^2 \rho_1 c_{p1}$ and the blocked force as $F_B = 2F_{inc}$, we obtain the transfer function

$$t_A(\omega) = 2R_{12} \exp(2ik_{p1}D) \Big[1 - \exp(ik_{p1}a^2/2D)$$
$$\cdot \Big\{ J_0(k_{p1}a^2/2D) - iJ_1(k_{p1}a^2/2D) \Big\} \Big]. \tag{5.18}$$

It is interesting to note that apart from the reflection coefficient Eq. (5.18) is identical to Eq. (5.12) if we replace the D in Eq. (5.12) by $2D$. This similarity occurs because we can view the reflected waves as arising from a fictitious "image" transmitting transducer located a distance $2D$ from the receiving transducer. Thus, for the pitch-catch response of two transducers of the same radius located co-axially in a fluid we have, from Eq. (5.12)

$$t_A(\omega) = \tilde{D}_p(k_p a^2/D) \exp(ik_p D) \tag{5.19}$$

and for the pulse-echo case, from Eq. (5.18)

$$t_A(\omega) = \tilde{D}_p(k_p a^2/2D) R_{12} \exp(2ik_p D) \tag{5.20a}$$

with

$$\tilde{D}_p(u) = 2\Big[1 - \exp(iu)\{ J_0(u) - iJ_1(u) \} \Big]. \tag{5.20b}$$

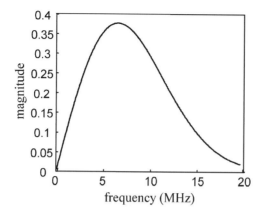

Fig. 5.6. The magnitude of the acoustic/elastic transfer function for two identical circular 3.175 mm radius planar piston transducers in water facing one another in a pitch-catch configuration as shown in Fig. 5.4 with the distance D = 444 mm. The effect of attenuation was included by using Eq. (5.22a) with the attenuation given by Eq. (5.21).

From Eq. (5.19) we can recognize the term without the $\tilde{D}_p$ function as just the transfer function for a plane wave that had traveled directly from the transmitter to the receiver, while in Eq. (5.20a) the terms without the $\tilde{D}_p$ function would be the transfer function describing a plane wave that had traveled from the transmitter to the interface, been reflected from the interface and then traveled back to the receiver. Thus, $\tilde{D}_p$ is just the diffraction correction term for these two cases that takes into account the deviations from a plane wave result. These deviations exist because the transducer produces a beam of sound rather than just a plane wave (see the discussion in Chapter 8 of diffraction corrections and the paraxial approximation). The factor of two in the $\tilde{D}_p$ expression arises simply because our transfer function is defined in terms of the blocked force rather than the force of the incident waves.

In using these transfer functions to model the propagation of waves in a real fluid, such as water, it is important to include the effects of material attenuation, which is absent in these transfer functions since they were developed under the assumption that the waves were propagating in an ideal (loss free) compressible fluid. Adding attenuation to these transfer functions can be done by including a term of the form $\exp\left[-\alpha(f)z\right]$,

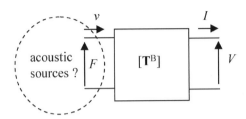

Fig. 5.7. A receiving transducer as a two port system. To use this model we need to know the nature of the acoustic sources driving the transducer.

where $\alpha(f)$ is a frequency dependent attenuation coefficient (measured in Nepers/unit length – see Appendix D) for the material the waves are traveling in and z is the distance traveled in that material. The attenuation coefficient for water at room temperature, for example, has been measured as [Fundamentals]

$$\alpha_w(f) = 25.3 \times 10^{-6} f^2 \text{ Nepers/mm} \tag{5.21}$$

where f is the frequency in MHz. Using this attenuation correction the transfer functions of Eq. (5.19) and (5.20) become

$$t_A(\omega) = \tilde{D}_p \left(k_p a^2 / D \right) \exp\left(i k_p D \right) \exp\left[-\alpha_w(f) D \right] \tag{5.22a}$$

and

$$t_A(\omega) = \tilde{D}_p \left(k_p a^2 / 2D \right) R_{12} \exp\left(2 i k_p D \right) \exp\left[-2\alpha_w(f) D \right] \tag{5.22b}$$

An example calculation to show the behavior of the transfer function in Eq. (5.22a) is given in Fig. 5.6.

There are other simple setups where one can develop explicit expressions for the transfer function $t_A(\omega)$ but we will not discuss those cases here. The two setups we have described will be particularly useful in setting up model-based measurements that allow us to characterize all the electrical and electromechanical components in an ultrasonic measurement system (see Chapter 7) and for determining material attenuation (see Appendix D).

5.4 The Acoustic Sources and Transducer on Reception

The elements of the sound reception process are the receiving transducer, the cabling, and the receiver portion of the pulser/receiver as shown in Fig. 5.2 (a). In this section we will model the receiving transducer while in the next section we will discuss models of the cabling and receiver. By combining all of those components we will obtain the transfer function that describes the entire reception process (Fig. 5.2 (b)).

First, consider a receiving transducer B. We can model this transducer as a two port system where the input port is the acoustic port and the output port is the electrical port, i.e. we have reversed the inputs and outputs from the transmitting case as shown in Fig. 5.7. Note that along with this reversal we have also changed the direction of the velocity at the acoustic port and the current at the electrical port of the transducer. By inverting the transducer transfer matrix $\left[\mathbf{T}^B \right]$ that describes B when it is used as a transmitter (see Eq. (4.5)), using the fact that $\det \left[\mathbf{T}^B \right] = 1$, and accounting for the sign changes on the velocity and current, we have

$$\begin{Bmatrix} F \\ v \end{Bmatrix} = \begin{bmatrix} T_{22}^B & T_{12}^B \\ T_{21}^B & T_{11}^B \end{bmatrix} \begin{Bmatrix} V \\ I \end{Bmatrix}, \tag{5.23}$$

i.e. the diagonal terms are interchanged but the elements of the transfer matrix in Eq. (5.23) are exactly the same elements defined for the case where the transducer acts as a transmitter. To make use of this two port system model we need to know how the force and velocity inputs are related at the acoustic port and define the "driving" sources at this port. For the receiving transducer, the "sources" at the acoustic port are obviously the waves incident on the transducer as well as the waves scattered from the transducer by the interaction of the incident waves with the transducer (see Fig. 5.8), generating a normal velocity on the face of the transducer. We will again assume that the receiving transducer behaves as a piston and let the normal velocity on its face be $v_n (\omega)$. To see how these waves generate the input force, F, and the input velocity, v, for our two port model, we break up our original problem into the sum of the two problems shown in Fig. 5.9 [5.2]. In Problem I, the face of the transducer is held rigidly fixed. In this case we have the pressure from the incident waves, p_{inc}, as well as the pressure of the waves scattered from the "blocked" transducer face, $p_{scatt}^{blocked}$. The integral of the sum of these two pressures over the transducer face is just the blocked force, $F_B (\omega)$, we

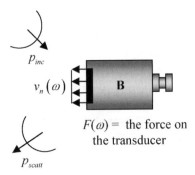

Fig. 5.8. The incident and scattered waves at a receiving transducer and the total force, $F(\omega)$, and normal velocity, $v_n(\omega)$, that those waves produce on the face of the transducer.

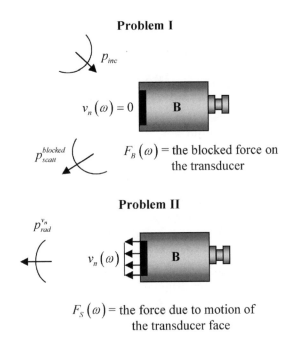

Fig. 5.9. The decomposition of the original problem shown in Fig. (5.8) into the sum of two auxiliary problems, labeled Problem I and Problem II.

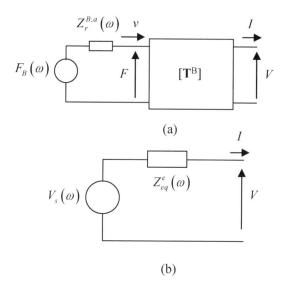

$$(a)$$

$$(b)$$

Fig. 5.10. (a) Representation of the waves received by a transducer as a blocked force source in series with the acoustic radiation impedance of the transducer, and **(b)** the representation of the acoustic sources and receiving transducer by a Thévenin equivalent voltage source and electrical impedance.

defined earlier. In Problem II the incident waves are absent and we have just the pressure of the radiated waves, $p_{rad}^{v_n}$, generated by the motion, $v_n(\omega)$, of the transducer face, which is taken as the same motion as in the original problem shown in Fig. 5.8. Let $F_s(\omega)$ be the force acting on the face of the transducer in Problem II due to this motion of the transducer face. However, Problem II is just the same form as if the transducer were radiating waves when the transducer is used as a transmitter so the force, $F_s(\omega)$, acting on the transducer in this case is related to $v_n(\omega)$ by $F_s(\omega) = Z_r^{B;a}(\omega)v_n(\omega)$, where $Z_r^{B;a}(\omega)$ is the acoustic radiation impedance of the receiving transducer B, the same impedance found when B acts as a transmitter. Since we have taken the velocity $v(\omega)$ in our two port system as flowing into the system (Fig. 5.7) and $v_n(\omega)$ is the normal velocity pointing outwards from the transducer (Fig. 5.8), we have $F_s(\omega) = -Z_r^{B;a}(\omega)v(\omega)$. The total force, $F(\omega)$, acting on the transducer, is the sum of the forces in Problems I and II, so:

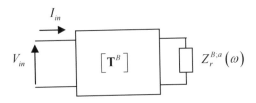

Fig. 5.11. A model of the receiving transducer when the acoustic sources are removed.

$$F(\omega) = F_B(\omega) - Z_r^{B;a}(\omega)v(\omega). \tag{5.24}$$

Equation (5.24) shows us explicitly how the force, F, and the velocity, v, are related at the acoustic port. This relationship is equivalent to the configuration shown in Fig. 5.10 (a), where a force "source", $F_B(\omega)$, is placed in series with an acoustic radiation impedance, $Z_r^{B;a}(\omega)$. Thus, we now have characterized the input side of the transducer. We see that the blocked force arises naturally in this model so that it is the quantity that makes sense to use in our transfer function definitions for both the acoustic/elastic processes and the reception process. From our previous discussion we see we could replace the blocked force source $F_B(\omega)$ by a source given by $2F_{inc}(\omega)$, where F_{inc} is the force due to the incident waves only (i.e. with the transducer absent).

Since there is at present no practical way to experimentally obtain the transfer matrix of the receiving transducer (see the discussion in Chapter 4), we need to replace the system shown in Fig. 5.10 (a) by an equivalent system whose elements we can determine. The system in Fig. 5.10 (a) is an active system (a system with a source) so Thévenin's theorem (Appendix B) allows us to replace that system with a single equivalent voltage source, $V_s(\omega)$, and an equivalent electrical impedance, $Z_{eq}^e(\omega)$, as shown in Fig. 5.10 (b). Recall from Appendix B that to obtain the equivalent impedance we can short out (remove) the sources and examine the ratio between the input voltage and current for this config-uration. When we do that for this system we find the configuration shown in Fig. 5.11, where the transducer is simply terminated at the acoustic port by its acoustic radiation impedance, $Z_r^{B;a}(\omega)$. This configuration is

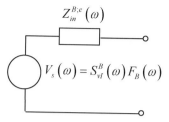

Fig. 5.12. The Thévenin equivalent circuit that characterizes a receiving transducer and its acoustic driving sources.

identical to the situation when this transducer is being used as a transmitter and so if we measured the voltage and current (V_{in}, I_{in}) shown in Fig. 5.11, we would find an equivalent impedance that is the same as that when the transducer is being used as a transmitter, i.e. $Z_{eq}^e(\omega) = Z_{in}^{B;e}(\omega)$ where

$$Z_{in}^{B;e}(\omega) = \frac{Z_r^{B;a}T_{11}^B + T_{12}^B}{Z_r^{B;a}T_{21}^B + T_{22}^B}. \tag{5.25}$$

To obtain the equivalent voltage source, we need to examine the system shown in Fig. 5.10 (a) under open circuit conditions. For this case, we have from Eq. (5.23)

$$\begin{aligned}F(\omega) &= T_{22}^B(\omega)V^\infty(\omega)\\ v(\omega) &= T_{21}^B(\omega)V^\infty(\omega),\end{aligned} \tag{5.26}$$

where $V^\infty(\omega)$ is the open circuit voltage and the source for our Thévenin equivalent circuit, i.e. $V_s(\omega) = V^\infty(\omega)$. Placing Eq. (5.26) into Eq. (5.24) we find

$$\frac{V^\infty(\omega)}{F_B(\omega)} = \frac{1}{T_{22}^B(\omega) + Z_r^{B;a}(\omega)T_{21}^B(\omega)}. \tag{5.27}$$

This ratio is a receiving sensitivity called the open-circuit, blocked force receiving sensitivity, $M_{VF_B}^{B;\infty}(\omega)$ [5.3]. However, comparing Eq. (5.27) with Eq. (4.20) where we defined the sensitivity, $S_{vI}^B(\omega)$, for this transducer when used as a transmitter, we see that:

$$M_{VF_B}^{B;\infty}(\omega) = S_{vI}^{B}(\omega) \tag{5.28}$$

and it follows that the Thévenin equivalent voltage is just

$$V_s(\omega) = S_{vI}^{B}(\omega) F_B(\omega), \tag{5.29}$$

which reduces the transducer and its driving sources to the simple circuit shown in Fig. 5.12. Since in Chapter 7 we will show that it is possible to obtain S_{vI}^{B} and $Z_{in}^{B;e}$ by purely electrical measurements, those measurements will determine completely the role of the transducer when acting as both a transmitter and receiver of sound.

The equality of the two sensitivities in Eq. (5.28) is not accidental. In fact, it is directly a consequence of the fact that the transducer is assumed to be a reciprocal device. Thus, Eq. (5.28) can be considered as a statement of *transducer reciprocity* (see [5.5] for further discussions of transducer reciprocity). This fact can be easily demonstrated by again starting from the transfer matrix of a transducer B when it is acting as a transmitter (Eq. (4.5)) and then obtaining the transfer matrix relationship of Eq. (5.23) but without assuming that the transducer is reciprocal (i.e. let $\det\left[\mathbf{T}^B\right] \neq 1$). In place of Eq. (5.23) we then find during reception that

$$\begin{Bmatrix} F \\ v \end{Bmatrix} = \frac{1}{\det\left[\mathbf{T}^B\right]} \begin{bmatrix} T_{22}^{B} & T_{12}^{B} \\ T_{21}^{B} & T_{11}^{B} \end{bmatrix} \begin{Bmatrix} V \\ I \end{Bmatrix}. \tag{5.30}$$

Thus, when we relate the force and velocity in Eq. (5.30) to the open-circuit receiving voltage, V^{∞}, in place of Eq. (5.26) we obtain

$$\begin{aligned} F(\omega) &= T_{22}^{B}(\omega) V^{\infty}(\omega) / \det\left[\mathbf{T}^B\right] \\ v(\omega) &= T_{21}^{B}(\omega) V^{\infty}(\omega) / \det\left[\mathbf{T}^B\right], \end{aligned} \tag{5.31}$$

which, when placed into Eq. (5.24), gives

$$\begin{aligned} M_{VF_B}^{B;\infty} &= \frac{V^{\infty}(\omega)}{F_B(\omega)} = \frac{\det\left[\mathbf{T}^B\right]}{T_{22}^{B}(\omega) + Z_r^{B;a}(\omega) T_{21}^{B}(\omega)} \\ &= \det\left[\mathbf{T}^B\right] S_{vI}^{B}, \end{aligned} \tag{5.32}$$

where we have also used the definition of the transmitting sensitivity S_{vI}^{B} given by Eq. (4.20). Equation (5.32) shows that the equality of the two

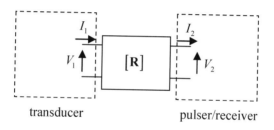

transducer   pulser/receiver

Fig. 5.13. A two port model of the receiving cable.

sensitivities as stated by Eq. (5.28) is then equivalent to requiring $\det\left[\mathbf{T}^B\right]=1$, i.e. the transducer must be reciprocal.

5.5 The Cable and the Receiver in the Reception Process

The role of the cable in the reception process is exactly the same as its role in the sound generation process. We can characterize the cable by a 2x2 reciprocal transfer matrix, $\left[\mathbf{R}\right]$, where (see Fig. 5.13)

$$\begin{Bmatrix} V_1 \\ I_1 \end{Bmatrix} = \begin{bmatrix} R_{11} & R_{12} \\ R_{21} & R_{22} \end{bmatrix} \begin{Bmatrix} V_2 \\ I_2 \end{Bmatrix} \tag{5.33}$$

and the reversing of the current directions does not affect this relationship if the cable is reciprocal ($\det[\mathbf{R}]=1$) and $R_{11} = R_{22}$ as found in a transmission line model of the cable. If the cable does not exactly satisfy these requirements of the transmission line model then we can take such behavior into account by replacing Eq. (5.33) by

$$\begin{Bmatrix} V_1 \\ I_1 \end{Bmatrix} = \frac{1}{\det[\mathbf{R}]} \begin{bmatrix} R_{22} & R_{12} \\ R_{21} & R_{11} \end{bmatrix} \begin{Bmatrix} V_2 \\ I_2 \end{Bmatrix}, \tag{5.34}$$

where $\left(R_{11}, R_{12}, R_{21}, R_{22}\right)$ are the measured transfer matrix of the cable when it is transferring signals from the pulser/receiver to the transducer during the sound generation process. These components of the receiving cable transfer matrix can again be found through the electrical measurements described in Chapter 3.

The receiver part of a pulser/receiver amplifies the received signals and can also filter them. Figure 2.1 shows these types of controls on the

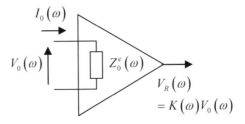

Fig. 5.14. Model of a receiver as an electrical impedance and an amplification factor.

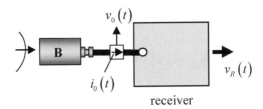

Fig. 5.15. A measurement setup where the waves driving a receiving transducer are used as inputs to the receiver. The input voltage, $v_0(t)$, and current, $i_0(t)$, are measured at the input port of the receiver, as is the receiver output voltage, $v_R(t)$.

right side of the front panel of a spike pulser and Fig. 2.4 shows similar gain and filtering settings that can be made on under computer control of a square wave pulser. Here, any filtering operations of the receiver will not be modeled as they can be easily applied to the unfiltered output at a later stage if desired. In many quantitative studies filtering may be detrimental because it removes frequency components that may contain useful information.

Since the receiver provides an electrical termination at one end of the cable, we will model the receiver as an electrical impedance, $Z_0^e(\omega)$ (Fig. 5.14). The amplifier action of the receiver will be modeled by an amplification (gain) factor, $K(\omega) = V_R(\omega)/V_0(\omega)$, as shown in Fig. 5.14, where $V_R(\omega)$ is the output voltage frequency components of the receiver and $V_0(\omega)$ is the corresponding voltage at the receiver's input port. By measuring the voltages and currents at the input and output of the receiver when it is receiving signals from a receiving transducer (see Fig. 5.15)

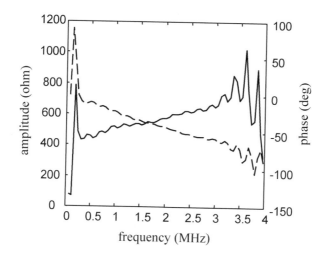

Fig. 5.16. The measured magnitude (solid line) and phase (dashed line) of the electrical impedance, $Z_0^e(\omega)$, of the receiver portion of a Panametrics 5052PR pulser/receiver when driven by a 2.25 MHz transducer in a pitch-catch mode.

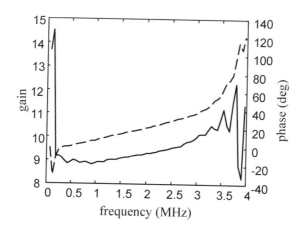

Fig. 5.17. The measured magnitude (solid line) and phase (dashed line) of the amplification (gain) factor, $K(\omega)$, of the receiver portion of a Panametrics 5052PR pulser/receiver when driven by a 2.25 MHz transducer in a pitch-catch mode.

and calculating their Fourier transforms, the quantities $V_0(\omega)$, $I_0(\omega)$, $V_R(\omega)$ can be found for a specific gain setting of the receiver. From these measurements both the impedance, $Z_0^e(\omega)$, and the amplification factor, $K(\omega)$, can be obtained since

$$Z_0^e(\omega) = \frac{V_0(\omega)}{I_0(\omega)}$$

$$K(\omega) = \frac{V_R(\omega)}{V_0(\omega)},$$

(5.35)

where a Wiener filter can be used to desensitize these divisions to noise (see Appendix C). Figure 5.16 shows the measured impedance of a Panametrics 5052PR pulser/receiver determined in this fashion when the pulser/receiver is operating in a pitch-catch mode. Fig. 5.17 gives the corresponding measured amplification (gain) factor. There is little structure seen in the impedance plot as a function of frequency. It is nearly a constant, having a value of approximately 500 ohms. This is consistent with the circuit diagrams of this particular instrument in a pitch-catch mode. The amplification factor also has little structure, having a value near 10 which corresponds well with the 20dB gain setting at which the measurements were taken. Since the 2.25 MHz receiving transducer used in these measurements band limits the received response the results shown in Figs. 5.16 and 5.17 can only be reliably estimated over the bandwidth present. If the transducer used in such a calibration is the same as the one used in an actual inspection, this may not be an issue since the same bandwidth constraints will also be present in the inspection. Otherwise, we may need to excite the receiver with a wider bandwidth source or combine the measurements made with several different transducers to obtain $Z_0(\omega)$, $K(\omega)$ over a larger range of frequencies.

In a pulse-echo mode the received signals must pass through some of the circuits of the pulser section so it is not surprising that in this case the properties of the receiver are affected by the pulser settings. Figure 5.18 (a) shows the behavior of the amplification factor, $K(\omega)$, of a spike pulser/receiver computed at two different damping settings and Fig. 5.18 (b) gives the receiving impedance, $Z_0^e(\omega)$, as measured over a range of different damping settings. The receiver was driven in these cases by waves received from a broadband 5 MHz transducer in a pulse-echo setup of the type shown in Fig. 5.5.

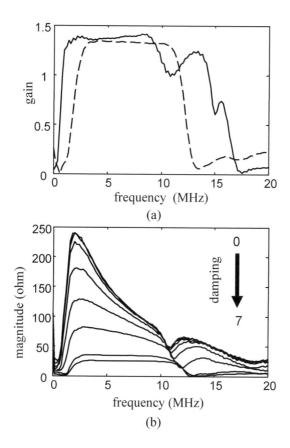

Fig. 5.18. (a) Magnitude of the amplification factor for the receiver section of a spike pulser/receiver in a pulse-echo mode obtained at a damping setting of 2 (solid line) and a damping setting of 9 (dashed line). **(b)** The equivalent impedance of the spike pulser/receiver at a range of damping settings from 0 to 7 (the arrow indicates the trend of the curves for changing damping settings).

Figure 5.19 shows the results of measurement of the amplification factor and receiving impedance of a square wave pulser/receiver when operated in a pitch-catch mode while Fig. 5.20 shows these same parameters when the square wave pulser is operated in a pulse-echo mode. In both cases the receiver was being driven by a broadband 5 MHz transducer. In the pulse-echo mode it can be seen that there is some dependency of the square wave receiver parameters on the pulse width setting in pulse-echo but these changes are not large.

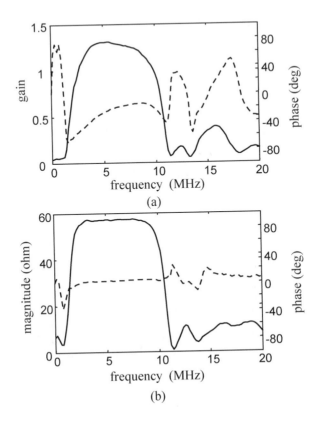

Fig. 5.19. (a) The magnitude (solid line) and phase (dashed line) of the amplification (gain) factor of the receiver section of a square wave pulser/receiver in a pitch-catch mode. **(b)** The magnitude and phase of the equivalent impedance of the receiver section of a square wave pulser/receiver in a pitch-catch mode.

5.6 A Complete Reception Process Model

By combining our transducer, cabling and receiver models we have the complete reception process shown in Fig. 5.21. From Fig. 5.21 we have

$$S_{vI}^B F_B - V_2 = Z_{in}^{B;e} I_2 \qquad (5.36)$$

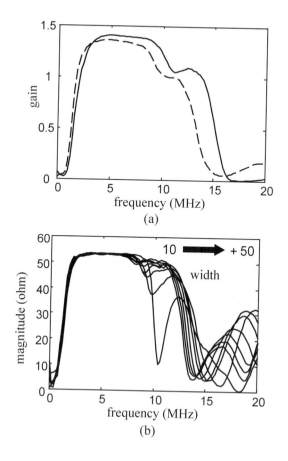

Fig. 5.20. (a) The magnitude of the amplification factor of the receiver section of a square wave pulser/receiver in a pulse-echo mode obtained at a pulse width setting of 10 (solid line) and a pulse width setting of 50 (dashed line). (b) The magnitude of the receiving impedance of the receiver section of a square wave pulser/receiver in a pulse-echo mode for a range of pulse width settings (the arrow indicates the trend of the curves for changing pulse widths).

$$\begin{Bmatrix} V_2 \\ I_2 \end{Bmatrix} = \begin{bmatrix} R_{22} & R_{12} \\ R_{21} & R_{11} \end{bmatrix} \begin{Bmatrix} V_0 \\ I_0 \end{Bmatrix} \tag{5.37}$$

$$V_R = K V_0 \tag{5.38}$$

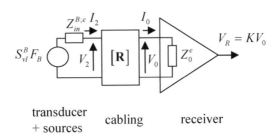

transducer
+ sources

cabling

receiver

Fig. 5.21. A model of the entire sound reception process.

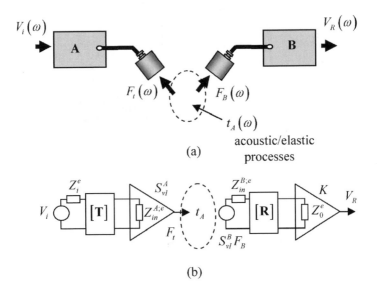

acoustic/elastic
processes

(a)

(b)

Fig. 5.22. (a) All the electrical and electromechanical elements of both the sound generation and sound reception parts of an ultrasonic measurement system, and **(b)** their representation by equivalent sources, impedances, sensitivities, amplification factors, and transfer matrix elements. All the wave propagation and scattering processes are shown in terms of the acoustic/elastic transfer function, $t_A(\omega)$.

$$V_0 = Z_0^e I_0, \tag{5.39}$$

where the components of the cabling transfer matrix are those obtained considering (V_0, I_0) as the input side of the cabling and we have assumed $\det[\mathbf{R}] = 1$ (i.e. the cable is reciprocal) but have not assumed that $R_{11} = R_{22}$ (see the discussion leading to Eq. (5.34)). Using Eqs. (5.36 - 5.39) it is easy to show that the transfer function for this entire reception process, $t_R(\omega)$, is given by [5.4]

$$t_R(\omega) = \frac{V_R(\omega)}{F_B(\omega)} = \frac{K Z_o^e S_{vI}^B}{\left(Z_{in}^{B;e} R_{11} + R_{12}\right) + \left(Z_{in}^{B;e} R_{21} + R_{22}\right) Z_o^e} \tag{5.40}$$

in terms of all the parameters defined earlier. Recall the transfer function for the sound generation process, $t_G(\omega)$, was given by Eq. (4.21) as

$$t_G(\omega) = \frac{F_t(\omega)}{V_i(\omega)} = \frac{Z_r^{A;a} S_{vI}^A}{\left(Z_{in}^{A;e} T_{11} + T_{12}\right) + \left(Z_{in}^{A;e} T_{21} + T_{22}\right) Z_i^e}. \tag{5.41}$$

All the electrical and electromechanical components in an ultrasonic measurement system are shown in Fig. 5.22 (a). The corresponding models are shown in Fig. 5.22 (b). It can be seen from Fig. 5.22 (b) that both the complex sound generation and reception processes models are combined in very similar ways, reflecting the close similarity between the sound generation and receptions transfer functions in Eqs. (5.40) and (5.41). Figure 5.23 shows an example where the magnitude and phase of a sound reception transfer function, $t_R(\omega)$, was experimentally determined by characterizing all the components contained in Eq. (5.40). In this case the receiver was the receiver section of a Panametrics 5052 PR pulser/receiver (measured at a specific gain setting). The cabling consisted of 1.83 m of flexible 50 ohm coaxial cable connected to a 0.76 m fixture rod. The rod also contained internal cabling and was terminated by a right-angle adapter to which the transducer was connected. The transducer was a relatively broadband 6.35 mm diameter, 5 MHz immersion transducer. The sensitivity and impedance of the transducer were obtained by the methods which will be discussed in Chapter 6.

In Chapter 7 it will be shown that these sound generation and reception transfer functions can be combined with the pulser voltage

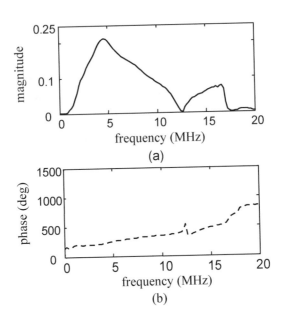

Fig. 5.23. A sound reception transfer function obtained experimentally. **(a)** Magnitude versus frequency and **(b)** phase versus frequency.

source term, $V_i(\omega)$, to form what is called the system function. It will also be shown in that Chapter that the system function can be obtained either by measuring of all its electrical and electromechanical components or by performing a single voltage measurement in a calibration setup. Thus, the acoustic/elastic transfer function, $t_A(\omega)$, shown in Fig. 5.22 is the only remaining part of the ultrasonic measurement system that is needed to completely characterize an entire ultrasonic measurement system. Since this acoustic/elastic transfer function involves the wave fields inside of solid components that are being inspected, it is not practical to measure this transfer function experimentally. Instead, accurate beam models and flaw scattering models are needed to describe $t_A(\omega)$ for an ultrasonic flaw measurement system. In Chapters 8-10 such ultrasonic beam models and flaw scattering models will be described in detail. In Chapter 11 these beam models and scattering models will be combined with a general reciprocity relationship to obtain the acoustic/elastic transfer function for many ultrasonic flaw measurement setups.

5.7 References

5.1 Beissner K (1981) Exact integral expression for the diffraction loss of a circular piston. Acustica 19: 21-217

5.2 Dang CJ, Schmerr LW, Sedov A (2002) Modeling and measuring all the elements of an ultrasonic nondestructive evaluation system. I: Modeling foundations. Research in Nondestructive Evaluation 14: 141-176

5.3 Dang CJ, Schmerr LW, Sedov A (2002) Ultrasonic transducer sensitivity and model-based transducer characterization. Research in Nondestructive Evaluation 14: 203-228

5.4 Dang CJ, Schmerr LW, Sedov A (2002) Modeling and measuring all the elements of an ultrasonic nondestructive evaluation system. II: Model-based measurements. Research in Nondestructive Evaluation 14: 177-201

5.5 Auld BA (1990) Acoustic fields and waves in solids, 2nd ed., Vols. 1, 2. Krieger Publishing Co., Malabar, FL

5.8 Exercises

1. Using Eqs. (5.21) and (5.22b) write a MATLAB function t_a that computes the acoustic/elastic transfer function for the pulse-echo setup shown in Fig. 5.5, where the fluid is water at room temperature. The calling sequence for this function should be:

>> t =t_a(f, a, d, d1, d2,c1,c2);

where f is the frequency (in MHz), a is the radius of the transducer (in mm), d is the distance from the transducer to the plane surface (in mm), d1 is the density of the fluid (in gm/cm^3), c1 is the compressional wave speed of the fluid (in m/sec), d2 is the density of the solid (in gm/cm^3), and c2 is the compressional wave speed of the solid (in m/sec).

   Using this function, obtain a plot of the magnitude of this transfer function versus frequency similar to Fig. 5.6 for a = 6.35 mm, d = 100 mm, d1 = 1.0 gm/cm^3, c1 = 1480 m/sec, d2 = 7.9 gm/cm^3, c2 = 5900 m/sec (steel). Let the frequencies range from 0 to 20 MHz. On the same plot, show the magnitude of this transfer function versus frequency when the attenuation of the fluid is neglected, so that the effects of attenuation on this function can be demonstrated.

6 Transducer Characterization

The sending and receiving transducers are some of the most important parts of an ultrasonic measurement system and also some of the most challenging components to completely characterize. To date there is no practical way to determine the complete transfer matrix components of a transducer, but as we have shown the role of the transducer as both a transmitter and a receiver in an ultrasonic measurement can be completely described in terms of its electrical impedance and sensitivity. In this Chapter we will describe methods to obtain a transducer's electrical imped-ance and sensitivity and also obtain a transducer's effective geometrical parameters such as effective radius and effective focal length.

6.1 Transducer Electrical Impedance

The transducer electrical impedance, $Z_{in}^{A;e}(\omega)$, of a given transducer A is relatively simple to determine in the calibration setup shown in Fig. 6.1. The transducer is connected by a short cable to the pulser and the input voltage, $v_1(t)$, and current, $i_1(t)$, are measured at point a as shown in Fig. 6.1 for the short time that the pulser is exciting the transducer and generating waves in the fluid but before any reflected waves have arrived back at the transducer. Taking the Fourier transform of these measurements to obtain $V_1(\omega), I_1(\omega)$ then gives the impedance directly since for a short cable the transfer matrix of the cable is just the unit matrix and $V_1(\omega) = V_{in}(\omega)$, $I_1(\omega) = I_{in}(\omega)$, where $V_{in}(\omega), I_{in}(\omega)$ are the voltage and current directly at the transducer electrical input port (point b in Fig. 6.1) and

$$Z_{in}^{A;e}(\omega) = \frac{V_{in}(\omega)}{I_{in}(\omega)}. \tag{6.1}$$

As discussed earlier for other measurements of this type, in implementing Eq. (6.1) it may be necessary to use a Wiener filter to desensitize the division

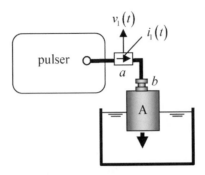

Fig. 6.1. A calibration setup for measurement of a transducer's electrical impedance.

process to noise (see Appendix C). The voltage measurement can be made by inserting a T-connector in the cable and measuring the voltage on the connector while the current can be measured directly by tapping the cable and using a commercial current probe (Tektronix CT-2, Tektronix, Inc., Wilsonville, OR) attached to the central conductor of the cable. A current probe of this type is shown in Fig. 6.2. If it is not practical to use a very short cable, then the measurements at point a must be compensated for cabling effects. This is easy to do since in this case

$$\begin{Bmatrix} V_{in} \\ I_{in} \end{Bmatrix} = \frac{1}{\det[\mathbf{T}]} \begin{bmatrix} T_{22} & -T_{12} \\ -T_{21} & T_{11} \end{bmatrix} \begin{Bmatrix} V_1 \\ I_1 \end{Bmatrix}, \tag{6.2}$$

where $[\mathbf{T}]$ is the transfer matrix for the cable between points a and b in Fig. 6.1 (considering a as the input port and b the output port). If the cabling acted as an ideal reciprocal device the determinant of the transfer matrix would be unity, i.e. $\det[\mathbf{T}] = 1$. In practice, the measured determinant is normally close to but not identically unity so those small differences are accounted for by using Eq. (6.2) with the determinant calculated directly from the measured component values. If the cable transfer matrix has been measured, we can use Eq. (6.2) to determine $V_{in}(\omega), I_{in}(\omega)$ from $V_1(\omega), I_1(\omega)$ and use Eq. (6.1) to obtain the impedance.

Figure 6.3 shows a measured transducer impedance plotted versus the frequency, f. To first order the magnitude of the impedance varies like $1/f$ and the phase is approximately 90 degrees. Figure 6.4 shows the corresponding frequency response of a capacitor, which we see has the

Fig. 6.2. A probe for measuring the current in a cable.

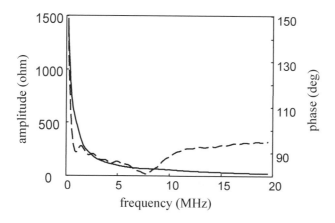

Fig. 6.3. The measured electrical impedance of a transducer showing the magnitude of the impedance (solid line) and the phase (dashed line) versus frequency.

same overall behavior. This is not surprising since a piezoelectric crystal that is plated on its faces will act to first order much like an ordinary capacitor. We cannot always expect to see purely a capacitor-like behavior

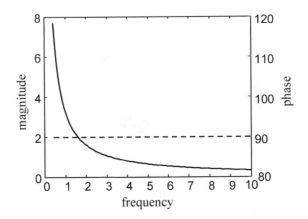

Fig. 6.4. The magnitude (solid line) and phase (dashed line) of the impedance, $Z^e = 1/(-2\pi i f C)$, of a capacitor versus frequency, f, where C is the capacitance.

for the impedance, however, if a commercial transducer contains additional internal electrical "tuning" elements.

6.2 Transducer Sensitivity

With a new pulse-echo technique that has been recently developed, determining the transducer sensitivity of transducer A, $S_{vI}^A(\omega)$, is only slightly more involved than finding the impedance [6.1]. In this case we use a calibration setup such as the one shown in both Figs. 6.5 and 6.6 where the waves from the transducer are reflected from a solid block at normal incidence and the acoustic/elastic transfer function, $t_A(\omega)$, is known (see Eq. (5.18)). We first measure the input voltage, $v_1(t)$, and current, $i_1(t)$, when the transducer is firing and before any reflected waves arrive at the transducer (Fig. 6.5). After a time delay of approximately $t = 2D/c_{p1}$, where c_{p1} is the wave speed in the water, we measure the received voltage, $v_2(t)$, and current, $i_2(t)$ generated by the waves reflected from the block (Fig. 6.6). In Fig. 6.7 we show the sound generation process model corresponding to Fig. 6.5, where the frequency components of $v_1(t), i_1(t)$ at point a are labeled $V_1(\omega), I_1(\omega)$ and the

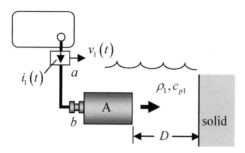

Fig. 6.5. Measurement of voltage and current when transducer A is radiating waves.

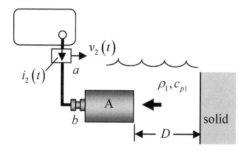

Fig. 6.6. Measurement of voltage and current when transducer A is receiving the waves reflected from the block.

frequency components of the voltage and current at the electrical input port are labeled $V_{in}(\omega), I_{in}(\omega)$. It is likely that the measurements of $v_1(t), i_1(t)$ must by physical necessity be made outside the water tank so that there may be a non-negligible length of cable between the measurement point a and the electrical port of the transducer (point b). Again, however, if the transfer matrix [**T**] of the cabling is known, the voltages and currents measured in these two setups can be related directly to the corresponding voltages and currents at the transducer electrical input port. During the sound generation process, we can again use Eq. (6.2). Note that $V_{in}(\omega)$ and $I_{in}(\omega)$ here are identical to those used in Eq. (6.1) so

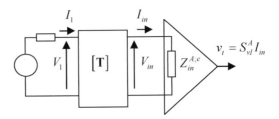

Fig. 6.7. The generation process model for the measurement of voltage and current when transducer A radiates waves.

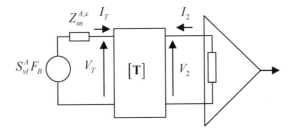

Fig. 6.8. The reception process model for the measurement of the voltage and current when transducer A receives waves reflected from the block.

that the impedance can also be calculated directly in the setup of Fig. 6.5 from $Z_{in}^{A;e}(\omega) = V_{in}(\omega)/I_{in}(\omega)$. In Fig. 6.8 we show the sound reception process model corresponding to Fig. 6.6 where the frequency components of $v_2(t), i_2(t)$ at point a are labeled $V_2(\omega), I_2(\omega)$ and the frequency components of the voltage and current at the electrical input port are labeled $V_T(\omega), I_T(\omega)$. To compensate for the cabling in this case we note that $(V_T, -I_T)$ in the reception process (Fig. 6.8) replaces (V_{in}, I_{in}) in the generation process (Fig. 6.7) and similarly (V_2, I_2) replaces (V_1, I_1) so we find

$$\begin{Bmatrix} V_T \\ -I_T \end{Bmatrix} = \frac{1}{\det[\mathbf{T}]} \begin{bmatrix} T_{22} & -T_{12} \\ -T_{21} & T_{11} \end{bmatrix} \begin{Bmatrix} V_2 \\ I_2 \end{Bmatrix}. \tag{6.3}$$

Note that I_1 and I_2 are taken to be in the same direction in both cases since these currents are both measured by the current probe in Fig. 6.2.

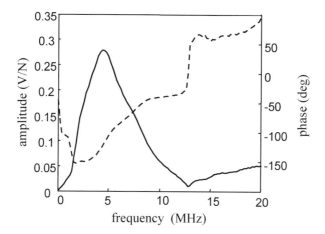

Fig. 6.9. The measured sensitivity of a 5MHz, 6.35 mm radius planar transducer. The magnitude of the sensitivity versus frequency (solid line) and phase versus frequency (dashed line).

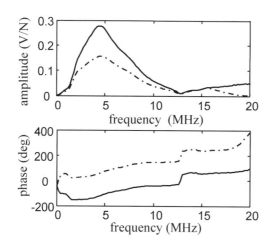

Fig. 6.10. The measured sensitivity of a transducer as determined with compensation for cabling effects (solid line) and where cabling effects are ignored (dash-dot line).

This probe is directional and is oriented so that it measures the current flowing into the cable at point a during both the sound generation and reception processes (see Figs. 6.5 and 6.6).

Now, consider determining the sensitivity from these measurements. From Fig. 6.8 we have

$$S_{vI}^A F_B = Z_{in}^{A;e} I_T + V_T \tag{6.4}$$

and also

$$F_B = \frac{F_B}{F_t}\frac{F_t}{v_t}\frac{v_t}{I_{in}}I_{in} \tag{6.5}$$

$$= t_A Z_r^{A;a} S_{vI}^A I_{in}$$

so that by combining these two relations and using $Z_{in}^{A;e} = V_{in}/I_{in}$ we obtain

$$S_{vI}^A = \sqrt{\frac{V_{in}I_T + V_T I_{in}}{t_A Z_r^{A;a} I_{in}^2}}. \tag{6.6}$$

Since we know the acoustic/elastic transfer function for this setup and we can take the acoustic radiation impedance as its high frequency value $Z_r^{A;a} = \rho_1 c_{p1} S_A$, measurements of $V_{in}(\omega), I_{in}(\omega), V_T(\omega), I_T(\omega)$ are suffi-cient to determine the transducer sensitivity. Since Eq. (6.6) involves division of frequency domain values, a Wiener filter can be used here also to handle noise issues.

Figure 6.9 shows a plot of a measured sensitivity. The dimensions of the sending sensitivity S_{vI}^A are velocity/current while the open-circuit receiving sensitivity, $M_{VF_B}^{A;\infty}$, has the dimensions of voltage/force. Since these two sensitivities are equal we can use either set of dimensions. We choose here to use Volts/Newton in the SI system to characterize these sen-sitivities. Figure 6.10 shows the differences in the measured sensitivity obtained when cabling effects are accounted for and when they are ignored. In most immersion setups such as the one used here there will likely be more than a meter of cable between where the voltages and currents are measured and the transducer electrical port, so that the cabling effects cannot be ignored, as shown in Fig. 6.10. It is important to realize that when the measured signals and modeled parameters are combined they determine the square of the transducer sensitivity, not the sensitivity itself. This can be seen from Eq. (6.6) if we rewrite it as

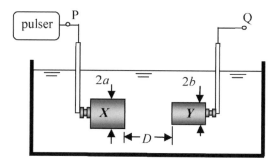

Fig. 6.11. A generic pitch-catch setup that can be used with three transducers (in various pairs) to determine the sensitivity of one of those transducers.

$$\left[S_{vI}^{A}(\omega) \right]^{2} = \frac{V_{in}(\omega)I_{T}(\omega) + V_{T}(\omega)I_{in}(\omega)}{t_{A}(\omega)Z_{r}^{A;a}(\omega)I_{in}^{2}(\omega)} \tag{6.7}$$

Thus, when the square root is taken of these values there is always an ambiguity about the sign that should be chosen. In a pulse-echo experiment, the sign is immaterial in predicting the measured voltage output of the system since the output voltage is proportional to the sensitivity squared (same transducer is both sender and receiver). In a pitch-catch experiment, however, two different transducers are used and this ambiguity in sign could affect the polarity of the predicted output voltage. There is no way to resolve the sign with the procedures discussed here, but there are two ways to deal with this issue. In a pitch-catch situation, the measured sensitivities of the two transducers involved could be combined with measurements of the other system components to predict the system function $s(\omega)$. If the transducers were placed in a measurement setup where the acoustic/elastic transfer function, $t_{A}(\omega)$ was known (such as the setup shown in Fig. 5.4) then the output voltage, $V_{R}(\omega) = s(\omega)t_{A}(\omega)$ could be obtained and Fourier transformed into the time domain and compared to the experimentally observed signal. If the predicted polarity of the time domain signal was correct (i.e. agreed with the experimental voltage), one could say that the signs of the two sensitivities were consistent. If the polarities did not agree, one could change the sign on one of the sensitivities to make them consistent. To determine the sign in a more fundamental manner one could instead place

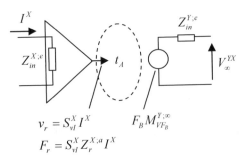

Fig. 6.12. A model for the generic pitch-catch setup of Fig. 6.11, showing the transmitting and receiving transducers and the acoustic/elastic transfer function that defines the wave processes occurring between them.

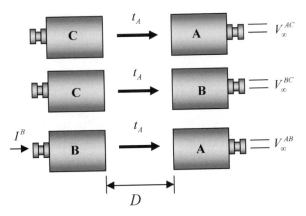

Fig. 6.13. Three separate pitch-catch setups and measurements for determining the sensitivity of transducer A. In this case we have assumed the transducers are all of the same diameter and the distance, D, is fixed for all three setups.

the transducer in a setup where the input current driving the transducer was measured as well as the pressure in the transducer wave field (such as the on-axis pressure measured with a separate calibrated probe). Such a measurement setup would only be needed, however, if it was essential to predict in an absolute sense the generated pressure wave field.

There exists another reciprocity-based measurement procedure to determine the open-circuit receiving sensitivity, $M_{VF_B}^{A;\infty}$, that is commonly described in the acoustics literature [6.2-6.10]. That method requires one to make measurements with three different transducers in three separate pitch-catch setups of the generic type shown in Fig (6.11) where the transmitting transducer is transducer X and the receiving transducer is transducer Y. The input current to transducer X measured at point P in Fig. 6.11 is labeled I^X and the open-circuit voltage measured at point Q received by transducer Y due to the waves generated by transducer X is labeled V_∞^{YX}. If the effects of cabling between point P and the transmitting transducer X and between transducer Y and point Q are both negligible, then the measured current at the input port of transducer X is the same as I^X and the open-circuit voltage at Q is the same as the open-circuit voltage directly at the receiving transducer electrical port. In this case the sound generation and reception model for the pitch-catch setup of Fig. 6.11 is as shown in Fig. 6.12, Note that the acoustic/elastic transfer function, t_A, for this pitch/catch configuration is known for a pair of circular, plane piston transducers (see Eq. (5.10) for the case where the transducers are of different size, or Eq. (5.12) when the transducers are of the same size). Since the open-circuit voltage at the receiving transducer electrical port is just the equivalent source term for transducer Y given by $F_B M_{VF_B}^{Y;\infty}$ (see Chapter 5) we find

$$
\begin{aligned}
\frac{V_\infty^{YX}}{I^X} &= \frac{F_B M_{VF_B}^{Y;\infty}}{I^X} \\
&= \frac{F_B}{F_t} \frac{F_t}{I^X} M_{VF_B}^{Y;\infty} \\
&= t_A Z_r^{X;a} S_{vI}^X M_{VF_B}^{Y;\infty}.
\end{aligned}
\tag{6.8}
$$

As shown in Chapter 5 the transmitting sensitivity S_{vI}^Z and the open-circuit receiving sensitivity, $M_{VF_B}^{Z;\infty}$, are the same for any reciprocal transducer Z (where $Z = X$ or Y), so we can express the voltage over current ratio in Eq. (6.8) in terms of either of these sensitivities. We will choose the open-circuit receiving sensitivity here, as that is the choice normally made in the acoustics literature. Then Eq. (6.8) becomes

$$
\frac{V_\infty^{YX}}{I^X} = t_A Z_r^{X;a} M_{VF_B}^{X;\infty} M_{VF_B}^{Y;\infty}.
\tag{6.9}
$$

Now, apply Eq.(6.9) to the three separate pitch-catch setups involving three transducers A, B, and C shown schematically in Fig. 6.13, where we have assumed that the distance, D, between transducers is held fixed for all three setups and the diameters of all three transducers are the same so that there is only one acoustic/elastic transfer function, t_A, for all three setups.

In setup one transducer $X = C$ is firing and transducer $Y = A$ is receiving while for setup two transducer $X = C$ again is firing and transducer $Y = B$ is receiving. In setup three, transducer $X = B$ is firing and transducer $Y = A$ is receiving. Applying Eq. (6.9) to each of these cases individually we have

$$\frac{V_{\infty}^{AC}}{I^C} = t_A Z_r^{C;a} M_{VF_B}^{A;\infty} M_{VF_B}^{C;\infty}$$

$$\frac{V_{\infty}^{BC}}{I^C} = t_A Z_r^{C;a} M_{VF_B}^{B;\infty} M_{VF_B}^{C;\infty} \qquad (6.10)$$

$$\frac{V_{\infty}^{AB}}{I^B} = t_A Z_r^{B;a} M_{VF_B}^{A;\infty} M_{VF_B}^{B;\infty}.$$

From Eq. (6.10) we see we can eliminate the sensitivities of transducers B and C by considering the particular combination of ratios

$$\frac{\left(\dfrac{V_{\infty}^{AB}}{I^B} \dfrac{V_{\infty}^{AC}}{I^C} \right)}{\left(\dfrac{V_{\infty}^{BC}}{I^C} \right)} = t_A Z_r^{B;a} \left[M_{VF_B}^{A;\infty} \right]^2 \qquad (6.11)$$

so solving for the open-circuit receiving sensitivity of transducer A we find:

$$M_{VF_B}^{A;\infty} = S_{vI}^A = \sqrt{\frac{V_{\infty}^{AB} V_{\infty}^{AC}}{V_{\infty}^{BC} I^B} \frac{1}{Z_r^{B;a} t_A}}. \qquad (6.12)$$

Equation (6.12), which is similar to the expression commonly found in the acoustics literature, is very much like Eq. (6.6) for our pulse-echo method. Instead of the two voltage and two current measurements needed for the pulse-echo method, Eq. (6.11) requires that we make three open-circuit voltage methods and one current measurement from the three pitch-catch setups of Fig. 6.13. For acoustic transducers operating at kHz frequencies or less, Eq. (6.12) has been commonly used in the acoustics community for many years to obtain transducer sensitivity. In fact, for transducers at those frequencies there exists a commercially available calibration system that can implement the measurements required in Eq.(6.12) and extract the sensitivity [6.11]. Dang. et al. [6.12] have also used this three transducer method to obtain the sensitivity of NDE transducers operating at MHz frequencies. However, Dang et al. [6.12] found that at MHz frequencies it was important to consider the effects of the cabling present. They defined a

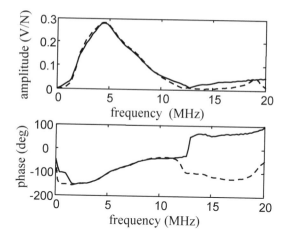

Fig. 6.14. The magnitude and phase of the sensitivity, S_{vI} , of a 5 MHz, 6.35 mm diameter planar transducer as calculated by the pulse-echo method (solid line) and the three transducer pitch-catch method (dashed line).

generalized sensitivity that took into account those cable effects and applied a modified version of Eq. (6.12).

The three transducer pitch-catch method is also a viable approach to obtaining sensitivity but the pulse-echo method has several advantages. First, the three-transducer method requires one to make measurements in three separate pitch-catch setups while only one setup is needed in the pulse-echo method. This makes the pulse-echo method faster and avoids any delicate re-alignment issues for the transducers. Second, we note that both the pulse-echo and the three transducer pitch-catch procedure for obtaining sensitivity are model-based approaches. This means that the model assumptions made on transducer behavior must be satisfied for all three transducers for the three transducer method but only for the transducer whose sensitivity is to be determined for the pulse-echo method. Figure 6.14 shows the sensitivity of a 5 MHz, 6.35 mm diameter planar transducer obtained via either the pulse-echo method or the three-transducer pitch-catch method. It can be seen that there is little difference between the results obtain with either method over the bandwidth of the transducer.

There is also a pulse-echo technique for determining sensitivity called the self-reciprocity method that has been developed in the acoustics literature [6.13-6.17]. The self-reciprocity method applies Eq. (6.9) to a

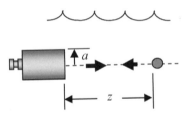

Fig. 6.15. A circular piston transducer of radius a receiving the waves reflected from the front surface of a spherical reflector located on the central axis of the transducer.

pulse-echo setup involving a single transducer, A, and solves for the sensitivity of A in the form

$$M_{VF_B}^{A;\infty} = S_{vI}^A = \sqrt{\frac{V_\infty^{AA}}{I^A} \frac{1}{t_A Z_r^{A;a}}}, \tag{6.13}$$

where V_∞^{AA} is the open-circuit voltage received by A due to the waves generated by A and I^A is the current driving transducer A when it is radiating into the fluid. Equation (6.13) is very similar to our pulse-echo expression, Eq. (6.6). In fact under open-circuit conditions $I_T = 0$ in Eq. (6.6) and that equation simply reduces to Eq. (6.13). However, in order to apply Eq. (6.13) directly one needs to measure the received voltage under open-circuit conditions. Since inherently in a pulse-echo setup the transducer will be loaded by the receiver and cabling on reception, this has forced some authors to use rather complicated measurement systems or special matching networks to infer the open-circuit response. Equation (6.6) can be applied directly from measurements taken under the actual conditions present in a pulse-echo setup, so it is significantly more convenient to use than Eq. (6.13).

6.3 Transducer Effective Radius and Focal Length

It would appear that geometrical parameters such as the transducer radius and focal length are parameters that are well-defined and need no experimental determination. In practice, however, it has been found that if one simply uses these parameters (as specified by the transducer manufacturer)

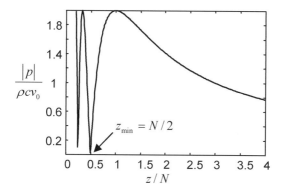

Fig. 6.16. The magnitude of the on-axis normalized pressure versus normalized distance z / N for a ½ inch diameter circular piston transducer radiating waves at 5 MHz into a fluid, where N is the near field distance given by $N = a^2 / \lambda$. As shown the last on-axis null occurs at one-half a near field distance.

in transducer beam models, one often does not get good agreement with theory when the behavior of the transducer beam is examined experimentally [6.18], [Fundamentals]. This is perhaps to be expected since, for example, a transducer crystal cannot have piston-like behavior over its entire face as the crystal is supported and constrained at its edges. Thus, one might define an *effective radius* for the transducer where a piston model agrees better with experiments. Similarly, the geometrical focal length of a focused transducer is determined in reality by a number of other unknown parameters such as the material properties and geometry of the focusing lens. Again, one might deal with these unknowns by defining an *effective focal length* that matches experiments.

First, consider the problem of determining the effective radius of a circular, planar immersion transducer. One configuration that can be used to determine the effective radius of this transducer is shown in Fig. 6.15. A spherical reflector is placed on the axis of the transducer and the transducer is scanned so that the sphere remains on the transducer's central axis at different distances, z. At each value of z, $z = z_i$ the received time domain voltage response, $v_R(t, z_i)$, from the front surface of the sphere is recorded and Fourier transformed to obtain its spectrum, $V_R(f, z_i)$. Then the magnitude of these frequency domain responses are plotted versus z at a single fixed frequency, f_0, which is usually taken near the center frequency of the transducer. Since the front surface reflection from the

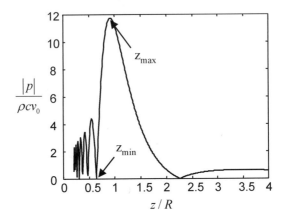

Fig. 6.17. The magnitude of the normalized on-axis pressure versus normalized distance z/R for a spherically focused piston transducer of radius a and geometrical focal length, R, radiating into water. The location of the null and maximum that are used in the determination of the effective focal length and radius are shown.

sphere is proportional to the square of the on-axis pressure of the transducer, the magnitude of the frequency domain plot of $V_R(f_0, z_i)$ has the same behavior as the on-axis pressure squared of the transducer when it is driven harmonically at frequency f_0 [Fundamentals]. In Chapter 8, an explicit expression for the on-axis pressure of a circular plane piston transducer at a fixed frequency is obtained analytically. This on-axis pressure is plotted in Fig. 6.16 versus the non-dimensional distance z/N, where $N = a^2/\lambda$ is called the near field distance and $\lambda = c_p/f_0$ is the wave length. It can be seen that in the region near the transducer there are a series of maxima and nulls. The last null (the one farthest from the transducer) can be shown to be located at the distance $z_{min} = a^2/2\lambda$. Since this is a null of the pressure field the squared pressure will also have a null at this position, as will $|V_R(f_0, z_i)|$. If, from the plot of $|V_R(f_0, z_i)|$ versus z one obtains an estimate of the distance to that null then one can define the corresponding effective radius, a_{eff}, as

$$a_{eff} = \sqrt{2\lambda z_{min}}. \tag{6.14}$$

This last on-axis null position is used because it is relatively simple to determine experimentally and does not require knowledge of the absolute amplitude of the on-axis pressure wave field. Some authors have used multiple on-axis nulls to obtain a better estimate of the effective radius or have used a least squares fitting to theory of many points, both on- and off-axis, in the transducer wave field to determine a_{eff}. All of these methods have the same goal – namely to obtain an estimate of a radius value that will match the theoretical wave field better than simply using the nominal radius. In principle the determination of a_{eff} in this fashion can be done at any fixed frequency and the result should not depend on the frequency chosen. In practice some variations of the effective radius value with frequency are found [Fundamentals]. Often these variations are not severe and a simple averaging of a_{eff} values over the bandwidth of the transducer gives good results.

For a spherically focused transducer one can use the same setup shown in Fig. 6.15 and the same procedures to obtain $\left|V_R\left(f_0, z_i\right)\right|$, which is proportional to the on-axis pressure squared wave field, but in this case we must obtain estimates of both the effective radius, a_{eff}, and the effective geometrical focal length, R_{eff} [6.19], [6.20]. Figure 6.17 shows a plot of a model prediction of the on-axis pressure of a circular, spherically focused piston transducer radiating into water. Again one sees nulls and maxima in the region close to the transducer and a very large peaked response due to focusing. Only the distance, z_{min}, to the last on-axis null can generally be obtained reliably, however, since at other nulls the response rapidly gets very small. One could also measure the distance, z_{max}, to the maximum value of $\left|V_R\left(f_0, z_i\right)\right|$, which also occurs when the magnitude of the pressure is a maximum. In this case, models show that the effective focal length is given in terms of z_{min} and z_{max} by [Fundamentals]

$$R_{eff} = z_{max}\left\{\frac{\pi - x}{\pi - x\left(z_{max} / z_{min}\right)}\right\}, \qquad (6.15)$$

where x is a solution of the transcendental equation

$$x\cos(x) = \frac{\pi - x\left(z_{max} / z_{min}\right)}{\pi - x}\sin(x). \qquad (6.16)$$

Once the effective focal length is found from these relations the effective radius is given by

$$a_{eff} = \sqrt{\frac{2\lambda z_{min} R_{eff}}{R_{eff} - z_{min}}}, \tag{6.17}$$

which we see reduces to the planar transducer case (Eq. (6.14)) when $R_{eff} \to \infty$. In practice it has been found that the location of the distance to the transducer peak response, z_{max}, is difficult to determine precisely and the results for R_{eff} are sensitive to those errors. It has been found better to use a range of estimates for z_{max} and choose the best combination of R_{eff} and a_{eff} values that match (in a least squares sense) the predicted and measured on-axis pressure values around the transducer focus. The details of these procedures can be found in [6.20]. There are other fitting methods that can be used to obtain these effective parameters but we will not discuss those alternatives here. As in the planar case, the effective parameters have been found to depend somewhat on the frequency one chooses, so one might need to take an average of their values over the bandwidth of the transducer.

Table 6.1. Effective radii and focal lengths found for some commercial transducers.

Transducers	Manufacturer's Specs		Effective Parameters		Center Frequency (MHz)
	R (mm)	a (mm)	R_{eff} (mm)	a_{eff} (mm)	
A	76.2	4.76	134.7	4.51	10
B	76.2	6.35	207.4	5.56	5
C	76.2	4.76	74.5	4.69	15

Equation (6.15) shows that the effective geometrical focal length, R_{eff}, is always larger than z_{max}. The distance z_{max}, which is the distance to the maximum on axis pressure, is often called the location of the "true focus". The difference between R_{eff} and z_{max} occurs because of wave diffraction effects at finite frequencies. It is only in the limit when the frequency goes to infinity that $z_{max} / z_{min} \to 1$, and one finds $R_{eff} = z_{max}$.

Table 6.1 gives some example values of the effective parameters obtained for several commercial transducers. It can be seen that in some cases the effective values are considerably different from the nominal values given by the transducer manufacturer. Those differences can lead to large errors if the nominal values are used in model calculations.

6.4 References

6.1 Lopez-Sanchez A, Schmerr LW (2006) Determination of an ultrasonic transducer's sensitivity and impedance in a pulse-echo setup. IEEE Trans. Ultrason., Ferroelect., Freq. Contr. 53: 2101-2112

6.2 Bobber RJ (1960) Underwater electroacoustic measurements. Naval Research Laboratory, Washington, DC

6.3 MacLean WR (1940) Absolute measurement of sound without a primary standard. J. Acoust. Soc. Am. 12: 140-146

6.4 Cook RK (1941) Absolute pressure calibration of microphone. J. Acoust. Soc. Am. 12: 415-420

6.5 DiMattia AL, Wiener FM (1946) On the absolute pressure calibration of condenser microphones by the reciprocity method. J. Acoust. Soc. Am. 18: 341-344

6.6 Ebaugh P, Mueser RE (1947) The practical application of the reciprocity theorem in the calibration of underwater sound transducers. J. Acoust. Soc. Am. 19: 695-700

6.7 Wathen-Dunn W (1949) On the reciprocity free-field calibration of microphones. J. Acoust. Soc. Am. 21: 542-546

6.8 Diestel HG (1961) Reciprocity calibration of microphones in a diffuse sound field. J. Acoust. Soc. Am. 33: 514-518

6.9 Hill EK, Egle DM (1980) A reciprocity technique for estimating the diffuse-field sensitivity of piezoelectric transducers. J. Acoust. Soc. Am. 67: 666-672

6.10 Vorlander M, Bietz H (1994) Novel broad-band reciprocity technique for simultaneous free-field and diffuse-field microphone calibration. Acustica 80: 365-377

6.11 Product Data Sheet, Reciprocity Calibration System – Type 9699 – Brüel & Kjaer

6.12 Dang CJ, Schmerr LW, Sedov A (2002) Ultrasonic transducer sensitivity and model-based transducer characterization. Research in Nondestructive Evaluation 14: 203-228

6.13 Carstensen EL (1947) Self-reciprocity calibration of electroacoustic transducers. J. Acoust. Soc. Am. 19: 961-965

6.14 White RM (1957) Self-reciprocity transducer calibration in a solid medium. J. Acoust. Soc. Am. 29: 834-836

6.15 Reid JM (1974) Self-reciprocity calibration of echo-ranging transducers. J. Acoust. Soc. Am. 55: 862-868

6.16 Widener MW (1980) The measurement of transducer efficiency using self reciprocity techniques. J. Acoust. Soc. Am, 67: 1058-1060

6.17 Brendel K, Ludwig G (1976/77) Calibration of ultrasonic standard probe transducers. Acustica 36: 203-208

6.18 Chivers RC, Bosselaar L, Filmore PR (1980) Effective area to be used in diffraction correction. J. Acoust. Soc. Am. 68: 80-84

6.19 Amin F, Gray TA, Margetan FJ (1991) A new method to estimate the effective geometrical focal length and radius of ultrasonic focused probes. In: Thompson DO, Chimenti DE (eds) Review of progress in quantitative nondestructive evaluation, 10A. Plenum Press, New York, NY, pp 861-865

6.20 Lerch T, Schmerr LW (1996) Characterization of spherically focused transducers using an ultrasonic measurement model approach. Res. Nondestr. Eval. 8: 1-21

6.5 Exercises

1. The MATLAB function transducer_x(z) returns the time-domain sampled voltage received from a spherical reflector in water (c = 1480 m/sec) located at a distance z (in mm) along the axis of a planar transducer as shown in Fig. (6.15). There are 1024 samples in this waveform, each separated by a sampling time interval Δt =.01 μsec. First, let z be the vector of values:

>> z = linspace (25, 400, 100);

Use this set of values in the transducer_x function, i.e. evaluate

>> V = transducer_x(z);

The matrix V will contain 100 waveforms calculated at each of these z-values. Use FourierT to generate the frequency spectra of these waveforms. Note that FourierT can operate on all of these waveforms at once as long as they are in columns (which is the case) and will return a matrix of the corresponding spectra, also in columns. Examine the magnitude of some of these spectra versus frequency to determine the range of frequencies over which there is a significant response. Pick one frequency value near the center frequency in this range and plot the magnitude of the spectra at that value versus the distance z.

Locate the last on-axis minimum in this plot and use Eq. (6.14) to determine the effective radius of this transducer. Try using a different frequency value within the transducer bandwidth to determine the effective radius. Does your answer vary with the frequency chosen?

7 The System Function and Measurement System Models

7.1 Direct Measurement of the System Function

In the previous Chapters we have obtained explicit expressions for the transfer functions $t_R(\omega), t_G(\omega)$ that define all the electrical and electro-mechanical components of an ultrasonic measurement system and we gave some examples of simple calibration setups where we can also obtain explicit expressions for the acoustic/elastic transfer function, $t_A(\omega)$. When all these transfer functions are combined with the Thévenin equivalent voltage of the pulser, $V_i(\omega)$, we have a model of the entire ultrasonic measurement system where the output voltage, $V_R(\omega)$, is given by

$$V_R(\omega) = t_G(\omega) t_R(\omega) t_A(\omega) V_i(\omega). \tag{7.1}$$

In section 7.3 we will give some examples of combining all of these models and measurements to synthesize the output voltage of an ultrasonic measurement system. Of course this type of synthesis requires a considerable number of measurements since we must obtain the equivalent voltage and electrical impedance of the pulser, the transfer matrices of the cabling, the impedances (electrical and acoustical) and sensitivities of the transducers, and the electrical impedance and amplification factor of the receiver. However, there is an alternative approach where we combine $t_R(\omega), t_G(\omega)$, and $V_i(\omega)$ into a single factor, $s(\omega)$, called the *system function*, where

$$s(\omega) = t_R(\omega) t_G(\omega) V_i(\omega). \tag{7.2}$$

In terms of the system function Eq. (7.1) reduces to simply:

$$V_R(\omega) = s(\omega)t_A(\omega). \tag{7.3}$$

For any calibration setup where we can model the transfer function $t_A(\omega)$ explicitly and where we measure the frequency components of the received voltage, $V_R(\omega)$, Eq. (7.3) shows that we can obtain the system function by deconvolution, i.e.

$$s(\omega) = \frac{V_R(\omega)}{t_A(\omega)}. \tag{7.4}$$

In practice, to reduce the sensitivity of the deconvolution to noise, we use a Wiener filter (see Appendix C) and obtain the system function from

$$s(\omega) = \frac{V_R(\omega)t_A^*(\omega)}{|t_A(\omega)|^2 + \varepsilon^2 \max\left\{|t_A(\omega)|^2\right\}}, \tag{7.5}$$

where ε is a constant that is used to represent the noise level present and $(\ )^*$ indicates the complex-conjugate.

The system function contains all the electrical and electromechanical components of the ultrasonic measurement system, so with one measurement of $V_R(\omega)$ in a well characterized calibration experiment, Eq. (7.5) allows us to characterize the effects of all those components at once. This is obviously a very convenient alternative to having to measure all the elements that make up $s(\omega)$. This method of determining the system function is done at a fixed set of system settings (e.g. energy and damping settings on a spike pulser, gain settings on the receiver) and with a given set of cables and transducers. If another experiment such as a flaw measurement is performed at exactly the same settings and with the same components the system function obtained from the calibration setup will be the same as for the flaw measurement. This fact allows us to quantitatively determine the effects that all the electrical and electromechanical parts of the measurement system have on a flaw measurement. Since $s(\omega)$ has nothing to do with the response of a flaw being measured, it is important to be able to characterize (and eliminate) those parts of the measured signals that are not flaw dependent so that we can determine a response more directly related to the flaw being examined.

Another way that we can use knowledge of $s(\omega)$ is to combine it with beam propagation and flaw scattering models that can model the

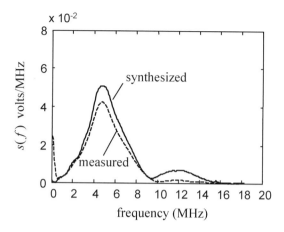

Fig. 7.1. A system function, $s(f)$, measured directly by deconvolution (dashed line) or synthesized by measuring all the electrical and electromechanical components contained in $s(f)$ (solid line).

acoustic/elastic transfer function, $t_A(\omega)$ explicitly. In later Chapters we will show just how to develop such detailed models. By combining a modeled $t_A(\omega)$ and a measured $s(\omega)$, Eq. (7.3) shows that we can predict the actual measured voltage, $V_R(\omega)$, in a flaw measurement setup in an absolute sense. This capability gives us a powerful engineering simulation tool to design and evaluate ultrasonic NDE inspections.

In using a directly measured system function, one must re-measure that function whenever a system setting or system component is changed and this approach does not permit us to determine the significance of individual changes, such as a replacement of a transducer, for example, without such a re-measurement. Determining $s(\omega)$ by combining a knowledge of $V_i(\omega)$ and all the components that make up $t_R(\omega)$ and $t_G(\omega)$, however, does allow us to examine the effects of such changes. Of course, either a directly measured system transfer function or one synthesized from its components should agree with each other. This is the case, as illustrated in Fig. 7.1, where a system function was both directly measured by deconvolution and constructed from individual measurements of all the electrical and electromechanical components [7.1].

7.2 System Efficiency Factor

In [Fundamentals] a quantity which is closely related to the system function was defined called the *system efficiency factor*, $\beta(\omega)$. This system efficiency factor is related to the measured voltage, $V_R(\omega)$, as follows:

$$V_R(\omega) = \beta(\omega)\frac{p_{ave}(\omega)}{\rho c v_0(\omega)},\qquad(7.6)$$

where $p_{ave}(\omega)$ is the average pressure generated by the incident waves at the receiving transducer, $v_0(\omega)$ is the output velocity of the transmitting transducer (which is assumed to act as a piston) and ρc is the specific acoustic impedance of the material into which the transmitting transducer radiates. The blocked force $F_B(\omega) = 2p_{ave}(\omega)S_R$, where S_R is the area of the receiving transducer, and the force transmitted by the sending transducer $F_t(\omega) = Z_r^{T;a}(\omega)v_0(\omega) = \rho c S_T v_0(\omega)$ for a piston transducer at high frequencies, where S_T is the area of the transmitting transducer. Thus, combining these relations with the two equivalent forms

$$V_R(\omega) = s(\omega)\frac{F_B(\omega)}{F_t(\omega)} = \beta(\omega)\frac{p_{ave}(\omega)}{\rho c v_0(\omega)}\qquad(7.7)$$

we see that the system function and the system efficiency factor are just proportional to one another, where

$$s(\omega) = \frac{S_T}{2S_R}\beta(\omega),\qquad(7.8)$$

so it makes no difference if we characterize our measurement system with either of these quantities.

In determining the system function or system efficiency factor experimentally by deconvolution in a reference experiment, the values of $s(\omega)$ or $\beta(\omega)$ should not depend on the choice of that reference experiment and it's corresponding transfer function, $t_A(\omega)$. Schmerr et al. [7.2] demonstrated this fact by using a number of different reference setups to calculate the system efficiency factor. Some of the simple calibration setups where the transfer function $t_A(\omega)$ is known are shown in Fig. 7.2.

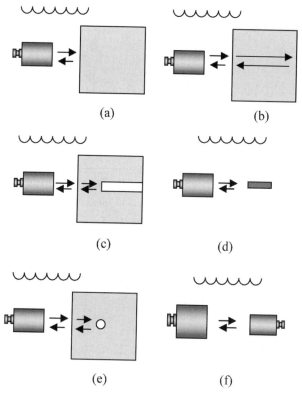

(a)   (b)

(c)   (d)

(e)   (f)

Fig. 7.2. Reference experiments that can be used to determine the system function or system efficiency factor where circular planar transducers are involved: **(a)** reflection from a plane front surface of a block at normal incidence, **(b)** reflection from the back surface of a block at normal incidence, **(c)** reflection from an on-axis flat-bottom hole at normal incidence, **(d)** reflection from an on-axis solid cylinder at normal incidence, **(e)** reflection from an on-axis side-drilled hole at normal incidence, and **(f)** two transducers (not necessarily the same) whose axes are aligned.

Cases (a) and (f) were discussed in Chapter 5. Cases (b), (c), (d) and (e) can be found in [7.2] and [Fundamentals]. All the cases shown in Fig. 7.2 are suitable for determining the system function for circular, planar transducers in pulse-echo immersion setups except Fig. 7.2 (f) which can be used for circular, planar transducers in immersion pitch-catch setups. In Chapter 8 we will develop an explicit expression for $t_A(\omega)$ in the setup shown in Fig. 7.2 (a) for a circular, spherically focused transducer that can be used to determine $s(\omega)$ for that type of transducer

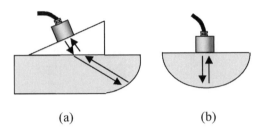

(a)   (b)

Fig. 7.3. Reference experiments that can be used to determine the system function for **(a)** an angle beam probe test setup where the waves are reflected from the curved surface of a calibration block, and **(b)**, a contact setup where the waves are reflected from a curved surface of a block.

as well as transfer functions for planar rectangular transducers and cylindrically focused rectangular transducers. In Chapter 13 we will show how a multi-Gaussian beam model can be used to numerically determine the transfer function $t_A(\omega)$ for the pulse-echo contact angle beam shear wave setup of Fig. 7.3 (a) where the waves are reflected from the cylindrical interface of a standard calibration block. That same approach can also be used for other contact testing setups such as the one shown in Fig. 7.3 (b) or in other contact setups with planar or curved surfaces. In contact problems, however, one must be aware of the fact that changes of the thin fluid couplant layer between the transducer and the component being inspected (or between the transducer wedge and the component) and non-uniform component surface conditions can produce measured response variabilities that must be carefully considered.

7.3 Complete Measurement System Modeling

The ultimate test of the ability of all these models and measurements to simulate an ultrasonic measurement system is to compare the measured output voltage of a particular setup with one that is synthesized from the models/measurements we have discussed in previous Chapters. Consider, for example, a calibration setup of the type shown in Fig. 7.2 (f) where two planar transducers of the same nominal radius are placed opposite to each other in an immersion tank with their axes aligned. An explicit acoustic/elastic transfer function for this configuration was given in Eq. (5.12) for an ideal lossless fluid. Adding attenuation into this ideal model as shown in Chapter 5 (see Eq. (5.22a)) we have a complete acoustic/elastic transfer function for this example. Combining this transfer function with a measured system function gives the frequency components of the measured output voltage (Eq. (7.3)). Finally, taking an inverse Fourier transform out this output voltage spectrum then yields a time domain A-scan signal for the entire system. We can simulate this A-scan signal using a system function that is calculated from Eq. (7.2), using measurements of all the components that make up $t_G(\omega)$ and $t_R(\omega)$ together with $V_i(\omega)$. Recall, these transfer functions were given by

$$t_R(\omega) = \frac{K Z_o^e S_{vl}^B}{\left(Z_{in}^{B;e} R_{11} + R_{12}\right) + \left(Z_{in}^{B;e} R_{21} + R_{22}\right) Z_o^e} \tag{7.9}$$

and

$$t_G(\omega) = \frac{Z_r^{A;a} S_{vl}^A}{\left(Z_{in}^{A;e} T_{11} + T_{12}\right) + \left(Z_{in}^{A;e} T_{21} + T_{22}\right) Z_i^e}. \tag{7.10}$$

The pulser used here was a Panametrics 5052 PR pulser/receiver operating at an energy setting of 1 and a damping setting of 7. The open-circuit voltage of the pulser was measured to obtain $V_i(\omega)$ and the pulser impedance, $Z_i^e(\omega)$, was measured by placing a 50 ohm resistor across the pulser output and measuring the resulting voltage across this resistance, as outlined in Chapter 2. The transfer matrix components, $[T_{11}, T_{12}, T_{21}, T_{22}]$ of the cabling between the pulser and transmitting transducer A and the cable components, $[R_{11}, R_{12}, R_{21}, R_{22}]$ for the cabling between the receiving transducer B and the receiver were both measured as functions of frequency using different cabling termination conditions as discussed in Chapter 3. The receiver gain, $K(\omega)$, and impedance, $Z_o^e(\omega)$, were obtained from measurements of voltage and current at the receiver inputs and outputs, as described in Chapter 5, at receiver gain and attenuation settings of 20 dB and 12 dB, respectively, and with the filter control of the receiver set to "off". The two transducers used in this pitch-catch setup were two nominally identical 5 MHz, 6.35 mm diameter planar transducers. Their electrical impedances, $Z_{in}^{A;e}(\omega)$ and $Z_{in}^{B;e}(\omega)$, and their sensitivities, S_{vl}^A and S_{vl}^B, were found using the electrical measurements which were discussed in Chapter 6. Finally, the acoustic radiation impedance of the transmitting transducer A, $Z_r^{A;a}(\omega)$, which appears in the sound generation transfer function, was computed from the high frequency

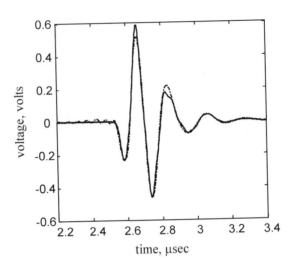

Fig. 7.4. Directly measured output voltage signal of an ultrasonic pitch-catch measurement system (solid line) and the voltage synthesized by measurement and modeling of all the ultrasonic components (dashed-dotted line) for a pair of 5 MHz, 6.35 mm diameter planar transducers in the configuration of Fig. 7.2 (f).

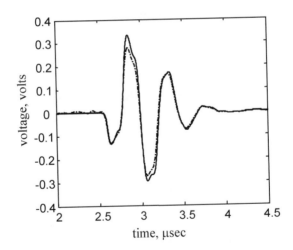

Fig. 7.5. Directly measured output voltage signal of an ultrasonic pitch-catch measurement system (solid line) and the voltage synthesized by measurement and modeling of all the ultrasonic components (dashed-dotted line) for a pair of 2.25 MHz, 12.7 mm diameter planar transducers in the configuration of Fig. 7.2 (f).

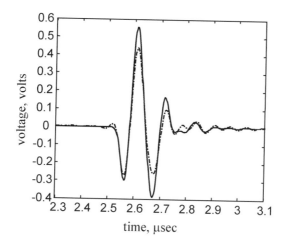

Fig. 7.6. Directly measured output voltage signal of an ultrasonic pitch-catch measurement system (solid line) and the voltage synthesized by measurement and modeling of all the ultrasonic components (dashed-dotted line) for a pair of 10 MHz, 6.35 mm diameter planar transducers in the configuration of Fig. 7.2 (f).

limit expression for a piston transducer, $Z_r^{A;a} = \rho c S_A$, using the density, $\rho = 1$ gm/cm^3, and measured wave speed, $c = 1481$ m/sec, of the water and a transducer area, $S_A = \pi a^2$, calculated from the nominal radius of the transducer, $a = 3.175$ mm. The distance, D, between the two transducers was set at $D = 67$ mm and the attenuation of the water (at room temperature) was taken as the value given by Eq. (5.21). Figure 7.4 shows a comparison of the directly measured output voltage for this configuration with the voltage synthesized from the measurement and modeling of all the system components. Figure 7.5 shows the corresponding results when a pair of 2.25 MHz, 12.7 mm diameter planar transducers were used instead in the same setup and Fig. 7.6 shows the results for a pair of 10 MHz, 6.35 mm diameter planar transducers. For the 5 MHz transducers a difference of −0.7 dB was observed between the peak-to-peak voltage response of the synthesized signal to that of the measured signal. The predicted waveform using 2.25 MHz transducers shows a difference of −1.1 dB in the peak-to-peak voltage with respect to that of the corresponding measured output voltage. For the 10 MHz transducers a somewhat larger difference (−2.5 dB) was observed. In all cases the predicted waveforms had very similar shapes to the measured ones.

Figure 7.7 shows some similar comparisons between a synthesized signal and a measured signal for the pulse-echo setup shown in Fig. 7.2 (a)

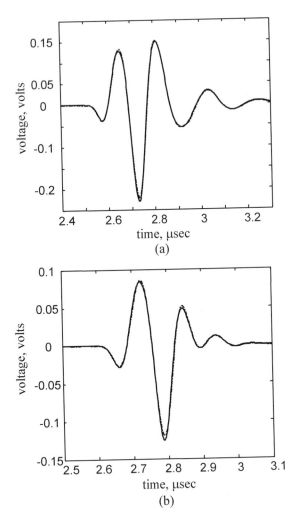

Fig. 7.7. Directly measured output voltage signal of an ultrasonic pulse-echo measurement system (solid line) and the voltage synthesized by measurement and modeling of all the ultrasonic components (dashed-dotted line) for **(a)** a 5 MHz, 6.35 mm diameter planar transducer in the configuration of Fig. 7.2 (a) ,and **(b)** a 10 MHz, 6.35 mm diameter planar transducer in the configuration of Fig. 7.2 (a).

where a planar transducer is receiving the signals reflected from the planar front surface of a solid. The acoustic/elastic transfer function is also available for this configuration (see Eq. (5.16)) in an explicit form. In this case a UTEX 320 square wave pulser/receiver was used in the measurements and again we compared the received measured signals with a voltage synthesized by combining the acoustic/elastic transfer function, the measured Thévenin equivalent source voltage of the pulser, and the sound generation and reception transfer functions obtained by measuring all the components contained in those functions. Figure 7.7 (a) shows a comparison of the measured and synthesized received voltage when a 5 MHz planar transducer was used in this setup. Figure 7.7 (b) shows the corresponding comparison for a 10 MHz transducer. In both cases the peak-to-peak values of the measured signals agreed with the synthesized wave forms to within about 0.2 dB.

7.4 References

7.1 Dang CJ, Schmerr LW, Sedov A (2002) Modeling and measuring all the elements of an ultrasonic nondestructive evaluation system. II: Model-based measurements. Research in Nondestructive Evaluation 14: 177-201
7.2 Schmerr LW, Song SJ, Zhang H (1994) Model-based calibration of ultrasonic system responses for quantitative measurements. In: Green RE Jr., Kozaczek KJ, Ruud CO (eds) Nondestructive characterization of materials, VI. Plenum Press, New York, NY, pp 111-118

7.5 Exercises

1. The beam of a planar immersion transducer is reflected off the front surface of a steel block (see Fig. 7.2 (a)) and this reference signal can be used to determine the system function. The file FBH_ref contains a sampled reference signal of this type and its corresponding sampled times. Place this file in your current MATLAB directory and then load it with the MATLAB command

```
>> load( ' FBH_ref ' )
```

This command will place in the MATLAB workspace 1000 sampled time values in the variable t_ref, and a 1000 point reference time domain waveform in the variable ref. Plot this waveform. Take the FFT of this reference waveform and keep only the first 200 values of the resulting 1000 point spectrum (from 0 to 20 MHz) in a variable, Vc. Plot the magnitude of Vc from 0 to 20 MHz. Use Vc and the data given below to determine the system function via deconvolution (using a Wiener filter) and plot the magnitude of this system function versus frequency from zero to 20 MHz. Compare this system function with Vc. Use the acoustic/elastic transfer function for this configuration as:

$$t_A = 2R_{12} \exp(-2\alpha_{p1}D)\left[1 - \exp(ik_{p1}a^2/2D)\right.$$
$$\left. \cdot \left\{J_0\left(k_{p1}a^2/2D\right) - iJ_1\left(k_{p1}a^2/2D\right)\right\}\right]$$

where we have dropped the phase term $\exp(2ik_pD)$ as it only produces a time delay and the plane wave reflection coefficient is:

$$R_{12} = \frac{\rho_2 c_{p2} - \rho_1 c_{p1}}{\rho_2 c_{p2} + \rho_1 c_{p1}}$$

The parameters for this setup are:

$\rho_1 = 1.0$, $\rho_2 = 7.86$: density of the water and steel, respectively (gm/cm^3)

$c_{p1} = 1484$, $c_{p2} = 5940$: P-wave speeds of the water and steel, respectively (m/sec)

$\alpha_{p1} = 24.79 \times 10^{-6} f^2$: water attenuation (Np/ mm) with f the frequency (in MHz)

$D = 50.8$: distance from the transducer to the block (mm)

$a = 6.35$: radius of the transducer (mm)

$e = 0.3$: noise coefficient for the Wiener filter

Note that the Bessel functions are available directly in MATLAB. The Bessel function of order zero, $J_0(x)$, is given by the MATLAB function besselj(0, x) and the Bessel function of order one, $J_1(x)$, is given by the MATLAB function besselj(1, x).

8 Transducer Sound Radiation

In this Chapter, we will examine models that can describe the radiated sound field generated by an ultrasonic transducer and some of the important parameters that govern the behavior of that field. We will demonstrate most of these results for immersion transducers but many of the concepts introduced also are valid for contact transducers as well. We will also discuss some of the major differences between immersion and contact transducers.

8.1 An Immersion Transducer as a Baffled Source

Figure 8.1 (a) shows a circular planar (non-focused) immersion transducer radiating into a fluid medium, where we have placed the face of the transducer in the x-y plane so that it is pointing in the positive z-direction. When this transducer is driven by the pulser the underlying piezoelectric crystal will move. That motion, in turn, will produce a transient velocity field on the face of the transducer which we will assume is a normal motion (in the z-direction). This velocity field we will write as $v_z(x,y,t)$. Since the pulser drives the transducer with a very short voltage pulse, the motion of the face of the transducer that is generated by this excitation will also be a short time duration pulse. However, we will not model this mechanical motion directly, but instead will deal with its Fourier transform, $v_z(x,y,\omega)$. Such a frequency domain response can alternately be viewed as the result of assuming that the velocity field on the face of the transducer has a harmonic motion given by $v_z = v_z(x,y,\omega)\exp(-i\omega t)$ which generates a radiated sound pressure field in the fluid given by $p(x,y,z,\omega)\exp(-i\omega t)$. Since all the variables for harmonic motion problems have the same common time factor, $\exp(-i\omega t)$, it is customary to drop this time factor and assume it implicitly, a convention we will often follow here.

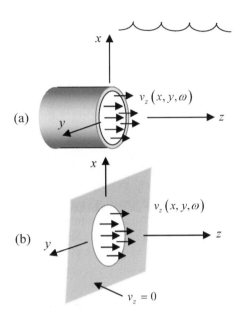

Fig. 8.1. (a) A planar immersion transducer radiating waves into a fluid produced by a harmonic velocity field $v_z(x,y,\omega)$ on its face, and **(b)** a transducer model consisting of the same velocity field in (a) surrounded by a motionless baffle on the z = 0 plane.

Most transducer models do not directly deal with the geometry of Fig. 8.1 (a) but instead consider the alternate geometry of Fig. 8.1 (b) where it is assumed that there is an infinite plane at z = 0 over which the velocity is specified [Fundamentals]. On the surface, S, of the transducer, which lies in this plane, the velocity is given as $v_z = v_z(x,y,\omega)$. For the remainder of the plane one takes $v_z = 0$. These conditions would correspond to having the transducer face embedded in an infinite, motionless, plane baffle. This modified geometry should still represent well our original problem, however, since the transducer will generate a sound field that is significant only in the region ahead of the transducer anyway and the actual fields in the fluid on the plane z = 0 outside of the surface S will be very small, if not identically zero. Mathematically it is more convenient to use the baffled geometry of Fig. 8.1 (b) rather than the original geometry since we then need only to find how a specified velocity field on z = 0 generates fields in the fluid half-space z > 0.

Determining what the velocity field distribution is on the face of a commercial transducer is not a trivial task. Although in principle it is

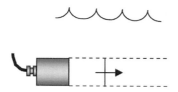

Fig. 8.2. A transducer radiating a perfectly collimated beam at high frequencies.

possible to determine this field experimentally, the measurements are time-consuming and require expensive equipment. Fortunately, for many commercial transducers we can avoid this difficulty by assuming a velocity distribution. The most common assumption is to treat the transducer as a *piston transducer* where the velocity is taken to be spatially uniform over the entire transducer face, i.e. $v_z(x,y,\omega) = v_0(\omega)$. This simple piston model has proven to work well as a basis for characterizing many commercial transducers so it is the model we will adopt here. One should be aware that the validity of this assumption, however, depends on the construction details of the transducer and may be violated in some cases.

If the frequency, ω, was infinitely large a transducer would emit a beam of sound that is confined only to the cylinder of fluid $z \geq 0, r \leq a$ ahead of the transducer as shown in Fig. 8.2. Such a beam is said to be perfectly *collimated*. In reality the frequency is not infinite so that the beam will spread beyond this cylinder, but at the MHz frequencies found in NDE testing a transducer beam will still remain fairly well collimated. This fact is demonstrated in Fig. 8.3 where the magnitude of the pressure

Fig. 8.3. A 5 MHz, ½ inch diameter circular piston transducer radiating sound into water.

field in the x-z plane is shown for a one half inch radius planar piston transducer radiating at 5 MHz into water. There are strong pressure variations in the pressure field, particularly in the region near the transducer. These variations show that one cannot consider the transducer beam to be a simple uniform and well collimated beam as seen, for example, in a flashlight beam. Modeling these pressure variations, therefore, is a nontrivial task.

8.2 An Angular Plane Wave Spectrum Model

Although a transducer does not generate only a plane wave, one way to model a transducer (as a baffled source) is to treat it as the superposition of an infinite number of plane waves, all traveling in the positive z-direction but with different x- and y- component directions. This is basic idea behind an *angular plane wave spectrum model*, where the pressure wave field at a point, $\mathbf{x} = (x, y, z)$, is represented in the form of a 2-D integral given by [Fundamentals], [8.1]

$$p(\mathbf{x}, \omega) = \left(\frac{1}{2\pi}\right)^2 \int_{-\infty}^{+\infty}\int_{-\infty}^{+\infty} P(k_x, k_y) \exp\left[i\left(k_x x + k_y y + k_z z\right)\right] dk_x dk_y. \quad (8.1)$$

Since the time-domain pressure, $p(\mathbf{x}, t)$, must satisfy the 3-D wave equation

$$\frac{\partial^2 p}{\partial x^2} + \frac{\partial^2 p}{\partial y^2} + \frac{\partial^2 p}{\partial z^2} - \frac{1}{c^2}\frac{\partial^2 p}{\partial t^2} = 0 \quad (8.2)$$

for $p(\mathbf{x}, t) = p(\mathbf{x}, \omega)\exp(-i\omega t)$ we must have $p(\mathbf{x}, \omega)$ satisfy

$$\frac{\partial^2 p}{\partial x^2} + \frac{\partial^2 p}{\partial y^2} + \frac{\partial^2 p}{\partial z^2} + k^2 p = 0, \quad (8.3)$$

which is called the *Helmholtz equation*. Clearly, $p(\mathbf{x}, \omega)$ will satisfy Eq. (8.3) if all of the exponential terms in Eq. (8.1) also satisfy that equation. Placing $\exp\left[i\left(k_x x + k_y y + k_z z\right)\right]$ into Eq. (8.3), we find as a requirement that $k_z = \pm\sqrt{k^2 - k_x^2 - k_y^2}$. In order to have waves traveling in the positive z-direction (as they must, physically, for our problem), only

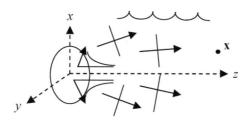

Fig. 8.4. Model of a transducer as a superposition of plane and inhomogeneous waves radiating into the region $z \geq 0$.

the positive value is acceptable and so we choose $k_z = \sqrt{k^2 - k_x^2 - k_y^2}$. Terms such as $p = \exp\left(ik_x x + ik_y y + i\sqrt{k^2 - k_x^2 - k_y^2}\, z\right)$ are just plane harmonic waves as long as $k^2 > k_x^2 + k_y^2$ is satisfied. In Eq. (8.1), however, all values of k_x, k_y are superimposed so that there will be values of those variables in the integrations where $k_x^2 + k_y^2 > k^2$ and k_z will be imaginary. For those cases if we take $k_z = i\sqrt{k_x^2 + k_y^2 - k^2}$ we will no longer have plane waves propagating into the half-space $z > 0$ but instead will have waves that propagate in the x- and y-directions from the transducer but that are exponentially decaying in the z-direction of the form $p = \exp\left(ik_x x + ik_y y - \sqrt{k_x^2 + k_y^2 - k^2}\, z\right)$ [note: $k_z = -i\sqrt{k_x^2 + k_y^2 - k^2}$ cannot be used since then we would obtain waves that grow exponentially in the z-direction away from the transducer, which is not physical]. Such waves are called *inhomogeneous waves*. Thus, strictly speaking, Eq. (8.1) represents the pressure wave fields as a superposition of both plane wave and inhomogeneous wave fields (see Fig. 8.4) where we must have

$$k_z = \begin{cases} \sqrt{k^2 - k_x^2 - k_y^2}, & k^2 \geq k_x^2 + k_y^2 \\ i\sqrt{k_x^2 + k_y^2 - k^2}, & k^2 < k_x^2 + k_y^2 \end{cases}. \tag{8.4}$$

Appendix D gives a discussion of inhomogeneous waves found when solving plane wave transmission/reflection problems.

In order for Eq. (8.1) to represent the solution to our baffled transducer model, we must determine the unknown $P(k_x, k_y)$ so that the velocity boundary conditions are satisfied on the plane $z = 0$. From the equation of motion for the fluid (see Appendix D) we have

$$v_z(x, y, z = 0, \omega) = \frac{1}{i\omega\rho} \frac{\partial p}{\partial z}(x, y, z = 0, \omega), \qquad (8.5)$$

where ρ is the density of the fluid. Placing Eq. (8.1) into this relationship we find

$$v_z(x, y, z = 0, \omega) = \left(\frac{1}{2\pi}\right)^2 \int_{-\infty}^{+\infty}\int_{-\infty}^{+\infty} \frac{ik_z P(k_x, k_y)}{i\omega\rho}$$
$$\cdot \exp\left[i(k_x x + k_y y)\right] dk_x dk_y \qquad (8.6)$$

To see what Eq. (8.6) means, let $V(k_x, k_y) = ik_z P(k_x, k_y)/i\omega\rho$. Then Eq. (8.6) becomes simply

$$v_z(x, y, z = 0, \omega) = \left(\frac{1}{2\pi}\right)^2 \int_{-\infty}^{+\infty}\int_{-\infty}^{+\infty} V(k_x, k_y)$$
$$\cdot \exp\left[i(k_x x + k_y y)\right] dk_x dk_y. \qquad (8.7)$$

Equation (8.7) is in the form of two inverse Fourier transforms where the t and ω parameters in the time and frequency domains (see Appendix A) are replaced by wave numbers and spatial parameters, i.e. $\omega \to k_x, -t \to x$ for one transform and $\omega \to k_y, -t \to y$ for the other transform. Thus, Eq. (8.7) is called an inverse 2-D *spatial* Fourier transform. By the properties of the Fourier transform it then follows that we must have

$$V(k_x, k_y) = \int_{-\infty}^{+\infty}\int_{-\infty}^{+\infty} v_z(x, y, z = 0, \omega)\exp\left[-i(k_x x + k_y y)\right] dxdy, \qquad (8.8)$$

which shows that $V(k_x, k_y)$ is just the 2-D spatial Fourier transform of the velocity field on the plane $z = 0$. For a circular piston transducer of radius a, for example, where

$$v_z(x, y, z = 0, \omega) = \begin{cases} v_0(\omega) & x^2 + y^2 \leq a^2 \\ 0 & x^2 + y^2 > a^2 \end{cases} \qquad (8.9)$$

the 2-D spatial Fourier transform in Eq. (8.8) can be obtained explicitly as

$$V(k_x,k_y) = 2\pi a^2 v_0(\omega)\frac{J_1\left(\sqrt{k_x^2+k_y^2}\,a\right)}{\sqrt{k_x^2+k_y^2}\,a},$$ (8.10)

where J_1 is a Bessel function of order one. Similarly, for a rectangular piston transducer of length l_x in the x-direction and length l_y in the y-direction we find

$$V(k_x,k_y) = l_x l_y v_0(\omega)\frac{\sin\left(\dfrac{k_x l_x}{2}\right)\sin\left(\dfrac{k_y l_y}{2}\right)}{\left(\dfrac{k_x l_x}{2}\right)\left(\dfrac{k_y l_y}{2}\right)}.$$ (8.11)

Thus, for any given velocity distribution on $z = 0$, the pressure wave field from the transducer can be found explicitly as

$$p(\mathbf{x},\omega) = \left(\frac{1}{2\pi}\right)^2 \int_{-\infty}^{+\infty}\int_{-\infty}^{+\infty} \frac{i\omega\rho\, V(k_x,k_y)}{ik_z}$$
$$\cdot \exp\left[i\left(k_x x + k_y y + k_z z\right)\right] dk_x dk_y$$ (8.12)

once the 2-D spatial Fourier transform of the velocity field at $z = 0$ is known. Equation (8.12) is an exact result that can be used directly for numerical modeling of transducer wave fields. However, it is a model that is numerically very challenging to implement since one still needs to perform two infinite integrations of rapidly varying functions. In practice, it has been found that the inhomogeneous waves contribute little to the pressure wave field except in a region very close to the transducer, which is usually not of great interest. Thus, most numerical evaluations of Eq. (8.12) simply ignore all the inhomogeneous waves and compute instead the finite integrals over all the plane wave terms

$$p(\mathbf{x},\omega) = \left(\frac{1}{2\pi}\right)^2 \iint_{k_x^2+k_y^2\le k^2} \frac{i\omega\rho V(k_x,k_y)}{ik_z}$$
$$\cdot \exp\left[i\left(k_x x + k_y y + k_z z\right)\right] dk_x dk_y.$$ (8.13)

Equation (8.13) is now a more tractable transducer model, but it still requires a significant amount of computation (i.e. many plane wave components need to be superimposed) in order to adequately simulate the transducer beam. Also, Eq. (8.13) does not explicitly show us much about the physics of the sound generation process. Thus, we will consider another transducer model that remedies some of these deficiencies.

8.3 A Rayleigh-Sommerfeld Integral Transducer Model

In discussing linear systems in Appendix C, we saw that the convolution theorem played a crucial role. In that case, we showed that a 1-D time domain convolution of two functions was equivalent to taking the inverse Fourier transform of a product of their Fourier transforms. Since here Eq. (8.12) is in the form of a 2-D inverse spatial Fourier transform of a product of 2-D transforms, we could expect that a 2-D form of the convolution theorem might play an equally important role here. This indeed turns out to be the case. First, we state the following 2-D (spatial) convolution theorem [8.2]:

If

$$f(x,y) = \left(\frac{1}{2\pi}\right)^2 \int_{-\infty}^{+\infty}\int_{-\infty}^{+\infty} H\left(k_x,k_y\right)G\left(k_x,k_y\right)\exp\left[i\left(k_x x + k_y y\right)\right]dk_x dk_y$$

then

$$f(x,y) = \int_{-\infty}^{+\infty}\int_{-\infty}^{+\infty} h(x',y')g(x-x',y-y')dx'dy'$$

where $H\left(k_x,k_y\right)$ is the 2-D spatial Fourier transform of $h(x,y)$ and $G\left(k_x,k_y\right)$ is the 2-D spatial Fourier transform of $g(x,y)$. We can use this theorem directly for Eq. (8.12) if we make the following definitions

$$H\left(k_x,k_y\right) = -i\omega\rho\, V\left(k_x,k_y\right)$$

$$G\left(k_x,k_y\right) \equiv G\left(k_x,k_y,z\right) = \frac{\exp\left(ik_z z\right)}{-ik_z}. \tag{8.14}$$

Then it follows that

$$h(x,y) = -i\omega\rho v_z(x,y,z=0,\omega)$$

$$g(x,y) \equiv g(x,y,z) = \frac{\exp\left[ik\sqrt{x^2+y^2+z^2}\right]}{2\pi\sqrt{x^2+y^2+z^2}}. \qquad (8.15)$$

The expression for h in Eq. (8.15) follows directly from the fact that $V(k_x,k_y)$ is the 2-D spatial Fourier transform of $v_z(x,y,z=0,\omega)$. The expression for g in Eq. (8.15) comes from Weyl's representation of a spherical wave in terms of an angular plane wave spectrum integral [Fundamentals]. In particular, Weyl showed that

$$\frac{\exp\left[ik\sqrt{x^2+y^2+z^2}\right]}{2\pi\sqrt{x^2+y^2+z^2}}$$

$$= \left(\frac{1}{2\pi}\right)^2 \int_{-\infty}^{+\infty}\int_{-\infty}^{+\infty} \frac{1}{-ik_z}\exp\left[i\left(k_x x + k_y y + k_z z\right)\right]dk_x dk_y \qquad (8.16)$$

$$= \left(\frac{1}{2\pi}\right)^2 \int_{-\infty}^{+\infty}\int_{-\infty}^{+\infty} \frac{\exp(ik_z)}{-ik_z}\exp\left[i\left(k_x x + k_y y\right)\right]dk_x dk_y.$$

From Eqs. (8.12), (8.14) and (8.15) and the 2-D convolution theorem then it follows that we have an alternate representation for the pressure wave field of a transducer given by

$$p(\mathbf{x},\omega) = \frac{-i\omega\rho}{2\pi}\int_{-\infty}^{+\infty}\int_{-\infty}^{+\infty} v_z(x',y',z=0,\omega)$$

$$\cdot \frac{\exp\left[ik\sqrt{(x-x')^2+(y-y')^2+z^2}\right]}{\sqrt{(x-x')^2+(y-y')^2+z^2}}dx'dy', \qquad (8.17)$$

which is called the *Rayleigh-Sommerfeld integral*. Just as Eq. (8.12) gave us a transducer model in terms of a superposition of plane (and inhomogeneous) waves traveling in different directions, the Rayleigh-Sommerfeld integral represents the transducer radiation as a superposition of spherical waves radiating from point sources distributed on the plane $z = 0$. Since any transducer only generates a non-zero velocity over some finite area, S, (see Fig. 8.5), we can rewrite Eq. (8.17) more compactly as

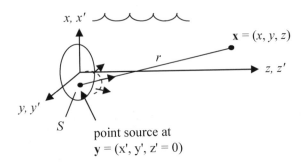

Fig. 8.5. A transducer modeled as a superposition of radiating point sources.

$$p(\mathbf{x},\omega) = \frac{-i\omega\rho}{2\pi} \iint_S v_z\left(x',y',z=0,\omega\right)\frac{\exp(ikr)}{r}dS \qquad (8.18)$$

where $r = \sqrt{\left(x-x'\right)^2 + \left(y-y'\right)^2 + z^2}$ (see Fig. 8.5) is the distance from an arbitrary point $\mathbf{y} = (x',y',0)$ on the transducer surface, S, to a point, $\mathbf{x} = (x,y,z)$, in the fluid and dS is an element of area on the transducer surface. For the particular case of a piston transducer the Rayleigh-Sommerfeld integral reduces to an even simpler form given by

$$p(\mathbf{x},\omega) = \frac{-i\omega\rho\, v_0(\omega)}{2\pi} \iint_S \frac{\exp(ikr)}{r}dS. \qquad (8.19)$$

The Rayleigh-Sommerfeld integral for a piston source, Eq. (8.19), is used in many texts to discuss transducer radiation in a fluid [Fundamentals]. In general, it still requires a significant amount of numerical effort to evaluate since although one now only has to integrate over the finite face of the transducer, the complex exponential term in the integrand of Eq. (8.19) has a rapidly varying phase for the frequencies and transducer sizes used in NDE tests that makes the 2-D numerical integrations lengthy. However, as we will see, the Rayleigh-Sommerfeld integral does allow us to examine more directly the physics of the transducer radiation problem than Eq. (8.12) permits and we can even extract exact results in some important special cases.

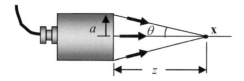

Fig. 8.6. Geometry for a circular planar piston transducer radiating direct and edge waves to a point **x** on the axis of the transducer.

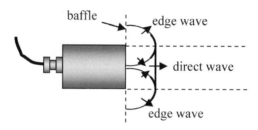

Fig. 8.7. The direct and edge waves generated by an impulsively excited circular piston transducer.

8.4 On-Axis Behavior of a Planar Circular Piston Transducer

Consider first the special case where we wish to obtain the pressure wave field on the central axis of a circular piston transducer of radius a as shown in Fig. 8.6. In this case because of symmetry we can take the area element as $dS = 2\pi\rho_0 d\rho_0$, where ρ_0 is the radial distance on the plane $z = 0$ from the center of the transducer to an arbitrary point on the transducer surface. Since $r^2 = \rho_0^2 + z^2$ it follows that $dS = 2\pi r dr$. Placing this result into Eq. (8.19) then allows us to integrate the remaining complex exponential term to obtain an exact expression for the on-axis pressure given by [Fundamentals]

$$p(z,\omega) = \rho c v_0(\omega)\left[\exp(ikz) - \exp\left(ik\sqrt{z^2 + a^2}\right)\right]. \tag{8.20}$$

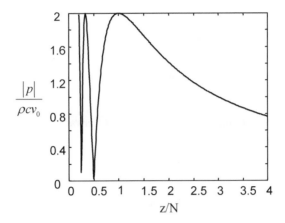

Fig. 8.8. On-axis normalized pressure versus normalized distance z/N for a 5 MHz, 1/2 inch diameter planar transducer radiating into water, where N is the near field distance.

The first term is a wave that has traveled a distance z directly from the face of the transducer to the point on the transducer axis while the second term is a wave that has traveled a distance $\sqrt{z^2 + a^2}$ so that it appears to have come from the edge (rim) of the transducer, as shown in Fig. 8.6. Indeed, if one examines the pulses which travel from an impulsively excited transducer, as shown in Fig. 8.7 , one sees a plane wave front (the "direct" wave) that travels normal to the face of the transducer and a doughnut-like wave front that comes from the transducer rim (the "edge" wave). Except very near the transducer and for very short pulses, however, we will likely not see these two waves separately. Indeed, at large distances from the transducer where $z \gg a$, an expansion of the edge wave term gives $\sqrt{z^2 + a^2} \approx z \left[1 + a^2 / 2z^2 \right]$. If we also assume $ka^2 / 2z \ll 1$ it follows to first order that

$$p(z,\omega) = \frac{-i\omega\rho a^2 v_0(\omega)}{2} \frac{\exp(ikz)}{z}, \qquad (8.21)$$

which now looks like a single, spherically spreading wave. This result is reasonable since at sufficiently large distances from the transducer the transducer should act like a point source. Distances that satisfy this criterion are said to be in the *transducer far field* or in the spherically spreading region of the transducer.

If one plots the magnitude of the on-axis pressure versus z that one obtains from Eq. (8.20), then one sees two distinct types of behavior for the on-axis response (Fig. 8.8). Near the transducer one sees a series of nulls and maxima. In this near field region, one can show from Eq. (8.20) that the maxima are located approximately at the distances $z = N/(2m+1)$ $m = 0,1,2,...$ while the nulls are at approximately $z = N/2n$ $n = 1,2,3,...$ where $N = a^2/\lambda$ (the ratio of the radius squared of the transducer to the wave length, λ) is called the *near field distance* and distances $z < N$ are said to be in the *transducer near field* [Fundamentals]. As the distance z increases, the last on-axis null occurs at $z = N/2$ and the last on-axis maximum occurs at $z = N$. Beyond $z = N$ the pressure field simply decays monotonically. At a distances greater than approximately three near field distances from the transducer the exact on-axis response begins to agree very well with the far field expression of Eq. (8.21) so that $z = 3N$ is generally taken as the start of the *transducer far field* region.

8.5 The Paraxial Approximation

Having the exact on-axis behavior of the transducer also enables us to discuss an important concept called the *paraxial approximation*. If we examine the direct and edge waves we see (Fig 8.6) that they are separated by the angle θ. At a distance z approximately equal to a transducer diameter $(2a)$, this angle begins to become small enough so that we can assume $\sqrt{z^2 + a^2} \approx z\left[1 + a^2/2z^2\right]$. However, unlike the far field case, we will not also assume $ka^2/2z \ll 1$ (which is equivalent to $z \gg \pi N$, i.e. under this condition we must be many near field distances away from the transducer), so that we are not necessarily in the transducer far field. This means that in the present case we are only assuming that the angle θ is small enough so that all the waves in the transducer beam can be considered to be traveling in approximately the same direction (which in this case is along the z-axis). This is the essence of the paraxial approximation. For this approximation we have

$$p(z,\omega) = \rho c v_0 \exp(ikz)\left[1 - \exp\left(\frac{ika^2}{2z}\right)\right].$$

$$(8.22)$$

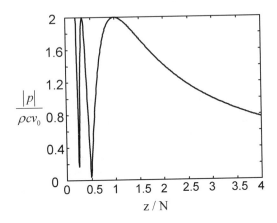

Fig. 8.9. On-axis normalized pressure versus normalized distance z/N for a 5 MHz, 1/2 inch diameter planar transducer radiating into water (paraxial approximation).

Equation (8.22) still contains the direct and edge waves of the original exact response but it is in the form of a quasi-plane wave since it can be written as

$$p(z,\omega) = C(z,a,\omega)\left[\rho c v_0 \exp(ikz)\right]. \qquad (8.23)$$

The term in the brackets in Eq. (8.23) is just a plane wave traveling in the z-direction. The coefficient $C(z,a,\omega)$ that multiplies this plane wave is called a *diffraction coefficient*. It accounts for all the deviations in amplitude and phase of the on-axis response in the actual transducer beam from that of a plane wave. In this case we simply have

$$C(z,a,\omega) = 1 - \exp(ika^2/2z). \qquad (8.24)$$

Figure 8.9 plots the on-axis response in the paraxial approximation (Eq. (8.22)) for the same case shown in Fig. 8.8. It can be seen from those figures that the paraxial approximation captures well both the near and far field on-axis behavior of the transducer. Only within approximately a transducer diameter, a region not shown in these figures, does the paraxial approximation begin to lose accuracy. This means that for most NDE testing situations where we are not concerned with the wave fields immediately adjacent to the transducer, the paraxial approximation should work well. The importance of the paraxial approximation is that it can also work well in much more general testing situations where we are considering

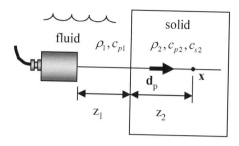

Fig. 8.10. An immersion transducer radiating at normal incidence through a planar fluid-solid interface.

off-axis transducer responses and where the transducer beam itself has been transmitted or reflected from various parts of a component's geometry. These types of complicated interactions occur frequently in NDE tests, so that if the paraxial approximation is valid, we may still treat the sound beam approximately as a quasi-plane wave and all the complicated interactions of the transducer sound beam with the component geometry can be treated approximately as interactions of a plane wave with that geometry. Plane wave interactions are much easier to deal with than interactions involving more general wave types so that the paraxial approximation gives us a powerful tool for accurately simulating many complex problems. The key, of course, is in being able to efficiently determine the diffraction coefficient (either analytically or numerically) for a given testing problem. Fortunately, this is possible, as we will see, in many cases. We will outline here one example where the paraxial approximation can be used in a more general testing setup to determine the transducer wave field. Consider a planar circular piston transducer of radius a radiating through a planar fluid-solid interface at normal incidence (see Fig. 8.10). In this case the compressional waves (P-waves) in the fluid generate primarily P-waves in the isotropic elastic solid and the on-axis velocity in the solid is given by [Fundamentals]

$$\mathbf{v}(z_2,\omega) = v_0 T_{12}^{P;P} \mathbf{d}_p \exp\left(ik_{p1}z_1 + ik_{p2}z_2\right)\left[1 - \exp\left(\frac{ik_{p1}a^2}{2\tilde{z}}\right)\right], \qquad (8.25)$$

where $k_{pj} = \omega / c_{pj}$ $(j = 1,2)$ are wave numbers for P-waves in the fluid and solid, respectively, $\mathbf{d}_p$ is a unit vector (polarization vector) along the

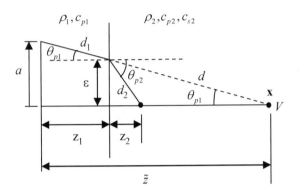

Fig. 8.11. Propagation of an edge wave through a fluid-solid interface to an on-axis point **x** in the solid and the corresponding "virtual" point V that the edge wave would travel to in the solid if it's angle was not changed upon refraction through the interface.

propagation direction, $T_{12}^{P;P}$ is a plane wave transmission coefficient for P-waves in the solid due to P-waves in the fluid (the ratio of the velocity at the interface on the solid side to the velocity on the fluid side) and $\tilde{z} = z_1 + c_{p2}z_2 / c_{p1}$. The combined leading terms multiplying the bracketed expression in Eq. (8.25) represent a plane wave that has traveled from the transducer to a depth, z_2, in the solid while the bracketed term itself is the diffraction coefficient for this problem. Interestingly, this diffraction coefficient is in exactly the same form as for the on-axis response for a single fluid medium so that all of the near and far field on-axis behavior we discussed previously for the single fluid case remain valid for this problem if we replace the z-distance in the fluid by the equivalent distance $z_1 + c_{p2}z_2 / c_{p1}$. This result can be explained by the behavior of the edge wave at the interface as shown in Fig. 8.11. From that figure we see that $\varepsilon = d_2 \sin\theta_{p2} = d\sin\theta_{p1}$ where d_2 is the path length of the edge wave in the solid and d is the distance from the interface to a "virtual" point, V, on the axis of the transducer in the solid, which is where the edge wave would arrive on the axis if it had not had its direction changed upon refraction. Solving for d, we find $d = d_2 \sin\theta_{p2} / \sin\theta_{p1}$. However, from Snell's law for refracted waves we have $\sin\theta_{p2} / \sin\theta_{p1} = c_{p2} / c_{p1}$ so $d = (c_{p2} / c_{p1})d_2$. If we now define the corresponding distance to the virtual point along the z-axis as $\tilde{z}$, in the paraxial approximation this virtual point distance is given

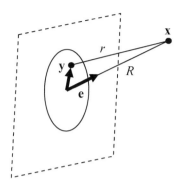

Fig. 8.12. Geometry parameters for defining the far field behavior of a transducer.

by $\tilde{z} \cong d_1 + d = d_1 + \left(c_{p2}/c_{p1}\right)d_2 \cong z_1 + \left(c_{p2}/c_{p1}\right)z_2$. We see that for the interface problem, in the paraxial approximation the refracted waves appear to go through a z-distance, $\tilde{z}$, to the virtual point on the axis in exactly the same manner as for a single medium problem where the interface is absent. In the diffraction correction for a single medium, therefore, one can simply replace the z-distance by the equivalent distance, $\tilde{z}$, to obtain the diffraction correction for this case.

8.6 Far field On-Axis and Off-Axis Behavior

In section 8.4 we obtained an explicit expression (Eq. (8.21)) for the on-axis far field wave field of a circular planar piston transducer. Here, we will show that it is possible to obtain an expression for the entire far field transducer behavior for both on- and off-axis points for planar transducers. This expression is often referred to as the *Fraunhoffer approximation* for the transducer wave field. First, we express the radius r in Eq. (8.18) in terms of the distance R and unit vector $\mathbf{e}$ pointing from the center of the transducer to point $\mathbf{x}$ as (see Fig. 8.12)

$$r = \sqrt{\left(\mathbf{x}-\mathbf{y}\right)\cdot\left(\mathbf{x}-\mathbf{y}\right)}$$
$$= \sqrt{\left(R\mathbf{e}-\mathbf{y}\right)\cdot\left(R\mathbf{e}-\mathbf{y}\right)}. \tag{8.26}$$

In the far field $|\mathbf{y}| \ll R$ so we can expand the square root in Eq. (8.26) to obtain:

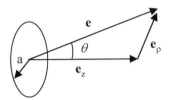

Fig. 8.13. The unit vector, **e**, and its cylindrical components, where $\mathbf{e}_z$ is along the z-axis and $\mathbf{e}_\rho$ is in a radial direction in a plane parallel to the circular transducer of radius a.

$$r \cong R\sqrt{1 - 2\mathbf{e}\cdot\mathbf{y}/R}$$
$$\cong R - \mathbf{e}\cdot\mathbf{y}. \tag{8.27}$$

Both terms in Eq. (8.27) are used to approximate r in the phase part of the spherical wave term in Eq. (8.18) while only the leading term is used to approximate the $1/r$ amplitude term. The reason for this difference in the number of terms retained is that the phase is much more sensitive to approximation than the amplitude since in the phase not only must a term that is neglected be smaller than those terms retained but the neglected term must also be much less than 2π. These approximations reduce Eq. (8.18) to the form

$$p(\mathbf{x}, \omega) = \frac{-i\omega\rho}{2\pi}\frac{\exp(ikR)}{R}\iint_S v_z(x', y', 0, \omega)\exp(-ik\mathbf{e}\cdot\mathbf{y})\,dS, \tag{8.28}$$

which can be rewritten as

$$p(\mathbf{x}, \omega) = \frac{-i\omega\rho}{2\pi}\frac{\exp(ikR)}{R}\iint_S \{v_z(x', y', 0, \omega)$$
$$\cdot\exp\left[-i\left(k_x x + k_y y\right)\right]dx'dy'\}, \tag{8.29}$$

where $k_x = ke_x, k_y = ke_y$. From Eq. (8.8) we recognize the integral in Eq. (8.29) as just the 2-D spatial Fourier transform of the velocity field, $V(k_x, k_y)$, so that we have, finally

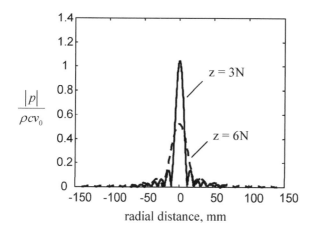

Fig. 8.14. The far field variation of the normalized pressure versus radial distance, ρ_0, for a circular transducer at three and six near field distances, showing the spreading of the angular lobes of the response and the decay in amplitude with increasing distance from the transducer.

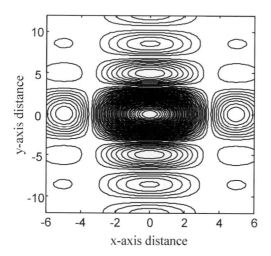

Fig. 8.15. The contours of the far field pressure distribution in a plane parallel to the face of a 3mm x 6 mm rectangular transducer radiating into water at 5 MHz and at a distance of 70 mm.

$$p(\mathbf{x},\omega) = \frac{-i\omega\rho V\left(k_x, k_y\right)}{2\pi} \frac{\exp(ikR)}{R}. \qquad (8.30)$$

For a circular piston transducer, we have, using Eq. (8.10) and $e_\rho = \sqrt{e_x^2 + e_y^2} = \sin\theta$ (see Fig. 8.13)

$$p(\mathbf{x},\omega) = -i\omega\rho a^2 v_0(\omega)\frac{J_1(ka\sin\theta)}{ka\sin\theta}\frac{\exp(ikR)}{R}, \qquad (8.31)$$

which represents a spherical pressure wave in the far field whose amplitude is angular dependent. Figure 8.14 plots the magnitude of the normalized pressure, $|p|/\rho c v_0$, at different fixed distances, z, from the transducer face as a function of the radial distance, ρ_0, from the transducer's central axis, where $\sin\theta = \rho_0/z$. For both $z = 3N$ and $z = 6N$ one sees the lobe structure generated by the $J_1(u)/u$ angular directivity term of the response in the far field. At $z = 6N$, however, the lobes are broader than at $z = 3N$ due to beam spreading and the amplitude is also smaller because of the $1/R$ spherical wave decay term.

For a rectangular transducer with length l_x in the x-direction and length l_y in the y-direction, Eqs. (8.11) and (8.30) give the far field behavior as

$$p(\mathbf{x},\omega) = \frac{-i\omega\rho l_x l_y v_0(\omega)}{2\pi}\frac{\sin(k_x l_x/2)\sin(k_y l_y/2)}{(k_x l_x/2)(k_y l_y/2)}\frac{\exp(ikR)}{R}. \qquad (8.32)$$

Figure 8.15 gives a 2-D cross sectional plot of the magnitude of the normalized pressure, $2\pi|p|/\rho c v_0$, as a function of the distances x and y for a given distance z, where $k_x = kx/R = kx/\sqrt{x^2 + y^2 + z^2}$ and $k_y = ky/R = ky/\sqrt{x^2 + y^2 + z^2}$. This figure shows the complex 2-D lobe structure present for a rectangular transducer.

8.7 A Spherically Focused Piston Transducer

Many commercial focused transducers produce a focused acoustic sound beam by incorporating an acoustic lens into the transducer design. Modeling

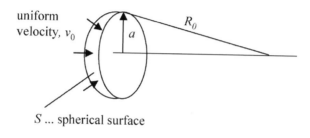

Fig. 8.16. The O'Neil model for a spherically focused piston transducer.

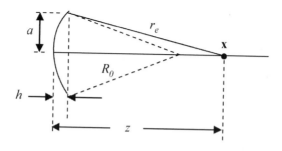

Fig. 8.17. Geometry parameters that appear in the on-axis response of a spherically focused transducer.

in detail such a configuration is very difficult but one can induce the same focusing effect by considering the transducer to be a piston transducer where a constant (radial) velocity is placed on a spherical surface instead of a plane one. In this case one still uses the Rayleigh-Sommerfeld integral (Eq. (8.19)) but now the integration is over a finite radius, a, of a spherical surface S whose radius of curvature is R_0, as shown in Fig. 8.16. This focused transducer model is due to O'Neil [8.3], [Fundamentals]. While the replacement of the integration over a plane surface in the Rayleigh-Sommerfeld integral by integration over a spherical surface is an ad-hoc approach that is not valid in a strict mathematical sense the O'Neil model has been shown to be accurate as long as the focusing is not too severe. Such severe focusing can be found in practice, for example, in acoustic microscopes. Most commercial focused NDE transducers, however, are not tightly focused so that the O'Neil model should work well in practice for most NDE applications. For a point $\mathbf{x}$ in the fluid on the axis of the spherically focused transducer one

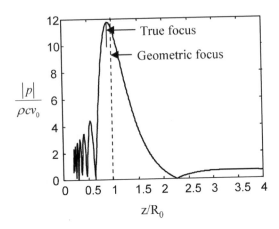

Fig. 8.18. The on-axis normalized pressure for a 6.35 mm radius spherically focused piston transducer radiating into water at 10 MHz with a geometric focal length of 76.2 mm.

can show that the element of area $dS = (2\pi / q_0) r dr$ where $q_0 = 1 - z / R_0$ [Fundamentals]. Thus, the O'Neil model, like the Rayleigh-Sommerfeld model, can be integrated exactly for this case. We find

$$p(z, \omega) = \frac{\rho c v_0}{q_0} \left[\exp(ikz) - \exp(ikr_e) \right], \qquad (8.33)$$

where $r_e = \sqrt{(z - h)^2 + a^2}$ and $h = R_0 - \sqrt{R_0^2 - a^2}$. These distances are shown in Fig. 8.17.

Figure 8.18 shows a plot of the normalized pressure, $|p| / \rho c v_0$, versus normalized distance, z / R_0, for a 6.35 mm radius transducer with geometrical focal length of 76.2 mm radiating into water at 10 MHz. It can be seen from that figure that for distances where $z < R_0$ the response has a series of nulls and maxima which eventually produce a single large peak near $z = R_0$ (the geometrical focal length). There is another null at approximately $z = 2.25 R_0$ and a very small response thereafter. It can be shown that the nulls are located approximately at distances z_n given by [Fundamentals]

$$z_n = R_0 \left(\frac{h}{h \pm n\lambda} \right), \tag{8.34}$$

where λ is the wave length and the plus sign is for nulls satisfying $z < R_0$ while the minus sign is for nulls where $z > R_0$. For nulls beyond the geometrical focus, however, there is an additional restriction $h \geq n\lambda$ that must be satisfied for those nulls so that in some cases such nulls may not exist at all. Unfortunately, one cannot write down a simple relationship for the location of the on-axis maxima as done for the planar transducer case. The most one can do is state that they are determined by the roots of a transcendental equation which is [Fundamentals]

$$\cos(k\delta/2) = \frac{2(\delta+z)\sin(k\delta/2)}{kR_0(\delta+h)q_0}, \tag{8.35}$$

where $\delta = r_e - z = \sqrt{(z-h)^2 + a^2} - z$.

Note that due to wave diffraction effects the maximum response (true focus) at finite frequencies occurs at a distance somewhat less than the geometric focal length (geometric focus), as shown for this case. It is only at infinitely large frequencies that the maximum on-axis response occurs at $z = R_0$.

With some algebra we can express the distance r_e also in the form $r_e = \sqrt{z^2 + (a^2 + h^2)q_0} - z$ [Fundamentals]. In the paraxial approximation we must have $h \ll a$ (not too severe focusing) and $z \gg a$ (not too near the transducer). In this approximation we find

$$\begin{aligned} r_e &\cong \sqrt{z^2 + a^2 q_0} - z \\ &\cong z\left(1 + \frac{a^2 q_0}{2z^2} + ...\right) - z \\ &= \frac{a^2 q_0}{2z} \end{aligned} \tag{8.36}$$

so that the on-axis response in Eq. (8.33) becomes

$$p(z,\omega) = \rho c v_0 \exp(ikz)\left\{ \frac{1}{q_0}\left[1 - \exp(ika^2 q_0 / 2z)\right] \right\}, \tag{8.37}$$

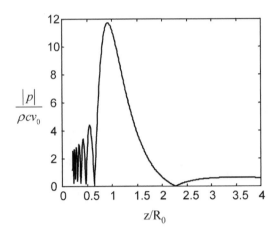

Fig. 8.19. The on-axis response calculated with the paraxial approximation for the same spherically focused transducer shown in Fig. 8.18.

which shows that the on-axis diffraction coefficient for a spherically focused piston transducer is given by

$$C(z,a,R_0,\omega) = \frac{1}{q_0}\left[1 - \exp\left(ika^2 q_0 / 2z\right)\right]. \tag{8.38}$$

For a planar transducer $q_0 \to 1$ and Eq. (8.38) reduces to Eq. (8.22). As in the planar case the paraxial approximation works very well in describing the ultrasonic beam from a spherically focused transducer as long as the focusing is not too severe and one is not too close to the transducer. Figure 8.19 shows the on-axis pressure plot predicted in the paraxial approximation for the same case shown in Fig. 8.18. It can be seen that the two responses are nearly identical.

The paraxial approximation also can be used as a means for illustrating a relatively simple way to incorporate focusing into a transducer beam model. Consider a planar circular piston transducer. Since the velocity is uniform over the face of the transducer, the phase of this velocity field is constant (zero) on this aperture. In contrast, if the transducer had generated a spherically converging wave which focuses at $z = R_0$ on the axis of the transducer the phase of the velocity field on the aperture would not be a constant (see Fig. 8.20). On the plane $z = 0$ we would instead have a phase term given, in the paraxial approximation, by

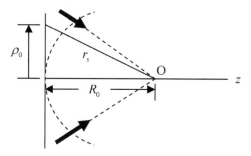

Fig. 8.20. Geometry for defining the phase variations on the plane $z = 0$ of a spherically converging wave that focuses at $z = R_0$.

$$\exp\left(-ik\left[r_s - R_0\right]\right) = \exp\left[-ik\left[\sqrt{\rho_0^2 + R_0^2} - R_0\right]\right]$$
$$\cong \exp\left(-ik\rho_0^2 / 2R_0\right). \tag{8.39}$$

[Note: we have included the ikR_0 term in Eq. (8.39) so that the phase of the wave is zero at the origin $\left(\rho_0 = z = 0\right)$, i.e. the wave starts out from that point at time $t = 0$. The $-ikr_s$ term has a negative sign because r_s decreases as the time t increases, i.e the wave is a spherical wave converging to point O on the axis]. Now, suppose we take the Rayleigh- Sommerfeld integral model of a planar piston transducer and simply include the phase term given in Eq. (8.39) over the planar transducer surface S. From Eq. (8.19) we would have

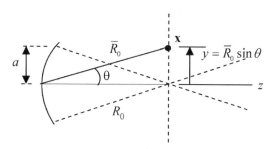

Fig. 8.21. Geometry variables for defining the field behavior at a plane located a distance from the transducer equal to the geometrical focal length.

$$p(\mathbf{x},\omega) = \frac{-i\omega\rho\, v_0(\omega)}{2\pi} \iint_S \exp\left(-ik\rho_0^2 / 2R_0\right) \frac{\exp(ikr)}{r} dS. \tag{8.40}$$

Consider now the on-axis response. For a circular transducer we can take $dS = 2\pi\rho_0 d\rho_0$. But $r = \sqrt{\rho_0^2 + z^2} \cong z + \rho_0^2/2z$ in the paraxial approximation so that we obtain an integral that can be done explicitly, giving

$$p(z,\omega) \cong \frac{-i\omega\rho\, v_0(\omega)\exp(ikz)}{z} \int_0^a \exp\left[ik\rho_0^2 q_0 / 2z\right]\rho_0 d\rho_0$$

$$= \frac{\rho c v_0 \exp(ikz)}{q_0}\left[1 - \exp\left(ika^2 q_0 / 2z\right)\right]. \tag{8.41}$$

Equation (8.41) is identical to the paraxial result of Eq. (8.38) obtained from the O'Neil model. Thus, in the paraxial approximation, the effect of spherical focusing can be modeled by including a phase term $\exp\left(-ik\rho_0^2 / 2R_0\right) = \exp\left[-ik\left(x^2 + y^2\right)/2R_0\right]$ on the aperture plane $z = 0$ of a planar transducer model. In a similar manner one could introduce bi-cylindrical focusing (different focal lengths R_x and R_y in the x- and y-directions, respectively) by including a phase term of the form $\exp\left[-ik\left(x^2/2R_x + y^2/2R_y\right)\right]$.

8.8 Wave Field in the Plane at the Geometrical Focus

The wave field of a spherically focused piston transducer in a plane located at a distance $z = R_0$ can also be obtained explicitly from the O'Neil model. One finds (see Fig. 8.21) that [Fundamentals]

$$p(\mathbf{x},\omega) = -i\omega\rho v_0 a^2 \frac{\exp\left(ik\bar{R}_0\right)}{\bar{R}_0} \frac{J_1\left(kay/\bar{R}_0\right)}{kay/\bar{R}_0}, \tag{8.42}$$

where $\bar{R}_0$ is the distance from the origin to a point $\mathbf{x}$ in the wave field. Since for most focused transducers the beam at the geometric focus is confined to a relatively small region near the transducer axis, in most cases

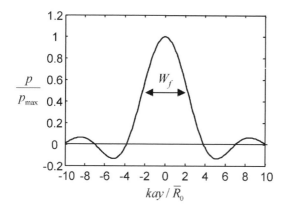

Fig. 8.22. The pressure distribution (due to the $J_1(u)/u$ function) on a plane parallel to the transducer face at a distance from the transducer equal to the geometric focal length.

we can take, approximately, $\overline{R}_0 = R_0$. It is interesting to note that the form of Eq. (8.42) is identical to that of the far field behavior of a circular planar piston transducer (see Eq. (8.31)). In this case, Eq. (8.42) gives us an explicit expression from which we can obtain an estimate of the beam width at the geometric focus. Usually that width is specified as the width of the main lobe when the magnitude of the response has dropped 6 dB from the maximum on-axis response, as shown in Fig. 8.22. Using Eq. (8.42), this beam width is given as [Fundamentals]

$$W_f\big|_{6\,dB} = 4.43\frac{R_0}{ka} = 1.41\,\lambda\,F \tag{8.43}$$

where $F = R_0/2a$ is called the *transducer F-number*.

8.9 Radiation of a Focused Transducer through an Interface

If one uses a focused transducer in an immersion setup, the transducer beam will be affected by the fluid-solid interface and focus at a shortened distance in the solid, as shown in Fig. 8.23 where a spherically focused piston transducer of radius a and focal length R_0 is radiating P-waves at

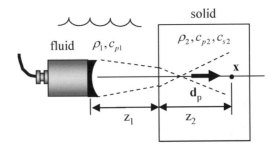

Fig. 8.23. A spherically focused piston transducer radiating a sound beam at normal incidence through a fluid-solid interface.

normal incidence to a planar fluid-solid interface. It can be shown that in the paraxial approximation the on-axis velocity wave field in the solid again can be expressed as a plane wave multiplied by a diffraction coefficient, C, i.e.[Fundamentals]

$$\mathbf{v}(\mathbf{x},\omega) = v_0 T_{12}^{P;P} \mathbf{d}_p \exp\left(ik_{p1}z_1 + k_{p2}z_2\right) C\left(z_1, z_2, a, R_0, \omega\right), \tag{8.44}$$

where

$$C\left(z_1, z_2, a, R_0, \omega\right) = \frac{1}{\tilde{q}_0}\left[1 - \exp\left(\frac{ik_{p1}a^2 \tilde{q}_0}{2\tilde{z}}\right)\right] \tag{8.45}$$

is of the same form as the diffraction coefficient for the single fluid medium, but with the distance z replaced by $\tilde{z} = z_1 + c_{p2}z_2/c_{p1}$ as in the planar transducer case and where $\tilde{q}_0 = 1 - \tilde{z}/R_0$.

8.10 Sound Beam in a Solid Generated by a Contact Transducer

All the examples discussed to this point have been for immersion transducers. In contact testing a P-wave transducer, like an immersion transducer, has an element whose motion is primarily normal to the face of the transducer. This transducer is placed in direct contact with the surface of the solid and a small layer of liquid couplant such as water, oil, or glycerin is placed between the transducer and the surface to ensure good coupling of the transducer to the solid. Under these conditions the transducer

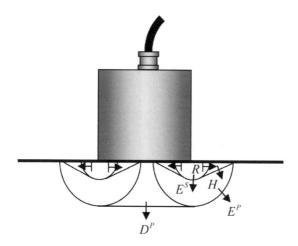

Fig. 8.24. The waves generated by a contact P-wave transducer radiating into a solid.

cannot drive the solid with a piston-like uniform velocity, since the solid is as stiff (or stiffer) than the transducer crystal and its wear plate. Instead, it is more reasonable to assume that the transducer generates a uniform pressure, p_0, over the transducer face. Even though this transducer is called a P-wave transducer, this pressure will actually launch a complicated set of waves of various types, as shown in Fig. 8.24 where a circular P-wave transducer is shown in contact with a stress-free planar surface of a solid. As in the fluid case, there will be a direct P-wave, D^P, that exists in a cylindrical region ahead of the transducer and an edge P-wave, E^P, that radiates from the transducer edge. However, there will also be an edge S-wave, E^S. When the edge P-wave grazes along the stress-free surface, it will generate a "Head" wave, H, (also called a von Schmidt wave) that radiates in a conical-like fashion from the interface and links up to the edge S-wave. Finally, the transducer also generates a surface Rayleigh wave, R, which moves radially from the transducer along the free surface at a wave speed slightly smaller than the shear wave velocity of the solid and is confined to a region between the free surface and the edge S-wave. Although it appears that the wave field of the contact transducer in Fig. 8.24 is considerably more complicated than the immersion transducer case, not all of the waves in Fig. 8.24 are of equal importance in determining the wave field below the transducer in the solid. The Rayleigh waves, for example, do not affect the wave field except in a region very close to the free surface. The head waves do travel into the solid but they

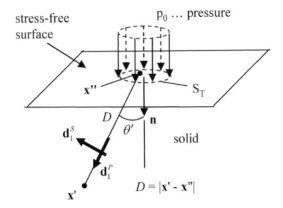

Fig. 8.25. A model of a contact P-wave transducer as a uniform pressure, p_0, acting on the free surface of an elastic solid.

radiate outwards at an angle from the transducer and generally are very weak. Thus, the predominant waves that one needs to consider are the direct P-wave and the edge P-waves and S-waves. A Rayleigh-Sommerfeld integral type of model can also be developed for these direct and edge waves, where the displacement vector, **u**, due to the waves in the solid is given by (see Fig. 8.25) [Fundamentals]

$$\mathbf{u}(\mathbf{x}',\omega) = \frac{p_0}{2\pi\rho_1 c_{s1}^2} \int_{S_T} K_s(\theta') \mathbf{d}_1^s \frac{\exp(ik_{s1}D)}{D} dS(\mathbf{x}'')$$

$$+ \frac{p_0}{2\pi\rho_1 c_{p1}^2} \int_{S_T} K_p(\theta') \mathbf{d}_1^p \frac{\exp(ik_{p1}D)}{D} dS(\mathbf{x}''),$$

(8.46)

where $D = |\mathbf{x}' - \mathbf{x}''|$, ρ_1 is the density of the solid, the compressional and shear wave speeds are c_{p1}, c_{s1}, respectively, and $\mathbf{d}_1^p, \mathbf{d}_1^s$ are the polarization vectors for the P-waves and S-waves. Unlike the immersion transducer case, the integrals also contain angular dependent directivity functions, $K_p(\theta'), K_s(\theta')$ for the P-waves and S-waves. These functions are given by the expressions [Fundamentals]

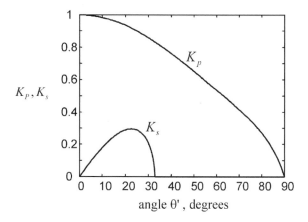

Fig. 8.26. The directivity functions for a contact P-wave transducer.

$$K_p(\theta') = \frac{\cos\theta'\kappa_1^2\left(\kappa_1^2/2 - \sin^2\theta'\right)}{2G(\sin\theta')}$$

$$K_s(\theta') = \frac{\kappa_1^3\cos\theta'\sin\theta'\sqrt{1-\kappa_1^2\sin^2\theta'}}{2G(\kappa_1\sin\theta')},$$

(8.47)

where $G(x) = \left(x^2 - \kappa_1^2/2\right)^2 + x^2\sqrt{1-x^2}\sqrt{\kappa_1^2 - x^2}$ and $\kappa_1 = c_{p1}/c_{s1}$. The directivities are plotted in Fig. 8.26. Near the central axis of the transducer $K_p \cong 1$, $K_s \cong 0$ so that Eq. (8.46) reduces to

$$\mathbf{u}(\mathbf{x}',\omega) = \frac{p_0\mathbf{n}}{2\pi\rho_1 c_{p1}^2}\int_{S_T}\frac{\exp(ik_{p1}D)}{D}dS,$$

(8.48)

which now only contains the direct and edge P-waves in a form almost identical to the expression for an immersion transducer. When such a transducer is used to interrogate a material for flaws, it is likely that the response will be "peaked up" by moving the transducer so that the flaw will be on or near the central axis of the transducer. In that case we see from Eq. (8.48) that a Rayleigh-Sommerfeld integral may also be an appropriate model.

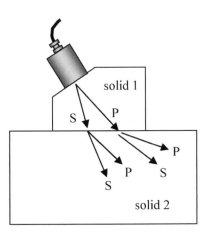

Fig. 8.27. A contact P-wave transducer on a wedge which is contact with another material that is to be inspected.

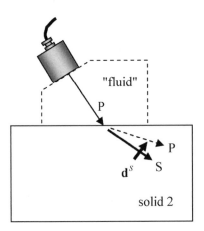

Fig. 8.28. An equivalent "fluid" model of an angle beam shear wave transducer. When the incident P-wave in the wedge is beyond the first critical angle, primarily a refracted S-wave only is generated in the solid with polarization $\mathbf{d}^s$, as shown. Since for the configuration shown $\mathbf{d}^s$ lies in a vertical plane, the S-wave in the solid is called a vertically polarized shear wave (SV-wave). There is a small transmitted P-wave as well in this configuration that can generally be neglected, as indicated by the dashed arrow in the figure.

8.11 Angle Beam Shear Wave Transducer Model

A contact P-wave can also be placed on a solid wedge and used to generate a shear wave in the solid by the process of mode conversion. In general, as shown in Fig. 8.27 the P-wave transducer generates in the wedge primarily the compressional and shear waves we have just discussed. These waves then mode convert to each generate compressional and shear waves in the solid, as shown. However, studies of this configuration have shown that again the only significant wave in the wedge is the compressional wave [Fundamentals]. If the angle of the wedge is chosen so that the compress-ional wave traveling along the central axis of the transducer is beyond the first critical angle, then primarily a shear wave is generated in the solid, a configuration in which the transducer is called an *angle beam shear wave transducer*. Since the only significant wave in the wedge is the P-wave, an angle beam shear wave transducer can be modeled by replacing the wedge by an equivalent fluid that has the same density and compressional wave speed of the wedge material, as shown in Fig. 8.28 and model the waves transmitted across the interface by using the transmission coefficients for two solids in smooth contact (see Appendix D). Thus, one can use an immersion transducer model as the basis for also modeling an angle beam shear wave transducer.

8.12 Transducer Beam Radiation through Interfaces

In immersion testing, the transducer sound beam inherently must pass through a fluid-solid interface. This causes the beam in the solid to be distorted from its behavior in the fluid. We have seen how at normal incidence to a plane interface we can model the on-axis behavior of these distortions in a simple manner for both planar and spherically focused transducers (see Eqs. (8.25) and (8.44)). For curved interfaces and oblique incidence, the models become much more complex. We can gain some understanding of these cases by using high frequency ray concepts. Consider, for example, a planar piston transducer radiating at oblique incidence to a curved interface, as shown in Fig. 8.29. If we model the wave field in the fluid by a Rayleigh-Sommerfeld integral, then in that model we are radiating a distribution of spherical waves to the interface. From an element of area dS at point $\mathbf{y}$ on the transducer surface a spherical wave generates a pressure at a general point $\mathbf{x}_1$ on the interface given by

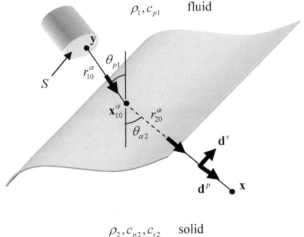

Fig. 8.29. An immersion transducer radiating a sound beam of type α $(\alpha = p, s)$ in a solid through a curved fluid-solid interface, showing a ray path from a point **y** on the transducer surface to a point **x** in the solid through the interface. In the transducer beam model, this ray path must satisfy generalized Snell's law. The polarization of the transmitted waves is defined by the unit vector $\mathbf{d}^{\alpha}$. For a transmitted P-wave, the polarization will be along the direction of propagation while for a transmitted S-wave it will be perpendicular to the direction of propagation. Both polarizations are shown along the refracted ray but for a given wave type only one will be present.

$$dp(\mathbf{x}_1, \omega) = \frac{-i\omega\rho v_0}{2\pi} \frac{\exp(ik_{p1}r_1)}{r_1} dS. \tag{8.49}$$

At high frequencies, the corresponding velocity in this spherical wave is given by

$$d\mathbf{v}(\mathbf{x}_1, \omega) = dv(\mathbf{x}_1, \omega)\mathbf{e}_{p1} = -\mathbf{e}_{p1} \frac{ik_{p1}v_0}{2\pi} \frac{\exp(ik_{p1}r_1)}{r_1} dS, \tag{8.50}$$

where $\mathbf{e}_{p1}$ is a unit vector along a line from point **y** on the transducer face to point $\mathbf{x}_1$ on the interface and $r_1 = |\mathbf{x}_1 - \mathbf{y}|$. By high frequency ray theory, this velocity is propagated into the solid as a bulk wave of type α, where

$\alpha = (p, s)$, to generate a velocity at point $\mathbf{x}$ in the solid of the form [Fundamentals]

$$dv^{\alpha}(\mathbf{x}, \omega) = \mathbf{d}^{\alpha} dv(\mathbf{x}_{10}, \omega) T_{12}^{\alpha;p}$$

$$\cdot \frac{\sqrt{|\rho_{v1}^{\alpha}|} \sqrt{|\rho_{v2}^{\alpha}|}}{\sqrt{|\rho_{v1}^{\alpha} + r_{20}^{\alpha}|} \sqrt{|\rho_{v2}^{\alpha} + r_{20}^{\alpha}|}} \exp\left(ik_{\alpha2} r_{20}^{\alpha} + i\phi^{\alpha}\right), \qquad (8.51)$$

where $r_{10}^{\alpha} = \left|\mathbf{x}_{10}^{\alpha} - \mathbf{y}\right|, r_{20}^{\alpha} = \left|\mathbf{x} - \mathbf{x}_{10}^{\alpha}\right|$ are distances from point $\mathbf{y}$ on the transducer surface to an interface point, $\mathbf{x}_{10}^{\alpha}$ and from that interface point to point $\mathbf{x}$ in the solid along a ray path that satisfies Snell's law for a wave of type α in the solid (see Fig. 8.29), i.e. we must have

$$\frac{\sin(\theta_{p1})}{c_{p1}} = \frac{\sin(\theta_{\alpha2})}{c_{\alpha2}}. \qquad (8.52)$$

We will assume that there is only one such path for the present argument, although that may not be true in general for complex curved interfaces. The term $T_{12}^{\alpha;p}$ is just the plane wave transmission coefficient (based on velocity ratios) for a wave of type α in the solid generated by the P-wave in the fluid traveling along this ray path. The factor

$$\frac{\sqrt{|\rho_{v1}^{\alpha}|} \sqrt{|\rho_{v2}^{\alpha}|}}{\sqrt{|\rho_{v1}^{\alpha} + r_{20}^{\alpha}|} \sqrt{|\rho_{v2}^{\alpha} + r_{20}^{\alpha}|}}$$

that appears in Eq. (8.51) involves two "virtual" source distances $\rho_{v1}^{\alpha}, \rho_{v2}^{\alpha}$ and represents the amplitude changes predicted by ray theory. Essentially this factor distorts the incident spherical wave fronts in the fluid to more general curved wave fronts in the solid. Ray theory also predicts that there are additional phase changes, ϕ^{α} in the wave traveling in the solid beyond the term, $k_{\alpha2} r_{20}^{\alpha}$ due to solely propagation in the solid. The vector $\mathbf{d}^{\alpha}$ in Eq. (8.51) is a unit vector that describes the polarization of the transmitted wave. It is identical to the polarization defined for a transmitted plane wave of type α generated by the interaction of a plane P-wave with a plane interface at point $\mathbf{x}_{10}^{\alpha}$ where the normal to the plane interface coincides with the actual interface normal of the curved interface at that point.

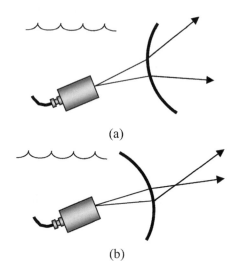

(a)

(b)

Fig. 8.30. (a) A planar transducer radiating through a curved fluid-solid interface that spreads (defocuses) the waves in the solid, and **(b)** a curved interface that focuses the waves in the solid.

By integrating the expression in Eq. (8.51) over the face of the transducer one then obtains a beam model for the total velocity in the transmitted waves:

$$\mathbf{v}^{\alpha}\left(\mathbf{x},\omega\right) = \int_{S} \Big[\, \mathbf{d}^{\alpha} T_{12}^{\alpha;p} \, \frac{\sqrt{\left|\rho_{v1}^{\alpha}\right|}\sqrt{\left|\rho_{v2}^{\alpha}\right|}}{\sqrt{\left|\rho_{v1}^{\alpha}+r_{20}^{\alpha}\right|}\sqrt{\left|\rho_{v2}^{\alpha}+r_{20}^{\alpha}\right|}}$$

$$\exp\left(ik_{\alpha 2}r_{20}^{\alpha}+i\phi^{\alpha}\right)dv\left(\mathbf{x}_{10},\omega\right)\Big]. \tag{8.53}$$

There are, however, some difficulties with this model [Fundamentals]. As long as the curved interface is of a defocusing type, as shown in Fig. 8.30 (a), where the rays from a point on the transducer surface traveling into the solid do not touch or cross, Eq. (8.53) is well-behaved and can be used, like the Rayleigh-Sommerfeld equation, to calculate the sound beam in the solid. However, if the curved interface is of a focusing type, as shown in Fig. 8.30 (b), the rays can touch or cross and the ray theory amplitude term becomes infinite. There are uniform ray theory approximations that can remove those singularities but the analysis and resulting expressions become much more complex. This difficulty arises mathematically because we have modeled the transducer beam as a superposition of spherical waves arising from point sources, and spherical waves can become singular, for example, when focused at a point by a curved interface. Similar focusing singularities can occur for plane waves incident on a curved interface so that an angular plane wave spectrum model will also have these same difficulties when focusing curved interfaces are present. In the next Chapter, we will show that these problems can be eliminated by expanding the transducer wave field in terms of Gaussian beams which always remain non-singular.

There is an important special case when Eq. (8.53) is always well-behaved [Fundamentals]. That case is when the planar piston transducer is incident at oblique incidence on a planar interface. In that case we have $\phi^\alpha = 0$ and

$$\rho_{v1}^\alpha = \frac{c_{p1} \cos^2\left(\theta_{\alpha 2}\right)}{c_{\alpha 2} \cos^2\left(\theta_{p1}\right)} r_{10}^\alpha$$

$$\rho_{v2}^\alpha = \frac{c_{p1}}{c_{\alpha 2}} r_{10}^\alpha$$

(8.54)

so Eq. (8.53) becomes, explicitly,

$$\mathbf{v}^\alpha\left(\mathbf{x},\omega\right) = \frac{-ik_{p1}v_0}{2\pi} \iint_S \left[T_{12}^{\alpha;p} \mathbf{d}^\alpha \right.$$

$$\left. \cdot \frac{\exp\left(ik_{p1}r_{10}^\alpha + ik_{\alpha 2}r_{20}^\alpha\right)}{\sqrt{r_{10}^\alpha + \left(c_{\alpha 2}^2/c_{p1}^2\right)r_{20}^\alpha}\sqrt{r_{10}^\alpha + \left(c_{\alpha 2}^2 \cos^2\theta_{p1}/c_{p1}^2 \cos^2\theta_{\alpha 2}\right)r_{20}^\alpha}} \right]dS.$$

(8.55)

Equation (8.55) is in a form very similar to the Rayleigh-Sommerfeld equation. Instead of superimposing spherical waves traveling directly from the transducer to the point in the fluid, we now need to superimpose a more general set of waves with elliptical wave fronts in the solid that travel along rays satisfying Snell's law and are modified by the plane wave transmission coefficient of the interface. Since both that transmission coefficient and the polarization vector depend on that ray path, they are both implicit functions of point y on the transducer surface and so must remain inside the integral. In general the integral in Eq. (8.55) must be performed numerically, so that like the Rayleigh-Sommerfeld integral the highly oscillatory complex exponentials in Eq. (8.55) make this evaluation a rather intensive computation. Fortunately, the Gaussian beam models discussed in the next Chapter will also be much more numerically efficient than these types of Rayleigh-Sommerfeld integral models.

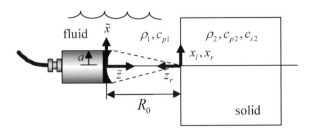

Fig. 8.31. An experimental setup for a spherically focused transducer of radius a and focal length R_0 where one can obtain the acoustic/elastic transfer function explicitly.

8.13 Acoustic/Elastic Transfer Function – Focused Transducer

In Chapter 7 it was shown that the acoustic/elastic transfer function is needed in order to determine experimentally the system function. In Chapter 6 the acoustic/elastic transfer function also played a key role in determining the transducer sensitivity. In Chapter 5 we obtained an acoustic/elastic transfer function for both a pitch-catch and a pulse-echo immersion setup. In Chapter 13 a general procedure is given for using a multi-Gaussian beam model to determine the acoustic/elastic transfer function in cases where the transfer function cannot be obtained analytically (angle beam testing and contact testing setups with curved surfaces, etc.). A number of other acoustic/elastic transfer functions can be derived from results given in [Fundamentals]. All of those cases, however, are for planar piston transducers. The acoustic/elastic transfer function for a spherically focused piston transducer in a pulse-echo immersion configuration is also available [8.4], [8.5], a case we will develop here as a simple application of the paraxial approximation and the use of the phase term discussed in Eq. (8.39). This approach will also lead to the transfer function for planar and cylindrically focused rectangular piston transducers in the following section.

The configuration we will consider is the pulse-echo setup shown in Fig. 8.31 where a spherically focused piston transducer of radius a and focal length, R_0, radiates waves into a fluid and receives the waves reflected from a plane fluid-solid interface. The distance from the transducer to interface is made equal to the geometrical focal length in this configuration.

As discussed in section 8.7, in the paraxial approximation we can use the Rayleigh-Sommerfeld equation to represent the wave field of a spherically focused transducer in the form (see Eq. (8.40))

$$p(\mathbf{x},\omega) = \frac{-i\omega\rho_1 v_0(\omega)}{2\pi} \iint_S \exp\left[-ik_{p1}\left(\tilde{x}^2 + \tilde{y}^2\right)/2R_0\right]$$

$$\cdot \frac{\exp(ik_{p1}r)}{r} dS, \tag{8.56}$$

where $(\tilde{x}, \tilde{y}, \tilde{z} = 0)$ are coordinates of a point on a plane at the transmitting transducer and $r = \sqrt{(\tilde{x} - x_I)^2 + (\tilde{y} - y_I)^2 + z_I^2}$ is the distance from that point to a point (x_I, y_I, z_I) in the fluid. Let the point in the fluid lie on the interface as shown in Fig. 8.31. Then $r = \sqrt{(\tilde{x} - x_I)^2 + (\tilde{y} - y_I)^2 + R_0^2}$. We also apply the paraxial approximation to this distance function to obtain $r \cong R_0 + \left[(\tilde{x} - x_I)^2 + (\tilde{y} - y_I)^2\right]/2R_0$ and Eq. (8.56) becomes

$$p(x_I, y_I, R_0, \omega) = \frac{-i\omega\rho_1 v_0(\omega)\exp(ik_{p1}R_0)}{2\pi R_0}$$

$$\cdot \iint_S \exp\left[-ik_{p1}\frac{\left(\tilde{x}^2 + \tilde{y}^2\right)}{2R_0}\right]\exp\left[ik_{p1}\frac{(\tilde{x} - x_I)^2 + (\tilde{y} - y_I)^2}{2R_0}\right]d\tilde{x}d\tilde{y}. \tag{8.57}$$

Equation (8.57) is in the form of a quasi-plane wave so at high frequencies the pressure in the reflected wave at the interface, $p_R(x_I, y_I, R_0, \omega)$, can be obtained by the plane wave relationship

$$p_R(x_I, y_I, R_0, \omega) = R_{12} p(x_I, y_I, R_0, \omega)$$

$$= \frac{\rho_2 c_{p2} - \rho_1 c_{p1}}{\rho_2 c_{p2} + \rho_1 c_{p1}} p(x_I, y_I, R_0, \omega), \tag{8.58}$$

where R_{12} is the reflection coefficient (based on a pressure ratio). The normal velocity at the interface in the z_r direction, v_r, (see Fig. 8.31) is also given by the plane wave relationship

$$v_r(x_I, y_I, R_0, \omega) = R_{12} p(x_I, y_I, R_0, \omega)/\rho_1 c_{p1}. \tag{8.59}$$

Using this velocity field as specified on the entire interface, we can again use the Rayleigh-Sommerfeld integral (with the paraxial approximation applied again to the radius, r, in that integral) to obtain the reflected waves that are incident on the transducer from the interface. We find

$$
p(x_r, y_r, z_r, \omega) = \frac{-ik_{p1} R_{12} \exp(ikz_r)}{2\pi z_r} \int_{-\infty}^{+\infty}\int_{-\infty}^{+\infty} p(x_I, y_I, R_0, \omega)
$$
$$
\cdot \exp\left[ik_{p1} \frac{(x_r - x_I)^2 + (y_r - y_I)^2}{2z_r} \right] dx_I dy_I.
$$
(8.60)

For a spherically focused transducer, this pressure is received not at the plane $z_r = R_0$ but instead over the curved spherical surface given by $z_r = R_0 - \left(x_r^2 + y_r^2\right)/2R_0$. Placing this distance into the plane wave phase term in Eq. (8.60) (and using $z_r = R_0$ elsewhere in Eq. (8.60)), the average pressure, p_{ave}, over the area, S, of the transducer is given by

$$
p_{ave} = \frac{-ik_{p1} R_{12} \exp(ikR_0)}{2\pi R_0 S} \iint_S \exp\left[-ik_{p1} \frac{\left(x_r^2 + y_r^2\right)}{2R_0} \right]
$$
$$
\left\{ \int_{-\infty}^{+\infty}\int_{-\infty}^{+\infty} p(x_I, y_I, R_0, \omega) \exp\left[ik_{p1} \frac{(x_r - x_I)^2 + (y_r - y_I)^2}{2R_0} \right] \right.
$$
(8.61)
$$
\left. \cdot dx_I dy_I \right\} dx_r dy_r.
$$

Substituting the expression for the pressure at the interface (Eq. (8.57)) into Eq. (8.61), we obtain an explicit expression for the average pressure acting on the transducer. Then from this average pressure we can find the blocked force, $F_B = 2p_{ave}S$, received by the transducer as

$$F_B = 2R_{12} \exp\left(2ik_{p1}R_0\right) \frac{-ik_{p1}}{2\pi R_0} \frac{-ik_{p1}\rho_1 c_{p1} v_0}{2\pi R_0}$$

$$\cdot \iint_S \left(\iint_S \exp\left[-ik_{p1}\frac{\left(x_r^2+y_r^2\right)}{2R_0}\right] \exp\left[-ik_{p1}\frac{\left(\tilde{x}^2+\tilde{y}^2\right)}{2R_0}\right] \right.$$

$$\cdot \left[\int_{-\infty}^{+\infty}\int_{-\infty}^{+\infty} \exp\left[ik_{p1}\frac{\left(x_r-x_l\right)^2+\left(y_r-y_l\right)^2}{2R_0}\right] \right.$$

$$\left. \exp\left[ik_{p1}\frac{\left(\tilde{x}-x_l\right)^2+\left(\tilde{y}-y_l\right)^2}{2R_0}\right] dx_l dy_l \right] dx_r dy_r \right) d\tilde{x}d\tilde{y}. \qquad (8.62)$$

Since $F_t = \rho_1 c_{p1} S v_0$ is the force transmitted by the transducer acting as a transmitter, the acoustic/elastic transfer function for our focused transducer, $t_A^{foc} = F_B / F_t$ is given by

$$t_A^{foc} = 2R_{12} \exp\left(2ik_{p1}R_0\right) \frac{-ik_{p1}}{2\pi R_0 S} \frac{-ik_{p1}}{2\pi R_0}$$

$$\cdot \iint_S \left(\iint_S \exp\left[-ik_{p1}\frac{\left(x_r^2+y_r^2\right)}{2R_0}\right] \exp\left[-ik_{p1}\frac{\left(\tilde{x}^2+\tilde{y}^2\right)}{2R_0}\right] \right.$$

$$\cdot \left[\int_{-\infty}^{+\infty}\int_{-\infty}^{+\infty} \exp\left[ik_{p1}\frac{\left(x_r-x_l\right)^2+\left(y_r-y_l\right)^2}{2R_0}\right] \right.$$

$$\left. \exp\left[ik_{p1}\frac{\left(\tilde{x}-x_l\right)^2+\left(\tilde{y}-y_l\right)^2}{2R_0}\right] dx_l dy_l \right] dx_r dy_r \right) d\tilde{x}d\tilde{y}. \qquad (8.63)$$

Equation (8.63) is a rather formidable looking expression, but we can proceed as follows. First, we note that the acoustic/elastic transfer function for a planar transducer of the same size as our spherically focused transducer, t_A^{planar} is given by exactly the same expression as Eq. (8.63) without the first two phase terms:

$$t_A^{planar} = 2R_{12}\exp\left(2ik_{p1}R_0\right)\frac{-ik_{p1}}{2\pi R_0 S}\frac{-ik_{p1}}{2\pi R_0}$$

$$\cdot \iint\limits_{S}\left(\iint\limits_{S}\left[\int\limits_{-\infty}^{+\infty}\int\limits_{-\infty}^{+\infty}\exp\left[ik_{p1}\frac{\left(x_r-x_l\right)^2+\left(y_r-y_l\right)^2}{2R_0}\right]\right.\right.\tag{8.64}$$

$$\left.\left.\cdot\exp\left[ik_{p1}\frac{\left(\tilde{x}-x_l\right)^2+\left(\tilde{y}-y_l\right)^2}{2R_0}\right]dx_l dy_l\right]dx_r dy_r\right)d\tilde{x}d\tilde{y}.$$

In Eqs. (8.63) and (8.64) the integrals over the interface are identical for the focused and planar cases. These integrals can be rewritten as

$$I=\int\limits_{-\infty}^{+\infty}\int\limits_{-\infty}^{+\infty}\exp\left[ik_{p1}\frac{\left(x_r-x_l\right)^2+\left(y_r-y_l\right)^2}{2R_0}\right]$$

$$\cdot\exp\left[ik_{p1}\frac{\left(\tilde{x}-x_l\right)^2+\left(\tilde{y}-y_l\right)^2}{2R_0}\right]dx_l dy_l$$

$$=\exp\left(ik_{p1}\frac{\left(x_r^2+y_r^2\right)}{2R_0}\right)\exp\left(ik_{p1}\frac{\left(\tilde{x}^2+\tilde{y}^2\right)}{2R_0}\right)\tag{8.65}$$

$$\cdot\int\limits_{-\infty}^{+\infty}\exp\left(ik_{p1}\frac{x_l^2}{R_0}\right)\exp\left(-ik_{p1}\frac{\left(\tilde{x}+x_r\right)x_l}{R_0}\right)dx_l$$

$$\cdot\int\limits_{-\infty}^{+\infty}\exp\left(ik_{p1}\frac{y_l^2}{R_0}\right)\exp\left(-ik_{p1}\frac{\left(\tilde{y}+y_r\right)y_l}{R_0}\right)dy_l$$

The remaining integrals can be performed exactly because we have [8.2]

$$\int\limits_{-\infty}^{+\infty}\exp\left(iAx^2\right)\exp\left(-iBx\right)dx=\sqrt{\frac{i\pi}{A}}\exp\left(\frac{-iB^2}{4A}\right),$$

$$\text{Im}\left[A\right]>0\tag{8.66}$$

where Im[ ] indicates "imaginary part of ". In Eq. (8.65) the corresponding A terms are purely real but if we add a small amount of "damping" by letting $A=A+i\varepsilon$ and then take the limit as $\varepsilon\rightarrow 0$, the result is the same as using Eq. (8.66) directly on the forms given in Eq. (8.65) and we find

$$I = \frac{i\pi R_0}{k_{p1}} \exp\left(ik_{p1} \frac{\left(x_r^2 + y_r^2\right)}{2R_0} \right) \exp\left(ik_{p1} \frac{\left(\tilde{x}^2 + \tilde{y}^2\right)}{2R_0} \right)$$
$$\cdot \exp\left(-ik_{p1} \frac{\left(\tilde{x} + x_r\right)^2}{4R_0} \right) \exp\left(-ik_{p1} \frac{\left(\tilde{y} + y_r\right)^2}{4R_0} \right). \tag{8.67}$$

In the focused case, we see that the first two phase terms in Eq. (8.67) simply cancel the first two phase terms in Eq. (8.63) and we obtain

$$t_A^{foc} = 2R_{12} \exp\left(2ik_{p1}R_0\right) \frac{-ik_{p1}}{4\pi R_0 S}$$
$$\iint_S \left(\iint_S \exp\left[-ik_{p1} \frac{\left(\tilde{x} + x_r\right)^2}{4R_0} \right] \right.$$
$$\left. \cdot \exp\left[-ik_{p1} \frac{\left(\tilde{y} + y_r\right)^2}{4R_0} \right] dx_r dy_r \right) d\tilde{x} d\tilde{y}. \tag{8.68}$$

However, we note that for a circular, spherically focused transducer the integrations in Eq. (8.68) are over symmetrical intervals in both x_r and y_r so that we can make the replacements $x_r \rightarrow -x_r$ and $y_r \rightarrow -y_r$ in Eq. (8.68) without affecting the end result. With those, replacements, we have, finally,

$$t_A^{foc} = 2R_{12} \exp\left(2ik_{p1}R_0\right) \frac{-ik_{p1}}{4\pi R_0 S} \iint_S \left(\iint_S \exp\left[-ik_{p1} \frac{\left(\tilde{x} - x_r\right)^2}{4R_0} \right] \right.$$
$$\left. \exp\left[-ik_{p1} \frac{\left(\tilde{y} - y_r\right)^2}{4R_0} \right] dx_r dy_r \right) d\tilde{x} d\tilde{y} \tag{8.69}$$

In the planar transducer case, we can place Eq. (8.67) into Eq. (8.64) to find

$$t_A^{planar} = 2R_{12} \exp\left(2ik_{p1}R_0\right)\frac{-ik_{p1}}{4\pi R_0 S} \iint_S \left(\iint_S \exp\left(ik_{p1}\frac{\left(x_r^2+y_r^2\right)}{2R_0}\right) \right.$$

$$\cdot \exp\left(ik_{p1}\frac{\left(\tilde{x}^2+\tilde{y}^2\right)}{2R_0}\right) \exp\left(-ik_{p1}\frac{\left(\tilde{x}+x_r\right)^2}{4R_0}\right) \qquad (8.70)$$

$$\left. \cdot \exp\left(-ik_{p1}\frac{\left(\tilde{y}+y_r\right)^2}{4R_0}\right) dx_r dy_r \right) d\tilde{x}d\tilde{y},$$

which, when the exponential terms are combined, gives

$$t_A^{planar} = 2R_{12} \exp\left(2ik_{p1}R_0\right)\frac{-ik_{p1}}{4\pi R_0 S}$$

$$\cdot \iint_S \left(\iint_S \exp\left(ik_{p1}\frac{\left(\tilde{x}-x_r\right)^2}{4R_0}\right) \right. \qquad (8.71)$$

$$\left. \cdot \exp\left(ik_{p1}\frac{\left(\tilde{y}-y_r\right)^2}{4R_0}\right) dx_r dy_r \right) d\tilde{x}d\tilde{y}.$$

In Chapter 5, we obtained an explicit expression for acoustic/elastic transfer function for the planar transducer case. For the geometry of Fig. 8.31 we can write the transfer function for a planar transducer in terms of the diffraction correction, $\tilde{D}_p$, used in Chapter 5 (see Eq. (5.20)) as

$$t_A^{planar}(\omega) = \tilde{D}_p\left(k_{p1}a^2/2R_0\right)R_{12}\exp\left(2ik_{p1}R_0\right), \qquad (8.72)$$

where

$$\tilde{D}_p(u) = 2\left[1-\exp(iu)\{J_0(u)-iJ_1(u)\}\right]. \qquad (8.73)$$

Comparing Eqs. (8.69) and (8.71) and using Eq. (8.72) for the planar case, we see that for the focused case we have

$$t_A^{foc}(\omega) = -\left[\tilde{D}_p\left(k_{p1}a^2/2R_0\right)\right]^* R_{12} \exp\left(2ik_{p1}R_0\right) \qquad (8.74)$$

where [ ]* denotes the "complex conjugate". Thus, by making the changes indicated by Eq. (8.74) one can simply use the same diffraction correction obtained for the planar case for this focused case as well. Note, however, that while in the planar transducer case the interface is not restricted to being at a particular distance from the transducer the interface *must* be placed at the geometrical focal length of the focused transducer in order to use Eq. (8.74).

8.14 Acoustic/Elastic Transfer Function – Rectangular Transducer

The results of the previous section can also be used to obtain the acoustic/ elastic transfer function for a rectangular piston transducer that is either planar or cylindrically focused and receiving the waves reflected from the front surface of a block (same setup as shown in Fig. 8.31). First, consider a planar rectangular transducer of length $2a$ in the $\tilde{x}$-direction and $2b$ in the $\tilde{y}$-direction and let the distance $R_0 = D$ (see Fig. 8.31). Then from Eq. (8.71) the acoustic/elastic transfer function, t_A^{rect}, is

$$t_A^{rect} = 2R_{12} \exp\left(2ik_{p1}D\right)\frac{-ik_{p1}}{16\pi Dab}$$

$$\cdot \int_{-b-a}^{+b+a}\left(\int_{-b-a}^{+b+a}\exp\left(ik_{p1}\frac{\left(\tilde{x}-x_r\right)^2}{4D}\right)\right.$$

$$\left.\cdot\exp\left(ik_{p1}\frac{\left(\tilde{y}-y_r\right)^2}{4D}\right)dx_r dy_r\right)d\tilde{x}d\tilde{y}. \tag{8.75}$$

But in this case we have

$$\int_{-a-a}^{+a+a}\exp\left(ik_{p1}\frac{\left(\tilde{x}-x_r\right)^2}{4D}\right)dx_r d\tilde{x} = \frac{4\pi D}{k_{p1}}\int_{0}^{\sqrt{2ka^2/\pi D}}F(x)dx, \tag{8.76}$$

where $F(x)$ is the Fresnel integral

$$F(x) = \int_{0}^{x}\exp\left(i\pi t^2/2\right)dt. \tag{8.77}$$

and similarly

$$\int_{-b-b}^{+b+b}\exp\left(ik_{p1}\frac{\left(\tilde{y}-y_r\right)^2}{4D}\right)dy_r d\tilde{y} = \frac{4\pi D}{k_{p1}}\int_{0}^{\sqrt{2kb^2/\pi D}}F(x)dx. \tag{8.78}$$

For the integral of the Fresnel function we can use the relationship [8.6] (which comes directly from integration by parts)

$$\int_{x_1}^{x_2} F(x)\,dx = \left[x F(x) + \frac{i}{\pi}\exp\left(i\pi x^2/2\right) \right]_{x_1}^{x_2}$$ (8.79)

to obtain

$$t_A^{rect} = R_{12}\exp\left(2ik_{p1}D\right)\frac{4}{i}\left\{ F\left(\sqrt{2k_{p1}a^2/\pi D}\right) + \right.$$

$$\frac{i}{\pi\sqrt{2k_{p1}a^2/\pi D}}\left[\exp\left(ik_{p1}a^2/D\right)-1\right]\right\}$$ (8.80)

$$\cdot\left\{ F\left(\sqrt{2k_{p1}b^2/\pi D}\right) + \frac{i}{\pi\sqrt{2k_{p1}b^2/\pi D}}\left[\exp\left(ik_{p1}b^2/D\right)-1\right]\right\}.$$

We can express Eq. (8.80) in terms of a diffraction correction term, $\tilde{D}_p^{rect}$, where

$$\tilde{D}_p^{rect} = \frac{4}{i}\left\{ F\left(\sqrt{2k_{p1}a^2/\pi D}\right) + \right.$$

$$\frac{i}{\pi\sqrt{2k_{p1}a^2/\pi D}}\left[\exp\left(ik_{p1}a^2/D\right)-1\right]\right\}$$ (8.81)

$$\cdot\left\{ F\left(\sqrt{2k_{p1}b^2/\pi D}\right) + \frac{i}{\pi\sqrt{2k_{p1}b^2/\pi D}}\left[\exp\left(ik_{p1}b^2/D\right)-1\right]\right\}$$

so that

$$t_A^{rect}(\omega) = \tilde{D}_p^{rect}\left(k_{p1}a^2/2D\right)R_{12}\exp\left(2ik_{p1}D\right).$$ (8.82)

Figure 8.32 shows a plot of $\tilde{D}_p^{rect}$ versus frequency for a rectangular transducer where $D = 50.8$ mm and $a = 12.7$ mm, $b = 6.35$ mm. For comparison the corresponding diffraction correction for a 12.7 mm radius circular transducer (Eq. 8.73) is also plotted in Fig. 8.32. It can be see that the rectangular transducer has a very similar behavior to the circular probe and that both diffraction corrections asymptotically approach a value of two for high frequencies.

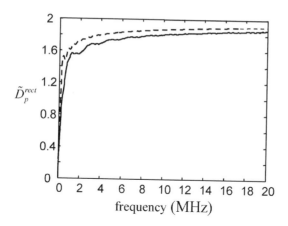

Fig. 8.32. The diffraction correction, $\tilde{D}_p^{rect}$, for a rectangular 25.4 x 12.7 mm rectangular transducer (solid line) and the corresponding diffraction correction, $\tilde{D}_p$, for a 12.7 mm radius circular transducer (dashed line). In both cases the distance $D = 50.8$ mm.

We can also consider a rectangular cylindrically focused transducer in the same fashion as done for the spherically focused transducer. For a transducer with cylindrical focusing of radius R in the $\tilde{y}$-direction, we can introduce the phase term $\exp\left(-ik_{p1}\tilde{y}^2/2R\right)$ into the Rayleigh-Sommerfeld equation and follow the same steps as in the spherically focused transducer case to obtain the acoustic/elastic transfer function, t_A^{cyl}, in the form

$$
t_A^{cyl} = 2R_{12}\exp\left(2ik_{p1}R\right)\frac{-ik_{p1}}{16\pi Rab}\int\limits_{-b-a}^{+b+a}\int\limits_{-b-a}^{+b+a}\left(\int\int\exp\left(ik_{p1}\frac{\left(\tilde{x}-x_r\right)^2}{4R}\right)\right.
$$

$$
\left.\cdot\exp\left(ik_{p1}\frac{\left(\tilde{y}+y_r\right)^2}{4R}\right)dx_r dy_r\right)d\tilde{x}d\tilde{y},
\tag{8.83}
$$

where we must set the distance, $D = R$, as in the spherically focused case. Again, we can express these integrations in terms of Fresnel integrals. Since the details are the same as for the planar case, we just give the end result, namely

$$
t_A^{cyl} = R_{12} \exp\left(2ik_{p1}R\right) \frac{4}{i} \left\{ F\left(\sqrt{2k_{p1}a^2/\pi R}\right) + \right.
$$

$$
\left. \frac{i}{\pi\sqrt{2k_{p1}a^2/\pi R}} \left[\exp\left(ik_{p1}a^2/R\right)-1\right] \right\}
$$

(8.84)

$$
\cdot\left\{ F\left(\sqrt{2k_{p1}b^2/\pi R}\right) + \frac{i}{\pi\sqrt{2k_{p1}b^2/\pi R}} \left[\exp\left(ik_{p1}b^2/R\right)-1\right] \right\}^* ,
$$

where again $\{\ \}^*$ indicates the complex conjugate.

8.15 References

8.1 Stamnes JJ (1986) Waves in focal regions. Institute of Physics Publishing, Bristol, England
8.2 Gaskill JD (1978) Linear systems, Fourier transforms and optics. John Wiley and Sons, New York, NY
8.3 O'Neil HT (1949) Theory of focusing radiators. J. Acoust. Soc. Am. 21: 516-526
8.4 Thompson RB, Gray TA (1982) Range of applicability of inversion algorithms. In: Thompson DO, Chimenti DE (eds) Review of progress in quantitative nondestructive evaluation 1, Plenum Press, New York, NY, pp 233-248
8.5 Chen X, Schwartz KQ (1994) Acoustic coupling from a focused transducer to a flat plate and back to the transducer. J. Acoust. Soc. Am. 95: 3049-3054
8.6 Abramowitz M, Stegun IA (1965) Handbook of mathematical functions. Dover Publications, New York, NY

8.16 Exercises

1. The exact on-axis pressure for a circular piston transducer was given by Eq. (8.20) and the far field approximation for this same pressure was given by Eq. (8.21). Using MATLAB, write a script that computes these two pressure expressions and plots the magnitude of the normalized pressure, $p/\rho c v_0$, versus the normalized distance, z/N, for both of these expressions on the same plot, where N is the near field distance. Let the transducer radius $a = 6.35$ mm, the frequency $f = 5$ MHz, and the wave speed of the fluid $c = 1480$ m/sec. Show both pressure plots over the range $z/N = 0.2$ to

$z/N = 4.0$. What can you conclude about when Eq. (8.21) is valid?

2. Equation (8.31) shows that the angular distribution of the far field radiation field of a circular planar piston transducer is controlled by the directivity function $J_1(ka\sin\theta)/(ka\sin\theta)$. Using MATLAB, write a function that calculates the angle where the amplitude of this directivity function drops by 6 dB from its maximum on-axis value. Use this function to determine the 6 dB angular spread of a 0.5 inch diameter piston transducer radiating into water at frequencies of 2.25, 5, and 10 MHz.

3. Equation (8.19) is the Rayleigh-Sommerfeld integral for a planar piston transducer radiating into a fluid. Consider this equation for a rectangular transducer with width $2a$ in the x-direction and width $2b$ in the y-direction. In the paraxial (Fresnel) approximation we can approximate the radius $r = \sqrt{z^2 + (x-x')^2 + (y-y')^2}$ appearing in the denominator of that equation as $r \cong R = \sqrt{x^2 + y^2 + z^2}$, where (x,y,z) is a point in the fluid and $(x',y',0)$ is a point on the transducer face. In the phase term of Eq. (8.19), however, we approximate the radius r instead as

$$r = z\sqrt{1 + \frac{(x-x')^2}{z^2} + \frac{(y-y')^2}{z^2}}$$

$$\cong z + \frac{(x-x')^2}{2z} + \frac{(y-y')^2}{2z}$$

Thus, with these approximations Eq. (8.19) for a rectangular transducer is:

$$p = \frac{-i\omega\rho v_0}{2\pi R}\exp(ikz)\int_{-a}^{+a}\exp\left[\frac{ik(x-x')^2}{2z}\right]dx'\int_{-b}^{+b}\exp\left[\frac{ik(y-y')^2}{2z}\right]dy'$$

Show that this expression can be written as the product of the difference of two Fresnel integrals in the form

$$\frac{p}{\rho c v_0} = \frac{-iz}{2R}\exp(ikz)\left[F\left(\sqrt{\frac{k}{\pi z}}(x+a)\right) - F\left(\sqrt{\frac{k}{\pi z}}(x-a)\right)\right]$$

$$\cdot\left[F\left(\sqrt{\frac{k}{\pi z}}(y+b)\right) - F\left(\sqrt{\frac{k}{\pi z}}(y-b)\right)\right]$$

where $F(x)$ is the Fresnel integral as defined in Eq. (8.77). Using the MATLAB function fresnel_int and the above expression, write a MATLAB function that computes this pressure wave field at any point (x, y, z) in the fluid. For a 6mm by 12mm rectangular transducer radiating into water (c = 1480 m/sec) at 5 MHz, plot the magnitude of the normalized on-axis pressure for distances $z = 6$ mm to $z = 100$ mm. For the same transducer plot cross-axis pressure profiles in the x- and y-directions at $z = 45, 70$ mm.

4. Write a MATLAB function that returns the normalized on-axis pressure, $p/\rho c v_0$, versus distance for a spherically focused piston transducer (see Eq. (8.37)). The input arguments of the function should be the distance values (in mm), the frequency (in MHz), the radius (in mm), the geometrical focal length (in mm), and the wave speed (in m/sec). Use this function to find the location of the true focus (i.e. the distance to the maximum pressure) for a 12.7 mm (0.5 inch) diameter, 101.6 mm (4 in.) focal length transducer radiating into water at 5, 10, and 20 MHz. What can you conclude about the relationship between the location of the true focus versus the geometrical focal length?

5. Equation (8.20) gives the exact on-axis pressure for a planar immersion transducer at a single frequency. Ultrasonic NDE transducers, however, do not normally operate at a single frequency but are driven by a voltage pulse and hence contain a spectrum of frequencies that generate a time domain pulse. The near field behavior of such a pulsed transducer does not show nearly the same strong near field structure as a single frequency model suggests.

   Write a MATLAB function that computes the normalized pressure, $p/\rho c v_0$, at a given on-axis distance at many frequencies and multiplies this pressure at each frequency by the MATLAB function spectrum1 written for exercise 1 in Appendix A. The function should evaluate this product at 1024 positive frequencies ranging from 0 to 100 MHz and then use the Fourier transform IFourierT defined in Appendix A to obtain the time-domain pulse generated by the transducer at the given location. Finally, the function should compute the peak-to-peak magnitude of this pulse and return that value. The inputs to the MATLAB function should be the distance (in mm), the transducer radius (in mm), the wave speed of the fluid (in m/sec), the center frequency, fc (in MHz), and the bandwidth, bw (in MHz).

Use this function to evaluate the peak-to peak response of a transducer radiating into water for 200 points ranging from 10 to 400 mm and plot this peak-to-peak response versus distance. Take the radius of the transducer to be 6.35 mm (0.25 in.), the center frequency fc = 5 MHz and the bandwidth bw = 2 MHz.

9 Gaussian Beam Theory and Transducer Modeling

As seen in the last Chapter and in Appendix D plane waves and spherical waves are important wave types. They can be used as a means to understand many aspects of wave propagation and scattering and they can serve as building blocks to form more complex waves such as the beam of ultrasound generated by an ultrasonic transducer. As building blocks, however, plane waves and spherical waves have some disadvantages. To adequately represent the high frequency beams found in ultrasonic NDE applications, many plane wave components or spherical wave sources are needed, leading to computational inefficiencies. Also, as discussed in the last Chapter, when these wave types are transmitted or reflected through certain geometries at high frequencies mathematical singularities in the resulting approximate wave fields can be encountered that must be eliminated. These wave types do have the virtue of being exact solutions to the equations of motion for both fluids and solids so that other wave fields formed from them also satisfy the equations of motion exactly as long as the wave fields are not obtained with the use of approximations.

Gaussian beams are another important wave type that can eliminate many of the disadvantages of plane waves and spherical waves. In this Chapter we will show that it is possible to accurately model the sound beam of an ultrasonic transducer with as few as ten Gaussian beams. Furthermore, we will see that it is possible to analytically define the propagation and transmission/reflection laws for these Gaussian beams even after they have undergone multiple interactions with curved interfaces. These properties of Gaussian beams will allow us to construct a multi-Gaussian transducer beam model that is computationally efficient and capable of simulating sound beams generated in very complex inspection geometries. Unlike plane waves and spherical waves, Gaussian beams are only approximate paraxial solutions to the governing equations of motion. Similarly, a multi-Gaussian transducer beam model will also be an approximate paraxial solution. Thus, there will be some situations where a multi-Gaussian beam model will lose accuracy. We will describe those special cases in some detail later. Fortunately, many of those special cases are not encountered in common testing setups so a multi-Gaussian beam model is a practical, powerful modeling tool for many NDE applications.

In Appendix F we have given an extensive discussion of Gaussian beam fundamentals for the special case of circularly symmetrical Gaussian beams to illustrate the important properties of Gaussian beams in a simple context. While circularly symmetrical Gaussian beams are very useful for describing many laser science problems, they are of limited use for the types of problems we need to model in ultrasonic inspections. In this Chapter we extend the treatment given in Appendix F to the more general Gaussian beams that are needed for ultrasonic NDE applications.

9.1 The Paraxial Wave Equation and Gaussian Beams in a Fluid

Consider first the case of wave propagation in a fluid. We know that the pressure, p, satisfies the wave equation. If we place a harmonic wave solution (of $\exp(-i\omega t)$ time dependency) into the wave equation in the form of a quasi-plane wave traveling in the x_3-direction given by:

$$p = P(x_1, x_2, x_3)\exp(ik_p x_3) \tag{9.1}$$

(Note - we will not write the time dependency explicitly here or in most subsequent expressions) then we find that P satisfies the equation

$$\frac{\partial^2 P}{\partial x_1^2} + \frac{\partial^2 P}{\partial x_2^2} + \frac{\partial^2 P}{\partial x_3^2} + 2ik_p\frac{\partial P}{\partial x_3} = 0. \tag{9.2}$$

If we use the solution of Eq. (9.1) to represent a wave which is propagating primarily in the x_3-direction, then we expect that at high frequencies the complex exponential term in Eq. (9.1) will capture most of the wave field variations in the x_3-coordinate so that the wave diffraction effects associated with the $\partial^2 P/\partial x_3^2$ term in Eq. (9.2) will be small in comparison to all the other terms in that equation, i.e. we make the *paraxial approximation* [9.1]

$$\frac{\partial^2 P}{\partial x_3^2} << \frac{\partial^2 P}{\partial x_1^2}, \frac{\partial^2 P}{\partial x_2^2}, 2ik_p\frac{\partial P}{\partial x_3} \tag{9.3}$$

which leads to the *paraxial wave equation* for P:

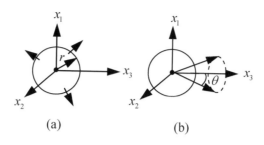

(a)   (b)

Fig. 9.1. (a) Propagation of a spherical wave from a point source and **(b)** the behavior of the spherical wave in a small region around the x_3-axis.

$$\frac{\partial^2 P}{\partial x_1^2} + \frac{\partial^2 P}{\partial x_2^2} + 2ik_p \frac{\partial P}{\partial x_3} = 0. \tag{9.4}$$

In Appendix F it is shown that the paraxial approximation of Eq. (9.3) places some physical limits on the properties of a propagating Gaussian beam.

We can also gain some physical understanding of the meaning of the paraxial approximation by considering the radiation of a spherical wave from a point source in a fluid as shown in Fig. 9.1 (a). The pressure in the fluid in this spherical wave is given by

$$p = A \frac{\exp(ik_p r)}{r}, \tag{9.5}$$

where $r = \sqrt{x_1^2 + x_2^2 + x_3^2}$ is the radial distance from the source and $k_p = \omega / c_p$ is the wave number, with ω the frequency in radians/sec and c_p the wave speed of the fluid.

Now, consider this spherical wave in the neighborhood of a fixed direction, which we will take as the x_3-axis (see Fig. 9.1 (b)). In a small angular region about this axis (where $x_1/x_3 \ll 1, x_2/x_3 \ll 1$) the spherical wave is traveling approximately in the x_3-direction and we have $r = \sqrt{x_3^2 + \rho_0^2} \cong x_3 + \rho_0^2/2x_3$, where $\rho_0 = \sqrt{x_1^2 + x_2^2}$. In this case, the spherical wave can be approximated by

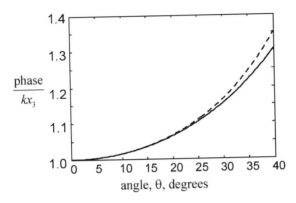

Fig. 9.2. Normalized phase for a spherical wave in the neighborhood of the x_3-axis. Solid line: exact normalized phase, dashed line: paraxial approximation for the normalized phase.

$$p \cong A \frac{\exp\left[ik_p\left(x_3 + \rho_0^2/2x_3\right)\right]}{x_3}, \tag{9.6}$$

which satisfies the paraxial wave equation exactly so that Eq. (9.6) is the paraxial approximation of the spherical wave in the neighborhood of the x_3-axis. How large of an angular neighborhood about the x_3-axis can we take before the paraxial approximation loses accuracy in describing the spherical wave? To answer this question, consider the phase term of the spherical wave $\left(k_p r\right)$ divided by the phase of a plane wave traveling in the x_3-direction $\left(k_p x_3\right)$ and let $\rho_0/x_3 = tan(\theta)$, where θ defines an angle about the x_3-axis (Fig. 9.1(b)). [Remark - a normalized phase term is considered here so that we can discuss phase differences in non-dimensional terms and we consider the phase differences, not the amplitude differences since it is the former that are most sensitive to approximation] Then we have:

exact spherical wave:
$$\frac{k_p r}{k_p x_3} = \sqrt{1 + tan^2\theta} \tag{9.7a}$$

paraxial approximation:

$$\frac{k_p r}{k_p x_3} \cong 1 + \frac{\tan^2 \theta}{2} \qquad (9.7b)$$

Figure 9.2 compares these two normalized phase terms versus θ where the solid line is the exact phase result and the dashed line is the paraxial approximation to this phase. As can be seen from that figure, the paraxial approximation for the phase begins to lose accuracy at an angle of approximately 30 degrees from the x_3-axis.

   Now, apply the results for this simple example to the case of Fig. 9.3 where a planar piston transducer radiates waves into water that travel from the transducer surface to a point on the transducer axis. As discussed in the last Chapter a Rayleigh-Sommerfeld integral model represents this transducer as a distribution of point sources over the face of the transducers, each of which generates a spherical wave of the type just discussed. Thus, if we apply the paraxial approximation to those distributed sources, we would expect that the paraxial approximation for the Rayleigh-Sommerfeld model also breaks down if the angle θ shown in Fig. 9.3 exceeds approximately 30 degrees. Typically, this means that the paraxial approximation should begin to lose accuracy when the distance from the face of the transducer to the point where the wave field being evaluated is less than about a transducer diameter. This can be demonstrated by comparing the magnitude of the exact on-axis pressure for a circular planar piston transducer radiating into a fluid, as found in Eq. (8.20):

$$p(z, \omega) = \rho c v_0 \left[\exp(ik_p z) - \exp\left(ik_p \sqrt{a^2 + z^2}\right) \right] \qquad (9.8)$$

with the same pressure in the paraxial approximation given by Eq. (8.22):

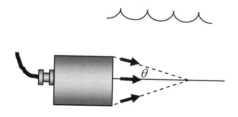

Fig. 9.3. A transducer radiating into a fluid.

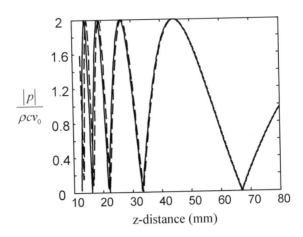

Fig. 9.4. A comparison of the magnitude of the normalized pressure versus on-axis z-distance for a 12.7 mm diameter, 5 MHz planar piston transducer radiating into water where the solid line is for the exact results and the dashed line is for the paraxial result.

$$p(z,\omega) \cong \rho c v_0 \exp(ik_p z)\left[1 - \exp(ik_p a^2 / 2z)\right]. \tag{9.9}$$

Figure 9.4 plots the magnitude of these exact and approximate pressures versus z for the case of a 5MHz, 12.7 mm diameter transducer radiating into water. It can be seen from that figure that even in the near field, where there are significant pressure variations, the paraxial approximation represents the pressure of this transducer very well but that the approximation begins to have a significant shift from the exact on-axis pattern at about one diameter distance from the transducer which is the smallest distance plotted in Fig. 9.4. This distance corresponds to an angle θ in Fig. 9.3 of 30 degrees.

Another way of viewing the paraxial approximation of Eq. (9.3) is to recall from the last Chapter that we can also use an angular spectrum of plane waves to represent the sound beam of a transducer. Thus, consider a plane wave component of this spectrum that is traveling in the $x_1 - x_3$ plane at an angle θ with respect to the x_3-axis (Fig. 9.5). This plane wave is given by

$$p = A\exp(ik_p x_1 \sin\theta + ik_p x_3 \cos\theta), \tag{9.10}$$

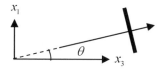

Fig. 9.5. A plane wave traveling in the $x_1 - x_3$ plane at an angle θ with respect to the x_3-axis.

which can be placed in the quasi-plane wave form of Eq. (9.1), $p = P\exp(ik_p x_3)$, where

$$P = A\exp\left[ik_p x_1 \sin\theta + ik_p x_3 (1 - \cos\theta)\right].\tag{9.11}$$

Then for small angles θ

$$\frac{\partial^2 P}{\partial x_1^2} = -k_p^2 P \sin^2\theta \cong -k_p^2 P\theta^2$$

$$2ik_p \frac{\partial P}{\partial x_3} = -2k_p^2 P(1 - \cos\theta) \cong k_p^2 P\theta^2 \tag{9.12}$$

$$\frac{\partial^2 P}{\partial x_3^2} = -k_p^2 P(1 - \cos\theta)^2 \cong -\frac{k_p^2 P\theta^4}{4}$$

so that we see that $\partial^2 P / \partial x_3^2$ will be at least an order of magnitude smaller than the other derivative terms if $\theta < 0.5$ rad, or approximately $\theta < 30°$. This shows that as long as the transducer beam is sufficiently well collimated so that the angular plane wave spectrum components needed to represent the beam are very small outside a cone angle of about 30 degrees about the x_3-axis, we expect the paraxial approximation will be valid.

There are number of exact solutions to the paraxial wave equation, Eq. (9.4). An ordinary plane wave where $P = A = $ constant is a solution. Also, as mentioned previously, the paraxial approximation of a spherical wave given by Eq. (9.6) is

$$P = A\frac{\exp\left[ik_p \rho_0^2 / 2x_3\right]}{x_3},\tag{9.13}$$

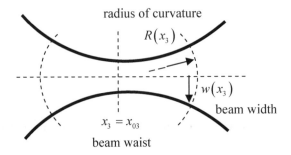

Fig. 9.6. A Gaussian beam of circular cross-section propagating in the x_3-direction, showing the wave front curvature and the beam width. The beam waist is located at $x_3 = x_{03}$.

which is an exact solution of the paraxial wave equation. We can also obtain a solution of Eq. (9.4) in the form of a Gaussian beam propagating along the x_3-axis. Here, we will consider a general form of a Gaussian beam given by:

$$P = P(x_3)\exp\left(\frac{i\omega}{2}\mathbf{X}^T\mathbf{M}_p(x_3)\mathbf{X}\right), \qquad \mathbf{X} = [x_1, x_2]^T \quad (9.14)$$

where $P(x_3)$ is a complex-valued scalar, and $\mathbf{M}_p$ is a 2×2 complex-valued symmetric matrix. A circular cross-section Gaussian beam of the type considered in Appendix F is then a special case of Eq. (9.14). This type of Gaussian beam is shown schematically in Fig. 9.6 along with some of its defining parameters (beam width, radius of curvature). For an in-depth discussion of these and other defining parameters, see Appendix F. We will also discuss later in this Chapter how the $\mathbf{M}_p$ matrix is related to these properties of the propagating beam. Substituting Eq. (9.14) into Eq. (9.4), we obtain

$$\frac{2}{c_p}\frac{dP}{dx_3} + P\,tr(\mathbf{M}_p) + iP\mathbf{X}^T\left(\frac{1}{c_p}\frac{d\mathbf{M}_p}{dx_3} + \mathbf{M}_p^2\right)\mathbf{X} = 0. \quad (9.15)$$

In order to satisfy Eq. (9.15) for all $\mathbf{X}$, we obtain the two equations

$$\frac{2}{c_p}\frac{dP}{dx_3} + P\,tr(\mathbf{M}_p) = 0 \quad (9.16)$$

and

$$\frac{1}{c_p}\frac{d\mathbf{M}_p}{dx_3}+\mathbf{M}_p^2=0,$$

(9.17)

where $\text{tr}(\mathbf{M}_p)$ is the trace of the matrix $\mathbf{M}_p$. In ray theory, Eq. (9.16) is usually called the *transport equation* [9.2]. Equation (9.17) is in the form of a non-linear matrix *Ricatti equation* [9.2].

We can manipulate both of these equations into alternative forms where we can solve them directly. Consider first Eq. (9.17). We start by differentiating the identity $\mathbf{M}_p\mathbf{M}_p^{-1}=\mathbf{I}$ with respect to x_3. We obtain

$$\left(d\mathbf{M}_p/dx_3\right)\mathbf{M}_p^{-1}+\mathbf{M}_p\left(d\mathbf{M}_p^{-1}/dx_3\right)=0.$$

(9.18)

If we use Eq. (9.17) in this result and pre-multiply by $\mathbf{M}_p^{-1}$, then we obtain

$$\frac{d\mathbf{M}_p^{-1}}{dx_3}-c_p\mathbf{I}=0,$$

(9.19)

where $\mathbf{I}$ is the 2×2 identity matrix. Equation (9.19) gives us a simple differential relationship that we will use shortly to obtain the solution of Eq. (9.17). Now, consider transforming the $\mathbf{M}_p$ part of Eq. (9.16). If we pre-multiply Eq. (9.19) by $\mathbf{M}_p$ we find

$$\mathbf{M}_p=\frac{1}{c_p}\mathbf{M}_p\left(d\mathbf{M}_p^{-1}/dx_3\right).$$

(9.20)

Using the relationship

$$\mathbf{M}_p=\left(\mathbf{M}_p^{-1}\right)^{-1}=\frac{\text{adj}\left(\mathbf{M}_p^{-1}\right)}{\det\left(\mathbf{M}_p^{-1}\right)}$$

(9.21)

(which comes directly from the definition of the inverse of a matrix) in Eq. (9.20) yields

$$\mathbf{M}_p=\frac{1}{c_p\det\left(\mathbf{M}_p^{-1}\right)}\text{adj}\left(\mathbf{M}_p^{-1}\right)\left(d\mathbf{M}_p^{-1}/dx_3\right),$$

(9.22)

where adj[ ] denotes the adjoint and det[ ] the determinant. Taking the trace of both sides of Eq. (9.22) and applying the general matrix relationship [9.3]

$$\frac{d\left[\det\left(\mathbf{M}_p^{-1}\right)\right]}{dx_3} = \text{tr}\left[\text{adj}\left(\mathbf{M}_p^{-1}\right)\left(d\mathbf{M}_p^{-1}/dx_3\right)\right], \tag{9.23}$$

it follows that

$$\begin{aligned}
\text{tr}\left(\mathbf{M}_p\right) &= \frac{1}{c_p \det\left(\mathbf{M}_p^{-1}\right)} \frac{d\left[\det\left(\mathbf{M}_p^{-1}\right)\right]}{dx_3} \\
&= \frac{1}{c_p} \frac{d\left\{\ln\left[\det\left(\mathbf{M}_p^{-1}\right)\right]\right\}}{dx_3}.
\end{aligned} \tag{9.24}$$

Placing Eq. (9.24) into Eq. (9.16) then gives

$$2\frac{dP}{dx_3} + P\frac{d}{dx_3}\left[\ln\left(\det\left[\mathbf{M}_p^{-1}\right]\right)\right] = 0. \tag{9.25}$$

The solutions of Eqs. (9.19) and (9.25) are now both easy to obtain. The solution of Eq. (9.19) by direct integration gives us the *propagation law:*

$$\begin{aligned}
\mathbf{M}_p^{-1}\left(x_3\right) &= c_p x_3 \mathbf{I} + \mathbf{M}_p^{-1}\left(0\right) \\
&= \left[c_p x_3 \mathbf{M}_p\left(0\right) + \mathbf{I}\right]\mathbf{M}_p^{-1}\left(0\right).
\end{aligned} \tag{9.26}$$

Taking the inverse of both sides of Eq. (9.26) gives the corresponding solution for $\mathbf{M}_p$:

$$\mathbf{M}_p\left(x_3\right) = \mathbf{M}_p\left(0\right)\left[\mathbf{I} + c_p x_3 \mathbf{M}_p\left(0\right)\right]^{-1}, \tag{9.27}$$

which can be rewritten as

$$\mathbf{M}_p\left(x_3\right) = \frac{1}{\Delta}\left(\mathbf{M}_p\left(0\right) + x_3 c_p \mathbf{I} \det\left[\mathbf{M}_p\left(0\right)\right]\right) \tag{9.28}$$

where

$$\Delta = 1 + \left(x_3 c_p\right)\text{tr}\left[\mathbf{M}_p\left(0\right)\right] + \left(x_3 c_p\right)^2 \det\left[\mathbf{M}_p\left(0\right)\right]. \tag{9.29}$$

The solution of Eq. (9.25) also follows directly, since we can write it in the equivalent form

$$d\left\{\ln\left[\frac{P(x_3)}{P(0)}\right]\right\}/dx_3 = d\left\{\ln\left[\det\left(\frac{\mathbf{M}_p^{-1}(x_3)}{\mathbf{M}_p^{-1}(0)}\right)^{-1/2}\right]\right\}/dx_3, \tag{9.30}$$

where $P(0)$ is $P(x_3)|_{x_3=0}$. Equation (9.30) can then also be integrated, leading to any one of the following equivalent forms:

$$\frac{P(x_3)}{P(0)} = \sqrt{\frac{\det\left[\mathbf{M}_p^{-1}(0)\right]}{\det\left[\mathbf{M}_p^{-1}(x_3)\right]}} = \sqrt{\frac{\det\left[\mathbf{M}_p(x_3)\right]}{\det\left[\mathbf{M}_p(0)\right]}}$$

$$= \frac{1}{\sqrt{\det\left[\mathbf{I}+c_p x_3 \mathbf{M}_p(0)\right]}}. \tag{9.31}$$

Using the second of these forms our Gaussian beam solution for the pressure, p, then can be written as

$$p(\mathbf{x},\omega) = P(0)\exp(ik_p x_3)\sqrt{\frac{\det\left[\mathbf{M}_p(x_3)\right]}{\det\left[\mathbf{M}_p(0)\right]}}$$

$$\cdot\exp\left(\frac{i\omega}{2}\mathbf{X}^T\mathbf{M}_p(x_3)\mathbf{X}\right) \tag{9.32}$$

with $\mathbf{X}=[x_1,x_2]^T$, which shows that both the amplitude and phase of the Gaussian beam are functions solely of the matrix $\mathbf{M}_p(x_3)$ and the starting values $P(0),\mathbf{M}_p(0)$ at $x_3 = 0$. The velocity in the Gaussian beam can also be obtained by differentiating this pressure. However, in the paraxial approximation the dominant term in such a differentiation comes from the $\exp(ik_p x_3)$ term so that the velocity is simply given by

$$\mathbf{v}^p(\mathbf{x},\omega) = V^p(0)\mathbf{d}^p\exp(ik_p x_3)\sqrt{\frac{\det\left[\mathbf{M}_p(x_3)\right]}{\det\left[\mathbf{M}_p(0)\right]}}$$

$$\cdot\exp\left(\frac{i\omega}{2}\mathbf{X}^T\mathbf{M}_p(x_3)\mathbf{X}\right), \tag{9.33}$$

where $V^p(0) = P(0)/\rho c_p$, ρ is the density of the fluid and $\mathbf{d}^p$ is a unit vector in the x_3-direction (the direction of propagation). For a proof that the other terms obtained when differentiating the pressure to obtain the velocity are indeed negligible, see the discussion in Appendix F leading up to Eq. (F.25).

Gaussian beams are often used to also represent the light beam in a laser. In the laser field the matrix $\mathbf{M}_p$ is usually taken to be a diagonal matrix of the form (see Appendix F and [9.1])

$$\mathbf{M}_p(x_3) = \begin{bmatrix} \dfrac{1}{c_p q(x_3)} & 0 \\ 0 & \dfrac{1}{c_p q(x_3)} \end{bmatrix},$$ (9.34)

where $q(x_3)$ is a complex scalar. In this case Eq. (9.32) becomes

$$p(\mathbf{x},\omega) = P(0)\exp\left(ik_p x_3\right)\frac{q(0)}{q(x_3)}\exp\left[\frac{ik\left(x_1^2 + x_2^2\right)}{2q(x_3)}\right]$$ (9.35)

and the propagation law for $\mathbf{M}_p$ (Eq. (9.26)) is simply

$$q(x_3) = q(0) + x_3.$$ (9.36)

Equation (9.35) represents a propagating Gaussian beam of circular cross section. As long as the imaginary part of the starting value at $x_3 = 0$, $q(0)$, has a negative imaginary part, the propagation law shows that $q(x_3)$ will also have a negative imaginary part so that Eq. (9.35) will represent a beam that is always localized near the axis of propagation. If we let

$$\frac{1}{q(x_3)} = \frac{1}{R(x_3)} + i\frac{\lambda}{\pi w^2(x_3)}$$ (9.37)

then Eq. (9.35) can be written as

$$p(\mathbf{x},\omega) = P(0)\exp\left(ik_p x_3\right)\frac{q(0)}{q(x_3)}$$

$$\cdot\exp\left[\frac{ik\left(x_1^2 + x_2^2\right)}{2R(x_3)}\right]\exp\left[-\frac{\left(x_1^2 + x_2^2\right)}{w^2(x_3)}\right]$$ (9.38)

which shows that $R(x_3)$ represents a wave front radius of curvature that varies as the Gaussian beam propagates while $w(x_3)$ represents a beam width parameter that defines the radial distance to which the beam amplitude drops by a factor e^{-1} from its on-axis value. Figure 9.6 illustrates these quantities for a propagating Gaussian beam. From the results given in Appendix F (see Eq. (F.14)) one can write down relatively simple expressions for $R(x_3), w(x_3)$:

$$R(x_3) = (x_3 - x_{03}) + x_{R3}^2 / (x_3 - x_{03})$$
$$w(x_3) = w_0 \sqrt{1 + (x_3 - x_{03})^2 / x_{R3}^2},$$

(9.39)

where w_0 is the beam width at the waist (see Fig. 9.6), located at $x_3 = x_{03}$ and $x_{R3} = \pi w_0^2 / \lambda$ is the *confocal parameter,* as discussed in Appendix F.

In the laser field, most of the discussion of Gaussian beams is for circular cross-section beams where Eq. (9.34) is valid. This is because in the interactions of the Gaussian light beam in a laser (reflection from mirrors, etc) the cross-section of the Gaussian beam often remains circular. Appendix F describes similar cases where a circular cross-section Gaussian beam propagates in a fluid and interacts with spherical interfaces, resulting in transmitted and reflected beams also of circular cross-section.

In NDE problems, although the Gaussians used to model a transducer may have circular cross-sections to begin with at the transducer face, after transmission and reflection from interfaces we must normally use the more general form of Eq. (9.14) and let $\mathbf{M}_p(x_3)$ be a complex 2x2 symmetrical matrix. As long as the two eigenvalues of $M_m^I(x_3) \equiv \text{Im}\{\mathbf{M}_p(x_3)\}$ $(m = 1, 2)$, satisfy $M_m^I(x_3) > 0$, where $\text{Im}\{\ \}$ indicates "imaginary part of", Eq. (9.14) will represent a wave which has an elliptical Gaussian profile with decay away from the x_3 axis and hence will be a localized beam traveling along that axis. If the general Gaussian beam of Eq. (9.14) starts out at $x_3 = 0$ with eigenvalues of $M_m^I(0) \equiv \text{Im}\{\mathbf{M}_p(0)\}$ $(m = 1, 2)$, that satisfy $M_m^I(0) > 0$, then during propagation the eigenvalues of $\text{Im}\{\mathbf{M}_p(x_3)\}$ will also satisfy $M_m^I(x_3) > 0$ since the propagation law, Eq. (9.26), shows that only the real parts of the eigenvalues of $\mathbf{M}_p^{-1}$ (and, hence, $\mathbf{M}_p$) are affected during propagation. Thus a localized Gaussian at $x_3 = 0$ always generates a localized propagating Gaussian beam, just as in the circular cross-section case. Note

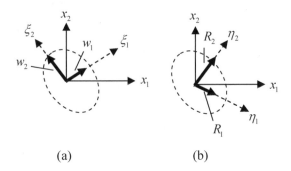

Fig. 9.7. The elliptical cross-section of a general propagating Gaussian beam, showing **(a)** the principal beam widths and principal beam width directions where the beam amplitude has fallen to 1/e of its value on the beam axis, and **(b)** the principal wave front radii of curvatures and their directions.

that the eigenvalues of $\mathrm{Re}\{\mathbf{M}_p(x_3)\}$, $M_m^R(x_3)$ $(m=1,2)$, are related to the principal wave front curvatures, where $\mathrm{Re}\{\ \}$ denotes "real part of". The directions of those principal curvatures, however, are different from the directions associated with the eigenvalues $M_m^I(x_3)$, which are related to the two principal beam widths for a Gaussian beam of elliptical cross-section (see Fig. 9.7). Thus, as an elliptical cross-section Gaussian beam represented by Eq. (9.32) propagates the angle between the major axes of that elliptical cross section and the principal wave front curvatures changes. If we let

$$M_m^I(x_3) = \frac{\lambda}{\pi c_p w_m^2(x_3)}$$

$$M_m^R(x_3) = \frac{1}{c_p R_m(x_3)},$$

(9.40)

where R_m, w_m are the principal radii of curvature and beam widths, respectively, the general Gaussian beam of Eq. (9.32) can be written as

$$p(\mathbf{x},\omega) = P(0)\exp\left(ik_p x_3\right)\sqrt{\frac{\det\left[\mathbf{M}_p(x_3)\right]}{\det\left[\mathbf{M}_p(0)\right]}}$$

$$\cdot \exp\left[\frac{ik_p}{2}\left(\frac{\eta_1^2}{R_1}+\frac{\eta_2^2}{R_2}\right)\right]\exp\left[-\left(\frac{\xi_1^2}{w_1^2}+\frac{\varepsilon_2^2}{w_2^2}\right)\right],$$

(9.41)

where (ξ_1, ξ_2) and (η_1, η_2) are the principal axes for the imaginary and real parts of the $\mathbf{M}_p$ matrix, respectively. The orientation of both these axes are functions of x_3.

Another difference between the circular cross-section case (Eq. (9.35)) and the more general case (Eq. (9.32)) is that square roots appear in the latter equation. Since the matrix $\mathbf{M}_p$ is complex and the principal curvature and beam width directions are not aligned in general, some care must be taken in evaluating those square roots. This issue appears to have received little attention in the literature as in many Gaussian beam problems discussed the $\mathbf{M}_p$ matrix is diagonal, i.e.

$$\mathbf{M}_p(x_3) = \begin{bmatrix} M_1(x_3) & 0 \\ 0 & M_2(x_3) \end{bmatrix}. \tag{9.42}$$

In this case it is easy to specify the roots since we can write Eq. (9.31) as

$$\frac{P(x_3)}{P(0)} = \frac{\sqrt{\det[\mathbf{M}_p(x_3)]}}{\sqrt{\det[\mathbf{M}_p(0)]}} = \frac{\sqrt{M_1(x_3)}\sqrt{M_2(x_3)}}{\sqrt{M_1(0)}\sqrt{M_2(0)}}. \tag{9.43}$$

Because the imaginary parts of $M_m(0), M_m(x_3)$ $(m = 1, 2)$ are always positive, the individual square roots in Eq. (9.43) also must be taken to have positive imaginary parts.

For the more general case where $\mathbf{M}_p$ is not diagonal, although the principal directions of the real and imaginary parts of $\mathbf{M}_p$ do not coincide, the real part of $\mathbf{M}_p$ is a real, symmetrical matrix and the imaginary part is a real, symmetrical and positive definite matrix. Under these conditions, matrix theory [9.5] shows that it is always possible to define a generalized eigenvalue problem where a real 2x2 transformation matrix, $\mathbf{T}$, can be found that *simultaneously* diagonalizes both the real and imaginary parts of $\mathbf{M}_p$. Knowing this transformation matrix we can then form up the term

$$\frac{\sqrt{\det[\mathbf{T}^T(x_3)\mathbf{M}_p(x_3)\mathbf{T}(x_3)]}}{\sqrt{\det[\mathbf{T}^T(0)\mathbf{M}_p(0)\mathbf{T}(0)]}} = \frac{\sqrt{\tilde{M}_1(x_3)}\sqrt{\tilde{M}_2(x_3)}}{\sqrt{\tilde{M}_1(0)}\sqrt{\tilde{M}_2(0)}} \tag{9.44}$$

and calculate the complex $\tilde{M}_m(0), \tilde{M}_m(x_3)$ terms which are the diagonal matrix terms obtained after applying the transformation matrices to $\mathbf{M}_p$ as shown in Eq. (9.44). We have placed the tilde over these diagonal terms to emphasize that these complex quantities are not the same as the complex values given in Eq. (9.43). However, the square roots on the right side of Eq. (9.44) can be found in the same fashion as done with Eq. (9.43). Then in terms of the remaining real determinants, we find

$$
\frac{P(x_3)}{P(0)} = \frac{\sqrt{\det[\mathbf{M}_p(x_3)]}}{\sqrt{\det[\mathbf{M}_p(0)]}}
$$

$$
= \frac{\sqrt{\det^2[\mathbf{T}(0)]}}{\sqrt{\det^2[\mathbf{T}(x_3)]}} \frac{\sqrt{\tilde{M}_1(x_3)}\sqrt{\tilde{M}_2(x_3)}}{\sqrt{\tilde{M}_1(0)}\sqrt{\tilde{M}_2(0)}}
$$

(9.45)

Many mathematical software packages such as MATLAB are available that obtain the transformation matrix $\mathbf{T}$, so that Eq. (9.45) is easy to implement in practice.

9.2 The Paraxial Wave Equation and Gaussian Beams in a Solid

For a homogeneous, elastic solid, the displacement potentials satisfy wave equations so that they also have paraxial Gaussian beam solutions of the form

$$
\phi = \Phi(x_3)\exp(ik_p x_3)\exp\left(\frac{i\omega}{2}\mathbf{X}^T\mathbf{M}_p(x_3)\mathbf{X}\right)
$$

$$
\psi = \Psi(x_3)\,\mathbf{t}\exp(ik_s x_3)\exp\left(\frac{i\omega}{2}\mathbf{X}^T\mathbf{M}_s(x_3)\mathbf{X}\right).
$$

(9.46)

At high frequencies we can obtain the velocity, $\mathbf{v}^\alpha$ $(\alpha = p,s)$, for a P-wave or S-wave by again just differentiating the $\exp(ik_\alpha x_3)$ terms in these equations to obtain

$$
\mathbf{v}^\alpha = V^\alpha(x_3)\mathbf{d}^\alpha\exp(ik_\alpha x_3)\exp\left(\frac{i\omega}{2}\mathbf{X}^T\mathbf{M}_\alpha(x_3)\mathbf{X}\right) \quad (\alpha = p,s)
$$

(9.47)

with $V^P = \omega^2\Phi/c_p$, $V^s = \omega^2\Psi/c_s$ and $\mathbf{d}^p = \mathbf{e}_3$, $\mathbf{d}^s = \mathbf{e}_3\times\mathbf{t}$, where $\mathbf{e}_3$ is a unit vector in the x_3-direction. Note that these relations are identical in form to those for a plane wave since in a plane wave $\exp(ik_\alpha x_3)$ is the only spatially varying term present.

Alternatively, we can show that a formal high frequency approximation of Navier's equations for the displacements in the quasi-plane wave form

$$u_i = \tilde{U}_i(x_1, x_2, x_3)\exp(ik_s x_3) \tag{9.48}$$

leads to the paraxial wave equation

$$\frac{\partial^2 \tilde{U}_3}{\partial x_1^2} + \frac{\partial^2 \tilde{U}_3}{\partial x_2^2} + 2ik_p \frac{\partial \tilde{U}_3}{\partial x_3} = 0 \tag{9.49}$$

with $\tilde{U}_1 = \tilde{U}_2 = 0$ for P-waves while for S-waves

$$\frac{\partial^2 \tilde{U}_I}{\partial x_1^2} + \frac{\partial^2 \tilde{U}_I}{\partial x_2^2} + 2ik_s \frac{\partial \tilde{U}_I}{\partial x_3} = 0 \quad (I = 1,2) \tag{9.50}$$

with $\tilde{U}_3 = 0$ [9.6]. Since both P-waves and S-waves in a homogeneous, isotropic elastic solid satisfy paraxial equations (Eqs. (9.49) and (9.50)), elastic wave Gaussian beam solutions can be written in vector form for the displacements of both wave types as

$$\mathbf{u}^\alpha = U^\alpha(x_3)\mathbf{d}^\alpha \, \exp\left(ik_\alpha x_3\right)\exp\left(\frac{i\omega}{2}\mathbf{X}^T\mathbf{M}_\alpha(x_3)\mathbf{X}\right) \quad (\alpha = p, s) \tag{9.51}$$

Then Eq. (9.47) again follows, where $\mathbf{v}^\alpha = -i\omega\mathbf{u}^\alpha$, $V^\alpha(x_3) = -i\omega U^\alpha(x_3)$.

In the solid these Gaussian beam solutions of the paraxial equation also must satisfy transport and Riccati equations given by [9.2]

$$\frac{2}{c_\alpha}\frac{dV^\alpha}{dx_3} + V^\alpha tr(\mathbf{M}_\alpha) = 0 \tag{9.52}$$

and

$$\frac{1}{c_\alpha}\frac{d\mathbf{M}_\alpha}{dx_3} + \mathbf{M}_\alpha^2 = 0. \tag{9.53}$$

Following exactly the same steps outlined for the fluid case, the solutions of Eqs. (9.52) and (9.53) are then

$$\mathbf{M}_\alpha(x_3) = \mathbf{M}_\alpha(0)\left[\mathbf{I} + c_\alpha x_3 \mathbf{M}_\alpha(0)\right]^{-1} \tag{9.54}$$

and

$$\frac{V^{\alpha}(x_3)}{V^{\alpha}(0)} = \sqrt{\frac{\det\left[\mathbf{M}_{\alpha}^{-1}(0)\right]}{\det\left[\mathbf{M}_{\alpha}^{-1}(x_3)\right]}} = \sqrt{\frac{\det\left[\mathbf{M}_{\alpha}(x_3)\right]}{\det\left[\mathbf{M}_{\alpha}(0)\right]}}$$
$$= \frac{1}{\sqrt{\det\left[\mathbf{I} + c_{\alpha}x_3\mathbf{M}_{\alpha}(0)\right]}}$$
(9.55)

so that the velocity in the solid for a Gaussian beam of type α $(\alpha = p, s)$ is

$$\mathbf{v}^{\alpha} = V^{\alpha}(0)\mathbf{d}^{\alpha}\frac{\sqrt{\det\left[\mathbf{M}_{\alpha}(x_3)\right]}}{\sqrt{\det\left[\mathbf{M}_{\alpha}(0)\right]}}$$
$$\cdot \exp\left(ik_{\alpha}x_3\right)\exp\left(\frac{i\omega}{2}\mathbf{X}^T\mathbf{M}_{\alpha}(x_3)\mathbf{X}\right) \quad (\alpha = p, s)$$
(9.56)

which shows that apart from the polarization vector the form of a Gaussian beam propagating in a solid is identical to that in a fluid (Eq. (9.33)).

9.3 Transmission/Reflection of a Gaussian Beam at an Interface

In the last section, we obtained explicit expressions for a Gaussian beam propagating in either a fluid or a solid. Here, we will obtain the transmission/reflection laws for a Gaussian beam incident on a curved interface between two solids (Fig. 9.8). A fluid-solid interface as found in immersion testing is then merely a special case of these relations. We will consider the case where the Gaussian beam may interact with an interface more than one time so the interface shown in Fig. 9.8 will be used to represent the Gaussian beam on the mth interface ($m = 1, 2, \ldots$).

When the incident Gaussian beam strikes the interface, transmitted and reflected Gaussian beams of various types will be generated. In Fig. 9.8 we show a Gaussian beam incident on a general curved interface Σ between two homogenous, isotropic media (solid or fluid) and only one other Gaussian beam that will be used to represent any one of the transmitted or reflected Gaussian beams generated. We will let the first medium be medium m and the second medium $m+1$. The wave speed of a Gaussian beam type β $(\beta = p, s)$ in medium m and the wave speed of a

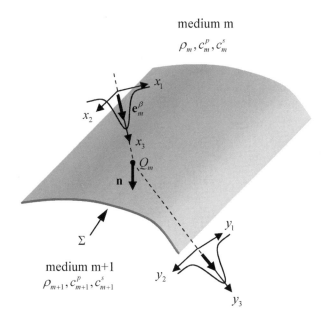

medium m

ρ_m, c_m^p, c_m^s

medium m+1

$\rho_{m+1}, c_{m+1}^p, c_{m+1}^s$

Fig. 9.8. A Gaussian beam incident on a curved interface between two elastic media and one of the transmitted or reflected Gaussian beams. The origins of the (x_1, x_2, x_3) and (y_1, y_2, y_3) axes are both at the point Q_m where the central axis of the incident Gaussian beam meets the interface, but these origins are shown displaced for clarity of illustration.

Gaussian beam of type α $(\alpha = p, s)$ in medium $m+1$ will be given by $c_m^\beta, c_{m+1}^\alpha$, respectively, and the corresponding wave numbers by $k_m^\beta, k_{m+1}^\alpha$. The velocity amplitude, polarization vector, and complex phase of a Gaussian beam of type β in medium m and of type α in medium $m+1$ will be designated as $V_m^\beta, \mathbf{d}_m^\beta, \mathbf{M}_m^\beta$ and $V_{m+1}^\alpha, \mathbf{d}_{m+1}^\alpha, \mathbf{M}_{m+1}^\alpha$, respectively. The propagation direction of the incident Gaussian beam will be along the x_3-axis in the (x_1, x_2, x_3) coordinate system and the propagation of the generated wave will be along the y_3-axis in the (y_1, y_2, y_3) coordinates (Fig. 9.8). Unit vectors along both of these propagation directions are given by $\mathbf{e}_m^\beta, \mathbf{e}_{m+1}^\alpha$, respectively, as shown. The normal to the interface at the point Q_m where the central axis of the incident Gaussian beam strikes the interface is the unit vector, $\mathbf{n}$. The origins of both the (x_1, x_2, x_3) and (y_1, y_2, y_3) axes will be taken to be at point Q_m. The origins are shown displaced from Q_m in Fig. 9.8 for clarity of illustration only.

In relating the Gaussian beams at the interface it will be necessary to perform some coordinate rotations in three dimensions [9.2]. Thus, we need to extend the definition of the 2x2 complex matrices involved to 3x3 matrices in a three-dimensional space. We will denote the 3-D version of matrix $\mathbf{M}_m^\beta$ as $\hat{\mathbf{M}}_m^\beta$, where

$$\hat{\mathbf{M}}_m^\beta = \begin{bmatrix} \left(\mathbf{M}_m^\beta\right)_{11} & \left(\mathbf{M}_m^\beta\right)_{12} & 0 \\ \left(\mathbf{M}_m^\beta\right)_{21} & \left(\mathbf{M}_m^\beta\right)_{22} & 0 \\ 0 & 0 & 0 \end{bmatrix} \tag{9.57}$$

with a similar definition for $\hat{\mathbf{M}}_{m+1}^\alpha$.

Using the notations just given, we will write the velocity components of the incident Gaussian beam in medium m as

$$\left(v_m^{\beta;(x)}\right)_j = V_{mj}^{\beta;(x)}\left(x_3\right)\exp\left[i\omega t_0 + ik_m^\beta x_3 + i\frac{\omega}{2}\mathbf{x}^T\hat{\mathbf{M}}_m^{\beta;(x)}\left(x_3\right)\mathbf{x}\right], \tag{9.58}$$

where $V_{mj}^{\beta;(x)}\left(x_3\right) = V_m^\beta\left(x_3\right)\left(d_m^{\beta;(x)}\right)_j$. Similarly for a Gaussian beam in medium $m+1$:

$$\left(v_{m+1}^{\alpha;(y)}\right)_j = V_{m+1j}^{\alpha;(y)}\left(y_3\right)\exp\left[i\omega t_0 + ik_{m+1}^\alpha y_3 + i\frac{\omega}{2}\mathbf{y}^T\hat{\mathbf{M}}_{m+1}^{\alpha;(y)}\left(y_3\right)\mathbf{y}\right] \tag{9.59}$$

where $V_{m+1j}^{\alpha;(y)}\left(y_3\right) = V_{m+1}^\alpha\left(y_3\right)\left(d_{m+1}^{\alpha;(y)}\right)_j$. In both Eq. (9.58) and Eq. (9.59) $\mathbf{x} = \left(x_1, x_2, x_3\right)$ and $\mathbf{y} = \left(y_1, y_2, y_3\right)$ are now full 3-D coordinates. We have placed an (x) or (y) in the notation for the vector and matrix terms appearing in Eqs. (9.58) and (9.59) to emphasize that the components involved in those quantities are being calculated in the x- and y-coordinates, respectively. This will be useful because it will become necessary to introduce several other coordinate systems when we solve the transmission/reflection problem. The term $\exp\left(i\omega t_0\right)$ appearing in both Eq. (9.58) and (9.59) corresponds to the time delay, t_0, it has taken for the incident beam to reach point Q_m on the interface from its starting location (which will be at the transducer face when this Gaussian beam is used to

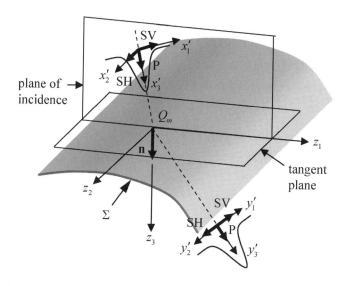

Fig. 9.9. Plane of incidence coordinates along the incident and T/R Gaussian beam directions and interface coordinates (z_1, z_2, z_3). The origin of all these coordinates is taken at point Q_m but the incident and T/R axes are shown displaced for clarity of illustration.

represent a transducer wave field). Point Q_m is at $x_3 = y_3 = 0$ for both coordinate systems, as mentioned previously.

Most transmission/reflection problems are solved in plane of incidence coordinates. The *plane of incidence* (POI) at interface m is the plane that contains both the incident wave direction, $\mathbf{e}_m^\beta$, and the normal, $\mathbf{n}$, to the interface at point Q_m where the central axis of the incident beam strikes the interface (see Fig. 9.9). The POI coordinates (x_1', x_2', x_3') for the incident beam are obtained from the (x_1, x_2, x_3) axes through a 2-D rotation about the x_3-axis, where $x_3 = x_3'$ is along the direction of propagation of the incident beam. The (x_1', x_3') axes lie in the POI while the x_2'-axis is perpendicular to the POI, as shown in Fig. 9.9. Similarly, the POI coordinates (y_1', y_2', y_3') for the transmitted/reflected beam are obtained from the (y_1, y_2, y_3) axes through a 2-D rotation about the y_3-axis, where $y_3 = y_3'$ is along the direction of propagation of the transmitted/reflected beam. The (y_1', y_3') axes lie in the POI while the y_2'-axis is perpendicular to the POI. As for the (x_1, x_2, x_3), (y_1, y_2, y_3) axes, the origin of both the (x_1', x_2', x_3') and (y_1', y_2', y_3') axes will be taken at point Q_m although those coordinates are shown displaced from Q_m in Fig. 9.9 for purposes of clarity of illustration. We also define the (z_1, z_2, z_3) coordinates shown in Fig. 9.9, where z_3 is along the unit normal, $\mathbf{n}$, (z_1, z_3) lie in the POI, and the z_2-axis is perpendicular to the POI, as shown. We can express both the incident and transmitted/reflected waves in these z-coordinates as

$$\left(v_m^{\beta;(z)} \right)_j = V_m^{\beta}(\mathbf{z}) \left(d_m^{\beta;(z)} \right)_j \exp\left[i\omega t_0 + ik_m^{\beta} \mathbf{e}_m^{\beta} \cdot \mathbf{z} + i\phi_m^{\beta}(\mathbf{z}) \right] \tag{9.60}$$

and

$$\left(v_{m+1}^{\alpha;(z)} \right)_j = V_{m+1}^{\alpha}(\mathbf{z}) \left(d_{m+1}^{\alpha;(z)} \right)_j \exp\left[i\omega t_0 + ik_{m+1}^{\alpha} \mathbf{e}_{m+1}^{\alpha} \cdot \mathbf{z} + i\phi_{m+1}^{\alpha}(\mathbf{z}) \right] \tag{9.61}$$

where, to simplify the notation we have lumped all the quadratic phase terms in the $\phi_m^{\beta}, \phi_{m+1}^{\alpha}$ terms.

We require that the velocity components of Eqs. (9.60) and (9.61) satisfy the conditions of velocity and traction matching on the interface Σ. These conditions are

$$\sum_{\gamma} \left(v_m^{\gamma;(z)}(\Sigma) \right)_j = \sum_{\delta} \left(v_{m+1}^{\delta;(z)}(\Sigma) \right)_j$$

$$\sum_{\gamma} n_i^{(z)} C_{ijkl}^{(m)} \frac{\partial \left(v_m^{\gamma;(z)}(\Sigma) \right)_k}{\partial z_l} = \sum_{\delta} n_i^{(z)} C_{ijkl}^{(m+1)} \frac{\partial \left(v_{m+1}^{\delta;(z)}(\Sigma) \right)_k}{\partial z_l}, \tag{9.62}$$

where $C_{ijkl}^{(m)}, C_{ijkl}^{(m+1)}$ are the elastic constants for the mth and $(m+1)$th media, respectively. The sums in Eq. (9.62) are taken over all the waves (incident, reflected, or transmitted) of type γ that are present in medium m, and of type δ that are present in medium $(m+1)$. We will not satisfy the conditions of Eq. (9.62) exactly over the entire interface Σ. Instead, consistent with the paraxial approximation which treats our Gaussian beams as propagating quasi-plane waves confined to a region near the central beam axis, we will match the "amplitude" parts of Eqs. (9.60) and (9.61) only at point Q_m, where the amplitudes are just the complex-valued coefficients of the complex exponentials appearing in those equations. The "phase" parts of Eqs. (9.60) and (9.61), however, we will match to second order in the z-coordinates at point Q_m, where the phase are the arguments of the complex exponentials in those equations [Note: we place quotes on the "amplitude" and "phase" terms considered here since they are in fact both complex quantities]. The amplitude matching conditions from Eqs. (9.60), (9.61) and Eq. (9.62) are then

$$
\sum_{\gamma} V_m^{\gamma} (Q_m)(d_m^{\gamma})_j = \sum_{\delta} V_{m+1}^{\delta} (Q_m)(d_{m+1}^{\delta})_j,
$$

$$
\sum_{\gamma} n_i^{(z)} C_{ijkl}^{(m)} (d_m^{\gamma})_k
$$

$$
\cdot \left[ik_m^{\gamma} \left(e_m^{\gamma;(z)} \right)_l V_m^{\gamma} (Q_m) + \left. \frac{\partial V_m^{\gamma}}{\partial z_l} \right|_{Q_m} + iV_m^{\gamma} \left. \frac{\partial \phi_m^{\gamma}}{\partial z_l} \right|_{Q_m} \right]
\tag{9.63}
$$

$$
= \sum_{\delta} n_i C_{ijkl}^{(m+1)} (d_{m+1}^{\delta})_k
$$

$$
\cdot \left[ik_{m+1}^{\delta} \left(e_{m+1}^{\delta} \right)_l V_{m+1}^{\delta} (Q_m) + \left. \frac{\partial V_{m+1}^{\delta}}{\partial z_l} \right|_{Q_m} + iV_{m+1}^{\delta} \left. \frac{\partial \phi_{m+1}^{\delta}}{\partial z_l} \right|_{Q_m} \right].
$$

The derivatives of the $\phi_m^{\gamma}, \phi_{m+1}^{\delta}$ terms appearing in Eq. (9.63) all vanish at Q_m since these terms are both quadratic functions in the z-coordinates. Also, at high frequencies the derivatives of the $V_m^{\gamma}, V_{m+1}^{\delta}$ terms in Eq. (9.63) are much smaller than the terms which only involve the $V_m^{\gamma}, V_{m+1}^{\delta}$ themselves, since the latter terms are multiplied by the frequency terms $k_m^{\gamma}, k_{m+1}^{\delta}$. Thus, Eq. (9.63) reduces to

$$
\sum_{\gamma} V_m^{\gamma} (Q_m)(d_m^{\gamma})_j = \sum_{\delta} V_{m+1}^{\delta} (Q_m)(d_{m+1}^{\delta})_j,
$$

$$
\sum_{\gamma} n_i^{(z)} C_{ijkl}^{(m)} (d_m^{\gamma})_k \left[ik_m^{\gamma} \left(e_m^{\gamma;(z)} \right)_l V_m^{\gamma} (Q_m) \right]
\tag{9.64}
$$

$$
= \sum_{\delta} n_i C_{ijkl}^{(m+1)} (d_{m+1}^{\delta})_k \left[ik_{m+1}^{\delta} \left(e_{m+1}^{\delta} \right)_l V_{m+1}^{\delta} (Q_m) \right].
$$

But the conditions of Eq. (9.64) are just the same as if we had applied the boundary conditions of Eq. (9.62) to a set of plane waves given by

$$
\left(v_m^{\gamma;(z)} \right)_j = V_m^{\gamma} \left(d_m^{\gamma;(z)} \right)_j \exp \left[i\omega t_0 + ik_m^{\gamma} \mathbf{e}_m^{\gamma} \cdot \mathbf{z} \right]
\tag{9.65}
$$

$$\left(v_{m+1}^{\delta;(z)} \right)_j = V_{m+1}^{\delta} \left(d_{m+1}^{\delta;(z)} \right)_j \exp\left[i\omega t_0 + ik_{m+1}^{\delta} \mathbf{e}_{m+1}^{\delta} \cdot \mathbf{z} \right] \qquad (9.66)$$

at a plane interface that coincides with the tangent plane to the interface Σ at point Q_m. The solution of Eq. (9.64), therefore, just yields the appropriate plane wave transmission/reflection (T/R) coefficients. Normally these T/R coefficients are found by assuming a P, SV, or SH polarization direction for the incident and transmitted/reflected (T/R) waves in the POI coordinates (x_1', x_2', x_3'), as shown in Fig. 9.9, and solving for a corresponding transmitted or reflected P, SV, or SH wave component in the POI coordinates (y_1', y_2', y_3'). However, our incident beam will not necessarily have a polarization that lies in the POI of the mth interface. Also, all the incident and T/R waves are not coupled to each other. For example, there is no coupling between plane P-waves or SV-waves and an SH-wave. Thus, it is necessary to define a procedure so that if we have an incident beam of specified type β $(\beta = p,s)$ and a T/R beam of specified type α $(\alpha = p,s)$ that the velocity components of these incident and T/R waves are related properly to each other. We can do this by defining a T/R matrix that transforms the (x_1', x_2', x_3')-components of velocity of the incident wave into the correct (y_1', y_2', y_3')-components of the T/R wave [9.2]. Specifically, consider the case when we have an incident wave of type S and are considering a T/R wave of the S-wave type $(S \rightarrow S)$. Then the (y_1', y_2', y_3') components of the T/R wave, $V_{m+1j}^{s;(y')} = V_{m+1}^s \left(d_{m+1}^{s;(y')} \right)_j$, and the (x_1', x_2', x_3') components of the incident wave, $V_{mj}^{s;(x')} = V_m^s \left(d_m^{s;(x')} \right)_j$ can be related to each other by

$$V_{m+1i}^{s;(y')} = \left(\tilde{T}_m^{s;s} \right)_{ij} V_{mj}^{s;(x')}, \qquad (9.67)$$

where the 3x3 T/R matrix, $\tilde{\mathbf{T}}_m^{s;s}$, is

$$\tilde{\mathbf{T}}_m^{s;s} = \begin{bmatrix} T_m^{sv;sv} & 0 & 0 \\ 0 & T_m^{sh;sh} & 0 \\ 0 & 0 & 0 \end{bmatrix}. \qquad (9.68)$$

whose components are the ordinary plane wave T/R coefficients, $T_m^{\delta;\gamma}$ (based on velocity ratios) for a T/R plane wave of type δ due to a plane wave of type γ at the mth interface, where the polarization of the incident and T/R waves used to define these coefficients are defined in Fig. 9.9. For example, the polarization of an incident SV-wave is assumed to be in the x_1' -direction. Note that this 3x3 matrix properly transforms the individual SV- and SH-components of the incident S-wave (i.e. components along the x_1' and x_2' -axes, respectively) into corresponding SV- and SH- components of the T/R wave (i.e. components along the y_1' and y_2' -axes, respectively) and does not generate any P-wave (y_3' component) of the T/R wave, as is required since the T/R wave is specified to be an S-wave . In a similar fashion we can define T/R matrices appropriate to the $(S \rightarrow P), (P \rightarrow S), (P \rightarrow P)$ cases as

$$\tilde{\mathbf{T}}_m^{p;s} = \begin{bmatrix} 0 & 0 & 0 \\ 0 & 0 & 0 \\ T_m^{p;sv} & 0 & 0 \end{bmatrix} \tag{9.69}$$

and

$$\tilde{\mathbf{T}}_m^{s;p} = \begin{bmatrix} 0 & 0 & T_m^{sv;p} \\ 0 & 0 & 0 \\ 0 & 0 & 0 \end{bmatrix} \tag{9.70}$$

and

$$\tilde{\mathbf{T}}_m^{p;p} = \begin{bmatrix} 0 & 0 & 0 \\ 0 & 0 & 0 \\ 0 & 0 & T_m^{p;p} \end{bmatrix}. \tag{9.71}$$

With these definitions, then for the general case we can write simply

$$V_{m+1i}^{\alpha;(y')} = \left(\tilde{T}_m^{\alpha;\beta}\right)_{ij} V_{mj}^{\beta;(x')} \tag{9.72}$$

We now need to transform this relationship back into our original (x_1, x_2, x_3) and (y_1, y_2, y_3) coordinates. Let $(\mathbf{u}_1, \mathbf{u}_2, \mathbf{u}_3)$ be unit vectors along the (x_1, x_2, x_3) axes, respectively, and similarly let $(\mathbf{u}_1', \mathbf{u}_2', \mathbf{u}_3')$ be unit vectors along the (x_1', x_2', x_3') POI axes. Also, let $(\mathbf{v}_1, \mathbf{v}_2, \mathbf{v}_3)$ be unit vectors along the (y_1, y_2, y_3) axes, and $(\mathbf{v}_1', \mathbf{v}_2', \mathbf{v}_3')$ be unit vectors along the (y_1', y_2', y_3') POI axes. We can choose the (y_1, y_2) axes to have any orientation we wish about the T/R wave direction, y_3, but a particularly convenient choice is to make the orientation (y_1, y_2) axes relative to the (y_1', y_2') axes the same as the orientation of the (x_1, x_2) axes relative to the (x_1', x_2') axes (this is called the "standard" choice in [9.2]). With this choice, when the material is the same on both sides of the interface Eq. (9.72) will simply leave the (y_1, y_2, y_3) components of the velocity unchanged from the original (x_1, x_2, x_3) components. Based on this choice, we can define a 3-D rotation matrix, $\hat{\mathbf{G}}^R$, with components $\hat{G}_{np}^R = \left(\mathbf{u}_n' \cdot \mathbf{u}_p \right) = \left(\mathbf{v}_n' \cdot \mathbf{v}_p \right)$ where

$$
\hat{\mathbf{G}}^R = \begin{bmatrix} \mathbf{u}_1' \cdot \mathbf{u}_1 & \mathbf{u}_1' \cdot \mathbf{u}_2 & 0 \\ \mathbf{u}_2' \cdot \mathbf{u}_1 & \mathbf{u}_2' \cdot \mathbf{u}_2 & 0 \\ 0 & 0 & 1 \end{bmatrix} = \begin{bmatrix} \mathbf{v}_1' \cdot \mathbf{v}_1 & \mathbf{v}_1' \cdot \mathbf{v}_2 & 0 \\ \mathbf{v}_2' \cdot \mathbf{v}_1 & \mathbf{v}_2' \cdot \mathbf{v}_2 & 0 \\ 0 & 0 & 1 \end{bmatrix}
$$
$$
= \begin{bmatrix} \cos \lambda & \sin \lambda & 0 \\ -\sin \lambda & \cos \lambda & 0 \\ 0 & 0 & 1 \end{bmatrix}
$$

(9.73)

and we have

$$
V_{mj}^{\beta;(x')} = \hat{G}_{jp}^R V_{mp}^{\beta;(x)}
$$
$$
V_{m+1i}^{\alpha;(y')} = \hat{G}_{ir}^R V_{m+1r}^{\alpha;(y)}
$$

(9.74)

which, when placed into Eq. (9.72) gives

$$
\hat{G}_{ir}^R V_{m+1r}^{\alpha;(y)} = \left(\tilde{T}_m^{\alpha;\beta} \right)_{ij} \hat{G}_{jp}^R V_{mp}^{\beta;(x)}.
$$

(9.75)

If we pre-multiply Eq. (9.75) by $\hat{G}_{ik}^R$ and use the fact that $\hat{G}_{ik}^R \hat{G}_{ir}^R = \delta_{kr}$, we find

$$
V_{m+1k}^{\alpha;(y)} = \hat{G}_{ik}^R \left(\tilde{T}_m^{\alpha;\beta} \right)_{ij} \hat{G}_{jp}^R V_{mp}^{\beta;(x)}
$$

(9.76)

or, equivalently

$$
V_{m+1}^{\alpha} \left(Q_m \right) \left(d_{m+1}^{\alpha;(y)} \right)_k = \left(\mathrm{T}_m^{\alpha;\beta} \right)_{kp} V_m^{\beta} \left(Q_m \right) \left(d_m^{\beta;(x)} \right)_p,
$$

(9.77)

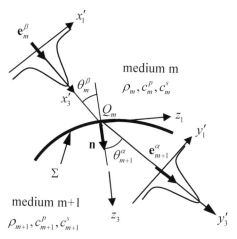

Fig. 9.10. The angles that an incident and transmitted Gaussian beam make at the interface in the POI coordinates.

where

$$\left(T_m^{\,\alpha;\beta} \right)_{kp} = \hat{G}_{jk}^R \left(\tilde{T}_m^{\,\alpha;\beta} \right)_{ij} \hat{G}_{jp}^R \tag{9.78}$$

Equation (9.77) gives the relationship needed to obtain the amplitude of the T/R Gaussian beam from the incident beam. In many NDE testing situations the POI for all the interfaces may be aligned and we can always assume that the $\left(x_1, x_2, x_3\right)$ and $\left(y_1, y_2, y_3\right)$ coordinates are the same as the POI coordinates. In that case the rotation matrix, $\hat{G}^R$, is not needed and the wave types α and β range over $\left(p, sv\right)$ only so we write

$$V_{m+1}^\alpha \left(Q_m\right) = T_m^{\alpha;\beta} V_m^\beta \left(Q_m\right) \tag{9.79}$$

without using a 3x3 T/R matrix. The polarization directions for the incident and T/R waves are then just the directions shown in Fig. 9.9. Figure 9.10 shows explicitly the acute angle θ_m^β that the direction of propagation an incident wave of type β makes with respect to the interface normal (z_3-axis) and the acute angle θ_{m+1}^α for a transmitted wave of type α. These angles will be needed as we now discuss the phase matching of the Gaussian beams at the interface.

The total phase of the complex exponential term in Eq. (9.58) for the incident wave in medium m is

$$\Phi_m^\beta = i\omega t_0 + ik_m^\beta x_3 + i\frac{\omega}{2}\mathbf{x}^T\hat{\mathbf{M}}_m^{\beta;(x)}(x_3)\mathbf{x} \tag{9.80}$$

Similarly the total phase for a transmitted wave in medium $m+1$ from Eq. (9.59) is

$$\Phi_{m+1}^\alpha = i\omega t_0 + ik_{m+1}^\alpha y_3 + i\frac{\omega}{2}\mathbf{y}^T\hat{\mathbf{M}}_{m+1}^{\alpha;(y)}(y_3)\mathbf{y}. \tag{9.81}$$

Here, we will match these phases at the interface, Σ, in the (z_1, z_2, z_3) coordinates in a neighborhood of the point Q_m. Consider first Eq. (9.80) for the incident wave. To transform from the (x_1, x_2, x_3) coordinates to the (z_1, z_2, z_3) coordinates, we first use the rotation matrix previously defined to transform from (x_1, x_2, x_3) to (x_1', x_2', x_3') POI coordinates (by rotating about the x_3-axis). We then transform from the (x_1', x_2', x_3') coordinates to the (z_1, z_2, z_3) coordinates through a rotation about an angle θ_m^β about the x_2'-axis, i.e. we let

$$x_i = \hat{G}_{ki}^R x_k' \tag{9.82}$$
$$x_k' = \hat{G}_{jk}^Z z_j,$$

where $\mathbf{G}^R$ has been given previously (Eq. (9.73)) and $\hat{\mathbf{G}}^Z$ is the rotation matrix

$$\hat{\mathbf{G}}^Z = \begin{bmatrix} \cos\theta_m^\alpha & 0 & \sin\theta_m^\alpha \\ 0 & 1 & 0 \\ -\sin\theta_m^\alpha & 0 & \cos\theta_m^\alpha \end{bmatrix}. \tag{9.83}$$

If we combine the two rotation matrices into a single matrix, $\hat{\mathbf{G}}^C$, where

$$\hat{G}_{ji}^C = \hat{G}_{ki}^R\hat{G}_{jk}^Z \tag{9.84}$$

then the total phase of the incident wave in the (z_1, z_2, z_3) coordinates can be written as

$$\Phi_m^\beta = i\omega t_0 + ik_m^\beta \left(z_1 \sin\theta_m^\beta + z_3 \cos\theta_m^\beta \right)$$
$$+ i\frac{\omega}{2} \left(\hat{\mathbf{M}}_m^{\beta;(x)}(Q_m) \right)_{ij} \hat{G}_{pi}^C \hat{G}_{rj}^C z_p z_r. \tag{9.85}$$

In the neighborhood of point Q_m for a curved interface we have to second order

$$z_3 = \frac{1}{2} h_{IJ}(Q_m) z_I z_J, \tag{9.86}$$

where the summation over capital subscripts such as the I and J in Eq. (9.86) is taken over the values (1,2) only. This is a convention we will also follow in subsequent expressions. The components of the 2x2 matrix h_{IJ} in Eq. (9.86) are the curvatures of the interface at Q_m as measured in the z-coordinates, with z_3 along the interface normal and z_1 in the plane of incidence, as shown in Figs. 9.9 and 9.10.

We now place Eq. (9.86) into Eq. (9.85), keeping only the terms which are at most quadratic in the (z_1, z_2) coordinates and use the fact that

$$\left(\hat{\mathbf{M}}_m^{\beta;(x)}(Q_m) \right)_{13} = \left(\hat{\mathbf{M}}_m^{\beta;(x)}(Q_m) \right)_{31} = \left(\hat{\mathbf{M}}_m^{\beta;(x)}(Q_m) \right)_{33} = 0. \quad (I = 1, 2) \tag{9.87}$$

Then Eq. (9.85) becomes

$$\Phi_m^\beta = i\omega t_0 + ik_m^\beta z_1 \sin\theta_m^\beta$$
$$+ i\frac{\omega}{2} \left(\left(\mathbf{M}_m^{\beta;(x)}(Q_m) \right)_{IJ} G_{PI}^C G_{RJ}^C + h_{PR}(Q_m) \frac{\cos\theta_m^\beta}{c_m^\beta} \right) z_P z_R. \tag{9.88}$$

In Eq. (9.88) all the capital subscripts take on the values $(1,2)$ only so the $\mathbf{M}_m^\alpha$ matrix in that equation is the 2x2 sub-matrix of $\hat{\mathbf{M}}_m^\alpha$ and similarly the rotation matrices in Eq. (9.88) only involve 2x2 sub-matrices of $\hat{\mathbf{G}}^C$, given by

$$\mathbf{G}^C = \mathbf{G}^Z \mathbf{G}^R = \begin{bmatrix} \cos\theta_m^\beta & 0 \\ 0 & 1 \end{bmatrix} \begin{bmatrix} \cos\lambda & \sin\lambda \\ -\sin\lambda & \cos\lambda \end{bmatrix}. \tag{9.89}$$

Equation (9.88) is an (approximate) expression for the total phase of a Gaussian beam for the incident beam in medium m. For a transmitted wave in medium $m+1$ in an entirely similar fashion one obtains:

$$\Phi^\alpha_{m+1} = i\omega t_0 + ik^\alpha_{m+1} z_1 \sin\theta^\alpha_{m+1}$$

$$+i\frac{\omega}{2}\left(\left(\mathbf{M}^{\alpha;(y)}_{m+1}(Q_m)\right)_{IJ} \tilde{G}^C_{PI}\tilde{G}^C_{RJ} + h_{PR}(Q_m)\frac{\cos\theta^\alpha_{m+1}}{c^\alpha_{m+1}}\right) z_P z_R, \tag{9.90}$$

where

$$\tilde{\mathbf{G}}^C = \tilde{\mathbf{G}}^Z \mathbf{G}^R = \begin{bmatrix} \cos\theta^\alpha_{m+1} & 0 \\ 0 & 1 \end{bmatrix}\begin{bmatrix} \cos\lambda & \sin\lambda \\ -\sin\lambda & \cos\lambda \end{bmatrix}. \tag{9.91}$$

Equating the total phases in Eq. (9.88) and Eq. (9.90) the terms involving t_0 cancel. We find from the term that is linear in z_1:

$$\frac{\sin\theta^\alpha_{m+1}}{c^\alpha_{m+1}} = \frac{\sin\theta^\beta_m}{c^\beta_m}, \tag{9.92}$$

which is a statement of generalized Snell's law. From equality of the quadratic terms it follows that

$$\left(\mathbf{M}^{\alpha;(y)}_{m+1}(Q_m)\right)_{IJ} \tilde{G}^C_{PI}\tilde{G}^C_{RJ} + h_{PR}(Q_m)\frac{\cos\theta^\alpha_{m+1}}{c^\alpha_{m+1}}$$

$$= \left(\mathbf{M}^{\beta;(x)}_m(Q_m)\right)_{IJ} G^C_{PI}G^C_{RJ} + h_{PR}(Q_m)\frac{\cos\theta^\beta_m}{c^\beta_m} \tag{9.93}$$

or, equivalently

$$\left(\mathbf{M}^{\alpha;(y)}_{m+1}(Q_m)\right)_{VU} = \left[\tilde{G}^C_{VP}\right]^{-1}\left[\tilde{G}^C_{UR}\right]^{-1}\left(\mathbf{M}^{\beta;(x)}_m(Q_m)\right)_{IJ} G^C_{PI}G^C_{RJ}$$

$$+\left[\tilde{G}^C_{VP}\right]^{-1}\left[\tilde{G}^C_{UR}\right]^{-1} h_{PR}(Q_m)\left(\frac{\cos\theta^\beta_m}{c^\beta_m} - \frac{\cos\theta^\alpha_{m+1}}{c^\alpha_{m+1}}\right). \tag{9.94}$$

Equation (9.94) gives the transformation law across the interface for the $\mathbf{M}$ matrix of a transmitted beam of type α in medium $m+1$ due to an incident wave of type β in medium m. The same equation also applies to a reflected beam of type α traveling in medium m if everywhere in Eq. (9.94) we simply replace c^α_{m+1} by c^α_m and $\cos\theta^\alpha_{m+1}$ by $-\cos\theta^\alpha_m$ where θ^α_m is then the acute angle between the negative z_3-axis and the propagation direction of the reflected beam. This process corresponds to making the replacement $\theta^\alpha_{m+1} \to \pi - \theta^\alpha_m$ and then interpreting θ^α_m as the acute angle that the direction of propagation the reflected wave makes with the negative z_3 axis.

Taking the real part of both sides of Eq. (9.94) gives the transformation of the wave front curvature of the Gaussian beam across the interface. That expression is the same as that found for wave front curvature changes from geometrical ray theory [9.2]. The imaginary parts of Eq. (9.94) relate the beam widths of the Gaussians on either side of the interface in terms of their projections on the interface [9.2]. To see this transformation law in a simple setting, consider a circularly symmetric Gaussian beam given by Eqs. (9.34) and (9.37) incident on a planar interface $\left(h_{PR} = 0 \right)$ where the $\left(x_1, x_2, x_3 \right)$ axes are aligned with the POI so that the rotation angle $\lambda = 0$. Then we find

$$\mathbf{M}_{m+1}^{\alpha;(y)}\left(Q_m \right) =$$

$$\begin{bmatrix} \dfrac{\cos^2 \theta_m^\beta}{c_m^\beta \cos^2 \theta_{m+1}^\alpha} \left(\dfrac{1}{R(Q_m)} + \dfrac{i\, c_m^\beta}{\pi f w^2 (Q_m)} \right) & 0 \\[4mm] 0 & \dfrac{1}{c_m^\beta} \left(\dfrac{1}{R(Q_m)} + \dfrac{i\, c_m^\beta}{\pi f w^2 (Q_m)} \right) \end{bmatrix}, \quad (9.95)$$

where f is the frequency. Equation (9.95) shows that the circularly symmetric incident beam is transformed into a T/R beam of elliptical cross section with wave front curvatures $\left(R_1, R_2 \right)$ and beam widths $\left(w_1, w_2 \right)$, both along the $\left(y_1, y_2 \right)$ directions, respectively, where

$$\mathbf{M}_{m+1}^{\alpha;(y)}\left(Q_m \right) =$$

$$\begin{bmatrix} \dfrac{1}{c_{m+1}^\alpha} \left(\dfrac{1}{R_1(Q_m)} + \dfrac{i\, c_{m+1}^\alpha}{\pi f w_1^2 (Q_m)} \right) & 0 \\[4mm] 0 & \dfrac{1}{c_{m+1}^\alpha} \left(\dfrac{1}{R_2(Q_m)} + \dfrac{i\, c_{m+1}^\alpha}{\pi f w_2^2 (Q_m)} \right) \end{bmatrix}. \quad (9.96)$$

Equating Eqs. (9.95) and (9.96) gives the transformation laws for the wave front curvatures and beam widths as:

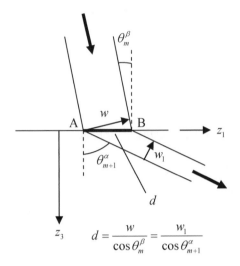

$$d = \frac{w}{\cos\theta_m^\beta} = \frac{w_1}{\cos\theta_{m+1}^\alpha}$$

Fig. 9.11. The transmission of an incident circular Gaussian beam at an interface showing the change of beam width in the POI, where the incident and transmitted beams both have a common width, d, between points A and B on the interface. The beam width in the direction normal to the POI is unchanged. This leads to a transmitted beam of elliptical cross-section.

$$w_1\left(Q_m\right) = \left|\frac{\cos\theta_{m+1}^\alpha}{\cos\theta_m^\beta}\right| w\left(Q_m\right)$$

$$w_2\left(Q_m\right) = w\left(Q_m\right)$$

$$R_1\left(Q_m\right) = \frac{c_m^\beta}{c_{m+1}^\alpha} \frac{\cos^2\theta_{m+1}^\alpha}{\cos^2\theta_m^\beta} R\left(Q_m\right) \qquad (9.97)$$

$$R_2\left(Q_m\right) = \frac{c_m^\beta}{c_{m+1}^\alpha} R\left(Q_m\right).$$

Thus, in the plane of incidence at a planar interface, Eq. (9.97) shows that the incident beam width w is changed to w_1 where the widths of incident and transmitted beams in the POI have the same projection on the interface, as shown in Fig. 9.11. In contrast $w_2 = w$ so the beam width in the direction normal to the POI is unchanged. The wave front curvatures are also changed but those changes depend on both the angles present and the wave speeds of the two materials. In Appendix F the special case when a circular Gaussian beam is normal to a planar interface was considered.

Setting $\theta^{\alpha}_{m+1} = \theta^{\beta}_m = 0$ in Eq. (9.97) then yields the same results found there (see Eqs. (F.44) and (F.46)).

Both the propagation and transmission laws do not affect the symmetry of the complex **M** matrix, as can be seen from Eq. (9.28) and Eq. (9.93). Thus, an **M** matrix that starts out symmetric remains symmetric after propagation in multiple media and after interactions with multiple interfaces. Like the propagation law, the interface transformation law of Eq. (9.94) also does not affect the sign on the eigenvalues of the imaginary part of the **M** matrix so that a localized Gaussian remains a localized Gaussian after interactions with the interface. For the simple case just discussed this is obvious since Eq. (9.97) shows that the beam widths on transmission or reflection remain positive, finite values. For the more general case, take the imaginary part of both sides of Eq. (9.93) and let $\mathbf{M}^I_{m+1} = \mathrm{Im}\left\{\mathbf{M}^{\alpha;(y)}_{m+1}\left(Q_m\right)\right\}, \mathbf{M}^I_m = \mathrm{Im}\left\{\mathbf{M}^{\beta;(x)}_m\left(Q_m\right)\right\}$. Then we find

$$\mathbf{M}^I_{m+1} = \left[\left(\mathbf{G}^R\right)^T\left(\tilde{\mathbf{G}}^Z\right)^{-1}\mathbf{G}^Z\mathbf{G}^R\right]\mathbf{M}^I_m\left[\left(\mathbf{G}^R\right)^T\left(\tilde{\mathbf{G}}^Z\right)^{-1}\mathbf{G}^Z\mathbf{G}^R\right]^T$$
$$= \mathbf{P}^T\mathbf{M}^I_m\mathbf{P},$$

(9.98)

where $\left(\mathbf{G}^Z\right)^T = \left(\mathbf{G}^Z\right), \left(\tilde{\mathbf{G}}^Z\right)^T = \left(\tilde{\mathbf{G}}^Z\right)$ and $\left(\tilde{\mathbf{G}}^Z\right)^{-1}\mathbf{G}^Z = \mathbf{G}^Z\left(\tilde{\mathbf{G}}^Z\right)^{-1}$.

The matrix $\mathbf{P} = \mathbf{P}^T = \left(\mathbf{G}^R\right)^T\left(\tilde{\mathbf{G}}^Z\right)^{-1}\mathbf{G}^Z\mathbf{G}^R$ is non-singular since for any angles $\theta^{\beta}_m, \theta^{\alpha}_{m+1} \neq \pi/2$ we have

$$\det[\mathbf{P}] = \det\left[\left(\mathbf{G}^R\right)^T\right]\det\left[\left(\tilde{\mathbf{G}}^Z\right)^{-1}\right]\det\left[\mathbf{G}^Z\right]\det\left[\mathbf{G}^R\right]$$
$$= (1)\left(\frac{1}{\cos\theta^{\alpha}_{m+1}}\right)\left(\cos\theta^{\beta}_m\right)(1) \neq 0.$$

(9.99)

Because the eigenvalues of $\mathbf{M}^I_m$ are positive, the matrix $\mathbf{M}^I_m$ is positive definite, i.e. $\mathbf{x}^T\mathbf{M}^I_m\mathbf{x} > 0$ for all real non-trivial vectors, $\mathbf{x}$. Then if we let $\mathbf{x} = \mathbf{P}\mathbf{y}$ where $\mathbf{y} \neq 0$ (which implies that $\mathbf{P}$ is non-singular) we have $\mathbf{y}^T\mathbf{P}^T\mathbf{M}^I_m\mathbf{P}\mathbf{y} > 0$ so it follows from Eq. (9.98) that $\mathbf{M}^I_{m+1}$ is also positive definite and its eigenvalues are positive.

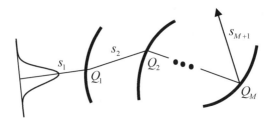

Fig. 9.12. The transmission/reflection of a Gaussian beam through multiple media with curved interfaces.

9.4 Gaussian Beams and ABCD Matrices

Since a Gaussian beam that starts out localized remains localized during propagation and after interaction with general curved surfaces, the beam is always well behaved, unlike plane waves or spherical waves that can lead to caustics or singularities. Furthermore, we can easily combine the propagation and interface transformation laws to obtain the form of the Gaussian beam after multiple interactions with curved surfaces or interfaces. To see this, consider now the general case where a Gaussian beam travels through or is reflected from M interfaces, as shown in Fig. 9.12. If we apply the propagation and transmission/reflection laws just derived to each medium and interface, the velocity of the Gaussian beam of type α in medium $M+1$ is then given by

$$
\mathbf{v}_{M+1}^{\alpha;(y)} = \frac{\sqrt{\det\left[\mathbf{M}_{M+1}^{\alpha}\left(s_{M+1}\right)\right]}}{\sqrt{\det\left[\mathbf{M}_{M+1}^{\alpha}\left(0\right)\right]}}
$$
$$
\cdot\left[\prod_{m=M}^{1}\mathbf{T}_{m}^{\gamma_{m+1};\gamma_{m}}\frac{\sqrt{\det\left[\mathbf{M}_{m}^{\gamma_{m}}\left(s_{m}\right)\right]}}{\sqrt{\det\left[\mathbf{M}_{m}^{\gamma_{m}}\left(0\right)\right]}}\right] \tag{9.100a}
$$
$$
\cdot V_{1}^{\gamma_{1}}\left(0\right)\mathbf{d}_{1}^{\gamma_{1}}\exp\left[i\omega\sum_{m=1}^{M+1}\frac{s_{m}}{c_{m}^{\gamma_{m}}}+i\frac{\omega}{2}\mathbf{y}^{T}\hat{\mathbf{M}}_{M+1}^{\alpha}\left(s_{M+1}\right)\mathbf{y}\right]
$$

or, equivalently

$$\mathbf{v}_{M+1}^{\alpha;(y)} = \frac{1}{\sqrt{\det\left[\mathbf{I} + s_{M+1}c_{M+1}^{\alpha}\mathbf{M}_{M+1}^{\alpha}(0)\right]}}$$

$$\cdot\left[\prod_{m=M}^{1}\frac{\mathbf{T}_{m}^{\,\gamma_{m+1};\gamma_{m}}}{\sqrt{\det\left[\mathbf{I} + s_{m}c_{m}^{\gamma_{m}}\mathbf{M}_{m}^{\gamma_{m}}(0)\right]}}\right] \qquad (9.100b)$$

$$\cdot V_{1}^{\gamma_{1}}(0)\mathbf{d}_{1}^{\gamma_{1}}\exp\left[i\omega\sum_{m=1}^{M+1}\frac{s_{m}}{c_{m}^{\gamma_{m}}} + i\frac{\omega}{2}\mathbf{y}^{T}\hat{\mathbf{M}}_{M+1}^{\alpha}(s_{M+1})\mathbf{y}\right],$$

where s_m is the distance the beam has traveled in medium m along its central axis and γ_m is the mode of the beam propagating in medium m and γ_{m+1} is the transmitted or reflected mode in medium $m+1$ after interaction with the m-th interface. The matrix $\mathbf{T}_m^{\,\gamma_{m+1};\gamma_m}$ is given from Eq. (9.78) by

$$\mathbf{T}_m^{\,\gamma_{m+1};\gamma_m} = \left(\hat{\mathbf{G}}^R\right)^T \tilde{\mathbf{T}}_m^{\gamma_{m+1};\gamma_m}\hat{\mathbf{G}}^R. \qquad (9.101)$$

As Eqs. (9.100a) and (9.100b) indicate the product of matrices is in the order

$$\mathbf{T}_M^{\,\gamma_{M+1};\gamma_M}\mathbf{T}_{M-1}^{\,\gamma_M;\gamma_{M-1}}...\mathbf{T}_1^{\,\gamma_2;\gamma_1}. \qquad (9.102)$$

Note that at the m-th interface $\mathbf{M}_m^{\gamma_m}(s_m) = \mathbf{M}_m^{\gamma_m}(Q_m)$ if s_m is the distance the Gaussian beam has traveled in medium m to the m-th interface and $\mathbf{M}_{m+1}^{\gamma_{m+1}}(0) = \mathbf{M}_{m+1}^{\gamma_{m+1}}(Q_m)$ since point Q_m is at the starting point for the Gaussian beam in medium $m+1$. Thus, we can use either of these notations interchangeably.

Both Eq. (9.100a) and Eq. (9.100b) are remarkably compact in form. The calculation of individual terms in those equations can be done in a highly modular and efficient way by the introduction of **A, B, C, D** matrices which are analogous to the scalar A, B, C, D terms discussed in Appendix F and commonly used in optics [9.7]. These matrices arise from the fact that both the propagation and transmission laws for $\mathbf{M}_m^{\gamma_m}$ can be written in a common form. First consider the propagation law (Eq. (9.27)). That law can be written as:

$$\mathbf{M}_m^{\gamma_m}(s_m) = \left[\mathbf{D}_m^d\mathbf{M}_m^{\gamma_m}(0) + \mathbf{C}_m^d\right]\left[\mathbf{A}_m^d + \mathbf{B}_m^d\mathbf{M}_m^{\gamma_m}(0)\right]^{-1}, \qquad (9.103)$$

where $\mathbf{A}_m^d, \mathbf{B}_m^d, \mathbf{C}_m^d, \mathbf{D}_m^d$ will denote the $\mathbf{A}, \mathbf{B}, \mathbf{C}, \mathbf{D}$ matrices that characterize propagation (displacement) of the beam in medium m. These matrices also depend on the wave type, γ_m, being considered in the m-th medium but for economy of notation we will not show that dependency explicitly. Comparing Eqs. (9.27) and (9.103) we find

$$\mathbf{A}_m^d = \mathbf{D}_m^d = \mathbf{I}, \qquad\qquad \mathbf{B}_m^d = c_m^{\gamma_m} s_m \mathbf{I}, \qquad\qquad \mathbf{C}_m^d = \mathbf{O}, \qquad (9.104)$$

where $\mathbf{O}$ is the zero matrix. Now, consider the transmission law (Eq. (9.94)). First we rewrite that equation as

$$\mathbf{M}_{m+1}^{\gamma_{m+1}}(Q_m) = \left\{ \left[\tilde{\mathbf{G}}^C \right]^{-1} \mathbf{G}^C \mathbf{M}_m^{\gamma_m}(Q_m) \right.$$
$$\left. + \left[\tilde{\mathbf{G}}^C \right]^{-1} \mathbf{h}(Q_m) \left(\frac{\cos\theta_m^{\gamma_m}}{c_m^{\gamma_m}} - \frac{\cos\theta_{m+1}^{\gamma_{m+1}}}{c_{m+1}^{\gamma_{m+1}}} \right) \left[\left[\mathbf{G}^C \right]^T \right]^{-1} \right\} \left[\left[\tilde{\mathbf{G}}^C \right]^{-1} \mathbf{G}^C \right]^T \qquad (9.105)$$

which is also of the form

$$\mathbf{M}_{m+1}^{\gamma_{m+1}}(Q_m) = \left[\mathbf{D}_m^t \mathbf{M}_m^{\gamma_m}(Q_m) + \mathbf{C}_m^t \right] \left[\mathbf{A}_m^t + \mathbf{B}_m^t \mathbf{M}_m^{\gamma_m}(Q_m) \right]^{-1}, \qquad (9.106)$$

where the transmission matrices $\mathbf{A}_m^t, \mathbf{C}_m^t$ are

$$\mathbf{A}_m^t = \left[\tilde{\mathbf{G}}^C \right]^T \left[\left[\mathbf{G}^C \right]^T \right]^{-1}$$
$$= \begin{bmatrix} \cos\lambda & -\sin\lambda \\ \sin\lambda & \cos\lambda \end{bmatrix} \begin{bmatrix} \dfrac{\cos\theta_{m+1}^{\gamma_{m+1}}}{\cos\theta_m^{\gamma_m}} & 0 \\ 0 & 1 \end{bmatrix} \begin{bmatrix} \cos\lambda & \sin\lambda \\ -\sin\lambda & \cos\lambda \end{bmatrix}, \qquad (9.107a)$$

and

$$\mathbf{C}_m^t = \left(\frac{\cos\theta_m^{\gamma_m}}{c_m^{\gamma_m}} - \frac{\cos\theta_{m+1}^{\gamma_{m+1}}}{c_{m+1}^{\gamma_{m+1}}} \right) \left[\tilde{\mathbf{G}}^C \right]^{-1} \mathbf{h}(Q_m) \left[\left[\mathbf{G}^C \right]^T \right]^{-1}$$
$$= \left(\frac{\cos\theta_m^{\gamma_m}}{c_m^{\gamma_m}} - \frac{\cos\theta_{m+1}^{\gamma_{m+1}}}{c_{m+1}^{\gamma_{m+1}}} \right) \begin{bmatrix} \cos\lambda & -\sin\lambda \\ \sin\lambda & \cos\lambda \end{bmatrix}$$
$$\cdot \begin{bmatrix} \dfrac{h_{11}}{\cos\theta_{m+1}^{\gamma_{m+1}} \cos\theta_m^{\gamma_m}} & \dfrac{h_{12}}{\cos\theta_{m+1}^{\gamma_{m+1}}} \\ \dfrac{h_{21}}{\cos\theta_m^{\gamma_m}} & h_{22} \end{bmatrix} \begin{bmatrix} \cos\lambda & \sin\lambda \\ -\sin\lambda & \cos\lambda \end{bmatrix} \qquad (9.107b)$$

and the $\mathbf{B}'_m$, $\mathbf{D}'_m$ matrices are

$$\mathbf{B}'_m = \mathbf{O} \tag{9.107c}$$

$$\mathbf{D}'_m = \left[\tilde{\mathbf{G}}^C\right]^{-1}\mathbf{G}^C = \begin{bmatrix} \cos\lambda & -\sin\lambda \\ \sin\lambda & \cos\lambda \end{bmatrix}$$
$$\cdot \begin{bmatrix} \dfrac{\cos\theta_m^{\gamma_m}}{\cos\theta_{m+1}^{\gamma_{m+1}}} & 0 \\ 0 & 1 \end{bmatrix} \begin{bmatrix} \cos\lambda & \sin\lambda \\ -\sin\lambda & \cos\lambda \end{bmatrix}. \tag{9.107d}$$

The t superscript indicates these matrices are transmission matrices. They are also functions of the wave types γ_{m+1}, γ_m but again for notational simplicity we will not show that dependency explicitly. Note that the equivalent A, B, C, D matrices for a reflected wave at the m-th interface can be obtained by replacing $c_{m+1}^{\gamma_{m+1}}$ by $c_m^{\gamma_{m+1}}$ and $\cos\theta_{m+1}^{\gamma_{m+1}}$ by $-\cos\theta_m^{\gamma_{m+1}}$ in all the matrices of Eqs. (9.107a - d).

We have shown that both the propagation and transmission laws for a Gaussian beam can be represented in identical forms in terms $\mathbf{A}, \mathbf{B}, \mathbf{C}$, and $\mathbf{D}$ matrices. This representation is important since if we propagate a Gaussian beam in medium m from $s_m = 0$ (where the $\mathbf{M}$ matrix is $\mathbf{M}_m^{\gamma_m}(0)$) over a distance s_m to interface m (where the $\mathbf{M}$ matrix is $\mathbf{M}_m^{\gamma_m}(Q_m)$) and then transmit that beam across interface m to obtain $\mathbf{M}_{m+1}^{\gamma_{m+1}}(Q_m)$, the relationship between the $\mathbf{M}_{m+1}^{\gamma_{m+1}}(Q_m)$ and $\mathbf{M}_m^{\gamma_m}(0)$ after both types of interactions can also be written as

$$\mathbf{M}_{m+1}^{\gamma_{m+1}}(Q_m) = \left[\mathbf{DM}_m^{\gamma_m}(0) + \mathbf{C}\right]\left[\mathbf{A} + \mathbf{BM}_m^{\gamma_m}(0)\right]^{-1}, \tag{9.108}$$

where the $\mathbf{A}, \mathbf{B}, \mathbf{C}$, and $\mathbf{D}$ matrices in Eq. (9.108) are given by matrix products of the propagation and transmission matrices in the 4x4 matrix form

$$\begin{bmatrix} \mathbf{A} & \mathbf{B} \\ \mathbf{C} & \mathbf{D} \end{bmatrix} = \begin{bmatrix} \mathbf{A}'_m & \mathbf{B}'_m \\ \mathbf{C}'_m & \mathbf{D}'_m \end{bmatrix}\begin{bmatrix} \mathbf{A}^d_m & \mathbf{B}^d_m \\ \mathbf{C}^d_m & \mathbf{D}^d_m \end{bmatrix} \tag{9.109}$$

This result can be obtained directly by placing Eq. (9.103) into Eq. (9.106) and rearranging the result in the form of Eq. (9.108).

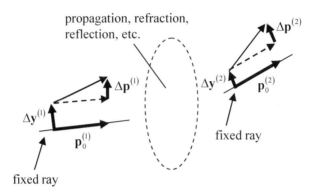

Fig. 9.13. Paraxial rays before and after a given wave process such as propagation, refraction, reflection, etc.

Obviously, this process can be continued for any additional materials and interfaces present. For example, in going from medium 1 to medium $M+1$ through M interfaces we can relate the $\mathbf{M}$ matrix in the final material at a distance s_{M+1} from the Mth interface directly to the starting $\mathbf{M}$ matrix values in medium 1 in terms of "global" matrices $\mathbf{A}^G, \mathbf{B}^G, \mathbf{C}^G, \mathbf{D}^G$ as

$$\mathbf{M}_{M+1}^{\gamma_{M+1}}\left(s_{M+1}\right) = \left[\mathbf{D}^G \mathbf{M}_1^{\gamma_1}\left(0\right) + \mathbf{C}^G\right]\left[\mathbf{A}^G + \mathbf{B}^G \mathbf{M}_1^{\gamma_1}\left(0\right)\right]^{-1}, \tag{9.110}$$

where $\mathbf{A}^G, \mathbf{B}^G, \mathbf{C}^G, \mathbf{D}^G$ are given by products of all the contributing propagation and transmission matrices, i.e.

$$\begin{bmatrix} \mathbf{A}^G & \mathbf{B}^G \\ \mathbf{C}^G & \mathbf{D}^G \end{bmatrix} = \begin{bmatrix} \mathbf{A}_{M+1}^d & \mathbf{B}_{M+1}^d \\ \mathbf{C}_{M+1}^d & \mathbf{D}_{M+1}^d \end{bmatrix} \begin{bmatrix} \mathbf{A}_M^t & \mathbf{B}_M^t \\ \mathbf{C}_M^t & \mathbf{D}_M^t \end{bmatrix}$$
$$\cdot \begin{bmatrix} \mathbf{A}_M^d & \mathbf{B}_M^d \\ \mathbf{C}_M^d & \mathbf{D}_M^d \end{bmatrix} \cdots \begin{bmatrix} \mathbf{A}_1^d & \mathbf{B}_1^d \\ \mathbf{C}_1^d & \mathbf{D}_1^d \end{bmatrix}. \tag{9.111}$$

Thus, all the $\mathbf{M}$ matrices appearing in either Eq. (9.100a) or Eq. (9.100b) can be obtained via the appropriate matrix multiplications of the type shown in Eq. (9.111). To compute the Gaussian beam in the final medium we need (1) the propagation and transmission/reflection $\mathbf{A}$, $\mathbf{B}$, $\mathbf{C}$, $\mathbf{D}$ matrices for a specified set of wave types and wave paths, (2) the plane wave transmission/reflection coefficients that allow us to compute the $\mathbf{T}_m^{\gamma_{m+1};\gamma_m}$ matrices, and (3) the velocity, $V_1^{\gamma_1}\left(0\right)\mathbf{d}_i^{\gamma_1}$, and phase matrix, $\mathbf{M}_1^{\gamma_1}\left(0\right)$, of the Gaussian beam at the starting point in the first medium.

The **A**, **B**, **C**, **D** matrices for propagation, transmission, and reflection appear in Eqs. (9.103) and (9.106) in exactly the same form because these equations are the consequence of some fundamental paraxial ray theory relations. To see this, consider a wave front moving in space defined at a given time by the function $T(\mathbf{x}) = \text{constant}$. As a simple example a plane wave traveling in the **e**-direction with wave speed c has the wave front $T(\mathbf{x}) = T_0 + \mathbf{e} \cdot \mathbf{x} / c = T_0 + \mathbf{p} \cdot \mathbf{x}$, with T_0 a constant and $\mathbf{p} \equiv \mathbf{e} / c$ the slowness vector. For a more general curved wave front we define the components of the slowness vector by $p_i = \partial T / \partial x_i$. We also define the curvatures of the wave front as the second derivatives, $\hat{M}_{ij} \equiv \partial^2 T / \partial x_i \partial x_j$ or, equivalently, $\hat{M}_{ij} = \partial p_i / \partial x_j$. In a homogeneous, isotropic medium the wave propagation rays are just straight lines along the slowness vector so we can examine a given fixed ray and some general process such as propagation, refraction, reflection, etc. as shown in Fig. 9.13. After such a process the slowness vector may be changed from $\mathbf{p}_0^{(1)}$ to $\mathbf{p}_0^{(2)}$. On a nearby (paraxial) ray, defined by its displacement vector $\Delta \mathbf{y}$ relative to our fixed ray, both the displacement and slowness will change during the process under consideration from $\Delta \mathbf{y}^{(1)}$ to $\Delta \mathbf{y}^{(2)}$ and $\mathbf{p}_0^{(1)} + \Delta \mathbf{p}^{(1)}$ to $\mathbf{p}_0^{(2)} + \Delta \mathbf{p}^{(2)}$, respectively. Since we are considering small deviations in going from the fixed ray to the nearby paraxial ray, we expect that these changes are linearly related to one another, i.e.

$$\begin{Bmatrix} \Delta \mathbf{y}^{(2)} \\ \Delta \mathbf{p}^{(2)} \end{Bmatrix} = \begin{bmatrix} \mathbf{A} & \mathbf{B} \\ \mathbf{C} & \mathbf{D} \end{bmatrix} \begin{Bmatrix} \Delta \mathbf{y}^{(1)} \\ \Delta \mathbf{p}^{(1)} \end{Bmatrix}, \tag{9.112}$$

where **A**, **B**, **C**, **D** are the "proportionality constants". To first order we also relate the changes in the slowness vectors to changes in the displacement vectors through wave front curvatures **M**, i.e.

$$\begin{aligned} \Delta \mathbf{p}^{(1)} &= \mathbf{M}_1 \Delta \mathbf{y}^{(1)} \\ \Delta \mathbf{p}^{(2)} &= \mathbf{M}_2 \Delta \mathbf{y}^{(2)}. \end{aligned} \tag{9.113}$$

If we use the expression for $\Delta \mathbf{p}^{(1)}$ in Eq. (9.113) in the expression for $\Delta \mathbf{y}^{(2)}$ in Eq.(9.112) we obtain

$$\Delta \mathbf{y}^{(2)} = \mathbf{A} \Delta \mathbf{y}^{(1)} + \mathbf{B} \mathbf{M}_1 \Delta \mathbf{y}^{(1)}. \tag{9.114}$$

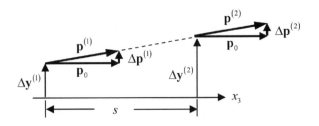

Fig. 9.14. Propagation of a paraxial ray.

Similarly, placing $\Delta\mathbf{p}^{(2)}$ from Eq. (9.113) into the expression for $\Delta\mathbf{p}^{(2)}$ in Eq. (9.112) gives

$$\mathbf{M}_2\Delta\mathbf{y}^{(2)} = \mathbf{C}\Delta\mathbf{y}^{(1)} + \mathbf{DM}_1\Delta\mathbf{y}^{(1)} \tag{9.115}$$

so that combining Eqs. (9.114) and (9.115) yields

$$\mathbf{M}_2\left(\mathbf{A} + \mathbf{BM}_1\right)\Delta\mathbf{y}^{(1)} = \left(\mathbf{C} + \mathbf{DM}_1\right)\Delta\mathbf{y}^{(1)}. \tag{9.116}$$

Since Eq. (9.116) must be true for all $\Delta y^{(1)}$ in the neighborhood of the fixed ray, we find

$$\mathbf{M}_2 = \left[\mathbf{DM}_1 + \mathbf{C}\right]\left[\mathbf{A} + \mathbf{BM}_1\right]^{-1}, \tag{9.117}$$

which has the same structure of Eqs. (9.103) and (9.106). Thus, our Gaussian beam relations can be thought of as the extension of ordinary paraxial ray theory relations for a real wave front curvature matrix $\mathbf{M}$ to a complex-valued $\mathbf{M}$ matrix that defines a Gaussian beam and we can view our $\mathbf{A}$, $\mathbf{B}$, $\mathbf{C}$, $\mathbf{D}$ matrices as the terms defining the paraxial changes of the ray parameters in Eq. (9.112). For example, consider propagation through a distance s at a wave speed c along a paraxial ray which is near a fixed ray along the x_3-axis as shown in Fig. 9.14. It is easy to see that $\Delta\mathbf{p}^{(2)} = \Delta\mathbf{p}^{(1)}$ and $\Delta\mathbf{y}^{(2)} = \Delta\mathbf{y}^{(1)} + sc\,\Delta\mathbf{p}^{(1)}$, where we have $\Delta\mathbf{p}^{(1)} = \left(\Delta p_1^{(1)}, \Delta p_2^{(1)}\right)$ and $\Delta\mathbf{p}^{(2)} = \left(\Delta p_1^{(2)}, \Delta p_2^{(2)}\right)$, and that this leads directly to

$$\begin{Bmatrix} \Delta\mathbf{y}_2 \\ \Delta\mathbf{p}_2 \end{Bmatrix} = \begin{bmatrix} \mathbf{I} & (sc)\mathbf{I} \\ 0 & \mathbf{I} \end{bmatrix} \begin{Bmatrix} \Delta\mathbf{y}_1 \\ \Delta\mathbf{p}_1 \end{Bmatrix}, \tag{9.118}$$

which corresponds to the propagation $\mathbf{A}$, $\mathbf{B}$, $\mathbf{C}$, $\mathbf{D}$ matrices defined by Eq. (9.104). Another more mathematical way to view the **ABCD** parameters is to recognize them as components of a *propagator matrix*. See Cerveny [9.2], who defines such propagator matrices and discusses their properties in detail.

Equations (9.108) and (9.110) can also lead to a further simplification of the amplitude terms in Eq. (9.100b), which can be rewritten in terms of propagation **A**, **B** matrices as

$$
\mathbf{V}_{M+1}^{\alpha;(y)} = \frac{1}{\sqrt{\det\left[\mathbf{A}_{M+1}^{d} + \mathbf{B}_{M=1}^{d}\mathbf{M}_{M+1}^{\alpha}(0)\right]}}
$$

$$
\cdot \left[\prod_{m=M}^{1} \frac{\mathbf{T}_{m}^{\gamma_{m+1};\gamma_{m}}}{\sqrt{\det\left[\mathbf{A}_{m}^{d} + \mathbf{B}_{m}^{d}\mathbf{M}_{m}^{\gamma_{m}}(0)\right]}}\right] \tag{9.119}
$$

$$
\cdot V_{1}^{\gamma_{1}}(0)\mathbf{d}_{1}^{\gamma_{1}} \exp\left[i\omega\sum_{m=1}^{M+1} \frac{s_{m}}{c_{m}^{\gamma_{m}}} + i\frac{\omega}{2}\mathbf{y}^{T}\hat{\mathbf{M}}_{M+1}^{\alpha}(s_{M+1})\mathbf{y}\right].
$$

To reduce this equation, we first use Eq. (9.106) and the fact that $\mathbf{B}_{m}' = \mathbf{0}$ to show directly that

$$
\left[\mathbf{A}_{m+1}^{d} + \mathbf{B}_{m+1}^{d}\mathbf{M}_{m+1}^{\gamma_{m+1}}(0)\right] = \left[\mathbf{A}' + \mathbf{B}'\mathbf{M}_{m}^{\gamma_{m}}(s_{m})\right]\left[\mathbf{A}_{m}'\right]^{-1}, \tag{9.120}
$$

where

$$
\begin{bmatrix}\mathbf{A}' & \mathbf{B}' \\ \mathbf{C}' & \mathbf{D}'\end{bmatrix} = \begin{bmatrix}\mathbf{A}_{m+1}^{d} & \mathbf{B}_{m+1}^{d} \\ \mathbf{C}_{m+1}^{d} & \mathbf{D}_{m+1}^{d}\end{bmatrix}\begin{bmatrix}\mathbf{A}_{m}^{t} & \mathbf{B}_{m}' \\ \mathbf{C}_{m}^{t} & \mathbf{D}_{m}'\end{bmatrix}. \tag{9.121}
$$

Also, using Eq. (9.103) it follows that

$$
\left[\mathbf{A}' + \mathbf{B}'\mathbf{M}_{m}^{\gamma_{m}}(s_{m})\right] = \left[\mathbf{A}^{G} + \mathbf{B}^{G}\mathbf{M}_{m}^{\gamma_{m}}(0)\right]\left[\mathbf{A}_{m}^{d} + \mathbf{B}_{m}^{d}\mathbf{M}_{m}^{\gamma_{m}}(0)\right]^{-1}, \tag{9.122}
$$

where $\mathbf{A}^{G}, \mathbf{B}^{G}$ are global matrices that combine the effects of propagation in media m and $m+1$ and transmission across the m-th interface, i.e.

$$
\begin{bmatrix}\mathbf{A}^{G} & \mathbf{B}^{G} \\ \mathbf{C}^{G} & \mathbf{D}^{G}\end{bmatrix} = \begin{bmatrix}\mathbf{A}_{m+1}^{d} & \mathbf{B}_{m+1}^{d} \\ \mathbf{C}_{m+1}^{d} & \mathbf{D}_{m+1}^{d}\end{bmatrix}\begin{bmatrix}\mathbf{A}_{m}^{t} & \mathbf{B}_{m}' \\ \mathbf{C}_{m}^{t} & \mathbf{D}_{m}'\end{bmatrix}\begin{bmatrix}\mathbf{A}_{m}^{d} & \mathbf{B}_{m}^{d} \\ \mathbf{C}_{m}^{d} & \mathbf{D}_{m}^{d}\end{bmatrix}. \tag{9.123}
$$

Then from Eq. (9.120) and Eq. (9.122) we obtain

$$
\left[\mathbf{A}_{m+1}^{d} + \mathbf{B}_{m+1}^{d}\mathbf{M}_{m+1}^{\gamma_{m+1}}(0)\right]
$$

$$
= \left[\mathbf{A}^{G} + \mathbf{B}^{G}\mathbf{M}_{m}^{\gamma_{m}}(0)\right]\left[\mathbf{A}_{m}^{d} + \mathbf{B}_{m}^{d}\mathbf{M}_{m}^{\gamma_{m}}(0)\right]^{-1}\left[\mathbf{A}_{m}^{t}\right]^{-1}. \tag{9.124}
$$

Now, examine two successive square root terms in Eq. (9.119) for medium m and $m+1$, i.e.

$$I = \frac{1}{\sqrt{\det\left[\mathbf{A}^d_{m+1} + \mathbf{B}^d_{m+1}\mathbf{M}^{\gamma_{m+1}}_{m+1}(0)\right]}} \frac{1}{\sqrt{\det\left[\mathbf{A}^d_{m} + \mathbf{B}^d_{m}\mathbf{M}^{\gamma_m}_{m}(0)\right]}}. \tag{9.125}$$

Placing Eq. (9.124) into Eq. (9.125) we find

$$I = \frac{\sqrt{\det\left[\mathbf{A}'_m\right]}}{\sqrt{\det\left[\mathbf{A}^G + \mathbf{B}^G\mathbf{M}^{\gamma_m}_m(0)\right]}}. \tag{9.126}$$

Since this same process can be repeated for all the other pairs of amplitude terms in Eq. (9.119), that equation reduces to

$$\mathbf{v}^{\alpha;(y)}_{M+1} = \frac{\breve{\mathbf{T}}_M}{\sqrt{\det\left[\mathbf{A}^G + \mathbf{B}^G\mathbf{M}^{\gamma_1}_1(0)\right]}} V^{\gamma_1}_1(0)\mathbf{d}^{\gamma_1}_1$$
$$\cdot \exp\left[i\omega\sum_{m=1}^{M+1}\frac{S_m}{c^{\gamma_m}_m} + i\frac{\omega}{2}\mathbf{y}^T\hat{\mathbf{M}}^\alpha_{M+1}\left(s_{M+1}\right)\mathbf{y}\right], \tag{9.127}$$

where $\mathbf{A}^G, \mathbf{B}^G$ are now the global matrices going from medium 1 to medium $M+1$ and

$$\breve{\mathbf{T}}_M = \prod_{m=M}^{1}\mathbf{T}^{\gamma_{m+1};\gamma_m}_m \sqrt{\det\left[\mathbf{A}'_m\right]}. \tag{9.128}$$

Equation (9.127) is in the form identical in structure to that of a Gaussian beam propagating in a single medium. Thus, use of the **ABCD** matrices can simplify our multiple media problems to an equivalent single medium expression. However, to use Eq. (9.127) one must be able to correctly evaluate the square root of the amplitude in that equation and at present we do not have a direct way to do that evaluation. The difficulty lies in that $\mathbf{A}^G, \mathbf{B}^G$ are no longer positive, real, diagonal matrices as they are for a single medium, so that the signs of the imaginary parts of the eigenvalues of $\mathbf{M}^{\gamma_1}_1$ are affected in a manner that is difficult to explicitly define. Thus, while Eq. (9.127) is in the most compact form possible either Eq. (9.100a) or Eq. (9.100b) appear to be needed in actual calculations. Of course the **ABCD** matrices can be conveniently used to obtain all the terms needed in those equations.

9.5 Multi-Gaussian Transducer Beam Modeling

We have seen in the previous sections how to analytically determine a Gaussian beam after it has propagated in multiple media and interacted with multiple interfaces. In NDE applications the value of those results would be limited if Gaussian beams were the only types of wave fields that we could consider since most ultrasonic transducers do not generate Gaussian-shaped beams. Instead, we would like to be able to model the wave fields from piston transducers. This is possible since in a seminal 1988 paper Wen and Breazeale showed that one can synthesize the sound beam from a circular piston transducer radiating into water using the superposition of as few as ten Gaussian beams [9.8]. On the face of a transducer of radius a located on the plane $x_3 = 0$ (and radiating into the region $x_3 > 0$) they let the normalized velocity field be given by a sum of Gaussians in the form

$$\frac{\mathbf{v}^P(x_1, x_2, 0, \omega)}{v_0(\omega)} = \sum_{r=1}^{10} A_r \mathbf{d}^P \exp\left(-B_r \rho^2 / a^2\right), \tag{9.129}$$

where $\rho^2 = x_1^2 + x_2^2$, $v_0(\omega)$ is the constant velocity on the transducer surface, and $\mathbf{d}^P = \mathbf{e}_3 = (0,0,1)$. These ten Gaussians will generate ten Gaussian beams having starting values of $\left[V_1^P(0)\right]_r, \left[\mathbf{M}_1^P(0)\right]_r$ $(r = 1,...10)$ given by

$$\left[V_1^P(0)\right]_r = A_r v_0(\omega)$$

$$\left[\mathbf{M}_1^P(0)\right]_r = \begin{bmatrix} \dfrac{iB_r}{c_{p1}D_R} & 0 \\ 0 & \dfrac{iB_r}{c_{p1}D_R} \end{bmatrix}, \tag{9.130}$$

where $D_R = k_{p1}a^2 / 2$ is the *Rayleigh distance*. A_r, B_r are complex-valued expansion coefficients that need to be determined to match the velocity field on the face of the transducer. For a circular planar piston transducer of radius a, the normalized velocity field in the x_3-direction, v_3 / v_0, is given by the *circ function*, where

$$\frac{v_3\left(x_1,x_2,0,\omega\right)}{v_0\left(\omega\right)} = \begin{cases} 1 & \rho^2/a^2 < 1 \\ 0 & otherwise \end{cases} \tag{9.131}$$

$$\equiv circ\left(\rho^2/a^2\right).$$

To obtain the A_r, B_r coefficients, Wen and Breazeale minimized an objective function, J, given by

$$J\left(A_r, B_r\right) = \int_0^\infty \left[circ\left(\rho^2/a^2\right) - \sum_{r=1}^{10} A_r \exp\left(-B_r \rho^2/a^2\right) \right]^2 d\rho \tag{9.132}$$

and they published the ten coefficients that they obtained. Although this is a non-linear optimization problem that is rather computationally intensive, once the A_r, B_r coefficients are calculated, they can be stored in a look-up table and used to synthesize the wave field in very complex problems simply by adding up the ten contributing Gaussians. Thus, we can use Eq. (9.100a) to write down the wave field from a piston transducer, after multiple propagations and interface interactions, in the form of a multi-Gaussian beam model, where

$$\mathbf{v}_{M+1}^{\alpha;(y)} = \sum_{r=1}^{10} \frac{\sqrt{\det\left[\mathbf{M}_{M+1}^{\alpha}\left(s_{M+1}\right)\right]_r}}{\sqrt{\det\left[\mathbf{M}_{M+1}^{\alpha}\left(0\right)\right]_r}} \left[\prod_{m=M}^{1} \mathbf{T}_m^{\gamma_{m+1};\gamma_m} \frac{\sqrt{\det\left[\mathbf{M}_m^{\gamma_m}\left(s_m\right)\right]_r}}{\sqrt{\det\left[\mathbf{M}_m^{\gamma_m}\left(0\right)\right]_r}} \right]$$

$$\cdot \left[V_1^{\gamma_1}\left(0\right)\right]_r \mathbf{d}_1^{\gamma_1} \exp\left[i\omega \sum_{m=1}^{M+1} \frac{s_m}{c_m^{\gamma_m}} + i\frac{\omega}{2}\mathbf{y}^T\left[\hat{\mathbf{M}}_{M+1}^{\alpha}\left(s_{M+1}\right)\right]_r \mathbf{y} \right]. \tag{9.133}$$

The success of this approach, of course, relies on how well the ten coefficients used here do represent a piston transducer wave field. Tests of the ten Wen and Breazeale coefficients show that they do a remarkably effective job of reproducing the piston transducer wave field to within the limits of the paraxial approximation. This means that they are accurate at distances of approximately one transducer diameter or greater from the transducer face. This can be easily tested by examining the normalized pressure wave field of the multi-Gaussian beam model for a single fluid medium given by

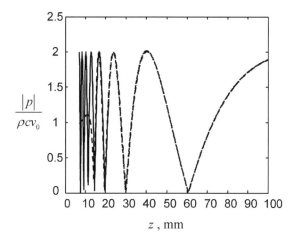

z , mm

Fig. 9.15. The magnitude of the normalized on-axis pressure for a 6 mm radius piston transducer radiating into water at 5 MHz modeled by the Rayleigh-Sommerfeld integral (solid line) and by a superposition of ten Gaussian beams which are defined by the ten coefficients of Wen and Breazeale.

$$\frac{p(x_1,x_2,x_3,\omega)}{\rho_1 c_{p1} v_0(\omega)} = \sum_{r=1}^{10} \frac{A_r}{1+iB_r x_3/D_R} \exp(ik_{p1}x_3)$$

$$\cdot \exp\left[i\omega\left(\frac{1}{2}\mathbf{x}^T\left[\mathbf{M}_1^p(x_3)\right]_r \mathbf{x}\right)\right], \tag{9.134}$$

where

$$\left[\mathbf{M}_1^p(x_3)\right]_r = \begin{bmatrix} \dfrac{iB_r/c_{p1}D_R}{1+iB_r x_3/D_R} & 0 \\[4mm] 0 & \dfrac{iB_r/c_{p1}D_R}{1+iB_r x_3/D_R} \end{bmatrix}. \tag{9.135}$$

Comparisons of Eq. (9.134) can be made with the exact solution obtained from the Rayleigh-Sommerfeld equation for any point in the transducer wave field but the results are very similar to comparisons done for on-axis wave fields, where we can write down an exact solution analytically (Eq. (9.8)). Figure 9.15 shows such an on-axis comparison for a 6 mm radius planar piston transducer radiating into water at 5 MHz. The multi-Gaussian beam model accurately models the on-axis near-field of the transducer down to approximately 15 mm from the transducer face.

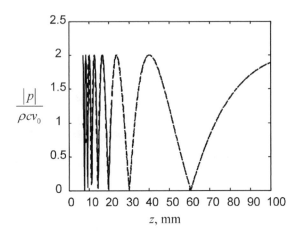

Fig. 9.16. The magnitude of the normalized on-axis pressure for a 6 mm radius piston transducer radiating into water at 5 MHz modeled by the Rayleigh-Sommerfeld integral (solid line) and by a superposition of fifteeen Gaussian beams defined by the fifteen "optimized" coefficients of Wen and Breazeale.

In a subsequent paper, Wen and Breazeale obtained even better results with a slightly larger number of optimized coefficients [9.9]. They used the normalized exact on-axis pressure, $\tilde{p}_{exact}$, defined as

$$\tilde{p}_{exact}\left(x_3,\omega\right)=\frac{p\left(0,0,x_3,\omega\right)}{\rho_1 c_{p1} v_0}=\exp\left(ik_{p1}x_3\right)-\exp\left(ik_{p1}\sqrt{a^2+x_3^2}\right) \quad (9.136)$$

and a 15 term multi-Gaussian beam model for this same wave field:

$$\tilde{p}_{MG}\left(x_3,\omega\right)=\sum_{r=1}^{15}\frac{A_r}{1+iB_r x_3/D_R}\exp\left(ik_{p1}x_3\right) \quad (9.137)$$

to define a modified objective function

$$J\left(A_r,B_r\right)=\int_0^\infty\left[circ\left(\rho^2/a^2\right)-\sum_{r=1}^{15}A_r\exp\left(-B_r\rho^2/a^2\right)\right]^2 d\rho$$
$$+\lambda_w\int_{z_1}^{z_2}\left[\left|\tilde{p}_{exact}\left(x_3,\omega\right)\right|-\left|\tilde{p}_{MG}\left(x_3,\omega\right)\right|\right]^2 dx_3 \quad (9.138)$$

where λ_w is a constant to weigh the on-axis matching conditions relative to the boundary matching conditions and z_1 and z_2 are near field limit values that define the range where matching to the on-axis field is to take

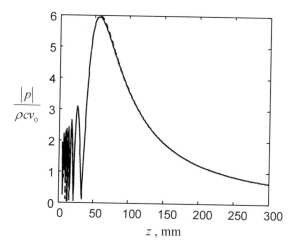

Fig. 9.17. The magnitude of the on-axis pressure for a 6 mm radius, 76 mm focal length spherically focused transducer radiating into water at 5MHz. Solid line – exact O'Neil theory, dashed line – multi-Gaussian beam model.

place. Fig. 9.16 shows the results of using these 15 optimized coefficients to calculate the on-axis field where now the multi-Gaussian beam model is accurate at distances from the transducer of approximately 10 mm or greater, matching the exact on-axis behavior for two additional near-field oscillations. These fifteen optimized coefficients are listed both in [9.9] and [9.10] and generated by a MATLAB function gauss_c15 given in Chapter 12 (Code Listing 12.2).

One of the nice properties of this multi-Gaussian beam model is that one can also model focused transducers by a simple modification of the B_r coefficient. As shown in Chapter 8, focusing in the paraxial approximation can be modeled by including a complex exponential term with a quadratic spatial variation for the velocity field over the face of a planar transducer. For a spherically focused circular transducer of radius a and geometrical focal length, F, this corresponds to specifying the velocity field at the transducer as

$$\frac{v_3(x_1,x_2,0,\omega)}{v_0(\omega)} = circ(\rho^2/a^2)\exp(-ik_{p1}\rho^2/2F).$$ (9.139)

Since we can view our multi-Gaussian beam model as approximating this *circ* function in the form

$$circ\left(\rho^2 / a^2\right) = \sum_{r=1}^{10} A_r \exp\left(-B_r \rho^2 / a^2\right) \qquad (9.140)$$

Eq. (9.139) shows that to include the effects of spherical focusing we need only to modify the B_r coefficients for the circular planar transducer case by making the replacement

$$B_r \rightarrow B_r + \frac{ik_{p1}a^2}{2F} \qquad (9.141)$$

Figure 9.17 shows the on-axis wave field predicted by a multi-Gaussian beam model obtained in this fashion for a 6 mm radius, 76 mm focal length transducer radiating into water at 5 MHz and the corresponding on-axis field obtained from the O'Neil model (Eq. (8.33)). It can be seen that there is very little discernable difference between the two results.

Recently, Ding et al. [9.11] made a clever use of the *circ* function and the ten coefficients of Wen and Breazeale to model rectangular piston transducers. For a rectangular piston transducer with sides of lengths $\left(2a_1, 2a_2\right)$ in the $\left(x_1, x_2\right)$ directions, respectively, the normalized velocity field on the transducer face is given by

$$\frac{v_3\left(x_1, x_2, 0, \omega\right)}{v_0\left(\omega\right)} = \begin{cases} 1 & \left|x_1 / a_1\right| < 1, \left|x_2 / a_2\right| < 1 \\ 0 & otherwise \end{cases} \qquad (9.142)$$
$$= circ\left(x_1^2 / a_1^2\right)circ\left(x_2^2 / a_2^2\right)$$

so if we use Eq. (9.140) in product form we have

$$circ\left(x_1^2 / a_1^2\right)circ\left(x_2^2 / a_2^2\right)$$
$$= \sum_{r=1}^{10}\sum_{q=1}^{10} A_r A_q \exp\left[i\omega\left(\frac{1}{2}\mathbf{x}^T\left[\mathbf{M}_1^p\left(0\right)\right]_{rq}\mathbf{x}\right)\right] \qquad (9.143)$$
$$= \sum_{r=1}^{10}\sum_{q=1}^{10} A_r A_q \exp\left[-B_r x_1^2 / a_1^2 - B_q x_2^2 / a_2^2\right],$$

where $\mathbf{x}^T = \left(x_1, x_2\right)$ and

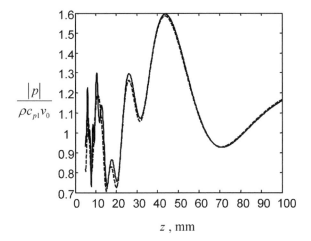

Fig. 9.18. Magnitude of the on-axis pressure for a 12x6 mm rectangular piston transducer radiating into water at 5 MHz using the Rayleigh-Sommerfeld integral (solid line) and a multi-Gaussian beam model based on the ten coefficients of Wen and Breazeale.

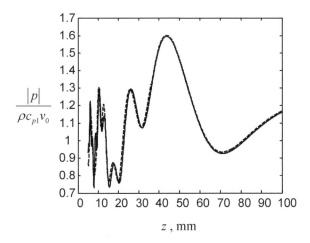

Fig. 9.19. Magnitude of the on-axis pressure for a 12x6 mm rectangular piston transducer radiating into water at 5 MHz using the Rayleigh-Sommerfeld integral (solid line) and a multi-Gaussian beam model based on the fifteen "optimized" coefficients of Wen and Breazeale.

$$\left[\mathbf{M}_1^p(0) \right]_{rq} = \begin{bmatrix} \dfrac{iB_r}{c_{p1}D_{R1}} & 0 \\ 0 & \dfrac{iB_q}{c_{p1}D_{R2}} \end{bmatrix} \tag{9.144}$$

and $D_{R1} = k_{p1}a_1^2/2, D_{R2} = k_{p1}a_2^2/2$.

Using these results we can write a multi-Gaussian beam model for a rectangular planar piston transducer radiating through multiple media and interacting with multiple interfaces as

$$\mathbf{v}_{M+1}^{\alpha;(y)} = \sum_{r=1}^{10}\sum_{q=1}^{10} \left[V_1^{\gamma_1}(0) \right]_{rq} \mathbf{d}_1^{\gamma_1} \frac{\sqrt{\det\left[\mathbf{M}_{M+1}^{\alpha}(s_{M+1}) \right]_{rq}}}{\sqrt{\det\left[\mathbf{M}_{M+1}^{\alpha}(0) \right]_{rq}}}$$

$$\cdot \left[\prod_{m=M}^{1} \mathbf{T}_m^{\gamma_{m+1};\gamma_m} \frac{\sqrt{\det\left[\mathbf{M}_m^{\gamma_m}(s_m) \right]_{rq}}}{\sqrt{\det\left[\mathbf{M}_m^{\gamma_m}(0) \right]_{rq}}} \right] \tag{9.145}$$

$$\cdot \exp\left[i\omega \sum_{m=1}^{M+1} \frac{s_m}{c_m^{\gamma_m}} + i\frac{\omega}{2}\mathbf{y}^T \left[\hat{\mathbf{M}}_{M+1}^{\alpha}(s_{M+1}) \right]_{rq} \mathbf{y} \right],$$

where $\left[V_1^{\gamma_1}(0) \right]_{rq} = A_r A_q v_0(\omega)$.

For a rectangular transducer radiating into a single fluid medium, the normalized on-axis pressure of this multi-Gaussian beam model is given by

$$\frac{p(0,0,x_3,\omega)}{\rho_1 c_{p1} v_0(\omega)} = \sum_{r=1}^{10}\sum_{q=1}^{10} \frac{A_r}{\sqrt{1+iB_r x_3/D_{R1}}} \frac{A_q}{\sqrt{1+iB_q x_3/D_{R2}}} \tag{9.146}$$

$$\cdot \exp\left(ik_{p1}x_3 \right).$$

This model has been compared to a highly accurate numerical integration of the Rayleigh-Sommerfeld equation for a 12x6 mm rectangular transducer $(a_1 = 6\,mm, a_2 = 3\,mm)$ radiating into water at 5 MHz as shown in Fig. 9.18. It can be seen from that figure that the multi-Gaussian beam model agrees well with the "exact" results at distances of 15 mm or greater from the transducer. Although the 15 optimized coefficients of Wen and Breazeale were specifically optimized for the circular transducer, it has been found that those coefficients also improve the results for rectangular transducers.

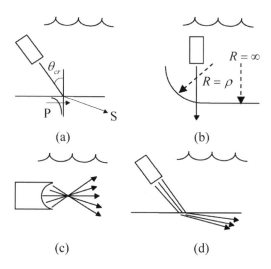

Fig. 9.20. Conditions under which the paraxial approximation can fail include: **(a)** transmission through an interface near a critical angle, **(b)** inspection through a surface with rapidly changing curvature, **(c)** for highly focused transducers, and **(d)** near grazing incidence to a surface.

For example, Figure 9.19 shows that using these 15 coefficients the multi-Gaussian beam model matched the exact results to within a distance less than 6 mm for the same transducer considered in Fig. 9.18.

As in the circular transducer case, it is easy to add focusing to the multi-Gaussian beam rectangular transducer model. For example, in the case of a bi-cylindrically focused rectangular transducer with geometrical focal lengths (F_1, F_2) in the (x_1, x_2) directions, respectively, one again merely has to modify the B_r, B_q coefficients by making the replacements

$$B_r \rightarrow B_r + \frac{ik_{p1}a_1^2}{2F_1}$$

$$B_q \rightarrow B_q + \frac{ik_{p1}a_2^2}{2F_2}.$$

(9.147)

It is possible to use the Wen and Breazeale coefficients to also model elliptical-shaped piston transducers [9.11]. There are also other fitting methods that can be used to obtain the coefficients such as the k-space method of Sha et al. [9.12]. However, the 15 optimized coefficients of

Wen and Breazeale generate about as accurate a wave field that is possible within the paraxial approximation [9.13].

Because the multi-Gaussian beam model relies on the paraxial approximation being valid, there are a number of testing situations where the model can degrade or fail. We have already seen that in the very near field the paraxial approximation will fail. Since most ultrasonic testing is not done under such very near field conditions, this limitation of the paraxial models may not be of practical importance. However, the breakdown of the paraxial approximation can also occur under other testing conditions. Figure 9.20 (a) shows the case when a transducer radiates through a planar fluid-solid interface at an angle near the first critical angle. In this case the waves reaching the point in the solid may be at very small angles relative to the central ray but the paraxial approximation can fail because, near a critical angle, the transmission coefficient that defines the amplitude of the waves in the solid varies rapidly for even small angular changes and such variations are neglected when paraxial beam models are used to treat transmission through interfaces (recall that we used the transmission coefficient along a central ray only in considering the interactions with an interface). Figure 9.20 (b) shows the inspection of a surface at the inter-section of a fillet and a plane surface. In this case the surface curvature, R, changes abruptly from $R = \infty$ to $R = \rho$ at the intersection and the paraxial approximation fails because of this rapid change of the surface curvature. Figure 9.20 (c) illustrates the case of a very tightly focused transducer. The paraxial approximation also fails in this case as the waves reaching the focus do not travel in approximately the same direction as required by that approximation. Finally, Fig. 9.20 (d) shows the case of the inspection through a plane interface at high angles or near grazing incidence on the interface. The paraxial approximation can fail in this case also since it cannot capture the strong beam distortions present at these high angles and there may be other waves besides bulk P-waves and S-waves (head waves, surface waves, etc.) present near the interface that are not considered by our beam model. Fortunately, many of these special testing situations are not encountered in practice so that the multi-Gaussian beam model is a fast and powerful tool and gives accurate results in many cases.

9.6 References

9.1   Siegman AE (1986) Lasers. University Science Books, Mill Valley, CA
9.2   Cerveny V (2001) Seismic ray theory. Cambridge University Press, Cambridge, United Kingdom

9.3 Jeffreys A (1995) Handbook of mathematical formulas and integrals. Academic Press, San Diego, CA, p 48

9.4 Heyman E, Felsen LB (2001) Gaussian beam and pulsed-beam dynamics: complex-source and complex-spectrum formulations within and beyond paraxial asymptotics. J. Opt. Soc. Am. 18: 1588-1611

9.5 Korn GA, Korn TM (1968) Mathematical handbook for scientists and engineers. McGraw-Hill Book Co., New York, NY

9.6 Cerveny V, Psencik I. (1983) Gaussian beams and paraxial ray approximation in three-dimensional inhomogeneous elastic media. Journal of Geophysics 53: 1-15

9.7 Goldsmith PF (1998) Quasioptical systems. IEEE Press, Piscataway, NJ

9.8 Wen JJ, Breazeale MA (1988) A diffraction beam field expressed as the superposition of Gaussian beams. J. Acoust. Soc. Am. 83: 1752-1756

9.9 Wen JJ, Breazeale MA (1990) Computer optimization of the Gaussian beam description of an ultrasonic field. In: Lee D, Cakmak A, Vichnevetsky R. (eds) Computational acoustics: scattering, Gaussian beams, and aeroacoustics, Vol. 2. Elsevier Science Publishers, Amsterdam, The Netherlands, pp 181-196

9.10 Huang D, Breazeale MA (1999) A Gaussian finite-element method for description of sound diffraction. J. Acoust. Soc. Am. 106: 1771-1781

9.11 Ding D, Zhang Y, Liu J (2003) Some extensions of the Gaussian beam expansion: radiation fields of rectangular and elliptical transducers. J. Acoust. Soc. Am. 113: 3043-3048

9.12 Sha K, Yang J, Gan WS (2003) A complex virtual source approach for calculating the diffraction beam field generated by a rectangular planar source. IEEE Trans. on Ultrasonics, Ferroelectrics, and Frequency Control 50: 890-896

9.13 Kim HJ, Schmerr LW, Sedov A (2006) Generation of the basis sets for multi-Gaussian ultrasonic beam models – an overview. J. Acoust. Soc. Am. 119: 1971-1978

9.7 Exercises

1. Equation (9.134) gives the normalized pressure of a circular planar piston transducer as computed by a multi-Gaussian beam model. Write a MATLAB function that takes as its inputs the frequency, f, the wave speed of the water (in m/sec), the radius, a, of the transducer (in mm), and the distances (x_1, x_2, x_3) (in mm) and computes this normalized pressure in the water. Use the fifteen Gaussian beam coefficients (instead of the ten terms indicated in Eq. (9.134)) which can be obtained from the MATLAB function gauss_c15. Verify that your function produces the magnitude of the on-axis pressure plot shown in Fig. 9.16 for a 6 mm radius transducer

radiating into water at 5 MHz. For this same transducer, plot the magnitude of the normalized pressure versus x_2 for $x_1 = 0, x_3 = 60$ mm.

2. Modify the MATLAB function written for the circular planar transducer of exercise 1 to model a spherically focused transducer of focal length, R, using the relationship of Eq. (9.141).Verify that your function produces the magnitude of the on-axis pressure plot for a 6 mm radius, 76 mm focal length focused transducer radiating into water at 5 MHz (Fig. 9.17). For this same transducer plot the magnitude of the cross-axis normalized pressure at the geometrical focal length $x_3 = 76$ mm and compare this pressure to the exact result given by Eq. (8.42).

3. The multi-Gaussian beam model for the normalized pressure of a planar rectangular transducer radiating into a fluid is given from Eq. (9.145) (for 15 Gaussian beams) by

$$\frac{p(x_1, x_2, x_3, \omega)}{\rho_1 c_{p1} v_0(\omega)} = \sum_{r=1}^{15} \sum_{q=1}^{15} \frac{A_r}{\sqrt{1 + iB_r x_3 / D_{R1}}} \frac{A_q}{\sqrt{1 + iB_q x_3 / D_{R2}}}$$

$$. \exp(ik_{p1} x_3) \exp\left[i\omega \left(\frac{1}{2} \mathbf{x}^T \left[\mathbf{M}_1^p(x_3) \right]_{rq} \mathbf{x} \right) \right],$$

where $\mathbf{x}^T = (x_1, x_2)$ and

$$\left[\mathbf{M}_1^p(x_3) \right]_{rq} = \begin{bmatrix} \dfrac{iB_r / c_{p1} D_{R1}}{1 + iB_r x_3 / D_{R1}} & 0 \\ 0 & \dfrac{iB_q / c_{p1} D_{R2}}{1 + iB_q x_3 / D_{R2}} \end{bmatrix}.$$

Write a MATLAB function that takes as its inputs the frequency, f, the wave speed of the water (in m/sec), the half-lengths (a_1, a_2) of the transducer (in mm), and the distances (x_1, x_2, x_3) (in mm) and computes this normalized pressure in the water. The Gaussian beam coefficients can be obtained from the MATLAB function gauss_c15. Verify that your function produces the magnitude of the on-axis pressure plot shown in Fig. 9.19 for a 12x6 mm transducer radiating into water at 5 MHz. Note that the half-lengths of the two sides are given here by $a_1 = 6$ mm, $a_2 = 3$ mm. For this same transducer, plot the magnitude of the normalized pressure versus distance x_2 for $x_1 = 0$, $x_3 = 20$ mm and versus distance x_1 for $x_2 = 0$, $x_3 = 20$ mm.

4. Modify the MATLAB function of exercise 3 to model a cylindrically focused rectangular transducer where the focusing is in the $x_1 - x_3$ plane, by using a relationship similar to Eq.(9.141), i.e.

$$B_r \to B_r + \frac{ik_{p1}a_1^2}{2F_1}$$

$$B_q \quad unchanged.$$

The Gaussian beam coefficients can again be obtained from the MATLAB function gauss_c15. Plot the magnitude of the on-axis pressure for a 12x6 mm transducer with cylindrical focal length $F_1 = 80$ mm radiating into water at 5 MHz.

5. Rewrite Eq. (9.134) for the normalized pressure wave field of a circular planar transducer radiating into a fluid (using 15 Gaussian beams) in terms of (x, y, z) coordinates as

$$\frac{p(x,y,z,\omega)}{\rho_1 c_{p1} v_0(\omega)} = \sum_{r=1}^{15} \frac{A_r}{1 + iB_r z / D_R}$$
$$\cdot \exp(ik_{p1}z) \exp\left[i\omega \left(\frac{1}{2} \mathbf{x}^T \left[\mathbf{M}_1^p(z) \right]_r \mathbf{x} \right) \right],$$

where $\mathbf{x}^T = (x, y)$ and

$$\left[\mathbf{M}_1^p(z) \right]_r = \begin{bmatrix} \dfrac{iB_r / c_{p1} D_R}{1 + iB_r z / D_R} & 0 \\ 0 & \dfrac{iB_r / c_{p1} D_R}{1 + iB_r z / D_R} \end{bmatrix}.$$

This expression can also be written as a quasi-plane wave in the form

$$p(x,y,z,\omega) = \rho_1 c_1 v_0 \exp(ik_{p1}z) C(a,x.y,z,\omega),$$

where C is a diffraction coefficient. We can use the paraxial approximation discussion of Chapter 8 (see section 8.5) and quickly obtain a multi-Gaussian beam model for a planar, circular P-wave transducer radiating at normal incidence to a fluid-solid interface by writing the normalized pressure in the solid in the quasi-plane wave form

$$\frac{p(x,y,z,\omega)}{\rho_1 c_1 v_0} = \exp\left(ik_{p1}z_1 + ik_{p2}z_2\right)C(a,x.y,\tilde{z},\omega),$$

where $\tilde{z} = z_1 + z_2 \dfrac{c_{p2}}{c_{p1}}$ and (z_1, z_2) are the distances traveled normal to the interface in the water and solid, respectively, as discussed in Chapter 8. Note that this normalized pressure is also the same as the normalized velocity v_z / v_0 (see Eq. (8.25)).

Use this result to write a MATLAB function that implements a multi-Gaussian beam model for a transducer radiating at normal incidence to a fluid-solid interface and evaluate and plot the magnitude of the normalized on-axis pressure for a 6.35 mm radius transducer radiating at 5 MHz through a water-aluminum interface where the water path distance $z_1 = 50.8$ mm, and the metal path distance, z_2, ranges from zero to 25.4 mm.

10 Flaw Scattering

Ultrasonic beam models can simulate the fields incident on a flaw in an ultrasonic inspection. Given those incident fields, we then must also determine the scattered waves produced by the interactions of those fields with the flaw. For complex flaw morphologies numerical methods are generally needed to solve for these scattered waves. For a number of simple flaw shapes and types, however, we can model some important characteristics of the flaw scattering process explicitly with approximate methods. In this Chapter we will describe two such approximations – the Kirchhoff approximation and the Born approximation – and also give a brief overview of a number of other flaw scattering methods.

10.1 The Far-Field Scattering Amplitude

To describe flaw scattering we will first consider the simple case shown in Fig. 10.1 where a plane wave in a fluid strikes an immersed object, generating scattered waves that travel from the "flaw" in all directions. At a distance of many wavelengths from the flaw, the flaw acts like a point source generating a spherical wave, as shown in Fig. 10.1. We can express the pressure in this spherical wave as

$$p^{scatt}\left(\mathbf{y},\omega\right) = p_0 A\left(\mathbf{e}_i;\mathbf{e}_s\right)\frac{\exp\left(ik_p r_s\right)}{r_s}, \tag{10.1}$$

where p_0 is the pressure amplitude of the incident wave, $A\left(\mathbf{e}_i;\mathbf{e}_s\right)$ is the *far-field scattering amplitude* of the flaw in the $\mathbf{e}_s$ direction due to an incident wave traveling in the $\mathbf{e}_i$ direction. The scattering amplitude is also a function of frequency but for economy of notation we will not show this frequency dependence explicitly. Note that we have implicitly assumed harmonic waves of $\exp\left(-i\omega t\right)$ time dependency, a factor that also will not be shown explicitly. The variable r_s is the distance from a

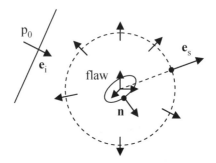

Fig. 10.1. The spherical P-wave scattered in the far-field from a "flaw" in a fluid due to an incident wave of pressure amplitude p_0.

fixed point at the flaw (usually taken to be the flaw "center") to the point in the fluid where the scattered pressure is being determined, and k_p is the wave number for compressional waves in the fluid.

It can be shown that the far-field scattering amplitude is related to the total fields (incident plus scattered fields) on the surface of the flaw through a surface integral given by [Fundamentals]:

$$A(\mathbf{e}_i;\mathbf{e}_s) = \frac{-1}{4\pi} \int_{S_f} \left[\frac{\partial \tilde{p}}{\partial n} + ik_p (\mathbf{e}_s \cdot \mathbf{n}) \tilde{p} \right] \exp\left(-ik_p \mathbf{x}_s \cdot \mathbf{e}_s\right) dS(\mathbf{x}_s), \qquad (10.2)$$

where $\mathbf{n}$ is the unit outwards normal to the surface of the flaw pointing into the fluid, $\mathbf{x}_s$ is a general point on the surface, S_f, and $\tilde{p} = p(\mathbf{x}_s, \omega)/p_0$ is the pressure normalized by the incident wave pressure amplitude. Note that the far-field scattering amplitude as defined here has a dimension of length. The unit vector $\mathbf{e}_i$ does not appear explicitly in Eq. (10.2) but the fields do depend on this direction so it is included as an argument of the scattering amplitude. Since $\partial p / \partial n = i\omega\rho v_n$ from the equation of motion for the fluid the scattering amplitude depends on both the pressure, p, and the normal component of the velocity, v_n, on the surface. It is possible to specify one of these variables. For example, for a void, we can set $p = 0$, while for a rigid, immobile scatterer we would have $v_n = 0$. For an elastic inclusion, we would have to instead specify conditions of continuity of the tractions and normal velocity at the surface. Given the incident waves and a set of boundary conditions of one of these types, it is then possible

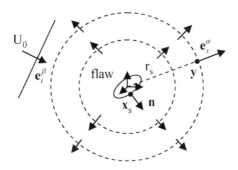

Fig. 10.2. The spherical P- and S-waves scattered in the far-field from a flaw in an elastic solid due to an incident wave of displacement amplitude U_0.

to formulate a boundary value problem and solve for the unknown fields on the surface of the scatterer [Fundamentals].

For ultrasonic NDE inspection problems the flaws of interest are located in an elastic solid. The scattering of elastic waves is more complex than the fluid case just considered, but again at a distance of many wavelengths from the flaw the scattered waves are just spherical waves, as shown in Fig. 10.2. In this case a flaw generates both scattered spherical P-waves and S-waves. The displacement of the solid produced by these scattered waves can be written as

$$\mathbf{u}^{scatt}(\mathbf{y},\omega) = U_0 \mathbf{A}\left(\mathbf{e}_i^\beta ;\mathbf{e}_s^p\right)\frac{\exp(ik_p r_s)}{r_s} + U_0 \mathbf{A}\left(\mathbf{e}_i^\beta ;\mathbf{e}_s^s\right)\frac{\exp(ik_s r_s)}{r_s}, \quad (10.3)$$

where U_0 is the displacement amplitude of the incident wave, $k_\alpha \, (\alpha = p,s)$ are the wave numbers for P- and S-waves, and $\mathbf{A}\left(\mathbf{e}_i^\beta ;\mathbf{e}_s^\alpha\right)$ is the *vector far-field scattering amplitude* for a scattered wave of type α $(\alpha = p,s)$ due to an incident wave of type β $(\beta = p,s)$. The vectors $\mathbf{e}_i^\beta$ and $\mathbf{e}_s^\alpha$ are unit vectors in the incident and scattered wave directions, respectively [Note: lower case p and s superscripts will be used here to denote P-waves and S-waves, respectively, while an s subscript will denote a "scattered" wave unit vector]. Far-field scattering amplitudes for both P-waves and S-waves can be written in terms of a single vector-valued function, $\mathbf{f}^{\alpha;\beta}$, where [Fundamentals]:

$$\mathbf{f}^{\alpha;\beta} = f_l^{\alpha;\beta}\mathbf{i}_l$$

$$= -\frac{\mathbf{i}_l}{4\pi\rho c_\alpha^2}\int_S\left[\tilde{\tau}_{lk}n_k + ik_\alpha C_{lkpj}e_{sk}^\alpha n_p\tilde{u}_j\right]$$ (no sum on s)   (10.4)

$$\cdot\exp\left(-ik_\alpha\mathbf{x}_s\cdot\mathbf{e}_s^\alpha\right)dS\left(\mathbf{x}_s\right)$$

and the vector far-field scattering amplitudes for P-waves and S-waves are given by

$$\mathbf{A}\left(\mathbf{e}_i^\beta;\mathbf{e}_s^p\right) = \left(\mathbf{f}^{p;\beta}\cdot\mathbf{e}_s^p\right)\mathbf{e}_s^p$$

$$\mathbf{A}\left(\mathbf{e}_i^\beta;\mathbf{e}_s^s\right) = \left[\mathbf{f}^{s;\beta} - \left(\mathbf{f}^{s;\beta}\cdot\mathbf{e}_s^s\right)\mathbf{e}_s^s\right]$$ (no sum on s)   (10.5)

for $\beta = (p,s)$. The vectors $\mathbf{i}_l$ are unit vectors along a set of Cartesian coordinate axes. The n_k terms in Eq. (10.4) are the components of the unit outward normal to the flaw surface (see Fig. 10.2) and C_{ijkl} is the fourth order elastic constants tensor, which here is taken to be for an isotropic elastic material. The stress and displacement components in Eq. (10.4) are normalized by the displacement amplitude of the incident wave, i.e. $\tilde{\tau}_{ij} = \tau_{ij}/U_0, \tilde{u}_j = u_j/U_0$. From Eq. (10.5) it can be seen that the polarization of the scattered P-wave is in the $\mathbf{e}_s^p$ direction while the polarization of the scattered S-wave is perpendicular to the $\mathbf{e}_s^s$ direction since $\mathbf{A}\left(\mathbf{e}_i^\beta;\mathbf{e}_s^s\right)\cdot\mathbf{e}_s^s = 0$.

In an ultrasonic flaw measurement system the output is a voltage which is a scalar quantity. Thus, if the scattering amplitude appears explicitly as part of a model for this measured voltage – which it does under certain conditions, as discussed in the next Chapter – there must be a specific scalar function of the vector scattering amplitude that is related to the output voltage. In the next Chapter it will be shown that the appropriate scalar function that appears in a model of the entire ultrasonic measurement system is the scalar component

$$A\left(\mathbf{e}_i^\beta;\mathbf{e}_s^\alpha\right) = \mathbf{A}\left(\mathbf{e}_i^\beta;\mathbf{e}_s^\alpha\right)\cdot\left(-\mathbf{d}^\alpha\right).$$ (10.6)

The unit vector $\mathbf{d}^\alpha$ is the polarization vector of a wave of type α (the same type as the scattered wave) that travels from the receiving transducer (acting like a transmitter) to the flaw along a completely reversed path from the path that the scattered waves take from the flaw to the receiving transducer (see Fig. 10.3). This polarization vector is defined when one

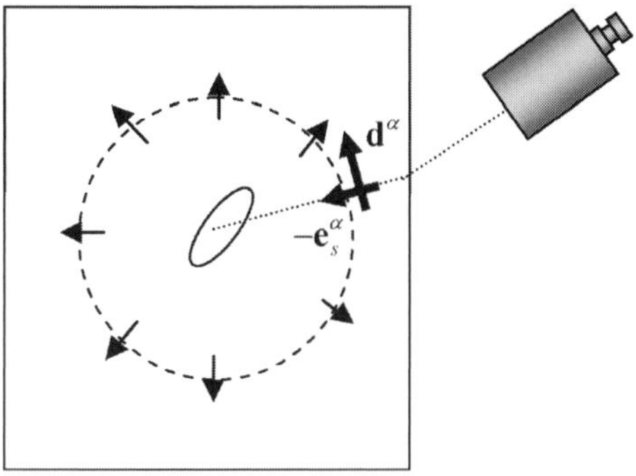

Fig. 10.3. The polarization, $\mathbf{d}^{\alpha}$, of the wave traveling from the receiver (acting as a transmitter) to the flaw along a path that is completely reversed from the actual received wave traveling from the flaw to the receiving transducer. For a scattered P-wave ($\alpha = \mathrm{p}$) we have $\mathbf{d}^{\alpha} = -\mathbf{e}_s^{\alpha}$ while for a scattered S-wave $\mathbf{d}^{\alpha}$ is perpendicular to $\mathbf{e}_s$.

solves for the waves propagated from the receiving transducer to the flaw. Note that the choice of sign of the polarization vector is arbitrary. For example, for a plane P-wave traveling in the $\mathbf{e}$-direction with velocity given by $\mathbf{v} = V\mathbf{e}$, we could take the polarization $\mathbf{d}^P = \mathbf{e}$ (as is normally done) and write $\mathbf{v} = V\mathbf{d}^P$ or we could choose $\mathbf{d}^P = -\mathbf{e}$ and write $\mathbf{v} = -V\mathbf{d}^P$ instead. The velocity of the wave is unaffected by this choice. Choosing a different sign on the polarization vector will affect the sign of the amplitude, as shown by this simple example, or it can affect individual parts of the total expression for the wave field such as transmission or reflection coefficients since those coefficients depend on the choice of the polarization direction (see Appendix D where the transmission coefficients were defined for specific choices of P-wave and S-wave polarizations). Sign changes of the transmission/reflection coefficients and polarizations, however, cancel so again the total wave field is unaffected by the choice for the direction of the polarization. However, with a given choice of the polarization vector we must be careful to use the transmission/reflection coefficients consistent with that choice.

Using Eqs. (10.4) and (10.5) in Eq. (10.6), the scalar scattering component, $A\left(\mathbf{e}_i^{\beta}; \mathbf{e}_s^{\alpha}\right)$ for both P-waves and S-waves is given by

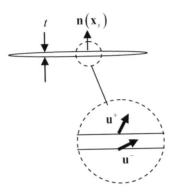

Fig. 10.4. A crack modeled as an open surface in a solid that is obtained by letting the thickness, t, go to zero of a thin volumetric shape, as shown, where the stress vector is zero on both sides of the crack but the displacement vector is allowed to have a displacement discontinuity given by $\Delta u(\mathbf{x}_s, \omega) = \mathbf{u}^+ - \mathbf{u}^-$, where $\mathbf{u}^+, \mathbf{u}^-$ are displacements on opposite sides of the crack at the same location on the open surface.

$$A\left(\mathbf{e}_i^\beta; \mathbf{e}_s^\alpha\right) = \frac{1}{4\pi\rho c_\alpha^2} \int_S d_i^\alpha \left[\tilde{\tau}_{lk}n_k + ik_\alpha C_{lkpj}e_{sk}^\alpha n_p \tilde{u}_j\right]$$

$$\cdot \exp\left(-ik_\alpha \mathbf{x}_s \cdot \mathbf{e}_s^\alpha\right) dS\left(\mathbf{x}_s\right).$$

(10.7)

(no sum on s, α)

Equation (10.7) gives the far-field scalar scattering response of a general volumetric flaw. One can also use this result and a limiting argument to obtain the response of a crack-like flaw where the crack is modeled as a zero volume open surface (Fig. 10.4). If one assumes that the faces of the crack are stress-free, we have on both faces of the crack $\tilde{\tau}_{lk}n_k = 0$. The displacement components, however, can be different from one face of the crack to the other, leading to *displacement discontinuities*, $\Delta\tilde{u}_j(\mathbf{x}_s, \omega)$, on the crack (see Fig. 10.4). The scattering amplitude of Eq. (10.7) then reduces to [Fundamentals]

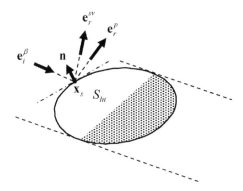

Fig. 10.5. The Kirchhoff approximation, where the fields on the "lit" surface of the flaw are assumed to be those obtained by plane wave interactions with a plane (dashed-dotted line) whose normal coincides with that of the flaw surface. On the remainder of the flaw surface (the shaded "shadow" region shown) the fields are assumed to be identically zero.

$$A\left(\mathbf{e}_i^\beta;\mathbf{e}_s^\alpha\right)=\frac{1}{4\pi\rho c_\alpha^2}\int_S\left[ik_\alpha C_{lkpj}d_l^\alpha e_{sk}^\alpha n_p\Delta\tilde{u}_j\right]$$

$$\cdot\exp\left(-ik_\alpha\mathbf{x}_s\cdot\mathbf{e}_s^\alpha\right)dS\left(\mathbf{x}_s\right),$$

(10.8)

(no sum on s, α)

where now S is the (open) surface of the crack and $\mathbf{n}$ is the unit normal to that open surface.

10.2 The Kirchhoff Approximation for Volumetric Flaws

One approximation that has been frequently used to describe the scattering of volumetric flaws or cracks is the Kirchhoff approximation [Fundamentals]. Consider first the volumetric flaw case. In this approximation, that part of the flaw surface where the incident wave (which is taken as a plane wave) can directly strike the surface is called the "lit" surface, S_{lit} (Fig. 10.5). On the lit surface it is assumed that the interaction of the incident plane wave with the surface is identical to that of the incident wave with a plane interface whose normal coincides locally with the surface normal, $\mathbf{n}$. Since we can solve for the interaction of a plane wave with a plane interface, we can write down explicit expressions for both the Kirchhoff approximation displacement components, $\tilde{u}_j^K$, and stresses, $\tilde{\tau}_{lk}^K$, on the lit surface as [Fundamentals]

$$\tilde{u}_j^K = d_{ij}^{\beta} \exp\left[ik_{\beta}\left(\mathbf{e}_i^{\beta} \cdot \mathbf{x}_s\right)\right] + \sum_{m=p,sv} R_{12}^{m;\beta} d_{rj}^m \exp\left[ik_m\left(\mathbf{e}_r^m \cdot \mathbf{x}_s\right)\right]$$

$$\tilde{\tau}_{lk}^K = C_{lkjp}\partial\tilde{u}_j^K / \partial x_p,$$

(10.9)

(no sum on r,i)

where $\mathbf{d}_i^{\beta}$ is the polarization vector for an incident wave (of type β) traveling in the $\mathbf{e}_i^{\beta}$ direction and $\mathbf{d}_r^m$ is the polarization of a reflected waves at the interface (of type m) traveling in the $\mathbf{e}_r^m$ direction. The reflection coefficients for a reflected wave of type m due to an incident wave of type β are the $R_{12}^{m;\beta}$. On the remaining part of the flaw surface where the incident wave cannot strike it directly, it assumed that the fields are totally absent and $\tilde{u}_j = \tilde{\tau}_{lk} = 0$. Then Eq. (10.7) becomes

$$A\left(\mathbf{e}_i^{\beta};\mathbf{e}_s^{\alpha}\right) = \frac{1}{4\pi\rho c_{\alpha}^2} \int_{S_{lit}} d_i^{\alpha}\left[\tilde{\tau}_{lk}^K n_k + ik_{\alpha}C_{lkpj}e_{sk}^{\alpha}n_p\tilde{u}_j^K\right]$$

$$\cdot \exp\left(-ik_{\alpha}\mathbf{x}_s \cdot \mathbf{e}_s^{\alpha}\right)dS\left(\mathbf{x}_s\right).$$

(10.10)

(no sum on s, α)

The Kirchhoff approximation is a high frequency approximation that allows us to avoid having to solve a boundary value problem in order to determine the far-field scattering amplitude. In general, the integrations in Eq. (10.10) must still be done numerically, but for the special case of the pulse-echo response of a void one can obtain some simple and explicit results. In that case we consider a scattered wave of the same type as the incident wave and let the scattered wave direction be opposite to that of the incident wave so that $\mathbf{e}_s^{\alpha} = -\mathbf{e}_i^{\beta}$. Since we are considering a void we also have $\tilde{\tau}_{lk}n_k = 0$ on the surface. Then Eq. (10.10) reduces to

$$A\left(\mathbf{e}_i^{\beta};-\mathbf{e}_i^{\beta}\right) = \frac{1}{4\pi\rho c_{\beta}^2} \int_{S_{lit}}\left[-ik_{\beta}C_{lkpj}d_l^{\beta}e_{ik}^{\beta}n_p\tilde{u}_j^K\right]\exp\left(ik_{\beta}\mathbf{x}_s \cdot \mathbf{e}_i^{\beta}\right)dS\left(\mathbf{x}_s\right). \quad (10.11)$$

(no sum on s, β)

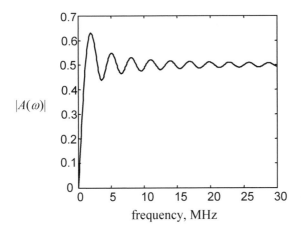

Fig. 10.6. Magnitude of the far-field pulse-echo P-wave scattering amplitude versus frequency for a 1 mm radius void in steel (c_p = 5900 m/s) in the Kirchhoff approximation.

Equation (10.11) can be simplified even further since it can be shown by a combination of analytical and numerical evaluations that [10.1]

$$\frac{C_{lkpj} d_l^\beta e_{ik}^\beta n_p u_j^K}{\rho c_\beta^2} = 2\left(\mathbf{e}_i^\beta \cdot \mathbf{n}\right)\exp\left[ik_\beta\left(\mathbf{e}_i^\beta \cdot \mathbf{x}_s\right)\right] \tag{10.12}$$

and the pulse-echo far-field scattering amplitude of the void becomes simply

$$A\left(\mathbf{e}_i^\beta; -\mathbf{e}_i^\beta\right) = \frac{-ik_\beta}{2\pi}\int_{S_{lit}}\left(\mathbf{e}_i^\beta \cdot \mathbf{n}\right)\exp\left(2ik_\beta \mathbf{x}_s \cdot \mathbf{e}_i^\beta\right)dS\left(\mathbf{x}_s\right). \tag{10.13}$$

Equation (10.13) is identical to the pulse-echo response of a void using a fluid model (see Eq. (10.2)) instead [Fundamentals]. It is a very important result since it shows that:

For any stress-free flaw in an isotropic elastic solid the Kirchhoff approximation for the pulse-echo far-field scattering amplitude component that appears in an ultrasonic measurement model is identical to the Kirchhoff approximation for the scalar scattering amplitude of a void in a fluid.

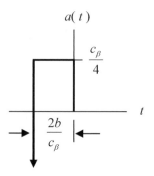

Fig. 10.7. The time domain pulse-echo impulse response of a spherical void in a solid in the Kirchhoff approximation, showing the leading edge delta function response followed by the response of the lit surface.

In this Chapter we will use this result to carry out the integrations in Eq. (10.13) explicitly for a number of important canonical scattering geometries including a spherical void, a flat elliptical crack (see Eq. (10.32)), and a side-drilled hole (see Eq. (10.53)). For a spherical void of radius b, for example, Eq. (10.13) gives [Fundamentals]

$$A\left(\mathbf{e}_i^\beta;-\mathbf{e}_i^\beta\right)=\frac{-b}{2}\exp\left(-ik_\beta b\right)\left[\exp\left(-ik_\beta b\right)-\frac{\sin\left(k_\beta b\right)}{k_\beta b}\right].$$

(10.14)

Figure 10.6 plots the magnitude of this scattering amplitude for a spherical void in steel. The characteristics of this plot can be better understood if we Fourier transform Eq. (10.14) into the time domain. This leads to the impulse response of the flaw, $a\left(\mathbf{e}_i^\beta;\mathbf{e}_s^\alpha,t\right)$, given by [Fundamentals]

$$a\left(\mathbf{e}_i^\beta;-\mathbf{e}_i^\beta,t\right)=\frac{-b}{2}\left[\delta\left(t+\frac{2b}{c_\beta}\right)-\frac{c_\beta}{2b}U\left(\frac{-2b}{c_\beta},0;t\right)\right],$$

(10.15)

where δ is a delta function (see Appendix A) and

$$U\left(t_1,t_2;t\right)=\begin{cases}1 & t_1<t<t_2\\0 & otherwise\end{cases}.$$

(10.16)

Figure 10.7 shows a plot of this time domain scattering amplitude. When the incident wave first reaches the flaw, there is a delta function response from the point where the incident wave first touches the flaw. This *leading*

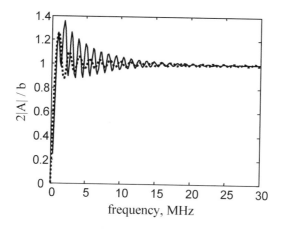

Fig. 10.8. The magnitude of the normalized scattering amplitude versus frequency for the pulse-echo P-wave response of a spherical void in the the Kirchhoff approximation (dotted line) and for the exact separation of variables solution (solid line).

edge response occurs at time $t = -2b/c_\beta$, where $t = 0$ is when the wave front reaches the center of the flaw, followed by a constant response that exists as the wave front sweeps across the lit surface. When the wave front reaches the boundary between the lit surface and the shadow zone of the flaw the response drops to zero. It is the interference of the leading edge response and the remaining lit surface response that causes the oscillations seen in Fig. 10.6. At very high frequencies, only the leading edge response remains, leading to the plateau seen in Fig. 10.6.

The sphere is one of the few shapes where we can obtain the exact far-field scattering amplitude by the method of separation of variables [Fundamentals]. Thus, we can compare the Kirchhoff approximation to the exact results for the spherical void just considered. Figure 10.8 shows this comparison made in the frequency domain for the normalized magnitude of the far-field scattering amplitude computed for the pulse-echo P-wave response of a spherical void. The two results agree at high frequencies, which show that the leading edge delta function response in the Kirchhoff approximation agrees with this same response in the exact solution. The frequency of oscillations in the exact solution is different from that in the Kirchhoff approximation because in the exact solution the oscillations are caused primarily by an interference of the leading edge response with a *creeping wave* that travels around the flaw and returns, as shown schematically in Fig. 10.9. This creeping wave can be seen explicitly in

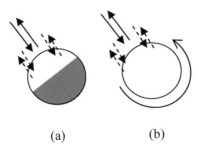

(a)   (b)

Fig. 10.9. The scattering from a spherical void in **(a)** the Kirchhoff approximation, where the response comes from a front surface leading edge response (solid arrows) and the response from the lit surface (dashed arrows), and in **(b)** the exact solution case where there are contributions from the leading edge and front surface but where there also exists a creeping wave that travels around the sphere as shown.

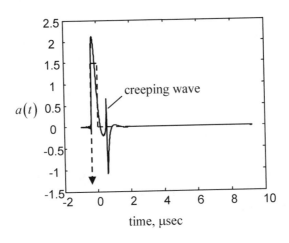

Fig. 10.10. The exact time domain pulse-echo impulse response (solid line) of a 1 mm radius spherical void in a solid as calculated from a separation of variables solution with the delta function removed, showing the response from the lit surface and a creeping wave. The same response in the Kirchhoff approximation (dashed line). Wave speeds: $c_p = 6000$ m/sec, $c_s = 3200$ m/sec.

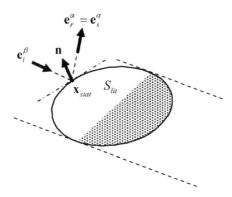

Fig. 10.11. The case where a stationary phase point, $\mathbf{x}_{stat}$, exists on the lit surface of a flaw, where the scattered wave direction coincides with one of the reflected wave directions.

Fig. 10.10 which shows the exact P-wave pulse-echo time domain impulse response of the spherical void obtained by Fourier transforming the exact separation of variables solution (after removal of the delta function leading edge response which is common to both the exact solution and the Kirchhoff approximation). The Kirchhoff solution is also shown in Fig. 10.10 for comparison purposes.

10.3 The Leading Edge Response of Volumetric Flaws

Although as we have seen the Kirchhoff approximation did not accurately represent the later arriving waves from a spherical void, it did model correctly the leading edge response of the flaw. This leading response is the dominant part of the solution at high frequencies, and in the time domain gives us a delta function signal from the front surface of the flaw. Since the delta function contains all frequencies equally whereas other parts of the flaw response typically go to zero as the frequency increases, even in real band-limited systems the leading edge response signal in the time domain is often the largest signal in the entire flaw response. Thus, it is useful to try to model this signal by itself. Fortunately, this is possible for general volumetric flaw types, not just voids. If we return to the Kirchhoff approximation (Eq. (10.10)) for a general volumetric flaw, we can approximate the integral in that equation at high frequencies by the method of stationary phase. The details are rather lengthy, but the end result is that in a general pitch-catch setup (which includes pulse-echo as a

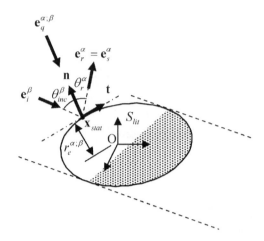

Fig. 10.12. The case where a stationary phase point, $\mathbf{x}_{stat}$, exists on the lit surface of a flaw, where the scattered wave direction coincides with one of the reflected wave directions.

special case) the major contribution to the integral for a scattered wave of type α traveling in the $\mathbf{e}_s^\alpha$ direction comes from a neighborhood of a point on the flaw surface, called a stationary phase point, $\mathbf{x}_{stat}$, where the direction of the reflected wave in the Kirchhoff approximation, $\mathbf{e}_r^\alpha$, coincides with $\mathbf{e}_s^\alpha$ (see Fig. 10.11). The contribution to the integral near this stationary phase point can then be calculated by the method of stationary phase to give, in the frequency domain [Fundamentals]

$$A\left(\mathbf{e}_i^\beta;\mathbf{e}_s^\alpha\right)=\frac{\left(\mathbf{e}_s^\alpha\cdot\mathbf{n}\right)R_{12}^{\alpha;\beta}\sqrt{R_1R_2}\left(\mathbf{d}_r^\alpha\cdot\mathbf{d}^\alpha\right)}{\left|\mathbf{g}^{\alpha;\beta}\cdot\mathbf{n}\right|}\exp\left(ik_\alpha\mathbf{g}^{\alpha;\beta}\cdot\mathbf{x}_{stat}\right),\qquad(10.17)$$

where $R_{12}^{\alpha;\beta}$ is the plane wave reflection coefficient (based on velocity ratios) between material 1 (the host material around the flaw) and material 2 (the flaw) for a reflected wave of type α due to an incident wave of type β. R_1,R_2 are the magnitudes of the principal curvatures of the flaw surface at the stationary phase point, $\mathbf{d}_r^\alpha$ is the polarization of the reflected wave and $\mathbf{d}^\alpha$ is the polarization of the wave coming from the receiving transducer (acting as a transmitter). The vectors $\mathbf{g}^{\alpha;\beta}$ are given by

$$\mathbf{g}^{\alpha;\beta} = \left(c_\alpha / c_\beta\right)\mathbf{e}_i^\beta - \mathbf{e}_s^\alpha, \tag{10.18}$$

where $\mathbf{e}_i^\beta$ is the incident wave direction for a wave of type β, $\mathbf{e}_s^\alpha = \mathbf{e}_r^\alpha$ is the reflected wave direction for a wave of type α, and c_p, c_s are the compressional and shear wave speeds for the host material surrounding the flaw, respectively. Note that one can always define the reflection coefficient so that the reflected wave polarization, $\mathbf{d}_r^\alpha$, coincides with the polarization $\mathbf{d}^\alpha$. In that case we have $\mathbf{d}_r^\alpha \cdot \mathbf{d}^\alpha = 1$. In all the subsequent results we will assume that this is true.

The vectors $\mathbf{g}^{\alpha;\beta}$ can be written in terms of their magnitudes and a unit vector, $\mathbf{e}_q^{\alpha;\beta}$ as $\mathbf{g}^{\alpha;\beta} = \left|\mathbf{g}^{\alpha;\beta}\right|\mathbf{e}_q^{\alpha;\beta}$. At the stationary phase point the unit vector $\mathbf{e}_q^{\alpha;\beta} = -\mathbf{n}$ so that $\left|\mathbf{g}^{\alpha;\beta} \cdot \mathbf{n}\right| = \left|\mathbf{g}^{\alpha;\beta}\right|$. We also have $\mathbf{g}^{\alpha;\beta} \cdot \mathbf{t} = 0$, where $\mathbf{t}$ is a unit vector in the tangent plane to the surface at the stationary phase point (see Fig. 10.12), which is just a statement of Snell's law. We can write the quantity $\mathbf{g}^{\alpha;\beta} \cdot \mathbf{x}_{stat} = -\left|\mathbf{g}^{\alpha;\beta}\right|r_e^{\alpha;\beta}$, where $r_e^{\alpha;\beta}$ is the distance in the direction $\mathbf{n}$ (or, equivalently, $-\mathbf{e}_q^{\alpha;\beta}$) at the stationary phase point from a fixed point (usually taken as the flaw "center") to the tangent plane of the surface at $\mathbf{n}$ (Fig. 10.12). The $r_e^{\alpha;\beta}$ distance is called the *equivalent radius* of the flaw in the $\mathbf{e}_q^{\alpha;\beta}$ direction. Thus, we can also write Eq. (10.17) in the form

$$A\left(\mathbf{e}_i^\beta;\mathbf{e}_s^\alpha\right) = \frac{\left(\mathbf{e}_s^\alpha \cdot \mathbf{n}\right)R_{12}^{\alpha;\beta}\sqrt{R_1 R_2}}{\left|\mathbf{g}^{\alpha;\beta}\right|}\exp\left(-ik_\alpha\left|\mathbf{g}^{\alpha;\beta}\right|r_e^{\alpha;\beta}\right). \tag{10.19}$$

If we Fourier transform Eq. (10.19) into the time domain, the leading edge impulse response of the flaw is given by

$$a\left(\mathbf{e}_i^\beta;\mathbf{e}_s^\alpha,t\right) = \frac{\left(\mathbf{e}_s^\alpha \cdot \mathbf{n}\right)R_{12}^{\alpha;\beta}\sqrt{R_1 R_2}}{\left|\mathbf{g}^{\alpha;\beta}\right|}\delta\left(t + \left|\mathbf{g}^{\alpha;\beta}\right|r_e^{\alpha;\beta}/c_\alpha\right). \tag{10.20}$$

For the special case of a pulse-echo (same mode) leading edge response, from Eqs. (10.19), (10.20) we have the even simpler expressions:

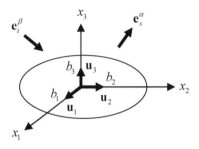

Fig. 10.13. Scattering geometry for an ellipsoidal flaw.

$$A\left(\mathbf{e}_i^\beta;-\mathbf{e}_i^\beta\right)=\frac{R_{12}^{\beta;\beta}\left(0^\circ\right)\sqrt{R_1R_2}}{2}\exp\left(-2ik_\beta r_e\right)$$

$$a\left(\mathbf{e}_i^\beta;-\mathbf{e}_i^\beta,t\right)=\frac{R_{12}^{\beta;\beta}\left(0^\circ\right)\sqrt{R_1R_2}}{2}\delta\left(t+2r_e/c_\beta\right),$$

(10.21)

where $r_e=r_e^{\beta;\beta}$ and now the reflection coefficient is just the normal incidence coefficient, as indicated in Eq. (10.21).

For a purely convex flaw shape such as an ellipsoid, as shown in Fig. 10.13, there can be at most only one stationary phase point on the lit surface. However, a stationary point may not exist on the lit surface at all for some combination of incident and scattered directions of a general pitch-catch setup. In that case, a leading edge response of the flaw is absent. For more general flaw shapes there may be multiple stationary phase points, in which case one must sum over all the leading edge responses.

For an ellipsoidal shaped flaw with semi-major axes $\left(b_1,b_2,b_3\right)$ along the $\left(\mathbf{u}_1,\mathbf{u}_2,\mathbf{u}_3\right)$ directions as shown in Fig. (10.13) we have the Gaussian curvature term [Fundamentals]

$$\sqrt{R_1R_2}=b_1b_2b_3/\left(r_e^{\alpha;\beta}\right)^2$$

(10.22)

and the equivalent radius is given by

$$r_e^{\alpha;\beta}=\sqrt{b_1^2\left(\mathbf{e}_q^{\alpha;\beta}\cdot\mathbf{u}_1\right)^2+b_2^2\left(\mathbf{e}_q^{\alpha;\beta}\cdot\mathbf{u}_2\right)^2+b_3^2\left(\mathbf{e}_q^{\alpha;\beta}\cdot\mathbf{u}_3\right)^2},$$

(10.23)

where, recall, $\mathbf{e}_q^{\alpha;\beta}=\mathbf{g}^{\alpha;\beta}/\left|\mathbf{g}^{\alpha;\beta}\right|$ (see Eq. (10.18)).

In this case the leading edge responses for the general pitch-catch setup (Eqs. (10.19) and (10.20)) become

$$A\left(\mathbf{e}_i^\beta;\mathbf{e}_s^\alpha\right)=\frac{\left(\mathbf{e}_s^\alpha\cdot\mathbf{n}\right)R_{12}^{\alpha;\beta}b_1b_2b_3}{\left|\mathbf{g}^{\alpha;\beta}\right|\left(r_e^{\alpha;\beta}\right)^2}\exp\left(-ik_\alpha\left|\mathbf{g}^{\alpha;\beta}\right|r_e^{\alpha;\beta}\right)$$

$$a\left(\mathbf{e}_i^\beta;\mathbf{e}_s^\alpha,t\right)=\frac{\left(\mathbf{e}_s^\alpha\cdot\mathbf{n}\right)R_{12}^{\alpha;\beta}b_1b_2b_3}{\left|\mathbf{g}^{\alpha;\beta}\right|\left(r_e^{\alpha;\beta}\right)^2}\delta\left(t+\left|\mathbf{g}^{\alpha;\beta}\right|r_e^{\alpha;\beta}/c_\alpha\right)$$

(10.24)

and for the pulse-echo case Eq. (10.21) reduces to

$$A\left(\mathbf{e}_i^\beta;-\mathbf{e}_i^\beta\right)=\frac{R_{12}^{\beta;\beta}\left(0^\circ\right)b_1b_2b_3}{2r_e^2}\exp\left(-2ik_\beta r_e\right)$$

$$a\left(\mathbf{e}_i^\beta;-\mathbf{e}_i^\beta,t\right)=\frac{R_{12}^{\beta;\beta}\left(0^\circ\right)b_1b_2b_3}{2r_e^2}\delta\left(t+2r_e/c_\beta\right).$$

(10.25)

For the particular case of a spherical void we have $b_1=b_2=b_3=b$, $R_{12}^{\beta;\beta}\left(0^\circ\right)=-1$, and $r_e=b$ so the pulse-echo results of Eq. (10.25) reduce to the leading edge results obtained previously as part of the full Kirchhoff solution for the sphere (see Eqs. (10.14) and (10.15)).

10.4 The Kirchhoff Approximation for Cracks

Our crack scattering model (Eq. (10.8)) considers the crack as a stress-free open surface on which there is a displacement discontinuity $\Delta\mathbf{u}\left(x_s,\omega\right)=\mathbf{u}^+-\mathbf{u}^-$. In the Kirchhoff approximation on the lit part of the front surface of the crack we would have $u_j^+=u_j^K$ and on the remainder of the front surface and the entire back surface (assuming the crack does not fold over so that part of the back surface can also be a "lit" surface) we would have zero displacements. Thus, the Kirchhoff approximation for a crack in an elastic solid gives

$$A\left(\mathbf{e}_i^\beta;\mathbf{e}_s^\alpha\right)=\frac{1}{4\pi\rho c_\alpha^2}\int_{S_{lit}}\left[ik_\alpha C_{lkpj}d_l^\alpha e_{sk}^\alpha n_p\tilde{u}_j^K\right]\exp\left(-ik_\alpha\mathbf{x}_s\cdot\mathbf{e}_s^\alpha\right)dS\left(\mathbf{x}_s\right).$$ (10.26)

(no sum on s, α)

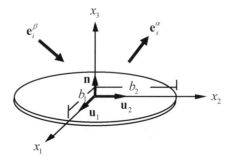

Fig. 10.14. Scattering geometry for a flat elliptical crack.

For the special case of pulse-echo we can again use Eq. (10.12) and write

$$A\left(\mathbf{e}_i^{\beta};-\mathbf{e}_i^{\beta}\right)=\frac{-ik_{\beta}}{2\pi}\int_{S_{lit}}\left(\mathbf{e}_i^{\beta}\cdot\mathbf{n}\right)\exp\left(2ik_{\beta}\mathbf{x}_s\cdot\mathbf{e}_i^{\beta}\right)dS\left(\mathbf{x}_s\right). \qquad (10.27)$$

Thus, the same Kirchhoff approximation expressions we used for the volumetric void can also be used for a crack. The only difference is that in Eqs. (10.26) and (10.27) we are integrating over the lit portion of an open surface of the crack rather than the lit part of a closed surface surrounding a volumetric flaw.

Now, consider the special case when the crack is a flat surface. Then Eq. (10.26) can be written as

$$A\left(\mathbf{e}_i^{\beta};\mathbf{e}_s^{\alpha}\right)=\frac{ik_{\alpha}C^{\alpha;\beta}}{2\pi}\int_{S}\exp\left[i\left(k_{\beta}\mathbf{e}_i^{\beta}-k_{\alpha}\mathbf{e}_s^{\alpha}\right)\cdot\mathbf{x}_s\right]dS\left(\mathbf{x}_s\right), \qquad (10.28)$$

where

$$C^{\alpha;\beta}=C_{kplj}e_{sj}^{\alpha}d_l^{\alpha}\left(d_{ip}^{\beta}+\sum_{m=p,s}R_{12}^{m;\beta}d_{rp}^{m}\right)n_k/2\rho c_{\alpha}^2 \qquad (10.29)$$

In this case the lit surface is now the entire surface, S, of the flaw. For the flat crack in pulse-echo, from Eq. (10.27)

$$A\left(\mathbf{e}_i^{\beta};-\mathbf{e}_i^{\beta}\right)=\frac{-ik_{\beta}\left(\mathbf{e}_i^{\beta}\cdot\mathbf{n}\right)}{2\pi}\int_{S}\exp\left(2ik_{\beta}\mathbf{x}_s\cdot\mathbf{e}_i^{\beta}\right)dS\left(\mathbf{x}_s\right). \qquad (10.30)$$

For the elliptical flat crack geometry shown in Fig. 10.14, the integrals in Eqs. (10.28) and (10.30) can be performed explicitly. We find for the pitch-catch case [Fundamentals]

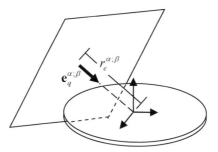

Fig. 10.15. The equivalent radius, $r_e^{\alpha;\beta}$ for the scattering by an elliptical crack shown as the distance from the center of the ellipse to a plane that is normal to $\mathbf{e}_q^{\alpha;\beta}$ and touches the crack edge at a single point.

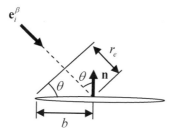

Fig. 10.16. The pulse-echo response of a circular flat crack of radius b showing that the equivalent radius $r_e = b\sin\theta$ where θ is the angle between the incident wave direction and the unit normal to the crack.

$$A\left(\mathbf{e}_i^\beta;\mathbf{e}_s^\alpha\right) = \frac{ib_1b_2 C^{\alpha;\beta}}{\left|\mathbf{g}^{\alpha;\beta}\right| r_e^{\alpha;\beta}} J_1\left(k_\alpha \left|\mathbf{g}^{\alpha;\beta}\right| r_e^{\alpha;\beta}\right) \tag{10.31}$$

and for the pulse-echo case

$$A\left(\mathbf{e}_i^\beta;-\mathbf{e}_i^\beta\right) = \frac{-ib_1b_2\left(\mathbf{e}_i^\beta\cdot\mathbf{n}\right)}{2r_e} J_1\left(2k_\alpha r_e\right), \tag{10.32}$$

where

$$r_e^{\alpha;\beta} = \sqrt{b_1^2\left(\mathbf{e}_q^{\alpha;\beta}\cdot\mathbf{u}_1\right)^2 + b_2^2\left(\mathbf{e}_q^{\alpha;\beta}\cdot\mathbf{u}_2\right)^2} \tag{10.33}$$

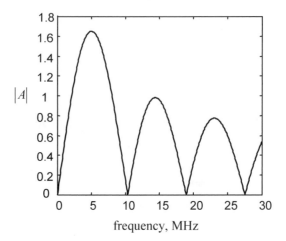

Fig. 10.17. The magnitude of the P-wave pulse-echo far-field scattering amplitude versus frequency calculated in the Kirchhoff approximation for a 1 mm radius circular crack in steel with an angle of incidence of $10°$ from the crack normal.

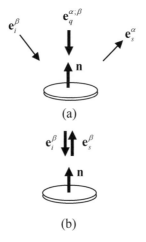

Fig. 10.18. (a) The "generalized normal incidence" case for pitch-catch where $\mathbf{e}_q^{\alpha;\beta}$ is parallel to the crack normal, and **(b)** the pulse-echo case where $\mathbf{e}_i^\beta = -\mathbf{e}_s^\beta$ is parallel to the normal.

and again we have $\mathbf{e}_q^{\alpha;\beta} = \mathbf{g}^{\alpha;\beta} / |\mathbf{g}^{\alpha;\beta}|$ (see Eq. (10.18)). For the pulse-echo case we have let $r_e = r_e^{\beta;\beta}$. As in the volumetric flaw case, we can interpret $r_e^{\alpha;\beta}$ as an "equivalent radius" for a given setup, as shown in Fig. 10. 15. In this case the equivalent radius is the distance in the $\mathbf{e}_q^{\alpha;\beta}$ direction from the center of the ellipse to a plane whose normal is $\mathbf{e}_q^{\alpha;\beta}$ and is touching the edge of the crack at a single point.

For the special case of the circular crack $b_1 = b_2 = b$ so that we find in the pitch-catch case

$$A\left(\mathbf{e}_i^{\beta};\mathbf{e}_s^{\alpha}\right) = \frac{ib^2 C^{\alpha;\beta}}{\left|\mathbf{g}^{\alpha;\beta}\right| r_e^{\alpha;\beta}} J_1\left(k_\alpha \left|\mathbf{g}^{\alpha;\beta}\right| r_e^{\alpha;\beta}\right) \tag{10.34}$$

and in the pulse-echo case

$$A\left(\mathbf{e}_i^{\beta};-\mathbf{e}_i^{\beta}\right) = \frac{-ib^2 \left(\mathbf{e}_i^{\beta} \cdot \mathbf{n}\right)}{2r_e} J_1\left(2k_\beta r_e\right). \tag{10.35}$$

In the pulse-echo response of the circular crack we have $\mathbf{e}_i \cdot \mathbf{n} = -\cos\theta$ and $r_e = b\sin\theta$ (see Fig. 10.16) so that

$$A\left(\mathbf{e}_i^{\beta};-\mathbf{e}_i^{\beta}\right) = \frac{ib\cos\theta}{2\sin\theta} J_1\left(2k_\beta b\sin\theta\right). \tag{10.36}$$

Figure 10.17 plots the behavior of the P-wave pulse-echo circular crack response (Eq. (10.36)) for a 1 mm radius crack in steel at an angle of incidence $\theta = 10°$. Unlike a spherical void the crack response has very strong oscillations that decrease with increasing frequency. At normal incidence, however, the crack scattering response is quite different. In the pitch-catch case we can have a similar situation. We will call either of these special cases "generalized normal incidence". At generalized normal incidence $\mathbf{e}_q^{\alpha;\beta}$ is parallel to the crack normal, $\mathbf{n}$. In the pulse-echo case this simply implies that the incident wave direction, $\mathbf{e}_i^{\beta}$, is parallel to $\mathbf{n}$ (Fig. 10.18). In either case we have $r_e \to 0$ and $\theta \to 0$ so that Eq. (10.34) for the pitch-catch case becomes

$$A\left(\mathbf{e}_i^{\beta};\mathbf{e}_s^{\alpha}\right) = \frac{ik_\alpha b^2 C^{\alpha;\beta}}{2} \tag{10.37}$$

and for the pulse-echo case

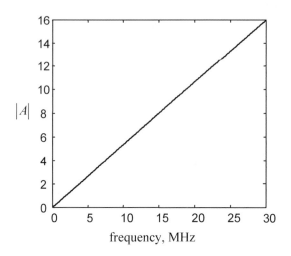

Fig. 10.19. The magnitude of the P-wave pulse-echo far-field scattering amplitude versus frequency calculated in the Kirchhoff approximation for a 1 mm radius circular crack in steel at normal incidence.

$$A\left(\mathbf{e}_i^\beta;-\mathbf{e}_i^\beta\right) = \frac{ik_\beta b^2}{2} \tag{10.38}$$

so that the crack response increases linearly with frequency as shown in Figure 10.19.

We can understand some of this frequency domain behavior if we Fourier transform our results back into the time domain to obtain the crack impulse response. From Eq. (10.35) for the pulse-echo response of the elliptical crack, for example, we find for the case when the incident wave direction is at oblique incidence to the crack normal [Fundamentals]

$$a\left(\mathbf{e}_i^\beta;-\mathbf{e}_i^\beta,t\right) = \begin{cases} \dfrac{-b_1 b_2 c_\beta \left(\mathbf{e}_i^\beta \cdot \mathbf{n}\right)}{4\pi\left(r_e\right)^2} \dfrac{t}{\sqrt{\left(2r_e/c_\beta\right)^2 - t^2}} & |t| \le 2r_e/c_\beta \\ 0 \quad otherwise \end{cases} \tag{10.39}$$

and for the normal incidence case, where

$$A\left(\mathbf{e}_i^\beta;-\mathbf{e}_i^\beta\right) = \frac{ik_\beta b_1 b_2}{2}, \tag{10.40}$$

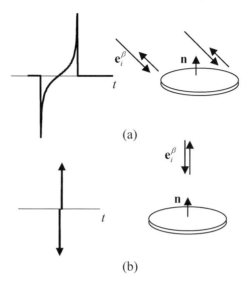

(a)

(b)

Fig. 10.20. (a) The time domain pulse-echo impulse response of an elliptical crack calculated in the Kirchhoff approximation at oblique incidence, and **(b)** at normal incidence.

we find

$$a\left(\mathbf{e}_i^{\beta};-\mathbf{e}_i^{\beta},t\right)=\frac{-b_1 b_2}{2c_\beta}\frac{d\delta(t)}{dt}.\tag{10.41}$$

These cases are both plotted in Fig. 10.20. One can see that in the oblique incidence case (Fig. 10.20 (a)) the crack signal has an anti-symmetrical form, with two distinct peaks. These peaks are called crack *flashpoint* responses. The first flashpoint occurs when the incident wave front first touches the crack and the second flashpoint occurs when the incident wave front last touches the crack. The interference of the frequency components of these two flashpoint responses is what causes the strong oscillations in the frequency domain response for non-normal incidence. At normal incidence, the two flash point signals merge to form a "doublet" signal as shown in Fig. 10.20 (b). The doublet is represented by the derivative of a delta function, as given in Eq. (10.41). Note that since the Fourier transform of the delta function is just unity, we can write formally

$$\delta(t)=\frac{1}{2\pi}\int_{-\infty}^{+\infty}(1)\exp(-i\omega t)\,d\omega\tag{10.42}$$

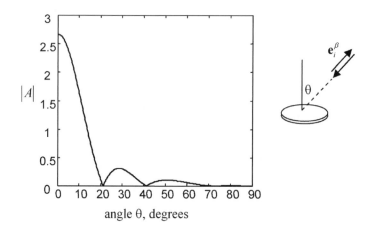

Fig. 10.21. The magnitude of the pulse-echo scattering amplitude response versus angle calculated in the Kirchhoff approximation for a 1 mm radius circular crack in steel at a frequency of 5 MHz.

from which we obtain

$$\frac{d\delta}{dt} = \frac{1}{2\pi} \int_{-\infty}^{+\infty} (-i\omega) \exp(-i\omega t) d\omega \qquad (10.43)$$

Equation (10.43) shows that the Fourier transform of the derivative of the delta function is just $-i\omega$ so that taking the Fourier transform of Eq. (10.41), we do indeed obtain Eq. (10.40). Thus, the linearly increasing frequency domain response is just a consequence of having a doublet time domain response for the crack at normal incidence.

A flat crack is a very specular scatterer since in pulse-echo its scattering response is large when the incident wave strikes a crack at normal incidence but decreases rapidly as a function of the angle, θ, that the incident wave makes with the crack normal, as shown in Fig. 10.21.

10.5 Validity of the Kirchhoff Approximation

The Kirchhoff approximation is a very useful tool for modeling the scattering of volumetric flaws and cracks. For the volumetric flaw case, the Kirchhoff approximation predicts a leading edge response that is in fact

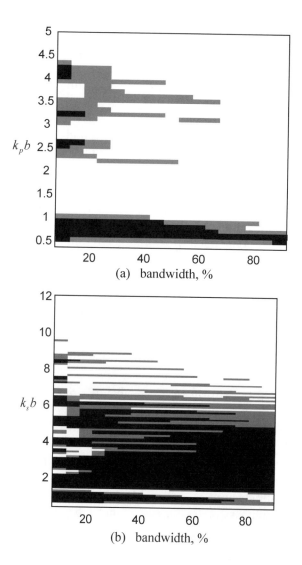

Fig. 10.22. A comparison of the peak-to-peak pulse-echo responses of a spherical void of radius b as calculated by the Kirchhoff approximation and the method of separation of variables where the non-dimensional wave number and bandwidth are varied. **(a)** Pulse-echo P-wave responses. **(b)** Pulse-echo SV-wave responses. White region: peak-to-peak differences < 1 dB, Gray region: differences > 1 dB and <1.5 dB, Black region: differences > 1.5 dB.

exact at high frequencies. Thus, as long as the flaw is not too small so that the later arriving waves can merge with this leading edge response or for special cases where the later arriving signals may be larger than the leading edge response, the Kirchhoff approximation will accurately model the amplitude of the flaw response, as measured, for example, by the maximum peak-to-peak amplitude of the time domain wave form. This fact can be demonstrated for the pulse-echo response of a spherical void by comparing wave forms synthesized by the method of separation of variables (discussed in section 10.8) and the Kirchhoff approximation. In this case the scattering amplitude was multiplied by a Gaussian window having a center frequency, f_c, and bandwidth, bw, (see Appendix A for an example). The result was then inverted into the time domain with a Fast Fourier Transform and the peak-to-peak value of the wave form was obtained. In order to compare the peak-to-peak values obtained in this fashion using either the method of separation of variables or the Kirchhoff approximation, it is necessary to have a practical criterion on when the Kirchhoff approximation is accurate. Since NDE inspection setups often have an uncertainty of 1-1.5 dB or greater in the amplitudes of the signals measured (due to experimental setup errors, noise, etc.) we will label the Kirchhoff approximate accurate if the peak-to-peak amplitude of the signal that it predicts is less than 1 dB different from the separation of variables result.

Figures 10.22 (a), (b) shows the results of simulating the peak-to-peak pulse-echo P-wave and SV-wave responses of a spherical void of radius b at different Gaussian window center frequencies and bandwidths. The white region in that figure is where the Kirchhoff and separation of variables solutions agree within 1 dB while the gray region is where the responses differ by more than 1 dB but less than 1.5 dB, and the black region is where the responses differ by 1.5 dB or more. The non-dimensional wave numbers, $k_p b = 2\pi f_c b / c_p$ and $k_s b = 2\pi f_c b / c_s$ shown in Fig. 10.22 were computed at the center frequency, f_c, of the Gaussian window and the bandwidth is given as a percentage of that center frequency value. For the P-wave case for values of $k_p b > 4.5$ it was found that the Kirchhoff approximation was accurate for all bandwidths but that below this value the bandwidth began to also play a role. However, for sufficiently large bandwidths Fig. 10.22 (a) shows that the Kirchhoff approximation remains accurate to wave numbers as small as $k_p b = 1$ in the P-wave case. At wave numbers $k_p b < 1$ there may be cases where the differences also are less than 1 dB but these only arise accidentally from

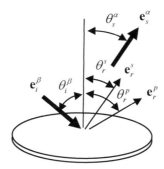

Fig. 10.23. In a general pitch-catch setup a flat crack generates a large specularly scattered signal when the scattered wave direction (and wave type) coincides with one of the reflected wave directions (and wave type) as determined by Snell's law, i.e. where $\theta_s^s = \theta_r^s$ or $\theta_s^p = \theta_r^p$ and where the reflected angles, θ_r^p, θ_r^s are given in terms of the incident angle, θ_i^β, by the relations $\sin\theta_r^p = \left(c_p/c_\beta\right)\sin\theta_i^\beta$, $\sin\theta_r^s = \left(c_s/c_\beta\right)\sin\theta_i^\beta$.

canceling errors since the Kirchhoff approximation and exact solution can be shown analytically to have different low frequency limits. Figure 10.22 (b) shows that in the case of shear waves, the wave number $k_s b$ must be greater than 10 for the Kirchhoff approximation to remain valid for all bandwidths and that from $k_s b = 10$ to $k_s b = 6$ approximately there are bandwidths effects.

Thus, while formally the Kirchhoff is a high frequency approximation where one assumes $kb \gg 1$, we see that this approximation remains useful and accurate in predicting pulse-echo peak-to-peak signal amplitudes for spherical voids at much lower frequencies and/or flaw sizes. It is also clear from Fig. 10.22 that bandwidth as well as frequency/size plays a role in how well the Kirchhoff approximation can perform.

For ideal flat cracks, the Kirchhoff approximation also accurately models the pulse-echo amplitude of the crack response when the incident wave direction is normal to the crack or in pitch-catch when the scattered wave direction is along a reflected wave direction as predicted by Snell's law (see Fig. 10.23). In either of these generalized normal incidence cases, as discussed previously, the vector $\mathbf{g}^{\alpha;\beta} = \left(c_\alpha/c_\beta\right)\mathbf{e}_i^\beta - \mathbf{e}_s^\alpha$ is parallel to the crack normal $\mathbf{n}$. The expression for the scattering of a flat crack, Eq. (10.28), can be written in terms of this vector as:

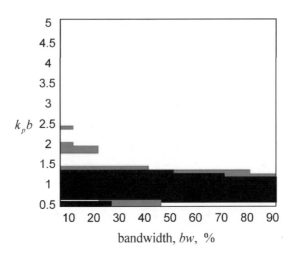

Fig. 10.24. A comparison of the normal incidence peak-to-peak pulse-echo P-wave responses of a circular crack of radius b as calculated by the Kirchhoff approximation and the method of separation of variables where the non-dimensional wave number, $k_p b$, and bandwidth, bw, are varied. White region: peak-to-peak differences < 1 dB, Light gray region: differences > 1 dB and < 1.5 dB, Black region: differences > 1.5 dB.

$$A\left(\mathbf{e}_i^\beta; \mathbf{e}_s^\alpha\right) = \frac{ik_\alpha C^{\alpha;\beta}}{2\pi} \int_S \exp\left[ik_\alpha \mathbf{g}^{\alpha;\beta} \cdot \mathbf{x}_s\right] dS(\mathbf{x}_s). \tag{10.44}$$

But $\mathbf{x}_s$ is a point lying in the plane of the crack and so we have $\mathbf{g}^{\alpha;\beta} \cdot \mathbf{x}_s = 0$ and Eq. (10.44) becomes for an arbitrarily shaped flat crack (see Eqs. (10.37), (10.38), (10.40) for the same result for different special shapes or setups)

$$A\left(\mathbf{e}_i^\beta; \mathbf{e}_s^\alpha\right) = \frac{ik_\alpha C^{\alpha;\beta} S_f}{2\pi}, \tag{10.45}$$

where S_f is the area of the flat crack. For all the cases where Eq. (10.45) holds we see a large specular response (like the doublet response shown earlier) that agrees with more exact scattering model predictions. Note that for a shear wave incident on the crack beyond the critical angle where the reflected P-wave disappears, the scattered S-wave response predicted by Eq. (10.45) will include pulse distortion since the coefficient $C^{\alpha;\beta}$ is then complex [Fundamentals]. Such pulse distortion, however, may be difficult to see in practice since at these angles the amplitude of the crack response will be small.

We can also examine the accuracy of the Kirchhoff approximation in predicting the pulse-echo normal incidence response of a circular crack in the same fashion as done for a spherical void. In this case, there is no separation of variables solution to compare to, but there have been numerical calculations done with the method of optimal truncation (MOOT) [10.34] for the pulse-echo P-wave response of a circular crack that can be used as an "exact" reference solution. Figure 10.24 shows the results of generating pulse-echo normal incidence P-wave peak-to-peak responses of a circular crack of radius b at different wave numbers, $k_p b = 2\pi f_c b / c_p$, and bandwidths, bw using the Kirchhoff approximation and MOOT. It can be seen from that figure that for wave numbers $k_p b > 2.5$ the Kirchhoff approximation is accurate for all bandwidths but bandwidth begins to play a role for smaller wave numbers. However, for sufficiently large bandwidths the Kirchhoff approximation remains accurate to non-dimensional wave numbers as small as $k_p b = 1.5$. At smaller wave numbers the Kirchhoff approximation is generally inaccurate although again there may be cases where canceling errors occur. These results demonstrate that, as in the spherical void case, the Kirchhoff approximation remains accurate for cases where the condition $kb \gg 1$ is not satisfied and that bandwidth also plays a role in determining when the Kirchhoff approximation is accurate but it is not as strong a factor as in the spherical void case.

For pulse-echo cases where the incident waves are not normal to the crack the Kirchhoff approximation predicts time domain flash point responses from the crack tips (see Fig. 10.20 (a)). It is commonly stated that the amplitudes of these signals do not agree with more exact scattering calculations except in a relatively small angular range (of about 20-30 degrees) from normal incidence [10.2]. However, this conclusion has been reached by considering either single frequency results or simulating narrow bandwidth time domain responses. It will be shown here that bandwidth plays an important role in determining the angular range over which the Kirchhoff approximation can accurately predict the pulse-echo peak-to-peak response of a circular crack. This fact is demonstrated by simulating oblique incident pulse-echo scattered P-wave responses of a circular crack with both the Kirchhoff approximation and MOOT and then comparing their predicted time domain peak-to-peak crack responses. In this study a Gaussian window having a center frequency of 10 MHz was

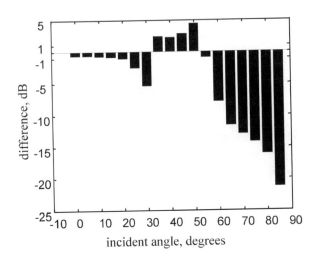

Fig. 10.25. Differences in dB between predicted peak-to-peak pulse-echo P-wave responses of a 0.381 mm radius circular crack in steel as calculated by the Kirchhoff approximation and by MOOT for a narrow band (2 MHz bandwidth, 10 MHz center frequency) system response.

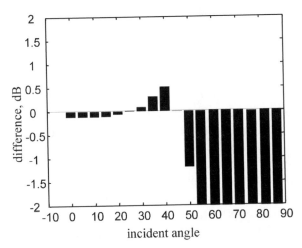

Fig. 10.26. Differences in dB between predicted peak-to-peak pulse-echo P-wave responses of a 0.381 mm radius circular crack in steel as calculated by the Kirchhoff approximation and by MOOT for a relatively wide band (6 MHz bandwidth, 10 MHz center frequency) system response. [For angles greater than 50° the differences are larger in magnitude than 2 dB so their values are off the scale of this figure.]

used with a flaw size of $b = 0.381$ mm. The compressional wave speed was taken as 6200 m/sec so that in all the cases considered the non-dimensional wave number was fixed at $k_p b = 3.86$. This wave number value is sufficiently large so that at normal incidence there were no bandwidth effects (see Fig. 10.24) but for oblique incidence this is not the case. For example, Fig. 10.25 shows the differences in dB between predicted peak-to-peak pulse-echo P-wave responses of the crack as calculated with the Kirchhoff approximation and MOOT and plotted versus angle of incidence for a narrow bandwidth (20%) window. In this case, differences exceeded 1 dB at an angle of incidence of about 20 degrees. However, if the bandwidth of the window is changed to 60%, holding all other variables fixed, the differences remain smaller than 1 dB for angles as large as 45 degrees, as shown in Fig. 10.26. For larger bandwidths, the range of angles where the Kirchhoff approximation is accurate is even larger. We have found that the precise way in which the angular range of accuracy of the Kirchhoff approximation varies is highly dependent on both the wave number and bandwidth so it is difficult to display comprehensible results for a wide range of cases on a single graph. Figure 10.27 shows a plot of the maximum incident angle at which the Kirchhoff approximation is accurate (i.e. within 1 dB of the MOOT solution) versus bandwidth for a

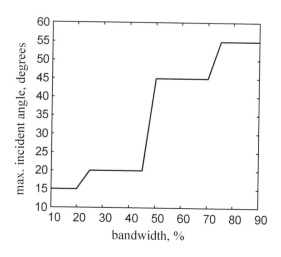

Fig. 10.27. The maximum incident angle at which the peak-to-peak pulse-echo flaw response predicted by the Kirchhoff approximation remains within 1 dB of the MOOT solution as a function of the bandwidth for the case of a P-wave obliquely incident on a 0.381 mm radius crack in steel ($k_p b = 5.0$).

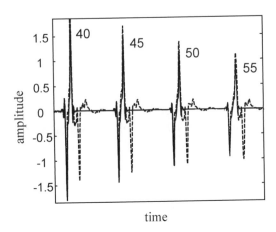

Fig. 10.28. Very wide-band simulated P-wave pulse-echo responses for a 0.381 mm radius crack in steel for angles of 40, 45, 50, and 55 degrees. Kirchhoff approximation (solid line), MOOT solution (dashed line).

non-dimensional wave number $k_p b = 5.0$. Although the curves at other wave numbers have different shapes, the trend shown in Fig. 10.27 remains the same for those other cases, namely the angular range where the Kirchhoff approximation is accurate can be as small as 15-20 degrees for very narrow bandwidth systems but as high as 55-60 degrees for very wide band systems.

To understand why the Kirchhoff approximation works better as the bandwidth increases consider Fig. 10.28 which shows a series of wave forms simulated by the Kirchhoff approximation and MOOT for the same 0.381 mm radius crack case examined in Fig. 10.27 but where all frequencies from 0-20 MHz were retained in calculating the time domain responses, yielding a very high bandwidth response. From Fig. 10.28 it can be seen that for angles from 40 to 55 degrees the flash point responses predicted by the Kirchhoff approximation agree well with those of the MOOT solution although the Kirchhoff approximation does predict a somewhat smaller trailing flashpoint signal than MOOT. Up to the 55 degree angle case the flashpoint signals are the largest signals present in the crack response but the MOOT solution also contains later arriving responses not predicted by the Kirchhoff approximation that grow as the angle increases. As the bandwidth decreases, these later arriving waves merge with the flashpoint responses, generating peak-to-peak responses that can differ significantly from the Kirchhoff approximation, which only contains the flashpoint signals.

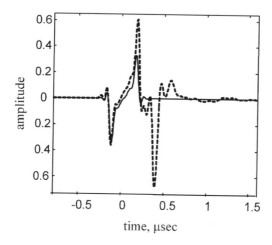

Fig. 10.29. A very wide-band simulated P-wave pulse-echo responses for a 0.381 mm radius crack in steel for an incident angle of 75 degrees. Kirchhoff approximation (solid line), MOOT solution (dashed line).

Ultimately the Kirchhoff approximation must fail at very high angles since in this approximation the flash point signals go to zero as the incident angle approaches 90 degrees while the exact solution remains finite. As an example of a very high angle case consider the flaw response at an angle of 75 degrees as shown in Fig. 10.29. At this angle the trailing flashpoint response predicted by the Kirchhoff approximation is much smaller than that given by the MOOT solution and the later arriving waves now are larger than either of the flashpoint responses so that the peak-to-peak response predicted by the Kirchhoff approximation is significantly in error. But as can be seen in Fig. 10.29 the Kirchhoff approximation continues to accurately model the first arriving flashpoint signal. It can also be seen from Fig. 10.29 that even for this angle the Kirchhoff approximation continues to model the arrival times of both flash point signals accurately. The arrival times of such crack tip signals are used in crack sizing methods such as the time-of-flight- diffraction (TOFD) method [10.3] and equivalent flaw sizing methods [Fundamentals], so the Kirchhoff approximation can be reliably used as the basis for those sizing methods.

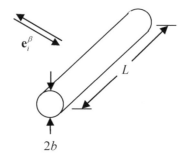

Fig. 10.30. The pulse-echo scattering of a cylindrical side-drilled hole of radius b and length L. The incident wave direction, $\mathbf{e}_i^\beta$, is assumed to lie in a plane perpendicular to the axis of the cylinder.

10.6 The Kirchhoff Approximation for Side-drilled Holes

Another scattering geometry that is commonly used in NDE calibration experiments is the side-drilled hole. This is a case where we can also obtain explicit results for the scattering amplitude in the Kirchhoff approximation. We will give the derivation here since it is not readily available in the literature. Consider the case of pulse-echo where the axis of a side-drilled hole of radius b and length L is normal to the plane of incidence (the plane containing the incident wave direction and the normal to the curved side of the side-drilled hole (Fig. 10.30). Equation (10.12) is again applicable to this case so the response of the side-drilled hole in the Kirchhoff approximation is given by

$$A\left(\mathbf{e}_i^\beta;-\mathbf{e}_i^\beta\right)=\frac{-ik_\beta}{2\pi}\int_{S_{lit}}\left(\mathbf{e}_i^\beta\cdot\mathbf{n}\right)\exp\left(2ik_\beta\mathbf{x}_s\cdot\mathbf{e}_i^\beta\right)dS\left(\mathbf{x}_s\right). \qquad (10.46)$$

Now, consider a surface S' that extends the lit surface to infinity in the $\mathbf{e}_i^\beta$ direction and a surface at infinity, S^∞, that closes this extended surface as shown in Fig. 10.31. Since $\mathbf{e}_i^\beta\cdot\mathbf{n}=0$ on S' and the integrand on S^∞ will

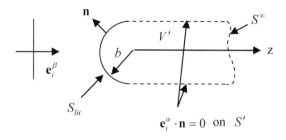

Fig. 10.31. The cross-section of the side-drilled hole showing the lit surface, S_{lit}, and the extension of that surface by the surfaces S' and S^{∞} to enclose the volume V'.

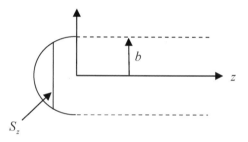

Fig. 10.32. The cross-sectional area, S_z, for the side-drilled hole geometry.

vanish if we add a small amount of damping to the plane wave term $\exp\left(2ik_{\beta}\mathbf{x}_s \cdot \mathbf{e}_i^{\beta}\right)$, we can write

$$A\left(\mathbf{e}_i^{\beta};-\mathbf{e}_i^{\beta}\right) = \frac{-ik_{\beta}}{2\pi} \int\limits_{S_{lit}+S'+S^{\infty}} \left(\mathbf{e}_i^{\beta}\cdot\mathbf{n}\right)\exp\left(2ik_{\beta}\mathbf{x}_s \cdot \mathbf{e}_i^{\beta}\right)dS\left(\mathbf{x}_s\right) \qquad (10.47)$$

and then use the divergence theorem to write the integral over the closed surface in Eq. (10.47) as a volume integral over the volume V' within this closed surface:

$$A\left(\mathbf{e}_i^{\beta};-\mathbf{e}_i^{\beta}\right) = \frac{k_{\beta}^2}{\pi} \int\limits_{V'} \exp\left(2ik_{\beta}\mathbf{e}_i^{\beta}\cdot\mathbf{x}\right)dV. \qquad (10.48)$$

If we take the z-axis as along the incident wave direction then $\mathbf{e}_i^\beta \cdot \mathbf{x} = z$ and we can write the volume element as $dV = S_z(z)dz$ where $S_z(z)$ is the cross-sectional area of the volume perpendicular to $\mathbf{e}_i^\beta$. For the volume V', however, we have directly from the geometry (see Fig. 10.32)

$$S_z(z) = \begin{cases} 0 & z < -b \\ 2L\sqrt{b^2 - z^2} & -b < z < 0 \\ 2Lb & z > 0 \end{cases} \tag{10.49}$$

so that Eq. (10.48) becomes

$$A\left(\mathbf{e}_i^\beta; -\mathbf{e}_i^\beta\right) = \frac{2Lk_\beta^2}{\pi} \int_{-b}^{0} \sqrt{b^2 - z^2} \, \exp\left(2ik_\beta z\right) dz$$
$$+ \frac{2Lbk_\beta^2}{\pi} \int_{0}^{\infty} \exp\left(2ik_\beta z\right) dz. \tag{10.50}$$

In the first integral in Eq. (10.50) let $x = -z/b$ and perform the integration explicitly for the second integral, again ignoring the limit at infinity by adding a small amount of damping to the complex exponential. We find

$$A\left(\mathbf{e}_i^\beta; -\mathbf{e}_i^\beta\right) = \frac{2L\left(k_\beta b\right)^2}{\pi} \int_{0}^{1} \sqrt{1 - x^2} \, \exp\left(-2ik_\beta bx\right) dx + \frac{iLbk_\beta}{\pi}. \tag{10.51}$$

But from Gradshteyn and Ryzhik [10.4]

$$\int_{0}^{1} \sqrt{1 - x^2} \, \cos\left(2kbx\right) dx = \frac{\sqrt{\pi}}{2}\left(\frac{1}{kb}\right)\Gamma\left(\frac{3}{2}\right)J_1\left(2kb\right)$$
$$\int_{0}^{1} \sqrt{1 - x^2} \, \sin\left(2kbx\right) dx = \frac{\sqrt{\pi}}{2}\left(\frac{1}{kb}\right)\Gamma\left(\frac{3}{2}\right)S_1\left(2kb\right) \tag{10.52}$$

and $\Gamma\left(\dfrac{3}{2}\right) = \dfrac{\sqrt{\pi}}{2}$ so that

$$A\left(\mathbf{e}_i^\beta; -\mathbf{e}_i^\beta\right) = \frac{\left(k_\beta b\right)L}{2}\left[J_1\left(2k_\beta b\right) - iS_1\left(2k_\beta b\right)\right] + \frac{i\left(k_\beta b\right)L}{\pi}, \tag{10.53}$$

where J_1 is a Bessel function of order one and S_1 is a Struve function of order one. Since these special functions can be easily calculated numerically,

Eq. (10.53) gives us an explicit expression for the pulse-echo scattering amplitude of the side-drilled hole. At high frequencies [10.5]

$$S_1(2kb) \cong Y_1(2kb) + \frac{2}{\pi} + O\left(\frac{1}{kb}\right)^2$$

$$J_1(2kb) - iY_1(2kb) = H_1^{(2)}(2kb) \tag{10.54}$$

$$\cong \sqrt{\frac{1}{\pi kb}} \exp\left[i(3\pi/4 - 2kb)\right],$$

where $H_1^{(2)}$ is a Hankel function of the second kind of order one and Y_1 is a Bessel function of the second kind of order one. Placing these approximations into Eq. (10.53), at high frequencies the pulse-echo scattering amplitude is given by

$$A\left(\mathbf{e}_i^\beta; -\mathbf{e}_i^\beta\right) \cong \frac{L}{2}\sqrt{\frac{k_\beta b}{\pi}} \exp\left[i(3\pi/4 - 2k_\beta b)\right]. \tag{10.55}$$

At low frequencies, we have instead

$$J_1(2kb) \cong \frac{kb}{\Gamma(2)} = kb$$

$$S_1(2kb) \cong \frac{8}{3\pi}(kb)^2 \tag{10.56}$$

so that the scattering amplitude becomes

$$A\left(\mathbf{e}_i^\beta; -\mathbf{e}_i^\beta\right) \cong \frac{ik_\beta bL}{\pi} \tag{10.57}$$

although we cannot expect the Kirchhoff approximation to be valid at these low frequencies.

Figures 10.33, 10.34 plot the magnitude and phase of the normalized pulse echo scattering amplitude versus wave number from the Kirchhoff solution, Eq. (10.53), and compares these results to the exact separation of variables solution for the two-dimensional pulse-echo P-wave scattering amplitude [10.6]. It can be seen that the Kirchhoff approximation agrees well with the separation of variables solution, particularly at the higher frequencies. In Fig. 10.33 both solutions approximately follow the high frequency square root behavior in frequency given by Eq. (10.55). The exact separation of variables solution has more oscillations than the Kirchhoff approximation since, like the spherical void case, the exact solution oscillations here come from the interference of the

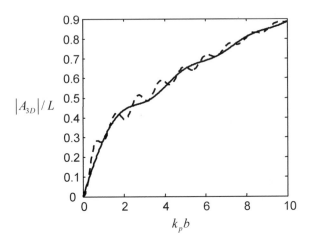

Fig. 10.33. The three-dimensional normalized pulse-echo P-wave scattering amplitude versus normalized wave number for a side drilled hole in the Kirchhoff approximation (solid line) and from the exact separation of variables solution (dashed line).

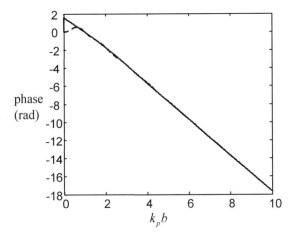

Fig. 10.34. The phase of the three-dimensional normalized pulse-echo P-wave scattering amplitude versus normalized wave number for a side drilled hole in the Kirchhoff approximation (solid line) and from the exact separation of variables solution (dashed line).

leading edge response of the side-drilled hole with a creeping wave that is not contained in the Kirchhoff approximation.

Since our Kirchhoff solution was obtained by considering the response of a three-dimensional cylinder of length L while the separation of variables solution is for the two-dimensional scattering from an infinitely long cylinder, some remarks are needed to describe how we made the comparison shown in Figs. 10.33 and 10.34. In two-dimensional scattering problems the waves in the far-field of the scatterer are not spherical waves but cylindrical waves and the two dimensional far-field scattering displacement $\mathbf{u}^{scatt} = \left(u_1^{scatt}, u_2^{scatt} \right)$ is given by [Fundamentals]

$$\mathbf{u}^{scatt}(\mathbf{y},\omega) = U_0 \frac{\tilde{\mathbf{A}}\left(\mathbf{e}_i^\beta;\mathbf{e}_s^p\right)}{\sqrt{R_s}} \exp\left(ik_p R_s\right) + U_0 \frac{\tilde{\mathbf{A}}\left(\mathbf{e}_i^\beta;\mathbf{e}_s^s\right)}{\sqrt{R_s}} \exp\left(ik_s R_s\right), \quad (10.58)$$

where the scattering amplitudes, $\tilde{\mathbf{A}}\left(\mathbf{e}_i^\beta;\mathbf{e}_s^\alpha\right)$, now are two-dimensional vectors and all the distances also are measured in a two dimensional space (y_1, y_2). In this case, if we compute the same component of the scattering amplitude as done for the three-dimensional case, we obtain

$$A_{2D}\left(\mathbf{e}_i^\beta;\mathbf{e}_s^\alpha\right) = \tilde{\mathbf{A}}\left(\mathbf{e}_i^\beta;\mathbf{e}_s^\alpha\right)\cdot\left(-\mathbf{d}^\alpha\right), \quad (10.59)$$

where we use the "2D" label to emphasize that the calculation is for the two-dimensional scattering amplitude. It can be shown that this component is given by [Fundamentals]

$$A_{2D}\left(\mathbf{e}_i^\beta;\mathbf{e}_s^\alpha\right) = \sqrt{\frac{i}{8\pi k_\alpha}} \frac{1}{\rho c_\alpha^2} \int_C d_\sigma^\alpha \left[\tilde{\tau}_{\gamma\sigma} n_\gamma + ik_\alpha C_{\sigma\delta\gamma\nu} e_{s\delta}^\alpha n_\gamma \tilde{u}_\nu \right]$$
$$\cdot \exp\left[-ik_\alpha e_{s\lambda}^\alpha y_\lambda \right] dc, \quad (10.60)$$

$$\text{(no sum on } s, \, \alpha)$$

where n_γ are the components of the outward normal to the flaw and the integration is a counterclockwise line integral around the edge of the two-dimensional scatterer. All the repeated Greek subscripts in Eq. (10.60) are summed over the values (1,2) only (no sum on s, α). Recall Eq. (10.7) for the same three-dimensional scattering amplitude component is given by

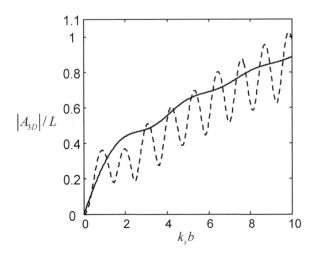

Fig. 10.35. The three-dimensional normalized pulse-echo SV-wave scattering amplitude versus normalized wave number for a side drilled hole in the Kirchhoff approximation (solid line) and from the exact separation of variables solution (dashed line).

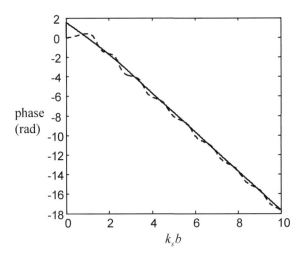

Fig. 10.36. The phase of the three-dimensional normalized pulse-echo SV-wave scattering amplitude versus normalized wave number for a side drilled hole in the Kirchhoff approximation (solid line) and from the exact separation of variables solution (dashed line).

$$A_{3D}\left(\mathbf{e}_i^\beta;\mathbf{e}_s^\alpha\right) = \frac{1}{4\pi\rho c_\alpha^2} \int\limits_S d_l^\alpha \left[\tilde{\tau}_{lk}n_k + ik_\alpha C_{lkpj}e_{sk}^\alpha n_p\tilde{u}_j\right]$$
$$\cdot \exp\left(-ik_\alpha \mathbf{x}_s \cdot \mathbf{e}_s^\alpha\right) dS\left(\mathbf{x}_s\right). \tag{10.61}$$

(no sum on s, α)

From Eqs. (10.60) and (10.61) we see that the two- and three-dimensional forms are very similar. In fact, if in the three-dimensional case the geometry and fields were all two-dimensional, i.e. if we set $n_3 = d_3^\alpha = e_{s3}^\alpha = \tilde{u}_3 = 0$ and assume $\tilde{\tau}_{\alpha\beta} = \tilde{\tau}_{\alpha\beta}\left(y_1,y_2,\omega\right)$, $\tilde{u}_\beta = \tilde{u}_\beta\left(y_1,y_2,\omega\right)$ we would obtain

$$A_{3D}\left(\mathbf{e}_i^\beta;\mathbf{e}_s^\alpha\right) = \frac{L}{4\pi\rho c_\alpha^2} \int\limits_S d_\sigma^\alpha \left[\tilde{\tau}_{\gamma\sigma}n_\gamma + ik_\alpha C_{\sigma\delta\gamma\nu}e_{s\delta}^\alpha n_\gamma\tilde{u}_\nu\right]$$
$$\cdot \exp\left(-ik_\alpha e_{s\lambda}^\alpha y_\lambda\right) dc. \tag{10.62}$$

(no sum on s, α)

Note that all these assumptions are fulfilled exactly by our three-dimensional solution for the side-drilled hole in the Kirchhoff approximation. These are also reasonable assumptions for more general scattering calculations if we assume the incident wave is a quasi-plane wave propagating in a plane which is perpendicular to the axis of the side-drilled hole. Comparing Eqs. (10.60) and (10.62) we find

$$A_{2D}\left(\mathbf{e}_i^\beta;\mathbf{e}_s^\alpha\right) = \left(\frac{2i\pi}{k_\alpha}\right)^{1/2} \frac{A_{3D}\left(\mathbf{e}_i^\beta;\mathbf{e}_s^\alpha\right)}{L}. \tag{10.63}$$

Equation (10.63) was used to transform the two-dimensional separation of variables scattering amplitude, A_{2D}, into an equivalent three-dimensional scattering amplitude, A_{3D}. In Figs. 10.33 and 10.34 the magnitude and phase of this exact three-dimensional amplitude was plotted and compared with the Kirchhoff solution. Thus, the quantities being plotted for both curves in those figures are based on A_{3D}/L. Figures 10.35 and 10.36 show the corresponding results for the pulse-echo scattering amplitude of the side-drilled hole calculated for shear (SV) waves. In this case the exact solution has deep oscillations since stronger SV- creep waves are generated than in the P-wave case. The Kirchhoff solution is unchanged in

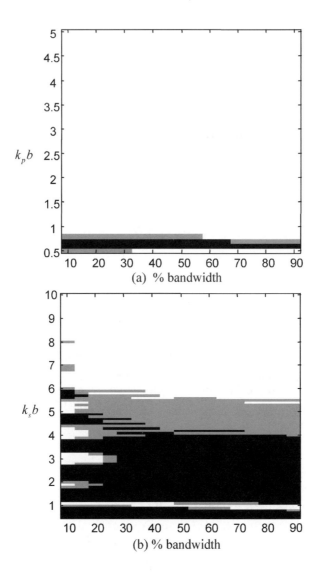

Fig. 10.37. A comparison of the peak-to-peak pulse-echo responses of a side drilled hole of radius b as calculated by the Kirchhoff approximation and the method of separation of variables where the non-dimensional wave number and bandwidth are varied. **(a)** Pulse-echo P-wave responses, **(b)** pulse-echo SV-wave responses. White region: peak-to-peak differences < 1 dB, Gray region: differences > 1 dB and <1.5 dB, Black region: differences > 1.5 dB.

form since Eq. (10.53) is applicable to both P- and SV-waves, but the normalized wave number appearing in the SV-wave case is $k_s b$.

The accuracy of the Kirchhoff approximation for the side drilled hole can also be studied as a function of the wave number and bandwidth as done with the spherical void. Like the spherical void, there is an exact separation of variables solution available for a cylindrical void that can be used to test the accuracy of the Kirchhoff approximation (see section 10.8). Figures 10.37 (a), (b) show the regions of validity of the Kirchhoff approximation for the side drilled hole that were obtained in the same fashion as Figs. 10.23 and 10.24 for the spheroid void and crack, respectively. Figure 10.37 (a) shows that for the pulse-echo P-wave case, the Kirchhoff approximation for the peak-to-peak response of the side drilled hole remains accurate (within 1 dB of the exact solution) for wave numbers even smaller than one and that there are virtually no bandwidth effects. In contrast the pulse-echo SV-wave response begins to show some small bandwidth effects at $k_s b = 8$ and the Kirchhoff approximation becomes inaccurate at all bandwidths for $k_s b < 4$, approximately.

10.7 The Born Approximation

Another approximation that is useful for simulating flaw scattering responses is the Born approximation [Fundamentals], [10.7-10.11]. This approximation is formally a low frequency, weak scattering approximation but we will show that with some modifications it may be applicable under a wider set of conditions. The Born approximation uses an exact volume integral representation of the far-field scattering amplitude given by [Fundamentals]

$$A\left(\mathbf{e}_i^\beta; \mathbf{e}_s^\alpha\right) = \frac{-d_q^\alpha}{4\pi\rho c_\alpha^2} \int_{V_f} \left[\Delta\rho\, \omega^2 \tilde{u}_q + ik_\alpha e_{sk}^\alpha \Delta C_{kqmj} \frac{\partial \tilde{u}_m}{\partial x_j} \right]$$
$$\cdot \exp\left(-ik_\alpha \mathbf{x} \cdot \mathbf{e}_s^\alpha\right) dV\left(\mathbf{x}\right) \tag{10.64}$$

(no sum on s, α)

where V_f is the volume of the flaw. In Eq. (10.64) $\Delta\rho = \rho_f(\mathbf{x}) - \rho$, $\Delta C_{kqmj} = C_{kqmj}^f(\mathbf{x}) - C_{kqmj}$, where $\rho_f(\mathbf{x}), C_{kqmj}^f(\mathbf{x})$ are the density and elastic constants of the flaw (both of which can vary with position in the

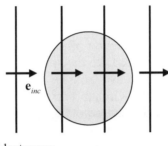

incident wave

Fig. 10.38. The Born approximation assumes that for a weakly scattering inclusion, the fields in the flaw are to first order the fields of the incident wave traveling in the host material as if the flaw was not present, as shown.

flaw) while ρ and C_{kqmj} are the density and elastic constants of the host material surrounding the flaw (both of which are assumed to be constants, i.e. the host material is taken to be homogeneous). The Born approximation assumes that the flaw is sufficiently similar to the surrounding host material that the fields appearing in Eq. (10.64) can be replaced approximately by those of the known incident wave. Physically, this means that to first order we are assuming that the incident wave passes through the flaw undisturbed, as shown in Fig. 10.38. For a pulse-echo setup and a homogeneous, isotropic flaw in a homogeneous, isotropic medium, for example, this results in a scattering amplitude expression given by [Fundamentals]

$$A\left(\mathbf{e}_i^\beta;-\mathbf{e}_i^\beta\right) = \frac{-\omega^2}{2\pi c_\beta^2}\left[\frac{\Delta\rho}{\rho} + \frac{\Delta c_\beta}{c_\beta}\right]\int_{V_f}\exp\left(2ik_\beta\mathbf{x}\cdot\mathbf{e}_i^\beta\right)dV\left(\mathbf{x}\right) \qquad (10.65)$$

that can, like the Kirchhoff approximation, be analytically evaluated for some simple flaw shapes. A similar expression can also be obtained for more general pitch-catch setups [Fundamentals]. An important feature of Eq. (10.65) is that the material properties of the flaw (contained in the coefficient of the integral) are completely separated from the flaw geometry information (contained in the integral itself). This separation has allowed the Born approximation to be successfully used in a number of flaw sizing applications [Fundamentals]. For a spherical inclusion Eq. (10.65) gives [Fundamentals]

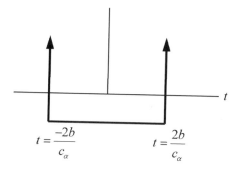

Fig. 10.39. The impulse response for the pulse echo scattering of a spherical inclusion of radius b in the Born approximation. The time $t = 0$ is when the incident wave front reaches the center of the flaw.

$$A\left(\mathbf{e}_i^{\beta};-\mathbf{e}_i^{\beta}\right) = -b\left[\frac{\Delta\rho}{\rho}+\frac{\Delta c_{\beta}}{c_{\beta}}\right]\left[\frac{\sin\left(2k_{\beta}b\right)-2k_{\beta}b\cos\left(2k_{\beta}b\right)}{2k_{\beta}b}\right], \qquad (10.66)$$

which can also be written in the alternate form

$$A\left(\mathbf{e}_i^{\beta};-\mathbf{e}_i^{\beta}\right) = -4k_{\beta}^2 b^3 F\frac{j_1\left(2k_{\beta}b\right)}{2k_{\beta}b}, \qquad (10.67)$$

where j_1 is a spherical Bessel function of order one and

$$F = \frac{1}{2}\left(\frac{\Delta\rho}{\rho}+\frac{\Delta c}{c_{\beta}}\right). \qquad (10.68)$$

If one inverts either Eq. (10.66) or (10.67) into the time domain, the impulse response of the spherical inclusion is the wave form shown in Fig. 10.39 [Fundamentals]. Like the Kirchhoff approximation, the Born approximation predicts a leading edge delta function response. This delta function is followed by a constant response as the wave passes through the entire flaw, and then one sees a trailing edge delta function response (which is equal to the leading edge delta function response) at the time when the wave has just finished passing through the flaw. Like the Kirchhoff approximation, the Born approximation is a single interaction type of approximation so that it neglects any other wave-flaw interactions such as creeping waves, multiple internal reflections, etc. Like the Kirchhoff approximation the Born approximation can also be applied to

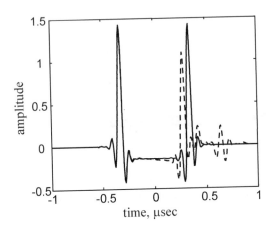

Fig. 10.40. The time domain pulse-echo P-wave response of a 1 mm radius spherical inclusion in steel where the density and compressional wave speed are both ten percent higher than the host steel. Solid line: Born approximation, dashed line: separation of variables solution.

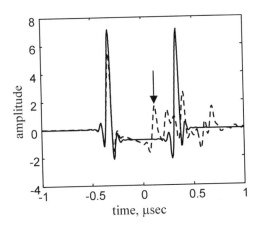

Fig. 10.41. The time domain pulse-echo P-wave response of a 1 mm radius spherical inclusion in steel where the density and compressional wave speed are both fifty percent higher than the host steel. Solid line: Born approximation, dashed line: separation of variables solution.

complex shaped flaws by performing the necessary integrations numerically.

Since the method of separation of variables can be used to obtain the "exact" solution for spherical inclusions, we can use that method to examine the accuracy of the wave form predictions of the Born approximation, just as we did with the Kirchhoff approximation for the spherical void (see Fig. 10.10). In the void case we used the separation of variables method to calculate the response to a relatively high frequency and then subtracted out (in the frequency domain) the known leading edge delta function response before inverting the result into the time domain with an FFT. Since the high frequency content of the other wave contributions (remainder of the lit surface response, creeping wave, etc.) is very small in pulse-echo for P-waves incident on a spherical void, we get in effect an infinite bandwidth time-domain response when we simply add the delta function back into the wave form symbolically, as done in Fig. 10.10. For weak-scattering inclusions the same process is not possible since the front and back surfaces are both delta functions. In the Born approximation these delta functions are always of equal amplitude but in comparing the Born approximation with the method of separation of variables it is found that the back surface delta function is only equal in amplitude to the front surface delta function in the very weak scattering limit and for all other flaws the back surface changes amplitude in an unknown fashion. Thus we cannot remove the delta functions from the Born response analytically, but we can still calculate the Born approximation and separation of variables responses over a range of frequencies and smoothly taper the high frequency response to zero with a cosine-squared windowing filter to reduce time domain "ringing". This is the method used here to compare the Born and separation of variables solutions in the following discussions. In all cases the cosine-squared filter began with values of one at 10 MHz and ended with a zero value at 20 MHz.

Figure 10.40 shows the pulse-echo P-wave response calculated in this fashion for a 1 mm radius spherical inclusion in steel where both the density and compressional wave speed of the inclusion was taken to be ten percent higher than the host steel. It can be seen for even these relatively small material changes the Born approximation does not accurately represent both the amplitude and time of arrival of the back surface response and there are other later arriving waves that are not predicted by the Born approximation. If one examines the same size inclusion in steel but takes the density and compressional wave speed to be 50% higher than the host then as seen in Fig. 10.41 the Born approximation is even more in error, with the back surface response located at a time well removed from

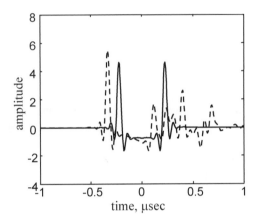

Fig. 10.42. The time domain pulse-echo P-wave response of a 1 mm radius spherical inclusion in steel where the density and compressional wave speed are both fifty percent higher than the host steel. Solid line: Doubly Distorted Born approximation, dashed line: separation of variables solution.

the actual back surface signal (located at the arrow in Fig. 10.41) and even the front surface leading edge response is significantly in error (Note: the specific changes in density and wave speed taken for this case and others that will be considered later are not intended to represent any particular real inclusion but are simply being used here to study the effects of large or small differences between the host and flaw materials). Having the leading edge response amplitude in error is particularly troublesome because it means that the Born approximation could not be reliably used to predict the detectability of inclusions except in the very weak scattering limit, which is not likely to be found in many real tests. A modified ad hoc approximation, called the *doubly distorted Born approximation* (DDBA) [10.12] was recently developed to try to remove some of these deficiencies of the ordinary Born approximation. In the DDBA, it was recognized that the wave field traveling in the flaw does not travel at the wave speed of the host material as assumed in the ordinary Born approximation. Instead, disturbances in the flaw should be traveling at the wave speed of the flaw material. Thus, the wave speeds appearing in both the integral kernel of the Born approximation expression and in the coefficient of that integral were changed in the DDBA from that of the host material to that of the flaw. For the pulse-echo response of a spherical inclusion, this change causes Eq. (10.67) to become instead

$$A\left(\mathbf{e}_i^\beta; -\mathbf{e}_i^\beta\right)_{DDBA} = -4k_{f\beta}^2 b^3 \tilde{F} \frac{j_1\left(2k_{f\beta}b\right)}{2k_{f\beta}b},$$ (10.69)

where $k_{f\beta} = \omega/c_{f\beta}$ is the wave number based on the flaw wave speed, $c_{f\beta}$, and the function, $\tilde{F}$, is given by

$$\tilde{F} = \frac{1}{2}\left(\frac{\Delta\rho}{\rho_f} + \frac{\Delta c}{c_{f\beta}}\right).$$ (10.70)

Figure 10.42 shows the result of using the DDBA on the same case shown in Fig. 10.40 (the pulse-echo P-wave response of a 1 mm radius inclusion in steel where the density and compressional wave speed are 50% higher than the host). Comparing the DDBA results of Fig. 10.42 with the Born approximation results of Fig. 10.41, we see that the DDBA amplitude of the leading edge response is closer to the separation of variables result than that of the Born approximation and also the time separation of the front and back surface responses in the DDBA now agrees with the separation of variables solution. However, the time of arrival of the front surface leading edge response is now incorrect, as seen in Fig. 10.41. The improvement in the leading edge response obtained with the DDBA can be understood by examining the behavior of the functions F and $\tilde{F}$. In [Fundamentals] it was noted that in pulse-echo the F function is just the weak scattering limit of the plane wave reflection coefficient, $R_{12}^{\beta;\beta}$ between the host and flaw materials, i.e.

$$\begin{aligned}
F &= \frac{1}{2}\left(\frac{\Delta\rho}{\rho} + \frac{\Delta c}{c_\beta}\right) \\
&= R_{12}^{\beta;\beta} + O\left(\left(\Delta\rho/\rho\right)^2, \left(\Delta c/c_\beta\right)^2\right) \\
&= \frac{\rho_f c_{f\beta} - \rho c_\beta}{\rho_f c_{f\beta} + \rho c_\beta} + O\left(\left(\Delta\rho/\rho\right)^2, \left(\Delta c/c_\beta\right)^2\right).
\end{aligned}$$ (10.71)

The functions F and $\tilde{F}$ together with $R_{12}^{\beta;\beta}$ are plotted versus $\Delta\rho/\rho = \Delta c/c_\beta = \lambda$ in Fig. 10.43. From that figure, we can see that the $\tilde{F}$ function does a much better job of following the behavior of $R_{12}^{\beta;\beta}$ for even large changes of density and wave speed. Since at high frequencies our previous discussions have shown that the pulse-echo leading edge response

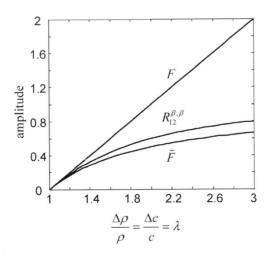

Fig. 10.43. A comparison of the F and $\tilde{F}$ functions and the plane wave reflection coefficient, $R_{12}^{\beta;\beta}$.

is controlled by $R_{12}^{\beta;\beta}$ (see Eq. (10.19)), the closer agreement of $\tilde{F}$ to this reflection coefficient is the reason for the improvements seen in Fig. 10.42. However, this fact also suggests that if one replaces the $\tilde{F}$ function in the DDBA by $R_{12}^{\beta;\beta}$ and includes a phase correction term to the DDBA to fix up its incorrect arrival time for the leading edge response, one should have a new model that agrees better with the separation of variables result. This new model we will call the *modified Born approximation* (MBA) [10.13]. In the MBA model Eq. (10.69) becomes

$$A\left(\mathbf{e}_i^\beta;-\mathbf{e}_i^\beta\right)_{MBA} = -4k_{f\beta}^2 b^3 R_{12}^{\beta;\beta} \exp\left[2ik_{f\beta}b\left(1-c_{f\beta}/c_\beta\right)\right]$$
$$\cdot \frac{j_1\left(2k_{f\beta}b\right)}{2k_{f\beta}b}. \tag{10.72}$$

Figure 10.44 shows the result of using the MBA model on the same case shown in Figs. 10.41 and 10.42 (the pulse-echo P-wave response of a 1 mm radius inclusion in steel where the flaw density and compressional wave speed are 50% higher than the host). It can be seen that the leading edge amplitude and time of arrival are now both correct as is the time of arrival of the back surface response. The MBA model will still model the amplitude of the back surface response as equal to the front surface amplitude and will not contain any of the other responses seen in the

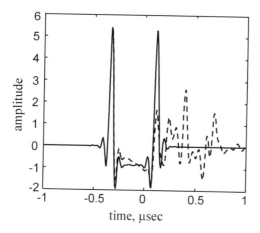

Fig. 10.44. The time domain pulse-echo P-wave response of a 1 mm radius spherical inclusion in steel where the density and compressional wave speed are both fifty percent higher than the host steel. Solid line: MBA model, dashed line: separation of variables solution.

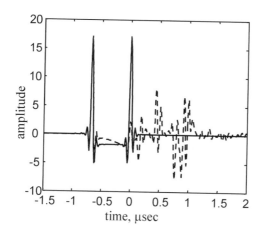

Fig. 10.45. The time domain pulse-echo P-wave response of a 1 mm radius spherical inclusion in steel where the density and compressional wave speed are both one hundred percent higher than the host steel. Solid line: MBA model, dashed line: separation of variables solution.

separation of variables solution, but in general Eq. (10.72) gives much better results than either the original Born or the DDBA models. Like the DDBA the MBA model is an ad hoc modification of the Born approximation but it appears to be a useful modification for dealing with inclusions that may be far from being weak scatterers. This can be seen in Figure 10.45 where the pulse-echo P-wave response for a 1 mm radius inclusion in steel is shown for a case with the flaw density and wave speed both 100% higher than that of the host material. Even in this extreme case the MBA model continues to capture the leading edge response correctly. For the pulse-echo response of a more general shaped inclusion, the MBA gives, from Eq. (10.65)

$$A\left(\mathbf{e}_i^\beta; -\mathbf{e}_i^\beta\right)_{MBA} = \frac{-\omega^2}{\pi c_{f\beta}^2} R_{12}^{\beta;\beta} \exp\left[2ik_{f\beta}r_e\left(1 - c_{f\beta}/c_\beta\right)\right]$$
$$\cdot \int_{V_f} \exp\left(2ik_{f\beta}\mathbf{x}\cdot\mathbf{e}_i^\beta\right)dV(\mathbf{x})$$

(10.73)

where r_e is the distance in the incident wave direction from a fixed point (usually the "center" of the flaw) to the point on the flaw surface where the incident wave front first touches the flaw.

One could of course use Eq. (10.19) to model just the leading edge response of an inclusion, even for more general pitch-catch setups. The advantage of using Eq. (10.73) is that although it is only valid for pulse-echo inspections it also captures the main features of the entire flaw response correctly in the weak scattering limit.

10.8 Separation of Variables Solutions

For spherical or cylindrical shaped scatterers in an elastic solid, one can use the method of separation of variables to express the exact scattering solution as an infinite sum of spherical Hankel functions and associated Legendre functions for the case of the sphere, and Hankel functions and complex exponential functions for the cylinder [10.14 – 10.25]. Even though both geometries are very simple shapes, they are useful for considering important scatterers such as pores or a side-drilled hole and they can serve as reference solutions for testing the accuracy of approximate methods. Although the separation of variables solutions are exact, they are expressed in terms of infinite sums that must be calculated numerically and more terms are needed as the scatterer becomes larger or the frequency becomes higher. Normally this is not a problem since with modern PCs it is possible to calculate scattering results for non-dimensional frequencies as high as, say, $kb \cong 100$.

In this section we will give the separation of variables solution for four cases: the pulse-echo response of a spherical void for both P-waves and S-waves, and the pulse echo response of a cylindrical void for P-waves and S-waves. These solutions have also been coded in MATLAB functions which are given in Appendix G.

First, consider the case of the pulse-echo P-wave response of a spherical void of radius b. Using the method of separation of variables, we find that [10.14], [Fundamentals]

$$A\left(\mathbf{e}_i^p; -\mathbf{e}_i^p\right) = \frac{-1}{ik_p} \sum_{n=0}^{\infty} (-1)^n A_n,$$ (10.74)

where

$$A_n = \frac{E_3 E_{42} - E_4 E_{32}}{E_{31} E_{42} - E_{41} E_{32}}$$ (10.75)

and we have

$$E_3 = (2n+1)\left\{\left[n^2 - n - \left(k_s^2 b^2 / 2\right)\right] j_n\left(k_{pb}\right) + 2k_p b j_{n+1}\left(k_p b\right)\right\}$$

$$E_4 = (2n+1)\left\{(n-1) j_n\left(k_{pb}\right) - k_p b j_{n+1}\left(k_p b\right)\right\}$$

$$E_{31} = \left[n^2 - n - \left(k_s^2 b^2 / 2\right)\right] h_n^{(1)}\left(k_p b\right) + 2k_p b h_{n+1}^{(1)}\left(k_p b\right)$$

$$E_{41} = (n-1) h_n^{(1)}\left(k_p b\right) - k_p b h_{n+1}^{(1)}\left(k_p b\right)$$ (10.76)

$$E_{32} = -n(n+1)\left[(n-1) h_n^{(1)}\left(k_s b\right) - k_s b h_{n+1}^{(1)}\left(k_s b\right)\right]$$

$$E_{42} = -\left[n^2 - 1 - \left(k_s^2 b^2 / 2\right)\right] h_n^{(1)}\left(k_s b\right) - k_s b h_{n+1}^{(1)}\left(k_s b\right).$$

For the SV-wave case, the separation of variables solution is of the form [10.18]

$$A\left(\mathbf{e}_i^s; -\mathbf{e}_i^s\right) = \frac{-1}{ik_s} \sum_{n=1}^{\infty} \frac{(-1)^n (2n+1) B_n}{2},$$ (10.77)

where

$$B_n = \frac{H_{13} J_{42} - H_{43} J_{12}}{H_{13} H_{42} - H_{43} H_{12}} - \frac{J_{41}}{H_{41}}$$ (10.78)

and

$$J_{12} = n(n+1)\left[(n-1)j_n(k_sb) - k_sb\,j_{n+1}(k_sb)\right]$$

$$H_{12} = n(n+1)\left[(n-1)h_n^{(1)}(k_sb) - k_sb\,h_{n+1}^{(1)}(k_sb)\right]$$

$$H_{13} = \left[n^2 - n - (k_s^2b^2/2)\right]h_n^{(1)}(k_pb) + 2k_pb\,h_{n+1}^{(1)}(k_pb)$$

$$J_{41} = (n-1)j_n(k_sb) - k_sb\,j_{n+1}(k_sb)$$

$$H_{41} = (n-1)h_n^{(1)}(k_sb) - k_sb\,h_{n+1}^{(1)}(k_sb) \tag{10.79}$$

$$J_{42} = \left[n^2 - 1 - (k_s^2b^2/2)\right]j_n(k_sb) + k_sb\,j_{n+1}(k_sb)$$

$$H_{42} = \left[n^2 - 1 - (k_s^2b^2/2)\right]h_n^{(1)}(k_sb) + k_sb\,h_{n+1}^{(1)}(k_sb)$$

$$H_{43} = (n-1)h_n^{(1)}(k_pb) - k_pb\,h_{n+1}^{(1)}(k_pb).$$

The pulse-echo P-wave time-domain response for a spherical void was shown previously in Fig. 10.10. There we could simply subtract off the leading edge response from the separation of variables solution in the frequency domain and then apply the inverse Fourier transform to the remaining portion of the response, which contains only low frequency signals. However, for the pulse-echo SV-wave response, the creeping waves are more significant and extend to very high frequencies. Figure 10.46 shows the magnitude of the pulse-echo SV-wave scattering amplitude of a 0.5 mm radius spherical pore in steel. The deep oscillations in the SV-wave response at high frequencies in comparison with the highly damped oscillations appearing in Fig. 10.8 for the P-wave response shows that a simple subtraction of the leading edge response will not lead to a response confined only to low frequencies. However, we can follow the procedure used in the Born approximation and apply a cosine-squared windowing filter to the frequency domain scattering amplitude before inverting the signal back into the time domain. Figure 10.47 shows the resulting time-domain SV-wave signal for the 0.5 mm radius pore in steel. For comparison purposes Fig. 10.48 shows the time-domain pulse-echo P-wave response for the 0.5 mm radius spherical pore in steel as calculated with the same filter function used in the Born approximation studies. It can be seen while the leading edge responses are almost identical in the two cases that the creeping wave in the P-wave case is indeed much smaller than in the SV-wave case.

For the case of a cylindrical void of radius b with the incident wave direction in a plane perpendicular to the axis of the cylinder, the 2-D separation of variables solution can be used to generate a normalized 3-D

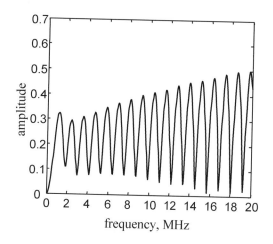

Fig. 10.46. The magnitude of the pulse-echo SV-wave response, $A(\mathbf{e}_i;-\mathbf{e}_i)$, versus frequency for a 0.5 mm radius spherical void in steel ($c_p = 5900$ m/s, $c_s = 3200$ m/sec) as calculated by the method of separation of variables.

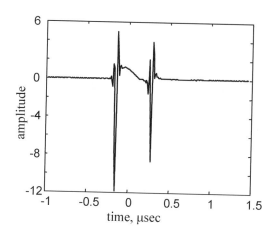

Fig. 10.47. The time domain response corresponding to Fig. 10.46 by applying a low-pass cosine-squared windowing filter between 10 and 20 MHz to the separation of variables solution and then inverting the result into the time domain with the inverse Fourier transform.

scattering amplitude for a cylinder of length L, as discussed previously. For an incident P-wave

$$\frac{A_{3D}\left(\mathbf{e}_i^p;-\mathbf{e}_i^p\right)}{L} = \frac{i}{2\pi}\sum_{n=0}^{\infty}\left(2-\delta_{0n}\right)\left(-1\right)^n F_n, \tag{10.80}$$

where

$$\delta_{0n} = \begin{cases} 1 & n=0 \\ 0 & otherwise \end{cases} \tag{10.81}$$

and

$$F_n = 1 + \frac{C_n^{(2)}\left(k_p b\right)C_n^{(1)}\left(k_s b\right) - D_n^{(2)}\left(k_p b\right)D_n^{(1)}\left(k_s b\right)}{C_n^{(1)}\left(k_p b\right)C_n^{(1)}\left(k_s b\right) - D_n^{(1)}\left(k_p b\right)D_n^{(1)}\left(k_s b\right)} \tag{10.82}$$

with

$$C_n^{(i)}\left(x\right) = \left(n^2 + n - \left(k_s b\right)^2/2\right)H_n^{(i)}\left(x\right) - \left(2n H_n^{(i)}\left(x\right) - x H_{n+1}^{(i)}\left(x\right)\right) \tag{10.83}$$

$$D_n^{(i)}\left(x\right) = n\left(n+1\right)H_n^{(i)}\left(x\right) - n\left(2n H_n^{(i)}\left(x\right) - x H_{n+1}^{(i)}\left(x\right)\right)$$

for $\left(i=1,2\right)$.

For S-waves, the polarization vector of the incident wave is assumed to lie in the plane perpendicular to the axis of the cylinder, so if we let that axis be horizontal, we are considering vertically polarized S-waves, i.e. SV-waves. Note that the 2-D scattering problem for horizontally polarized shear (SH) waves is just equivalent to a purely scalar scattering problem with no mode conversion while the SV-case does involve a coupling between P-waves and SV-waves. Here, we will only consider the SV-wave case as that is the one most commonly encountered in NDE setups. Again, transforming the 2-D separation of variables solution to a normalized 3-D scattering amplitude we have

$$\frac{A_{3D}\left(\mathbf{e}_i^{sv};-\mathbf{e}_i^{sv}\right)}{L} = \frac{i}{2\pi}\sum_{n=0}^{\infty}\left(2-\delta_{0n}\right)\left(-1\right)^n G_n \tag{10.84}$$

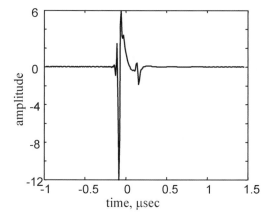

Fig. 10.48. The time-domain pulse-echo P-wave response of a 0.5 mm radius spherical void in steel ($c_p = 5900$ m/s, $c_s = 3200$ m/sec) obtained by applying a low-pass cosine-squared windowing filter between 10 and 20 MHz to the separation of variables solution and then inverting the result into the time domain with the inverse Fourier transform.

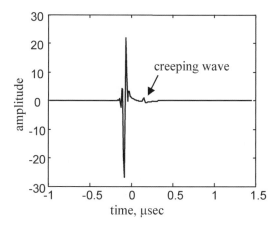

Fig. 10.49. The time-domain pulse-echo P-wave response of a 0.5 mm radius cylindrical void in steel ($c_p = 5900$ m/s, $c_s = 3200$ m/sec) obtained by applying a low-pass cosine-squared windowing filter between 10 and 20 MHz to the separation of variables solution and then inverting the result into the time domain with the inverse Fourier transform.

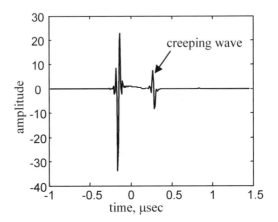

Fig. 10.50. The time-domain pulse-echo SV-wave response of a 0.5 mm radius cylindrical void in steel ($c_p = 5900$ m/s, $c_s = 3200$ m/sec) obtained by applying a low-pass cosine-squared windowing filter between 10 and 20 MHz to the separation of variables solution and then inverting the result into the time domain with the inverse Fourier transform.

where

$$G_n = 1 + \frac{C_n^{(2)}(k_s b) C_n^{(1)}(k_p b) - D_n^{(2)}(k_s b) D_n^{(1)}(k_p b)}{C_n^{(1)}(k_p b) C_n^{(1)}(k_s b) - D_n^{(1)}(k_p b) D_n^{(1)}(k_s b)}. \quad (10.85)$$

These solutions were used in Figs. 10.33-36 to calculate the "exact" solutions that were compared with the Kirchhoff approximation in the frequency domain. We can calculate these pulse-echo separation of variables solutions and then use a cosine-squared window again, as done for the spherical void, that allows us to invert these frequency domain values back into the time domain. The results for the P-wave response of a 0.5 mm radius cylindrical void in steel are shown in Fig. 10.49, and the corresponding SV-wave response is shown in Fig. 10.50. It can be seen that the creeping wave in the P-wave case is very small while it is much larger in the SV-wave case. In both cases, the early time response of the cylinder (Note: this early time response is not the leading edge response calculated earlier for 3-D scatterers), which is predicted well by the Kirchhoff approximation, is the dominant part of the overall pulse-echo response, which demonstrates that the Kirchhoff approximation works well except for very small flaws where the creeping wave and early time responses merge.

10.9 Other Scattering Models and Methods

In addition to the Born and Kirchhoff approximations there are other approximate methods that have been used to model flaw scattering problems, including elastodynamic ray theory (used to model the scattering of cracks) [10.26], [10.27] and approximate low frequency expansions [10.28 -10.30]. The high frequency ray methods lead to rather complex expressions and in certain cases singularities appear that invalidate the approximation. Low frequency expansions can produce some explicit analytical results but these have been of limited use in NDE applications because the responses of flaws of interest often are well beyond the range where such expansions are valid. This is unfortunate since it has been shown that there is much useful information in the low frequency response of flaws [10.30].

Besides the method of separation of variables there are five other numerical methods that have been commonly used for solving flaw scattering problems: the T-matrix method or the closely related method of optimal truncation (MOOT), the method of finite differences, the finite element method, the boundary element method, and the elastodynamic finite integration technique (EFIT).

The T-matrix method and the method of optimal truncation (MOOT) express the scattering solution for shapes other than spheres and cylinders in terms of the same special functions used in the method of separation of variables [10.31-10.35]. These two methods differ only in the way they approximately satisfy the boundary conditions. Both spheroidal and circular crack-like geometries have been considered with these methods. Like the separation of variables methods, these solutions are expressed in terms of infinite series and it is necessary to keep a sufficiently large number of terms in order to guarantee convergence of the solutions.

The method of finite differences has also been applied to flaw scattering problems [10.36], [10.37]. Unlike the separation of variables or T-matrix methods, this approach approximates the governing differential equations of motion directly in the time domain for the elastic material surrounding the flaw and replaces those equations with corresponding difference equations which are then solved numerically for field values defined on a mesh (or "grid") of discrete points as a function of time. This method can in principle handle rather general problems but in practice there are a number of issues that have limited the use of this method. First, it is convenient to use regular shaped meshes (such as rectangular meshes) with this method, but such meshes do not readily allow one to satisfy the boundary conditions at the flaw surface if that surface is rather complex. Second, since all of the material exterior to the flaw must be meshed, to keep the computational burden manageable the mesh must eventually be artificially truncated, leading to "fictitious" boundaries. Extraneous waves are generated at such fictitious boundaries that must be suppressed. This is often done by the application of special absorbing boundary conditions at the fictitious boundaries that minimize the extra waves generated or by keeping the fictitious surfaces sufficiently far from the flaw so that the extraneous waves do not contaminate the solution for the time interval considered. Finally, another problem inherent to the finite difference method is the large amount of computations needed, particularly for 3-D scattering problems. Thus even on modern computers, many of the finite difference solutions one sees are for 2-D problems. One nice feature of this method, however, is that it yields the solutions at all points in the solid directly, which allows one to view the complex wave/flaw interactions present in graphical form, including movies of those interactions. The mass-spring lattice model of Yim is a recent model of the finite difference type that has been used in this manner [10.38].

The finite element method, like the finite difference method, solves for the scattered fields on a mesh of discrete points [10.39], [10.40]. In the finite element method, however, the mesh is generated by an assemblage of small elements which can have different shapes so that it is not difficult to adapt the mesh to even complicated flaw shapes. Unlike the finite difference method, however, the finite element method does not directly approximate the equations of motion but instead it minimizes an energy functional for the assemblage of elements where the fields and material properties in each element are approximated by relatively simple functions such as polynomials. This allows the finite element method to model very complex materials, including both inhomogeneous and anisotropic materials. Ultimately, the finite element method generates a large, banded system of simultaneous equations that must be solved numerically. Like the finite difference method, the finite element method must deal with fictitious boundaries and suppress the extra waves generated by those boundaries. Perhaps the greatest challenge faced with the finite element method is the "curse of small wavelength", i.e. in order to maintain accurate solutions the finite element method must keep the element size very small (on the order of 5-10 elements per wavelength). Since most NDE applications use very high frequency waves (and hence the corresponding wavelengths are very small) this limitation makes it very computationally intensive to simulate general 3-D problems, so that (like with finite differences) one often sees finite element solutions applied to simpler 2-D or axisymmetric situations.

The boundary element method is an attractive method for solving flaw scattering problems [10.41-10.45] since it uses a fundamental solution for waves in the solid to generate integral equations for the displacements and tractions on the flaw surface, and these are precisely the fields needed to calculate the scattered waves and the far-field scattering amplitude of the flaw. These integral equations are solved by breaking the flaw surface into small elements and assuming some simple form for the fields in each element, which leads to a large set of simultaneous equations. Unlike the finite element system, however, the boundary element system of equations is not banded. Because the boundary element method deals only with the fields on the surface of the flaw, there are no fictitious surfaces in the solid that need to be considered with this method. In fact, the boundary element method can easily handle flaw scattering problems in infinite regions. However, when dealing with volumetric flaws the boundary element solution can be contaminated by fictitious resonances that render the solution inaccurate at certain frequencies so that special procedures need to be taken to suppress this unwanted behavior. Like the finite element method the boundary element method is affected by the curse of small wavelength. Thus, it is generally very computationally expensive in terms of both computer storage and calculation time to consider 3-D scattering problems with the boundary element method for, say, $kb > 20$. Perhaps the most important limitation of this method is the need for a fundamental solution to generate the requisite integral equations. Although a fundamental solution for a homogeneous, isotropic solid is available in exact analytical form [Fundamentals], for homogeneous anisotropic materials the fundamental solution is only known in an integral form that must be calculated numerically [10.46] and such fundamental solutions are not available for general inhomogeneous materials.

The elastodynamic finite integration technique (EFIT) is similar in some respects to the finite difference method in that it works with an approximation of the equations of motion, but, unlike the finite difference method, EFIT uses an integral form of those equations of motion [10.47]. Like the finite difference and finite element methods, EFIT can serve as both a beam model and a flaw scattering model since both of those aspects of the wave-flaw interactions are treated simultaneously. Also, like the other numerical methods, the generality of the EFIT approach, which in principle can handle quite complex inhomogeneous and anisotropic media problems, has to be weighed against its overall numerical costs.

There is still considerable opportunity for improving the state of the art in flaw scattering modeling for NDE applications. Although approximate methods such as the Kirchhoff and Born approximation are very valuable, they are limited in the features of the scattering process that they can simulate, while more exact numerical methods suffer from computational inefficiencies. Surface breaking cracks, porosity, and multiple distributed cracks (as found in stress-corrosion cracking problems), are examples where simple, efficient, and accurate scattering models are not currently available.

10.10 References

10.1 Schmerr LW, Sedov A (2003) Modeling ultrasonic problems for the 2002 benchmark session. In: Thompson DO, Chimenti DE (eds) Review of progress in quantitative nondestructive evaluation 22B. American Institute of Physics, Melville, NY, pp 1776-1783

10.2 Harker AH (1988) Elastic waves in solids. Institute of Physics Publishing Ltd., Bristol, England

10.3 Charlesworth JP, Temple JAG (2001) Engineering applications of ultrasonic time-of-flight diffraction. Research Studies Press, Philadelphia, PA

10.4 Gradshteyn IS, Ryzhik IM (1980) Table of integrals, series, and products. Academic Press, New York, NY

10.5 Abramowitz M, Stegun IA (1965) Handbook of mathematical functions. Dover Publications, New York, NY

10.6 Brind RJ, Achenbach JD, Gubernatis JE (1984) High-frequency scattering of elastic waves from cylindrical cavities. Wave Motion 6: 41-60

10.7 Gubernatis JE, Domany E , Krumhansl JA, Hubermann M (1977) The Born approximation in the theory of scattering of elastic waves by flaws. J. Appl. Phys. 48: 2812-2819

10.8 Hudson JA, Heritage JR (1981) The use of the Born approximation in seismic scattering problems. Geophys. J. Royal Astron. Soc. 66: 221-240

10.9 Gubernatis JE, Domany E , Krumhansl JA (1977) Formal aspects of the theory of scattering of ultrasound by flaws in elastic materials. J. Appl. Phys. 48: 2804-2811

10.10 Rose JH, Richardson JM (1982) Time domain Born approximation. Journ. Nondestr. Eval. 3: 45-53

10.11 Rose JH (1989) Elastic wave inverse scattering in nondestructive evaluation. Pure Appl. Geophys. 131: 715-739

10.12 Darmon M, Calmon P, Bele B (2004) An integrated model to simulate the scattering of ultrasound by inclusions in steels. Ultrasonics 42: 237-241

10.13 Huang R, Schmerr LW, Sedov A (2006) A modified Born approximation for scattering in isotopic and anisotropic elastic solids. Journ. Nondestr. Eval. 25: 139-154

10.14 Ying CF, Truell R (1956) Scattering of a plane compressional wave by a spherical obstacle in an isotropically elastic solid. J. Appl. Phys. 27: 1086-1097

10.15 Pao YH, Mow CC (1963) Scattering of a plane compressional wave by a spherical obstacle. J. Appl. Phys. 34: 493-499

10.16 Varadan VV, Ma Y, Varadan VK, Lakhtakia A (1991) Scattering of Waves by Spheres and Cylinders. In: Varadan VV, Lakhtakia A, Varadan VK (eds) Field representations and introduction to scattering. Elsevier Science publishers, Amsterdam, The Netherlands, Chapter 5

10.17 Pao YH, Mow CC (1973) Diffraction of elastic waves and dynamic stress concentrations. Crane, Russak and Co., New York, NY

10.18 Einspruch N, Witterholt E, Truell R (1960) Scattering of a plane transverse wave by a spherical obstacle in an elastic medium. J. Appl. Phys. 31: 806-818

10.19 McBride RJ, Kraft DW (1972) Scattering of a transverse elastic wave by an elastic sphere in a solid medium. J. Appl. Phys. 43: 4853-4861

10.20 Knopoff L (1959) Scattering of shear waves by spherical obstacles. Geophysics 24: 209-219

10.21 Morse PM, Feshbach H (1953) Methods of theoretical physics, parts I and II. McGraw-Hill, New York, NY

10.22 Einspruch N, Truell R (1960) Scattering of a plane longitudinal wave by a spherical fluid obstacle in an elastic medium. J. Acoust. Soc. Am. 32: 214-220

10.23 Truell R, Elbaum C, Chick BB (1969) Ultrasonic methods in solid state physics. Academic Press, New York, NY

10.24 Kraft DW, Franzblau M (1971) Scattering of elastic waves from a spherical cavity in a solid medium. J. Appl. Phys. 42: 3019-3024

10.25 Knopoff L (1959) Scattering of compressional waves by spherical obstacles. Geophysics 24: 30-39

10.26 Achenbach JD, Gautesen AK, McMaken H (1982) Ray methods for waves in elastic solids. Pitman Books Ltd., Boston, MA

10.27 Langenberg KJ, Schmitz V (1986) Numerical modeling of ultrasonic scattering by cracks. Nuclear Eng. and Design 94: 427-445

10.28 Richardson JM (1984) Scattering of elastic waves from symmetric inhomogeneities at low frequencies. Wave Motion 6: 325-336

10.29 Rose JH (1987) Elastodynamic long wavelength phase. Ultrasonics 25: 141-146

10.30 Kohn W, Rice JR (1979) Scattering of long-wavelength elastic waves from localized defects in solids. J. Appl. Phys. 50: 3346

10.31 Varadan VK, Varadan VV (eds.) (1980) Acoustic, electromagnetic and elastic wave scattering – focus on the T-matrix approach. Pergamon Press, New York, NY

10.32 Varatharajulu V, Pao YH (1976) Scattering matrix for elastic wave I. theory. J. Acoust. Soc. Am. 60: 556-566

10.33 Varadan VV (1978) Scattering matrix for elastic waves II. application to elliptical cylinders. J. Acoust. Soc. Am. 63: 1014-1024

10.34 Opsal JL, Visscher WM (1985) Theory of elastic wave scattering: applications of the method of optimal truncation. J. Appl. Phys. 58: 1102-1115

10.35 Visscher WM (1981) Calculation of the scattering of elastic waves from a penny-shaped crack by the method of optimal truncation. Wave Motion 3: 49-69

10.36 Bond LJ, Punjani M, Safari N (1988) Ultrasonic wave propagation and scattering using explicit finite difference methods. In: Blakemore M, Georgiou GA, (eds) Mathematical modelling in nondestructive testing. Clarendon Press, Oxford, England

10.37 Yamawaki H, Saito T (2000) Numerical calculation of ultrasonic propagation with anisotropy. Nondestr. Test. & Eval. Int'l 33: 489-497

10.38 Yim H, Baek E (2002) Two-dimensional numerical modeling and simulation of ultrasonic testing. J. Korean Soc. Nondestr. Test. 22: 649 -658

10.39 Ludwig R, Lord W (1988) A finite-element formulation for the study of ultrasonic NDT systems. IEEE Trans. Ultrason., Ferro., and Freq. Control UFFC-35: 809-820

10.40 Kishore NN, Sridhar I, Iyengar NGR (2000) Finite element modeling of the scattering of ultrasonic waves by isolated flaws. Nondestr. Test. & Eval. Int'l 33: 297-305

10.41 Dominguez J (1994) Boundary elements in dynamics. Elsevier Applied Science, Amsterdam, The Netherlands

10.42 Kitahara M (1985) Boundary integral equation methods in eigenvalue problems in elastodynamics and thin plates. Studies in applied mechanics Vol. 10, Elsevier, Amsterdam, The Netherlands

10.43 Do Rega Silva J (1994) Acoustic and elastic wave scattering using boundary elements. W.I.T. Press, Southhampton, United Kingdom

10.44 Krishnasamy G, Schmerr LW, Rudolphi T J, Rizzo FJ (1990) Hypersingular boundary integral equations: some applications in acoustic and elastic wave scattering. Trans. ASME, J. Appl. Mech 57: 404-414

10.45 Bonnet M (1995) Boundary integral equation methods for solids and fluids. J. Wiley and Sons, New York, NY

10.46 Wang CY, Achenbach JD (1995) Three-dimensional time-harmonic elastodynamic Green's functions for anisotropic solids. Proc. Roy. Soc. London, Ser. A 449: 441-458

10.47 Fellinger P, Marklein R, Langenberg KJ, Klaholz S (1995) Numerical modeling of elastic wave propagation and scattering with EFIT – elastodynamic finite integration technique. Wave Motion 21: 47-66

10.11 Exercises

1. Equations (10.14) and (10.36) give pulse-echo far-field scattering amplitude component versus frequency for a spherical pore and a circular crack, respectively, in the Kirchhoff approximation. In Chapter 12 a

MATLAB function A_void is given that implements Eq. (10.14). The function A_void has as its argument a setup structure that contains all the necessary parameters for ultrasonic flaw response simulations. Rewrite that function as a new MATLAB function, A_pore, that requires only the parameters needed to calculate the scattering amplitude and returns this scattering amplitude with the function call:

>> A = A_pore(f, b, c);

where f contains the frequencies (in MHz) at which the scattering amplitude is to be evaluated, b is the pore radius (in mm), and c is the wave speed of the surrounding material (in m/sec). By similarly modifying the function A_crack given in Chapter 12, write a new MATLAB function, A_circ, for the pulse-echo response of a circular crack of radius b which has a calling sequence:

>> A = A_circ(f, theta, b, c);

where theta is the angle (in degrees) that the incident wave makes with the normal to the crack.

In MATLAB generate a vector of 200 frequency values ranging from 0 to 30 MHz and using these two MATLAB functions plot the magnitude of the P-wave pulse-echo scattering amplitude component versus frequency for a 1mm radius pore in steel (c = 5900 m/sec) and a 1 mm radius crack in steel at an incident angle of 10 degrees. Compare your plots to Figs. (10.8) and (10.17).

2. The time-domain pulse-echo scattering amplitude responses of a spherical pore and a circular crack in the Kirchhoff approximation were given by Eqs. (10.15) and (10.39), respectively. These time-domain signals were computed by performing the inverse Fourier transforms of Eqs. (10.14) and (10.36) exactly so that they are for an infinite bandwidth system. Here we want to examine these time-domain responses for finite bandwidths.

(a) In MATLAB generate a vector, f, of 1024 frequencies ranging from 0 to 100 MHz using the function s_space (see Appendix G). Compute the scattering amplitude components of a spherical pore and circular crack for the parameters given in the previous exercise. Multiply these scattering amplitudes (element by element) with the output of the MATLAB function system_f(f, amp, fc, bw) which generates a Gaussian-shaped window of amplitude amp, centered at a frequency, fc, and having a 6 dB bandwidth, bw. Take amp = 1.0, fc = 5.0, bw = 4.0 (see Appendix G). Invert these products into the time domain using IFourierT and plot the time domain signal versus the time, t. These results show the flaw response as would be typically measured in a relatively wideband ultrasonic system.

(b) Repeat part (a) but replace the function system_f by the function lp_filter (f, fstart, f, end) which is a low-pass filter that is unity for frequencies below fstart and is tapered smoothly to zero at the frequency fend (see Appendix G). For frequencies above fend the filter is zero. Use this function with fstart = 20 MHz, fend = 30 MHz to generate and plot the same time-domain signals found in part (a). Comparing these results with part (a), what can you conclude?

11 Ultrasonic Measurement Models

In the previous Chapters we have shown that in order to predict the measured signals in an ultrasonic test one needs to know the system function, $s(\omega)$, and the acoustic/elastic transfer function, $t_A(\omega)$ of the system. Then the frequency components of the measured voltage, $V_R(\omega)$, are given by

$$V_R(\omega) = s(\omega) t_A(\omega). \tag{11.1}$$

We have seen how to obtain the system function, either by measurement of all the electrical and electromechanical components that it contains, or by a direct measurement in a calibration setup. In either case, if a flaw measurement is made with the same components and under the same conditions that the system function, $s(\omega)$, is measured, this same system function can be used in Eq. (11.1) for the flaw measurement. We have also given explicit expressions for the acoustic/elastic transfer function in some simple calibration setups. For a flaw measurement we need also to be able to describe this transfer function in terms of quantities that can be modeled or measured. Once such a transfer function is known, Eq. (11.1) provides a complete *ultrasonic measurement model* of the flaw measurement system. This Chapter will describe how to construct models of the acoustic/ elastic transfer function and the types of overall measurement models that result.

11.1 Reciprocity-based Measurement Model

It will be shown in this section that the acoustic/elastic transfer function can be modeled with reciprocity relations for fluid and elastic media. These reciprocity relations are very general, relying primarily on the assumption of linearity of the media involved. We have already seen reciprocity play a role in defining the electrical and electromechanical components of a measurement system. For purely electrical components, like the cable, reciprocity was given in terms of the electrical input and output voltages and currents of a two port system in the form (see Eq. (3.3)):

$$V_1^{(1)}I_1^{(2)} - V_1^{(2)}I_1^{(1)} = V_2^{(1)}I_2^{(2)} - V_2^{(2)}I_2^{(1)}.$$

(11.2)

Similarly the transducer satisfied a reciprocity relation between electrical and mechanical quantities (see Eq. (4.4)):

$$V^{(1)}I^{(2)} - V^{(2)}I^{(1)} = F^{(1)}v^{(2)} - F^{(2)}v^{(1)}.$$

(11.3)

To model the wave propagation and scattering processes contained in the acoustic/elastic transfer function, one needs to state similar reciprocity relations for the 3-D acoustic and elastic fields involved. For a fluid, for example, if one has a volume, V, of a fluid and two different wave fields (identified as states (1) and (2)) in that volume that satisfy the same homogeneous wave equation (no body force sources), then on the closed surface, S, of V we must have satisfied the reciprocity relationship [Fundamentals]

$$\int_S \left(p^{(1)}\mathbf{v}^{(2)} - p^{(2)}\mathbf{v}^{(1)} \right) \cdot \mathbf{n}dS = 0,$$

(11.4)

where $p^{(1)}, \mathbf{v}^{(1)}$ are the pressure and velocity fields for state (1), $p^{(2)}, \mathbf{v}^{(2)}$ are the pressure and velocity fields for state (2), and $\mathbf{n}$ is a unit vector normal to S.

Similarly for a linear, elastic solid, one has a reciprocity relationship between two stress and velocity fields acting in the same volume, V, of the same elastic material. If those two fields both satisfy Navier's equations in V for no body force sources, then [Fundamentals]

$$\int_S \left(\mathbf{t}^{(1)} \cdot \mathbf{v}^{(2)} - \mathbf{t}^{(2)} \cdot \mathbf{v}^{(1)} \right) dS = 0,$$

(11.5)

where $\mathbf{t}^{(1)}, \mathbf{v}^{(1)}$ are the stress (traction) vector and velocity vector for state (1) and $\mathbf{t}^{(2)}, \mathbf{v}^{(2)}$ are the stress vector and velocity vector for state (2). The Cartesian components of the stress vector are given in terms of the Cartesian stresses, τ_{ij}, in the solid by

$$t_i = \tau_{ji}n_j, \qquad (i = 1,2,3)$$

(11.6)

where n_j are the components of the normal to the surface, S, surrounding the volume.

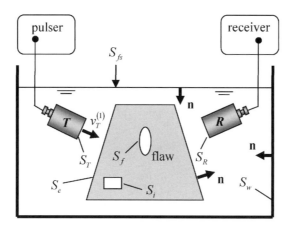

Fig. 11.1. An ultrasonic immersion flaw measurement system. This setup is designated as state (1) in our reciprocity relations.

We will demonstrate the application of these reciprocity relations to the flaw measurement system shown in Fig. 11.1. We will call this setup state (1). In this state we have the transmitting piston transducer, T, firing and generating a normal velocity, $v_T^{(1)}(\omega)$, on its surface, S_T, while the receiving transducer, R, is picking up the signals received from the flaw and other reflectors over its surface, S_R. The surface of the flaw itself is denoted as S_f. We have also labeled other surfaces in Fig. 11.1 as follows: Surface S_{fs} is the free surface of the fluid, S_w is the surface of the tank wall in contact with the fluid, S_e is the surface of the elastic solid being inspected, and S_i is the surface of one or more internal surfaces of the solid (other than the flaw). The unit normals to the various surfaces in contact with the fluid are also shown in Fig. 11.1. State (2) is shown in Fig. 11.2. In this state, we drive the "receiving" transducer R with a normal velocity, $v_R^{(2)}(\omega)$, on its surface, S_R, and we have the flaw in the component absent. We will also use a state (3), shown in Fig. 11.3. This state is identical to state (1), as shown in Fig. 11.3, except that the flaw is also absent in this state.

First, apply the reciprocity relationship to the common fluid region in states (1) and (2). There are no sources inside the fluid so we have:

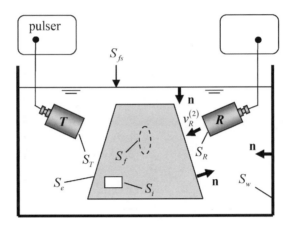

Fig. 11.2. The same measurement configuration as in Fig. 11.1 but where transducer R is assumed to be driven as a transmitter (the pulser driving T is quiescent) and the flaw is absent. The surface, S_f, is defined in this setup as the same surface that was occupied by the flaw in state (1). This configuration is designated as state (2) in the reciprocity relations.

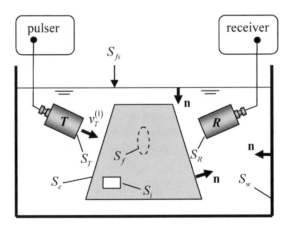

Fig. 11.3. The same configuration as shown in Fig. 11.1 except the flaw is now absent. The surface, S_f, is defined in this setup as the same surface that was occupied by the flaw in state (1). This configuration is designated as state (3) in the reciprocity relations.

$$\int\limits_{S_{fs}+S_w+S_T+S_R+S_e+S_{other}} \left(p^{(1)}\mathbf{v}^{(2)} - p^{(2)}\mathbf{v}^{(1)}\right)\cdot\mathbf{n}dS = 0. \tag{11.7}$$

For both states (1) and (2) at the free surface we have $p^{(1)} = p^{(2)} = 0$ so that the integral over S_{fs} is zero and can be eliminated from Eq. (11.7). Similarly for the tank wall, which is assumed to be rigid, we have $\mathbf{v}^{(1)}\cdot\mathbf{n} = \mathbf{v}^{(2)}\cdot\mathbf{n} = 0$ so the integral over S_w can also be eliminated. The surface S_{other} includes all other surfaces in contact with the fluid not shown explicitly in Fig. 11.1. These other surfaces would be the surfaces of the cables, the parts of the transducer surfaces other than the active surfaces S_T and S_R and the surfaces of the supports (not shown) of the component being inspected. We will assume these other surfaces, like the tank wall, are rigid so they also can be eliminated from Eq. (11.7). [Note: strictly speaking, the assumption that the tank wall and other surfaces are rigid is not needed to eliminate them from Eq. (11.7). The integrals over those surfaces can be eliminated by simply using the fact that they do not themselves contain any acoustic sources.] These results then reduce Eq. (11.7) to the form

$$\int\limits_{S_T+S_R} \left(p^{(1)}\mathbf{v}^{(2)} - p^{(2)}\mathbf{v}^{(1)}\right)\cdot\mathbf{n}dS = -\int\limits_{S_e} \left(p^{(1)}\mathbf{v}^{(2)} - p^{(2)}\mathbf{v}^{(1)}\right)\cdot\mathbf{n}dS. \tag{11.8}$$

Since we have assumed the transducers are acting as pistons, we can remove the velocity terms from the integrals in Eq. (11.8), which leaves the remaining integrals of the pressure over the transducer faces as just force terms, and we obtain

$$\left(F_T^{(1)}v_T^{(2)} - F_T^{(2)}v_T^{(1)}\right) + \left(F_R^{(1)}v_R^{(2)} - F_R^{(2)}v_R^{(1)}\right) = -\int\limits_{S_e} \left(p^{(1)}\mathbf{v}^{(2)} - p^{(2)}\mathbf{v}^{(1)}\right)\cdot\mathbf{n}dS \tag{11.9}$$

where $F_T^{(m)}, v_T^{(m)}$ $(m = 1, 2)$ are the compressive forces and normal velocities at the surface, S_T, of the transmitting transducer in states (1) and (2), and $F_R^{(m)}, v_R^{(m)}$ $(m = 1, 2)$ are the corresponding forces and normal velocities acting on the receiving transducer for those states. The directions of the normal velocities all are positive when pointing outwards from the transducer face into the fluid. On the surface, S_e, of the component being inspected the traction and normal velocity must be continuous, i.e. we have

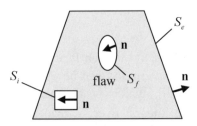

Fig. 11.4. The geometry of the solid component being inspected, showing the surfaces involved in the reciprocity relations and the directions of the unit normal on those surfaces.

$$-p^{(m)}\mathbf{n}\Big|_{fluid} = \mathbf{t}^{(m)}\Big|_{solid}$$

$$\mathbf{v}^{(m)} \cdot \mathbf{n}\Big|_{fluid} = \mathbf{v}^{(m)} \cdot \mathbf{n}\Big|_{solid} \tag{11.10}$$

for $m = 1,2$ so that Eq. (11.9) can also be written in terms of the surface fields on the solid as

$$\left(F_T^{(1)}v_T^{(2)} - F_T^{(2)}v_T^{(1)}\right) + \left(F_R^{(1)}v_R^{(2)} - F_R^{(2)}v_R^{(1)}\right)$$

$$= \int_{S_e}\left(\mathbf{t}^{(1)} \cdot \mathbf{v}^{(2)} - \mathbf{t}^{(2)} \cdot \mathbf{v}^{(1)}\right)dS. \tag{11.11}$$

Now, consider the volume of solid contained between the external surface, S_e, which is in contact with the fluid, and the internal surfaces consisting of the flaw surface, S_f, and other internal surfaces, S_i (Fig. 11.4). Since there are no sources of sound inside this volume, we must have

$$\int_{S_e+S_f+S_i}\left(\mathbf{t}^{(1)} \cdot \mathbf{v}^{(2)} - \mathbf{t}^{(2)} \cdot \mathbf{v}^{(1)}\right)dS = 0. \tag{11.12}$$

The surfaces S_i are present in both states (1) and (2). If those surfaces are traction free in both states (e.g. if there are holes in the component being inspected), or if those surfaces are source-free inclusions of other materials, the integral over S_i in Eq. (11.12) will vanish. We then find

$$\int_{S_e}\left(\mathbf{t}^{(1)} \cdot \mathbf{v}^{(2)} - \mathbf{t}^{(2)} \cdot \mathbf{v}^{(1)}\right)dS = -\int_{S_f}\left(\mathbf{t}^{(1)} \cdot \mathbf{v}^{(2)} - \mathbf{t}^{(2)} \cdot \mathbf{v}^{(1)}\right)dS \tag{11.13}$$

which, when placed into Eq. (11.11), gives

$$\left(F_T^{(1)} v_T^{(2)} - F_T^{(2)} v_T^{(1)} \right) + \left(F_R^{(1)} v_R^{(2)} - F_R^{(2)} v_R^{(1)} \right)$$
$$= - \int_{S_f} \left(\mathbf{t}^{(1)} \cdot \mathbf{v}^{(2)} - \mathbf{t}^{(2)} \cdot \mathbf{v}^{(1)} \right) dS. \tag{11.14}$$

In Eq. (11.14), the unit normal to the flaw is directed inwards, as shown in Fig. 11.4. If we express the stress vector in terms of the stresses and let the components of this inward normal be n'_j Eq. (11.14) becomes

$$\left(F_T^{(1)} v_T^{(2)} - F_T^{(2)} v_T^{(1)} \right) + \left(F_R^{(1)} v_R^{(2)} - F_R^{(2)} v_R^{(1)} \right)$$
$$= - \int_{S_f} \left(\tau_{ji}^{(1)} v_i^{(2)} - \tau_{ji}^{(2)} v_i^{(1)} \right) n'_j dS. \tag{11.15}$$

It is convenient, however, to switch the direction of the normal so that it points outwards from the flaw into the surrounding material. In that case, Eq. (11.15) becomes

$$\left(F_T^{(1)} v_T^{(2)} - F_T^{(2)} v_T^{(1)} \right) + \left(F_R^{(1)} v_R^{(2)} - F_R^{(2)} v_R^{(1)} \right)$$
$$= \int_{S_f} \left(\tau_{ji}^{(1)} v_i^{(2)} - \tau_{ji}^{(2)} v_i^{(1)} \right) n_j dS, \tag{11.16}$$

where now n_j are the components of the *outward normal*.

We can follow exactly the same steps outlined here for states (1) and (2) but use states (3) and (2) instead. On the left hand side of Eq. (11.16) the force and velocity terms for state (1) will be replaced by those for state (3). The right hand side of equation (11.16) will be zero since the surface S_f is itself merely a fictitious surface surrounding a source free region in both states (2) and (3). Thus, we find

$$\left(F_T^{(3)} v_T^{(2)} - F_T^{(2)} v_T^{(3)} \right) + \left(F_R^{(3)} v_R^{(2)} - F_R^{(2)} v_R^{(3)} \right) = 0. \tag{11.17}$$

Now, we subtract Eq. (11.17) from Eq. (11.16) to obtain

$$\left(\Delta F_T v_T^{(2)} - F_T^{(2)} \Delta v_T \right) + \left(\Delta F_R v_R^{(2)} - F_R^{(2)} \Delta v_R \right)$$
$$= \int_{S_f} \left(\tau_{ji}^{(1)} v_i^{(2)} - \tau_{ji}^{(2)} v_i^{(1)} \right) n_j dS, \tag{11.18}$$

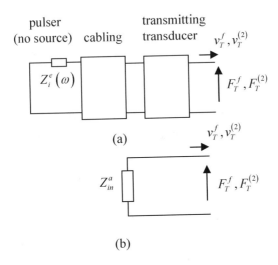

Fig. 11.5. (a) The force and velocity at the transmitting transducer due to the waves from the flaw or due to the waves in state (2), and **(b)** the corresponding equivalent acoustic impedance of the passive system shown in (a).

where

$$\Delta F_T = F_T^{(1)} - F_T^{(3)}$$
$$\Delta v_T = v_T^{(1)} - v_T^{(3)}$$
$$\Delta F_R = F_R^{(1)} - F_R^{(3)}$$
$$\Delta v_R = v_R^{(1)} - v_R^{(3)}.$$

(11.19)

The quantities in state (1) appearing in Eq. (11.19) are due to all the waves received at either the transmitter or receiver from 1) either the flaw directly (or interactions that involve the flaw) and 2) with other interactions that do not involve the flaw at all. We will call the first type of contribution the flaw response and the second type of contribution the non-flaw response. The quantities in state (3) appearing in Eq. (11.19), however, come from exactly the same non-flaw response as in state (1). Thus, all the differences in Eq. (11.19) are only due to waves received from the flaw directly or due to interactions involving the flaw (such as a bounce of the incident wave from a surface to the flaw and then to the receiver). We will call these differences, therefore, the *flaw responses* and define them as

$$F_T^f \equiv \Delta F_T$$
$$v_T^f \equiv \Delta v_T$$
$$F_R^f \equiv \Delta F_R \qquad (11.20)$$
$$v_R^f \equiv \Delta v_R.$$

In terms of these flaw responses then Eq. (11.18) becomes

$$\left(F_T^f v_T^{(2)} - F_T^{(2)} v_T^f \right) + \left(F_R^f v_R^{(2)} - F_R^{(2)} v_R^f \right)$$
$$= \int_{S_f} \left(\tau_{ji}^{(1)} v_i^{(2)} - \tau_{ji}^{(2)} v_i^{(1)} \right) n_j dS \qquad (11.21)$$

Since the same incident, driving waves (and the same corresponding voltage sources) are present in both states (1) and (3), F_T^f is the force at the transmitting transducer due to waves coming from the flaw in the absence of any voltage sources at the transmitter. The same is true for $F_T^{(2)}$ since by definition the only voltage sources active in that state are driving the receiving transducer (see Fig. 11.5 (a)). Thus, as shown in Fig. 11.5 (b), for both the flaw force response at the transmitter T and for the force at T in state (2) one can replace the passive electrical and electromechanical components of the sound generation process by the same equivalent acoustic impedance, Z_{in}^a, where

$$F_T^f = -Z_{in}^a v_T^f$$
$$F_T^{(2)} = -Z_{in}^a v_T^{(2)} \qquad (11.22)$$

[the minus signs are due to the fact that the velocities appearing in Eq. (11.22) were both defined as the normal velocities directed outwards from the transducer, as shown in Fig. 11.5 (b)]. Placing Eq. (11.22) into Eq. (11.21), the terms at the transmitter all cancel, leaving

$$\left(F_R^f v_R^{(2)} - F_R^{(2)} v_R^f \right) = \int_{S_f} \left(\tau_{ji}^{(1)} v_i^{(2)} - \tau_{ji}^{(2)} v_i^{(1)} \right) n_j dS \qquad (11.23)$$

In state (2), we have at the receiving transducer (which is firing as a transmitter) $F_R^{(2)} = Z_r^{R;a} v_R^{(2)}$, where $Z_r^{R;a}$ is the acoustic radiation impedance of the receiving transducer so Eq. (11.22) becomes:

$$\left(F_R^f - Z_r^{R;a}v_R^f\right)v_R^{(2)} = \int_{S_f}\left(\tau_{ji}^{(1)}v_i^{(2)} - \tau_{ji}^{(2)}v_i^{(1)}\right)n_j dS. \tag{11.24}$$

But the term in parentheses on the left side of Eq. (11.24) is just the total force at the receiver due to the waves from the flaw minus the force $F_s^f \equiv Z_r^{R;a}v_R^f$ due to the motion of the receiving transducer from the flaw response. By definition, this is just the blocked force, F_B, at the receiver due to the waves contained in the flaw response (see Eq. (5.24)), i.e.

$$F_B = F_R^f - Z_r^{R;a}v_R^f \tag{11.25}$$

and Eq. (11.24) can be rewritten as

$$F_B = \frac{1}{v_R^{(2)}}\int_{S_f}\left(\tau_{ji}^{(1)}v_i^{(2)} - \tau_{ji}^{(2)}v_i^{(1)}\right)n_j dS. \tag{11.26}$$

[Note that this blocked force is a force due only to wave interactions with the flaw, but we have dropped the superscript "f" and labeled the force simply F_B in order to be compatible with the notation used in previous Chapters.] Since the force at the transmitting transducer in our measurement setup is $F_t^{(1)} = Z_r^{T;a}v_T^{(1)}$, the acoustic/elastic transfer function, $t_A(\omega) = F_B/F_t^{(1)}$, for the measured flaw signals is given by

$$t_A(\omega) = \frac{1}{Z_r^{T;a}v_T^{(1)}v_R^{(2)}}\int_{S_f}\left(\tau_{ji}^{(1)}v_i^{(2)} - \tau_{ji}^{(2)}v_i^{(1)}\right)n_j dS. \tag{11.27}$$

When Eq. (11.27) is placed into Eq. (11.1), we have, finally a complete measurement model for the voltage received from the flaw, which can be written more explicitly as:

$$\begin{aligned}V_R(\omega) = &\frac{s(\omega)}{Z_r^{T;a}v_T^{(1)}(\omega)v_R^{(2)}(\omega)}\\ &\cdot\int_{S_f}\left[\tau_{ji}^{(1)}(\mathbf{x},\omega)v_i^{(2)}(\mathbf{x},\omega) - \tau_{ji}^{(2)}(\mathbf{x},\omega)v_i^{(1)}(\mathbf{x},\omega)\right]n_j(\mathbf{x})dS,\end{aligned} \tag{11.28}$$

where $\mathbf{x}$ is a general point on the surface of the flaw and recall that the n_j are the components of the outward normal.

Equation (11.28) is a significant result. It shows that if we can measure the system function and model the stress and velocity fields present at the flaw in states (1) and (2), we can predict the measured voltage response of the flaw in virtually any flaw measurement system.

We obtained this result for the pitch-catch immersion setup of Fig. 11.1 but it can be equally applied to pulse-echo and contact testing setups as well. Since the fields in states (1) and (2) in Eq. (11.28) are divided by the driving normal velocities $v_T^{(1)}, v_R^{(2)}$ in states (1) and (2), we only need to model the fields in both states due to driving transducers having a unit normal velocity on their faces. Thus, we do not need to explicitly know those normal velocities. This fact is important since if we had to model the absolute beam fields we would need a way to determine the "driving" normal velocities at the transmitting and receiving transducers in states (1) and (2), respectively. The normal velocities at the acoustic ports of these transducers are not easy quantities to determine experimentally, so it is fortunate that we do not need to know them to apply Eq. (11.28).

In state (2), the flaw is absent so that the stress and velocity fields appearing in that state in Eq. (11.28) are due to just the waves incident of the flaw surface. Those fields only require that we have an ultrasonic beam model in order to predict them. In state (1), however, the flaw is present, so that we must have both a beam model to predict the incident fields on the flaw in that state and a flaw scattering model that can predict the waves generated by the interaction of the incident waves with the flaw.

Equation (11.28) in a slightly different form was originally derived in 1979 by Bert Auld [11.1]. Because it is a very general result it has been frequently used as the basis for many ultrasonic modeling efforts world-wide. The main difference between Eq. (11.28) and Auld's original form is that Eq. (11.28) is an expression for the measured output voltage in an ultrasonic measurement system while Auld's result gave the measured flaw response in terms of a change of the fields present in a cable at the receiver. While this difference does not change the basic form of Eq. (11.28), it is an important difference when one wants to examine the elements in $s(\omega)$ as done in previous Chapters.

We can express Eq. (11.28) in a slightly different form that is also very useful. In both states (1) and (2), we assume the incident velocity field can be expressed as an incident plane wave modified by a spatially varying "amplitude" coefficient. Then we can write the incident velocity field in state (1) as (omitting the $\exp(-i\omega t)$ term):

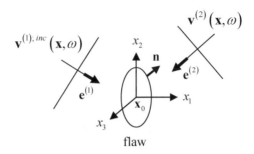

Fig. 11.6. The velocity fields incident on a flaw in states (1) and (2), respectively, which for small flaws can be treated locally as quasi-plane waves.

$$v_j^{(1);inc} = v_T^{(1)} \hat{V}^{(1)} (\mathbf{x}, \omega) d_j^{(1)} \exp\left[ik_{\beta 2} e_n^{(1)} x_n\right]$$ (11.29)

and the incident velocity in state (2) as

$$v_j^{(2)} = v_R^{(2)} \hat{V}^{(2)} (\mathbf{x}, \omega) d_j^{(2)} \exp\left[ik_{\alpha 2} e_n^{(2)} x_n\right]$$ (11.30)

where $\hat{V}^{(m)} (m = 1, 2)$ are the velocity field "amplitudes" (note that they are complex quantities) of the incident waves in states (1) and (2) normalized by the driving velocities on the faces of the transmitting transducers in those states. The $d_j^{(m)} (m = 1, 2)$ are polarizations of the incident waves in the two states, $k_{\alpha 2}$ and $k_{\beta 2}$ are the wave numbers for the incident waves in states (1) and (2) in the solid surrounding the flaw, respectively, and where α and β denote the incident wave type (P or S). The $e_n^{(m)} (m = 1, 2)$ terms are the components of the unit vectors in the direction of propagation for the incident waves in the two states. In Eqs. (11.29) and (11.30) the coordinates of the point $\mathbf{x} = (x_1, x_2, x_3)$ are measured from an origin located at point $\mathbf{x}_0$, which is a fixed point near the flaw, usually taken at the flaw "center" (see Fig. 11.6). Note that in state (1) the total velocity, $\mathbf{v}^{(1)}$ is given by $\mathbf{v}^{(1)} = \mathbf{v}^{(1);inc} + \mathbf{v}^{(1);scatt}$, where $\mathbf{v}^{(1);inc}$ is given by Eq. (11.29) and $\mathbf{v}^{(1);scatt}$ is the velocity field due to the waves scattered from the flaw, while in state (2), the total velocity field is only the incident field given by Eq. (11.30) since the flaw is absent in that state.

Using these quasi-plane wave forms in Eq. (11.28), we can write that equation in the form

$$V_R(\omega) = s(\omega) \left[\frac{4\pi\rho_2 c_{\alpha 2}}{-ik_{\alpha 2} Z_r^{T;a}} \right]$$
$$\cdot \int_{S_f} \hat{V}^{(1)}(\mathbf{x},\omega) \hat{V}^{(2)}(\mathbf{x},\omega) A(\mathbf{x},\omega) \exp\left[ik_{\alpha 2} e_n^{(2)} x_n \right] dS \qquad (11.31)$$

with

$$A(\mathbf{x},\omega) = \frac{1}{4\pi\rho_2 c_{\alpha 2}^2} \left[\tilde{\tau}_{ji}^{(1)} d_i^{(2)} + C_{ijkl} d_k^{(2)} \left(e_l^{(2)} / c_{\alpha 2} \right) \tilde{v}_i^{(1)} \right] n_j. \qquad (11.32)$$

(no sum on α)

where $\rho_2, c_{\alpha 2}$ are the density and wave speed, respectively, for the material surrounding the flaw. The normalized velocity and stress terms $\tilde{v}_j^{(1)}, \tilde{\tau}_{ij}^{(1)}$ are defined as

$$\tilde{\tau}_{ij}^{(1)} = \frac{-i\omega\tau_{ij}^{(1)}}{v_T^{(1)} \hat{V}^{(1)}}$$

$$\tilde{v}_j^{(1)} = \frac{-i\omega v_j^{(1)}}{v_T^{(1)} \hat{V}^{(1)}}. \qquad (11.33)$$

Physically, these normalized fields are the actual fields in state (1) normalized by an incident wave displacement amplitude term $U^{(1)}(\mathbf{x},\omega) = v_T^{(1)} \hat{V}^{(1)}(\mathbf{x},\omega)/(-i\omega)$.

Equation (11.31) begins to reveal some of the structure of the integral term that was not evident in Eq. (11.28). The $\hat{V}^{(1)}(\mathbf{x},\omega), \hat{V}^{(2)}(\mathbf{x},\omega)$ terms are quasi-plane wave incident field amplitudes at the flaw for states (1) and (2) due to the transmitting and receiving transducers radiating with a unit velocity on their surfaces. These terms can be modeled explicitly if one has a beam model such as the multi-Gaussian beam model of Chapter 9 and combines the beam model with terms that take into account the material attenuation present (see Appendix D). The remaining A term is closely related to the scattering properties of the flaw. If the displacement amplitude term $U^{(1)}$ was a constant or if it did not vary significantly over the surface of the flaw then the normalized velocity and stress fields in Eq. (11.32) would be those due to an incident plane wave of unit displacement amplitude on the flaw. But recall the component of the plane wave far-field scattering amplitude taken in the $-\mathbf{d}^{(2)}$ direction, $A(\omega) \equiv \mathbf{A}^{\alpha;\beta}\left(\mathbf{e}^{(1)}; \mathbf{e}^{(2)}\right) \cdot \left(-\mathbf{d}^{(2)}\right)$ is just given by (see Eqs. (10.6) and (10.7))

$$A(\omega) = \int_S A(\mathbf{x}, \omega) \exp\left(-ik_{\alpha 2} \mathbf{e}_s^\alpha \cdot \mathbf{x}\right) dS(\mathbf{x}_s). \tag{11.34}$$

[Note: to compare with the results in Chapter 10, in our current notation, $\mathbf{e}_i^\beta = \mathbf{e}^{(1)}$, $\mathbf{e}_s^\alpha = -\mathbf{e}^{(2)}$]. Thus, A contains the total fields on the surface of the flaw that in principle could be obtained by solving the flaw scattering problem and can be used to compute the far-field scattering amplitude component, $A(\omega)$. Of course, in Eq. (11.31) the far-field scattering amplitude component of Eq. (11.34) itself does not appear explicitly because of the beam variations contained in that equation but as we will see in the following section there are cases where the frequency spectrum of the received voltage is proportional directly to $A(\omega)$.

11.2 The Thompson-Gray Measurement Model

If we write the incident fields in states (1) and (2) as quasi-plane waves and if in addition we assume that the flaw is small enough so that the variations of the velocity field amplitudes $\hat{V}^{(m)}$ $(m = 1, 2)$ are negligible over the surface of the flaw, we have

$$\begin{aligned}
v_j^{(1);inc} &= v_T^{(1)} \hat{V}_0^{(1)} d_j^{(1)} \exp\left[ik_{\beta 2} e_n^{(1)} x_n \right] \\
v_j^{(2)} &= v_R^{(2)} \hat{V}_0^{(2)} d_j^{(2)} \exp\left[ik_{\alpha 2} e_n^{(2)} x_n \right],
\end{aligned} \tag{11.35}$$

where $\hat{V}_0^{(m)}(\omega) \equiv \hat{V}^{(m)}(\mathbf{x}_0, \omega)$ $(m = 1, 2)$, i.e. the velocity field amplitudes are now constants evaluated at a fixed point, $\mathbf{x}_0$, in the vicinity of the flaw, which is usually taken at the "center" of the flaw (Fig. 11.6). We see from Eq. (11.35) that under this assumption these incident fields are now indeed treated as simply plane waves. Then Eq. (11.31) becomes

$$\begin{aligned}
V_R(\omega) &= s(\omega) \hat{V}_0^{(1)} \hat{V}_0^{(2)} \left[\frac{4\pi \rho_2 c_{\alpha 2}}{-ik_{\alpha 2} Z_r^{T;a}} \right] \\
&\cdot \int_{S_f} A(\mathbf{x}, \omega) \exp\left[ik_{\alpha 2} e_n^{(2)} x_n \right] dS
\end{aligned} \tag{11.36}$$

and the far-field scattering amplitude component does appear explicitly so we find, finally

$$V_R(\omega) = s(\omega)\hat{V}_0^{(1)}(\omega)\hat{V}_0^{(2)}(\omega)A(\omega)\left[\frac{4\pi\rho_2 c_{\alpha 2}}{-ik_{\alpha 2}Z_r^{T;a}}\right].$$

(11.37)

A form similar to Eq. (11.37) was first obtained by Thompson and Gray in 1983 [11.2]. As we have seen, this Thompson-Gray measurement model is based on the general reciprocity-based measurement model (Eq. (11.28) and only two assumptions: 1) the incident waves can be expressed in a quasi-plane wave form, and 2) the flaw is small enough so that the amplitude of this quasi-plane wave does not vary significantly over the flaw surface [11.3]. With those two assumptions we obtain a modular measurement model where the flaw response, $A(\omega)$, is explicitly separated from all the other measurement system terms, including the system function, $s(\omega)$, and the normalized incident fields $\hat{V}_0^{(1)}, \hat{V}_0^{(2)}$ at the flaw from the transducers in states (1) and (2), respectively.

This separation of terms allows us to examine a ultrasonic measurement system in a variety of ways. For example, if the voltage response of an unknown flaw is measured and we also measure $s(\omega)$ and model the transducer beam fields $\hat{V}_0^{(1)}, \hat{V}_0^{(2)}$, we can write Eq. (11.37) in the form

$$V_R(\omega) = G(\omega)A(\omega),$$

(11.38)

where both $V_R(\omega)$ and $G(\omega)$ are known. In this case we can divide the measured voltage by the known G (using a Wiener filter) to obtain a measured flaw far-field scattering amplitude of the flaw. An example of this approach is given in Chapter 13. The flaw far-field scattering amplitude is related to the properties of the flaw only, so that we can use it in quantitative flaw characterization and sizing studies. Alternatively, we could model the beam fields and the scattering amplitude for an assumed flaw and measure the system function for a given measurement setup. In this case we could use Eq. (11.38) to predict the amplitude of the received signals from the known flaw directly. This information might be used, for instance, to optimize the transducer orientation during a scan so that the signals received from a given flaw are large. Engineering studies of these and other types can be done easily with the Thompson-Gray measurement model, so that it has been used for many practical industrial applications

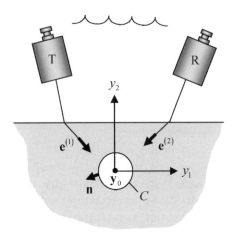

Fig. 11.7. A pitch-catch measurement of the scattering from a cylindrical reflector of length L where the axis of the cylinder is normal to planes of incidence for states (1) and (2).

(many examples are available in past volumes of Review of Progress in Quantitative Nondestructive Evaluation [11.4]).

11.3 A Measurement Model for Cylindrical Reflectors

The Thompson-Gray measurement model describes the ultrasonic response of a flaw where the variation of the incident fields can be neglected over the entire flaw surface. In some experiments cylindrical reflectors are used where the beam variations over the cross-section of the reflector may be neglected, but the variations over the length of the reflector are significant. An example is where a side-drilled hole is used as a reference reflector in a calibration block. Here, we will develop a measurement model suitable for these type of setups [11.5].

Consider the pitch-catch case shown in Fig. 11.7. We will assume that the reflector has a cylindrical geometry and is of length, L. We will also assume that the axis of the cylinder is normal to the planes of incidence of the incident waves in states (1) and (2), which are defined to be the planes that contain both the incident wave direction and the unit normal, **n**, to the reflector. This is a reasonable assumption since in most setups where a 2-D reflector such as a side-drilled hole is used, the transducers are usually oriented so that this condition holds. With this assumption it is reasonable to also assume that all the scattering occurs only from the cylindrical surface, i.e. the scattering from the ends of the reflector is neglected. For a side-drilled hole, for example, which is often drilled entirely through a reference calibration block, the length of the hole is generally larger than the axial extent of the interrogating transducer fields, so that this assumption is well satisfied.

Like the Thompson-Gray measurement model, we will also assume the incident fields in states (1) and (2) can be represented by the quasi-plane waves given by Eqs. (11.29) and (11.30). Thus, we can use as our starting point Eq. (11.31), which we write as:

$$V_R(\omega) = s(\omega) \left[\frac{4\pi \rho_2 c_{\alpha 2}}{-ik_{\alpha 2} Z_r^{T;a}} \right]$$
$$\cdot \int_{S_c} \hat{V}^{(1)}(\mathbf{y}, z, \omega) \hat{V}^{(2)}(\mathbf{y}, z, \omega) A \exp\left[ik_{\alpha 2} e_n^{(2)} x_n \right] dS,$$
(11.39)

where the integration is now over only the cylindrical surface, S_c, and the velocity field amplitudes over this surface are expressed as $\hat{V}^{(1)} = \hat{V}^{(1)}(\mathbf{y}, z, \omega)$, $\hat{V}^{(2)} = \hat{V}^{(2)}(\mathbf{y}, z, \omega)$ where $\mathbf{y} = (y_1, y_2)$ is a point in a plane normal to the cylinder axis and z is along the axis. If the cylinder is small enough to neglect the variations of these velocity fields over its cross-sectional area (but not neglecting these variations over its length) we let $\hat{V}^{(1)} = \hat{V}_0^{(1)}(z, \omega)$, $\hat{V}^{(2)} = \hat{V}_0^{(2)}(z, \omega)$, where

$$\hat{V}_0^{(1)}(z, \omega) \equiv \hat{V}^{(1)}(\mathbf{y}_0, z, \omega)$$
$$\hat{V}_0^{(2)}(z, \omega) \equiv \hat{V}^{(2)}(\mathbf{y}_0, z, \omega)$$
(11.40)

and $\mathbf{y}_0$ is a fixed point, usually taken as the center of the reflector (Fig. 11.7). Then we can write Eq. (11.39) as

$$V_R(\omega) = s(\omega) \left[\frac{4\pi \rho_2 c_{\alpha 2}}{-ik_{\alpha 2} Z_r^{T;a}} \right]$$
$$\cdot \int_C \int_L \hat{V}_0^{(1)}(z, \omega) \hat{V}_0^{(2)}(z, \omega) A \exp\left[ik_{\alpha 2} e_n^{(2)} x_n \right] dc dz,$$
(11.41)

where the 2-D surface integration has been decomposed into a counter-clockwise line integral over the cross-section, C, and a 1-D integral over the length, L.

Now, consider the normalized fields appearing in the A term in Eq. (11.41). They are:

$$\tilde{v}_i^{(1)} = \frac{-i\omega v_i^{(1)}(\mathbf{y}, z, \omega)}{\hat{V}_0^{(1)}(z, \omega)}$$

$$\tilde{\tau}_{ji}^{(1)} = \frac{-i\omega \tau_{ji}^{(1)}(\mathbf{y}, z, \omega)}{\hat{V}_0^{(1)}(z, \omega)}. \tag{11.42}$$

We will also assume that these normalized fields are functions of $\mathbf{y}$ only, i.e. we assume the z-variations of the non-normalized fields are identical to those in the incident waves. In this case we then also have $A = A(\mathbf{y}, \omega)$ only. This assumption is equivalent to breaking the cylindrical surface into small elements of length dz and at each z treating the scattering of each element as if it were a purely two-dimensional scattering process due to a plane wave whose displacement amplitude is given by $U^{(1)}(z, \omega) = v_T^{(1)}(\omega)\hat{V}_0^{(1)}(z, \omega)/(-i\omega)$. In Chapter 10, where the scattering amplitude for a side-drilled hole was calculated via the Kirchhoff approximation, this assumption was satisfied exactly. Here, we will make the assumption regardless of how the scattering problem for the cylinder is to be solved. With this assumption, then Eq. (11.41) becomes

$$V_R(\omega) = s(\omega)\left[\frac{4\pi\rho_2 c_{\alpha 2}}{-ik_{\alpha 2}Z_r^{T;a}}\right]\int_L \hat{V}_0^{(1)}(z, \omega)\hat{V}_0^{(2)}(z, \omega)\,dz$$
$$\cdot \int_C A(\mathbf{y}, \omega)\exp\left[ik_{\alpha 2}e_\lambda^{(2)}y_\lambda\right]dc(\mathbf{y}), \tag{11.43}$$

where the summation over the repeated λ subscript is taken over values (1,2) only.

But the far-field scattering amplitude of the cylindrical reflector is given by

$$A_{3D}(\omega) = L\int_C A(\mathbf{y}, \omega)\exp\left[ik_{\alpha 2}e_\lambda^{(2)}y_\lambda\right]dc(\mathbf{y}) \tag{11.44}$$

where we have used the "3D" label to emphasize that the reflector is still being treated as a three-dimensional scatterer even though under our assumptions the fields in Eq. (11.44) are all two-dimensional. With this definition, Eq. (11.43) can be reduced to

$$V_R(\omega) = s(\omega)\left[\int_L \hat{V}_0^{(1)}(z, \omega)\hat{V}_0^{(2)}(z, \omega)\,dz\right]\left[\frac{A_{3D}(\omega)}{L}\right]\left[\frac{4\pi\rho_2 c_{\alpha 2}}{-ik_{\alpha 2}Z_r^{T;a}}\right]. \tag{11.45}$$

Equation (11.45) is the measurement model for the cylindrical reflector that is the counterpart of Eq. (11.37) for the small three-dimensional flaw. In fact, if the incident velocity fields do not vary significantly also in the z-direction, we see that Eq. (11.45) simply reduces to Eq. (11.37).

Since under our assumptions the fields in Eq. (11.44) only depend on y and from our other assumptions we also have $n_3 = e_3^\alpha = d_3^\alpha = \tilde{v}_3^\alpha = 0$, Eq. (11.44) can be rewritten more explicitly as

$$A_{3D}(\omega) = \frac{L}{4\pi \rho_2 c_{\alpha 2}^2} \int_C \left[\tilde{\tau}_{\gamma\sigma}^{(1)} d_\sigma^{(2)} + C_{\sigma\gamma\nu\delta} d_\nu^{(2)} \left(e_\delta^{(2)} / c_{\alpha 2} \right) \tilde{v}_\sigma^{(1)} \right]$$
$$\cdot n_\gamma \exp\left[ik_{\alpha 2} e_\lambda^{(2)} y_\lambda \right] dc,$$

(11.46)

(no sum on α)

where all the Greek indices are summed over the values $(1,2)$ only (no sum on α). As shown in Chapter 10 this scattering amplitude component can be related to the corresponding far-field scattering amplitude component, $A_{2D}(\omega)$, in a purely two-dimensional scattering problem where both the incident fields and the geometry of the reflector are independent of the z-coordinate. That relationship was given as (see Eq. (10.63)):

$$A_{2D}(\omega) = \left(\frac{2i\pi}{k_{\alpha 2}} \right)^{1/2} \frac{A_{3D}(\omega)}{L}.$$

(11.47)

Thus, we can express our ultrasonic measurement model for the cylinder either in terms of the three-dimensional far-field scattering amplitude of the reflector or its two-dimensional far-field scattering amplitude counterpart.

11.4 References

11.1 Auld BA (1979) General electromechanical reciprocity relations applied to the calculation of elastic wave scattering coefficients. Wave Motion 1: 3-10
11.2 Thompson RB, Gray TA (1983) A model relating ultrasonic scattering measurements through liquid-solid interfaces to unbounded medium scattering amplitudes. J. Acoust. Soc. Am. 74: 140-146
11.3 Schmerr LW (2004) Fundamentals of ultrasonic measurement models. In: Lee SS, Yoon DJ, Lee JH, Lee S (eds) Advances in nondestructive evaluation. Trans Tech Publications Ltd, USA pp 402-409

11.4 Thompson DO, Chimenti DE (eds) Review of progress in quantitative non-destructive evaluation. Plenum Press, New York, NY/ American Institute of Physics, Melville, NY. Past Volumes (1981-present)

11.5 Schmerr LW, Sedov A (2003) Modeling ultrasonic problems for the 2002 ultrasonic benchmark session. In: Thompson DO, Chimenti DE (eds) Review of progress in quantitative nondestructive evaluation 22B. American Institute of Physics, Melville, NY pp 1776-1783

11.5 Exercises

1. In Chapter 12, a multi-Gaussian beam model is used in conjunction with the Kirchhoff approximation to implement the Thompson-Gray measurement model (Eq. (11.37)) for the pulse-echo P-wave response of a spherical void, as shown in Fig. 11.8. However, one can also implement this measurement model directly using the results of Chapter 8 and Chapter 10. First note that for this pulse-echo setup Eq. (11.37) becomes

$$V_R(\omega) = s(\omega)\left[\hat{V}_0^{(1)}(\omega)\right]^2 A(\omega)\left[\frac{4}{-ik_{p2}a^2}\frac{\rho_2 c_{p2}}{\rho_1 c_{p1}}\right], \qquad (11.48)$$

where a is the radius of the transducer, ρ_1, c_{p1} are the density and compressional wave speed of the fluid, ρ_2, c_{p2} are the density and wave speed of the solid, and $k_{p2} = \omega/c_{p2}$ is the wave number for compressional waves in the solid. The normalized on-axis velocity, $\hat{V}_0^{(1)}(\omega)$, can be obtained from Eq. (8.25) as

$$\hat{V}_0^{(1)}(\omega) = T_{12}^{P;P}\exp\left(-\alpha_{p1}z_1\right)\exp\left(ik_{p1}z_1 + ik_{p2}z_2\right)\left[1-\exp\left(\frac{ik_{p1}a^2}{2\tilde{z}}\right)\right], \quad (11.49)$$

where $\tilde{z} = z_1 + c_{p2}z_2/c_{p1}$ and $T_{12}^{P;P}$ is the plane wave transmission coefficient at normal incidence (see Eq. (D.36)):

$$T_{12}^{P;P} = \frac{2\rho_1 c_{p1}}{\rho_1 c_{p1} + \rho_2 c_{p2}}. \qquad (11.50)$$

An attenuation factor has been included in Eq. (11.49) to account for the water attenuation. The attenuation of the solid (which is glass) has been neglected here. In implementing Eq. (11.49), omit the propagation terms

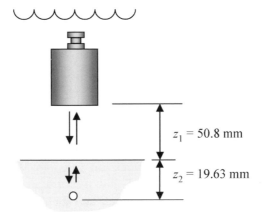

Fig. 11.8. Measurement of the pulse-echo P-wave response of an on-axis spherical pore.

$\exp\left(ik_{p1}z_1 + ik_{p2}z_2\right)$ as these only generate large phases that represent a time delay factor $\Delta t = z_1/c_{p1} + z_2/c_{p2}$ that can always be added in later, if necessary.

For the scattering amplitude term, $A(\omega)$, one can use the Kirchhoff approximation for the pulse-echo response of a void of radius b given by (Eq. (10.14)):

$$A(\omega) = \frac{-b}{2}\exp\left(-ik_{p2}b\right)\left[\exp\left(-ik_{p2}b\right) - \frac{\sin\left(k_{p2}b\right)}{k_{p2}b}\right]. \tag{11.51}$$

Write a MATLAB script that implements the Thompson-Gray measurement model of Eq. (11.48) for this on-axis spherical pore. The pertinent data for this problem are:

$a = 6.35\ mm$

$b = 0.34605\ mm$

$z_1 = 50.8\ mm$

$z_2 = 19.63\ mm$

$\rho_1 = 1.0\ gm/cm^3$

$$c_{p1} = 1484 \ m/\sec$$

$$\rho_2 = 2.2 \ gm/cm^3$$

$$c_{p2} = 5969.4 \ m/\sec$$

$$\alpha_{p1} = 0.02479 \times 10^{-3} f^2 \ \ Np/mm \ (f \ in \ MHz)$$

For the system function, $s(\omega)$, use the simulated MATLAB function system_f(f, amp, fc, bw) described in Appendix G. Take amp=0.08, fc = 5, bw = 4. These parameters simulate a system containing a broad band 5 MHz transducer. The frequency, f, in this function is measured in MHz.

The MATLAB script should calculate the received voltage from the void as a function of frequency, perform an inverse FFT to obtain the corresponding time-domain signal, and then plot this signal versus time. Verify that you results agree with Fig. 12.11 which is the solution of this same problem using a multi-Gaussian beam model instead of Eq. (11.49).

12 Ultrasonic Measurement Modeling with MATLAB

In this Chapter we will implement complete ultrasonic measurement models in a series of MATLAB functions and scripts for the pulse-echo setup of Fig. 12.1. These measurement models will be used to simulate a number of measurement setups where a reference reflector such as a spherical pore, a flat-bottom hole, or a side drilled hole is present. Reference reflectors are commonly used in NDE tests to serve as calibration standards and they are also used to measure system performance. Here we will demonstrate the ability of the measurement models to simulate experimentally determined signals from these types of reference reflectors [12.1]. Similar demonstrations have been carried out worldwide by a number of researchers in a recent series of benchmark studies (see [12.2] for an overview of these activities from 2001- 2005). In those studies a variety of beam models and flaw scattering models were employed. Here, we will use the multi-Gaussian beam model of Chapter 9 and two of the flaw scattering models discussed Chapter 10 (the Kirchhoff approximation and the method of separation of variables) in conjunction with the various measurement models described in Chapter 11.

The MATLAB models of this Chapter can be used by the reader as the basis for implementing and studying many of the concepts and results discussed in this book in a more hands-on fashion, where the parameters can be readily changed and the results easily illustrated. Although the models are implemented for a simple pulse-echo configuration (Fig. 12.1) they can be used for a number of advanced purposes, such as examining ultrasonic beam behavior at curved interfaces, for example, and they can serve as the starting point for developing more complex simulation models.

12.1 A Summary of the Measurement Models

In the previous Chapter we developed measurement models suitable for several different testing situations. These included a general model that

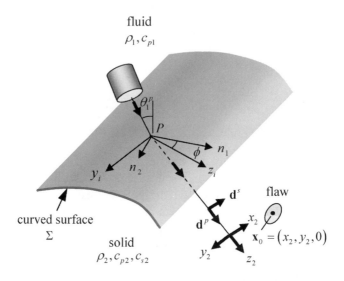

Fig. 12.1. Parameters for defining the problem of pulse-echo inspection of a flaw in a solid through a fluid-solid interface.

only relied on linearity and reciprocity and assumed the incident beam could be written in quasi-plane wave form. For that model the frequency components of the measured voltage were given by

$$
V_R(\omega) = s(\omega) \left[\frac{4\pi \rho_2 c_{\alpha 2}}{-ik_{\alpha 2} Z_r^{T;a}} \right]
$$
$$
\cdot \int_{S_f} \hat{V}^{(1)}(\mathbf{x}, \omega) \hat{V}^{(2)}(\mathbf{x}, \omega) A(\mathbf{x}, \omega) \exp\left[ik_{\alpha 2} e_n^{(2)} x_n \right] dS,
$$

(12.1)

where, recall,

$$
A(\mathbf{x}, \omega) = \frac{1}{4\pi \rho_2 c_{\alpha 2}^2} \left[\tilde{\tau}_{ji}^{(1)} d_i^{(2)} + C_{ijkl} d_k^{(2)} \left(e_l^{(2)} / c_{\alpha 2} \right) \tilde{v}_i^{(1)} \right] n_j
$$

(12.2)

involves the stresses and velocity on the surface of the flaw normalized by the incident wave displacement amplitude at the flaw, i.e.

$$\tilde{\tau}_{ij}^{(1)} = \frac{-i\omega\tau_{ij}^{(1)}}{v_T^{(1)}\hat{V}^{(1)}}$$

$$\tilde{v}_j^{(1)} = \frac{-i\omega v_j^{(1)}}{v_T^{(1)}\hat{V}^{(1)}}. \tag{12.3}$$

The terms $\hat{V}^{(\alpha)}(\mathbf{x},\omega)$ $(\alpha = 1,2)$ are the incident velocity field amplitudes on the flaw surface for states (1) and (2), where in state (1) the transmitting transducer is firing with a unit velocity on its face and for state (2) the receiving transducer is firing with a unit velocity on its face. Both of these amplitude terms, therefore, can be calculated with appropriate ultrasonic beam and attenuation models. The remaining fields in the $A(\mathbf{x},\omega)$ term are the total fields on the surface of the flaw normalized by the displacement of the incident wave. Those fields can also be modeled with an appropriate flaw scattering model. This measurement model is quite general and should apply to most testing situations. Note that in this form the flaw far-field scattering amplitude does not appear directly but, as shown in the last Chapter, $A(\mathbf{x},\omega)$ is closely related to the component of the scattering amplitude that appears in other measurement models (see Eq. (11.34)).

The second model developed assumed that the flaw was small enough so that the incident fields did not vary significantly over the surface of the flaw. In that case we found

$$V_R(\omega) = s(\omega)\hat{V}_0^{(1)}(\omega)\hat{V}_0^{(2)}(\omega)A(\omega)\left[\frac{4\pi\rho_2 c_{\alpha 2}}{-ik_{\alpha 2}Z_r^{T;a}}\right], \tag{12.4}$$

where

$$\hat{V}_0^{(1)} = \hat{V}^{(1)}(\mathbf{x}_0,\omega)$$

$$\hat{V}_0^{(1)} = \hat{V}^{(1)}(\mathbf{x}_0,\omega)$$

are the now the velocity amplitude terms evaluated at the "center" of the flaw and a flaw far-field scattering amplitude term, $A(\omega)$, is directly a part of the measurement model.

For a cylindrical scatterer where beam variations are not negligible we can again apply the measurement model of Eq. (12.1). For a small cylindrical scatterer, however, where beam variations over the scatterer cross section are negligible we found

$$V_R(\omega) = s(\omega)\left[\int_L \hat{V}_0^{(1)}(z,\omega)\hat{V}_0^{(2)}(z,\omega)dz\right]\left[\frac{A(\omega)}{L}\right]\left[\frac{4\pi\rho_2 c_{\alpha 2}}{-ik_{\alpha 2}Z_r^{T;a}}\right], \qquad (12.5)$$

where, recall,

$$\hat{V}_0^{(1)}(z,\omega) \equiv \hat{V}^{(1)}(\mathbf{y}_0,z,\omega)$$
$$\hat{V}_0^{(2)}(z,\omega) \equiv \hat{V}^{(2)}(\mathbf{y}_0,z,\omega)$$

are now the incident velocity amplitude terms calculated at the "center" of the scatterer and at any axial position along its length. The far-field scattering amplitude of the flaw appearing in Eq. (12.5) is the same 3-D scattering amplitude in Eq. (12.4), but as mentioned in the last Chapter we also can use a 2-D scattering amplitude calculation in Eq. (12.5) if we use the relationship of Eq. (11.48).

Each of the measurement models described above has three components: 1) the system function, $s(\omega)$, describing all the electrical and electromechanical elements of the measurement system, 2) the velocity fields $\hat{V}^{(1)}, \hat{V}^{(2)}$ that characterize the incident fields on the flaw from the transmitting transducer or receiving transducer, respectively, when there is a unit driving velocity on those transducer faces, and 3) the scattering properties of the flaw itself, described in terms of $A(\mathbf{x},\omega)$ or $A(\omega)$. In this Chapter we will develop a series of MATLAB functions that model each of these three components and implement the measurement models described above.

The problem we will consider is shown in Fig. 12.1 where a transducer is performing a pulse-echo inspection of a flaw in an immersion setup. First, assume that the flaw is small enough so that the beam variations over its surface can be neglected and the measurement model of Eq. (12.4) can be used. The distances along a ray (a path satisfying Snell's law) extending normally from the center of the transducer are z_1, z_2 for the fluid and solid, respectively, and the center of the flaw is located at a point (x_2, y_2) relative to that central ray as shown in Fig. 12.1, where the y_2-axis is normal to the plane of incidence. The acute angle of the central ray in the fluid and the normal to the interface (at point P where that ray intersects the interface) is the angle θ_1^p. The (y_i, z_i) coordinates are in the tangent plane of the interface and y_i is normal to the plane of incidence. The angle of the z_i-axis from one of the principal directions, n_1, of the surface is the angle ϕ. [Important: note that these definitions are different from some of those used in Chapter 9 and Chapter 11 so that in the MATLAB measurement models of this Chapter one should relate the quantities in those models back to Fig. 12.1].

We can express the measurement model of Eq. (12.4) more explicitly by examining the various pieces that contribute to the velocity terms. Since we are considering a pulse-echo setup here, our measurement model can be written as

$$V_R(\omega) = s(\omega) \left[\hat{V}_0^{(1)}(\omega) \right]^2 A(\omega) \left[\frac{4\pi\rho_2 c_{\alpha 2}}{-ik_{\alpha 2} Z_r^{T;a}} \right] \qquad (12.6)$$

and the incident velocity field, $\hat{V}_0^{(1)}$, can be written as

$$\hat{V}_0^{(1)} = \exp\left[-\alpha_{p1}(\omega) z_1 - \alpha_{\gamma 2}(\omega) z_2 \right] \left[V_i^\gamma / v_0 \right] \qquad (12.7)$$

where z_1, z_2 are the distances the sound beam from the transducer has propagated in the fluid and the solid, respectively, and $\alpha_{p1}(\omega), \alpha_{\gamma 2}(\omega)$ are the frequency dependent attenuation coefficients for the compressional wave in the fluid and the wave of type γ in the solid, respectively. The term V_i^γ / v_0 is the ideal velocity field (for a material with no losses) at the flaw normalized by the normal velocity, v_0, on the face of the transducer. This ideal field will be described by a multi-Gaussian beam model of the type discussed in Chapter 9. The types of transducer we will consider in the setup of Fig. 12.1 with a multi-Gaussian beam model are circular planar and spherically focused piston transducers. In the following section we will use the general formulation of Chapter 9 to derive a multi-Gaussian beam model that is directly applicable to a setup of the type given in Fig. 12.1.

12.2 The Multi-Gaussian Beam Model

In developing the multi-Gaussian beam model the interface will assumed to be either planar or curved, with the plane of incidence of the transducer aligned with one of the principal curvatures of the interface (i.e. $\phi = 0$ in Fig. 12.1). For a single fluid-solid interface on transmission through the interface it is not necessary to rotate axes and the angle $\lambda = 0$ in Eqs. (9.89)-(9.91). Also, we do not need to put the transmission coefficient in matrix form, but can use the simpler scalar relation of Eq. (9.79). The ideal normalized velocity for a wave of type γ in the solid as computed by the multi-Gaussian beam model (with 15 coefficients) for this case is then given by (see Fig. 12.1)

$$
V_i^\gamma \mathbf{d}^\gamma = \sum_{r=1}^{15} T_{12}^{\gamma;p} \mathbf{d}^\gamma \left[V_1^p(0) \right]_r \frac{\sqrt{\det\left[\mathbf{M}_2^\gamma(z_2) \right]_r}}{\sqrt{\det\left[\mathbf{M}_2^\gamma(0) \right]_r}} \frac{\sqrt{\det\left[\mathbf{M}_1^p(z_1) \right]_r}}{\sqrt{\det\left[\mathbf{M}_1^p(0) \right]_r}}
$$
$$
\cdot \exp\left[ik_{p1}z_1 + ik_{\gamma 2}z_2 + i\frac{k_{p1}}{2}\mathbf{y}^T \left[c_{p1}\mathbf{M}_2^\gamma(z_2) \right]_r \mathbf{y} \right],
$$

$$(12.8)$$

$$(\gamma = p,s)$$

where, $\mathbf{y}^T = (x_2, y_2)$ and at the face of the transducer

$$
\left[V_1^p(0) \right]_r = A_r v_0(\omega)
$$

$$
\left[\mathbf{M}_1^p(0) \right]_r = \begin{bmatrix} \dfrac{iB_r}{c_{p1}D_R} & 0 \\[2ex] 0 & \dfrac{iB_r}{c_{p1}D_R} \end{bmatrix}
$$

$$(12.9)$$

in terms of the Wen and Breazeale coefficients A_r, B_r. The polarization vector, $\mathbf{d}^\gamma$, is shown in Fig. 12.1 for both P-waves and SV-waves. The plane wave transmission coefficient, $T_{12}^{\gamma;p}$ is based on a velocity ratio. The parameter $D_R = k_{p1}a^2/2$ is the Rayleigh distance, where the radius of the transducer is a and k_{p1} is the wave number for P-waves in medium one. Similarly $k_{\gamma 2}$ $(\alpha = p,s)$ are wave numbers for P- or S-waves in medium two. From the propagation law for medium one, from Eq. (9.28) we have

$$
\left[\mathbf{M}_1^p(z_1) \right]_r = \begin{bmatrix} \dfrac{1}{c_{p1}(z_1 - iD_R/B_r)} & 0 \\[2ex] 0 & \dfrac{1}{c_{p1}(z_1 - iD_R/B_r)} \end{bmatrix}.
$$

$$(12.10)$$

From Eq. (12.9) and Eq. (12.10) then it follows that

$$
\frac{\sqrt{\det\left[\mathbf{M}_1^p(z_1) \right]_r}}{\sqrt{\det\left[\mathbf{M}_1^p(0) \right]_r}} \left[V_1^p(0) \right]_r = \frac{A_r v_0}{1 + iz_1 B_r/D_R}.
$$

$$(12.11)$$

The transmission law across the interface also gives (Eq. (9.94))

$$
\left[M_2^y(0) \right]_r = \begin{bmatrix} \dfrac{M_1}{c_{p1}\left(z_1 - iD_R/B_r\right)} & 0 \\ 0 & \dfrac{M_2}{c_{p1}\left(z_1 - iD_R/B_r\right)} \end{bmatrix},
\tag{12.12}
$$

where

$$
M_1 = \left(\cos^2\theta_1^p + Kh_{11}\right)/\cos^2\theta_2^y
\tag{12.13}
$$

and

$$
M_2 = 1 + Kh_{22}
\tag{12.14}
$$

are given in terms of the principal interface curvatures $\left(h_{11}, h_{22}\right)$ and

$$
K = \left(z_1 - iD_R/B_r\right)\left(\cos\theta_1^p - \dfrac{c_{p1}}{c_{y2}}\cos\theta_2^y\right).
\tag{12.15}
$$

Finally, from the propagation law (Eq. (9.28)) for the propagation in medium two we have

$$
\left[M_2^y(z_2) \right]_r = \begin{bmatrix} \dfrac{\left[\left\{M_2^y(0)\right\}_{11}\right]_r}{1 + z_2 c_{y2}\left[\left\{M_2^y(0)\right\}_{11}\right]_r} & 0 \\ 0 & \dfrac{\left[\left\{M_2^y(0)\right\}_{22}\right]_r}{1 + z_2 c_{y2}\left[\left\{M_2^y(0)\right\}_{22}\right]_r} \end{bmatrix}.
\tag{12.16}
$$

Thus, we have

$$
\frac{\sqrt{\det\left[\mathbf{M}_2^y(z_2)\right]_r}}{\sqrt{\det\left[\mathbf{M}_2^y(0)\right]_r}} = \frac{1}{\sqrt{1 + z_2 \dfrac{c_{y2}}{c_{p1}}\dfrac{M_1}{\left(z_1 - iD_R/B_r\right)}}} \cdot \frac{1}{\sqrt{1 + z_2 \dfrac{c_{y2}}{c_{p1}}\dfrac{M_2}{\left(z_1 - iD_R/B_r\right)}}}
\tag{12.17}
$$

and

$$c_{p1}\left[\mathbf{M}_2^{\gamma}\left(z_2\right)\right]_r =$$

$$\begin{bmatrix} \dfrac{1}{\dfrac{\left(z_1 - iD_R / B_r\right)}{M_1} + z_2 \dfrac{c_{\gamma 2}}{c_{p1}}} & 0 \\ 0 & \dfrac{1}{\dfrac{\left(z_1 - iD_R / B_r\right)}{M_2} + z_2 \dfrac{c_{\gamma 2}}{c_{p1}}} \end{bmatrix}. \tag{12.18}$$

To put the final expressions in a more compact form, let

$$Z_1^r = \frac{\left(z_1 - iD_R / B_r\right)}{M_1}$$

$$Z_2^r = \frac{\left(z_1 - iD_R / B_r\right)}{M_2}. \tag{12.19}$$

[Note: Z_1^r, Z_2^r are distances, not impedances here]. Then the multi-Gaussian beam model becomes, finally

$$V_i^{\gamma}\mathbf{d}^{\gamma} = T_{12}^{\gamma;P}\mathbf{d}^{\gamma} v_0 \sum_{r=1}^{15} \frac{A_r}{1 + i z_1 B_r / D_R}$$

$$\cdot \frac{\sqrt{Z_1^r}}{\sqrt{Z_1^r + z_2\left(c_{\gamma 2} / c_{p1}\right)}} \frac{\sqrt{Z_2^r}}{\sqrt{Z_2^r + z_2\left(c_{\gamma 2} / c_{p1}\right)}} \tag{12.20}$$

$$\cdot \exp\left[ik_{p1}z_1 + ik_{\gamma 2}z_2 + i\frac{k_{p1}}{2}\mathbf{y}^T\left[c_{p1}\mathbf{M}_2^{\gamma}\left(z_2\right)\right]_r \mathbf{y}\right]$$

with

$$c_{p1}\left[\mathbf{M}_2^{\gamma}\left(z_2\right)\right]_r = \begin{bmatrix} \dfrac{1}{Z_1^r + z_2\left(c_{\gamma 2} / c_{p1}\right)} & 0 \\ 0 & \dfrac{1}{Z_2^r + z_2\left(c_{\gamma 2} / c_{p1}\right)} \end{bmatrix}. \tag{12.21}$$

The square roots appearing in Eq. (12.21) are unambiguous so that they can be calculated directly.

12.3 Measurement Model Input Parameters

In order to model the single interface problem shown in Fig.12.1, there are a significant number of input parameters that need to be defined. Here we will outline those parameters and the manner in which they will be represented in MATLAB. First, there are several general parameters that we will call setup parameters:

Setup Parameters
f....the frequencies at which the response will be calculated (MHz)
type1....the type of wave ('p' or 's') in medium one (a string)
type2....the type of wave ('p' or 's') in medium two (a string)

Although we will initially only consider problems where medium one is a fluid where type1 = 'p', we will leave type1 arbitrary to show the structure of input parameters in a more general setting.
    Next, we need to define parameters that will allow us to determine the system function:

System Parameters
sysf....the name of a function that will either model the system function or calculate it experimentally (a string).
amp....the amplitude of a modeled system function (volts/MHz)
fc....the center frequency of a modeled system function (MHz)
bw....the bandwidth of a modeled system function
z1r....the distance in the fluid used in a reference scattering configuration to calculate the system function experimentally
en....the noise constant used in a Wiener filter when obtaining the system function experimentally
ref_file....the name of a MAT-file (a string). This file must contain the time axis and measured waveform obtained from the reference scattering configuration. These measured values are used in the function whose name is contained in sysf

In an ultrasonic system the system function determines the effects of all the electrical and electromechanical components. The sysf parameter allows us to use either an experimentally determined system function in the measurement model or a model-based function. If this value is the string 'systf' then the model-based function systf (which is defined later) will be used. The function systf obtains the amplitude, center frequency, and bandwidth to be used in calculating the system function from the amp, fc, and bw parameters, respectively. Otherwise the user must supply the name of a compatible function that calculates the system function experimentally. Examples of the use of both types of these functions will be given. The function that calculates the system function experimentally needs to have as one of its inputs a measured waveform from a reference scattering configuration. This waveform and its time axis is contained in a MATLAB MAT-file whose filename is given by the contents of ref_file. In this MAT-file the time axis is a MATLAB vector named t_ref and the reference waveform is a MATLAB vector named ref. The function that calculates the system function experimentally also must use the same transducer parameters, pulser/receiver settings, etc. as in a flaw measurement so that a system function can be determined that is also appropriate to the flaw measurement. However, in a reference experiment where the waves received from the front surface of an immersed block can be used to calculate the system function, as described in Chapter 6, the water path length might be different from that of a flaw measurement setup. Thus, this water path length is given by the parameter z1r. If there are other parameters that are different in the reference experiment from those used in the flaw measurement (such as the material properties of the block, etc.) then they must also be included as additional setup system parameters.

There are also parameters associated with the transducer. For circular piston probes we need to specify:

Transducer parameters
d....the transducer diameter (mm)
fl....the transducer geometrical focal length (mm)

There are also a number of geometry parameters:

Geometry Parameters
z1....the distance traveled by the sound in medium one along a central ray path (mm)
z2....the distance traveled by the sound in medium two along a central ray path (mm)
x2....the perpendicular distance from the central ray axis to the center of the flaw (see Fig. 12.1) in the plane of incidence (mm)
y2....the perpendicular distance from the central ray axis to the center of the flaw (see Fig. 12.1) in a plane perpendicular to the plane of incidence (mm)
i_ang....the acute angle between the normal to the transducer and the normal to the interface at the point where the central ray from the

transducer strikes the interface (deg) [This is the angle θ_1^p shown in Fig. 12.1].

R1....the principal radius of curvature (Fig. 12.1) in the n_1 direction (mm)

R2....the principal radius of curvature (Fig. 12.1) in the n_2 direction (mm)

p_ang....the angle between the plane of incidence and the n_1 direction (deg) [This is the angle ϕ shown in Fig. 12.1].

The present study will assume that the plane of incidence and the n_1 direction are aligned so that p_ang = 0, but this parameter has been included for generality.

Not surprisingly, there are also quite a number of material parameters to specify:

Material Parameters
d1....the density of medium one (gm/cm^3)
d2....the density of medium two (gm/cm^3)
cp1....the P-wave speed of medium one (m/sec)
cs1....the S-wave speed of medium one (m/sec)
cp2....the P-wave speed of medium two (m/sec)
cs2....the S-wave speed of medium two (m/sec)
p1....P-wave attenuation fitting coefficients for medium one
s1....S-wave attenuation fitting coefficients for medium one
p2....P-wave attenuation fitting coefficients for medium two
s2....S-wave attenuation fitting coefficients for medium two

Again, for generality, we will leave the possibility of medium one having shear properties. The attenuation fitting coefficients will be used to define the attenuation coefficients in terms of powers of frequency. These will be discussed when we develop the attenuation model term.

The "flaw" cases we will consider in these examples will be of simple shapes (e.g. spherical voids, cylindrical holes, circular cracks) so that only several parameters will be needed in addition to the name of the function that will calculate the scattering amplitude:

Flaw Parameters
b.... radius of the flaw (mm)
f_ang....acute angle of the flaw with respect to the central ray (deg) (see Fig. 12.1)
Afunc....the name of the function that will calculate the far-field scattering amplitude of the flaw (a string)

Finally, we have a number of parameters associated with the particular types of waves we are considering in medium one and two. They are the wave speeds in medium one and two for the specified wave types in those media and the corresponding plane wave transmission coefficient. We have labeled these parameters wave parameters:

Wave Parameters
$c1$....the wave speed for the wave of type1 in medium one (m/sec)
$c2$....the wave speed for the wave of type2 in medium two (m/sec)
$T12$....the plane wave transmission coefficient (based on velocity or displacement ratios) appropriate to waves of type1 and type2

There is one difference between the wave parameters and the other parameters in that the wave parameters are *derived* parameters so that if the wave types and/or wave speeds are changed these wave parameter values will not be consistent with those choices unless they also are appropriately changed. *Thus, it is necessary to update these wave parameters with the current values present in the setup before using them.*

Because there are a considerable number of parameters, it is essential to have a flexible method to examine, retrieve, and change them and to pass them to other functions. Thus we have placed all of these parameters in a MATLAB structure named setup. This setup structure has a number of fields called system (for the system function), trans (for transducer), geom (for geometry), matl (for material), flaw (for flaw), and wave (for wave parameters). These fields in turn have fieldnames that are associated with the parameters just listed. A MATLAB function called setup_maker defines a complete set of the default parameters needed for a measurement model suitable for problems of the type shown in Fig. 12.1 and generates the setup structure (Code Listing 12.1). In setup_maker all the setup parameters are first defined and then placed into the setup structure. Both of these operations could have been performed in one step but they have been separated here strictly to make them more explicit for the reader.

Code Listing 12.1. The MATLAB function for generating a default structure, setup, that contains all the parameters needed for a measurement model of the case shown in Fig 12.1

```
function setup =setup_maker( )

%setup parameters
f = 5;
type1 = 'p';
type2 ='p';
% system function parameters
sysf ='systf';
amp = 5.0E-02;
fc = 5;
bw = 3;
z1r = 0.0;
en =0.01;
ref_file ='empty';
% transducer parameters
d = 12.7;
fl= inf;
%geometry parameters
z1 = 0;
z2 = linspace(0,200,512);
x2 = 0.0;
y2 =0.0;
i_ang = 0;
R1 = inf;
R2 = inf;
p_ang = 0;
% material parameters
d1 = 1.0;
d2 = 1.0;
cp1 =1480;
cs1 = 0;
cp2 =1480;
cs2 = 0;
p1 = zeros(1,5);
s1 = zeros(1,5);
p2 = zeros(1,5);
s2 = zeros(1,5);
%flaw parameters
b =0.0;
f_ang = 0.0;
```

```
Afunc = 'empty';
%wave parameters
c1 =1480;
c2 = 1480;
T12 =1.0;

% put setup in a structure
setup.f = f;
setup.type1 = type1;
setup.type2 = type2;
setup.system.sysf = sysf;
setup.system.amp =amp;
setup.system.fc = fc;
setup.system.bw = bw;
setup.system.z1r =z1r;
setup.system.en =en;
setup.system.ref_file = ref_file;
setup.trans.d = d;
setup.trans.fl =fl;
setup.geom.z1 = z1;
setup.geom.z2 = z2;
setup.geom.x2 = x2;
setup.geom.y2 = y2;
setup.geom.i_ang = i_ang;
setup.geom.R1 =R1;
setup.geom.R2 = R2;
setup.geom.p_ang = p_ang;
setup.matl.d1 =d1;
setup.matl.d2 = d2;
setup.matl.cp1 =cp1;
setup.matl.cs1 = cs1;
setup.matl.cp2 = cp2;
setup.matl.cs2 =cs2;
setup.matl.p1 = p1;
setup.matl.s1 =s1;
setup.matl.p2 = p2;
setup.matl.s2 = s2;
setup.flaw.b = b;
setup.flaw.f_ang = f_ang;
setup.flaw.Afunc = Afunc;
setup.wave.c1 = c1;
setup.wave.c2 = c2;
setup.wave.T12 = T12;
```

It can be seen from Code Listing 12.1 that the default parameters are for a 5 MHz center frequency, 3 MHz bandwidth system function and a 12.7 mm diameter planar transducer radiating a P-wave directly into a single medium (water), since the material properties for water are used for both materials. The P-wave response is to be calculated at a single frequency of 5 MHz at 512 points along the transducer central axis from zero to 200 mm, with no attenuation and with the flaw parameters initially set to zero. It can be seen that the wave parameters are also made consistent with the other setup parameters in this default case. However, to remain consistent these wave parameters must be recomputed whenever the wave types or materials are changed, as mentioned previously.

This default set of parameters would be suitable for generating, for example, a central axis transducer beam response similar to those shown in Chapter 8 (see, for example, Fig. 8.9). We will demonstrate the use of this default set of parameters (and others) after we have developed the necessary MATLAB multi-Gaussian beam model.

The setup structure makes it easy to manipulate all the problem parameters and to set up various cases. Examples of using this structure will be given when we begin to discuss specific case studies later in this Chapter. A MATLAB function display_setup has also been defined that allows one to examine all these setup parameters.

12.4 A Multi-Gaussian Beam Model in MATLAB

To generate a complete multi-Gaussian beam model that can simulate the ideal normalized velocity field, V_i^γ / v_0 of Eq. (12.8), in addition to a subset of the setup parameters (attenuation parameters and flaw parameters, for example, are not needed for this beam model) we need the Gaussian coefficients and we must calculate the appropriate plane wave transmission coefficient. The Wen and Breazeale fifteen complex coefficients, (A_r, B_r), have been placed in a MATLAB function gauss_c15 that returns their values. This function is given in the following listing:

Code Listing 12.2. A MATLAB function that returns the fifteen Wen and Breazeale coefficients. These coefficients are used to generate a multi-Gaussian beam model of a circular piston transducer.

```
function [a, b] = gauss_c15

a = zeros(15,1);
b = zeros(15,1);
a(1) = -2.9716 + 8.6187*i;
a(2) = -3.4811 + 0.9687*i;
a(3) = -1.3982 - 0.8128*i;
a(4) = 0.0773 - 0.3303*i;
a(5) = 2.8798 + 1.6109*i;
a(6) = 0.1259 - 0.0957*i;
a(7) = -0.2641 - 0.6723*i;
a(8) = 18.019 + 7.8291*i;
a(9) = 0.0518 + 0.0182*i;
a(10) = -16.9438 - 9.9384*i;
a(11) = 0.3708 + 5.4522*i;
a(12) = -6.6929 + 4.0722*i;
a(13) = -9.3638 - 4.9998*i;
a(14) = 1.5872 - 15.4212*i;
a(15) = 19.0024 + 3.6850*i;
b(1) = 4.1869 - 5.1560*i;
b(2) = 3.8398 - 10.8004*i;
b(3) = 3.4355 - 16.3582*i;
b(4) = 2.4618 - 27.7134*i;
b(5) = 5.4699 + 28.6319*i;
b(6) = 1.9833 - 33.2885*i;
b(7) = 2.9335 - 22.0151*i;
b(8) = 6.3036 + 36.7772*i;
b(9) = 1.3046 - 38.4650*i;
b(10) = 6.5889 + 37.0680*i;
b(11) = 5.5518 + 22.4255*i;
b(12) = 5.4013 + 16.7326*i;
b(13) = 5.1498 + 11.1249*i;
b(14) = 4.9665 + 5.6855*i;
b(15) = 4.6296 + 0.3055*i;
```

The plane wave transmission coefficient must be calculated consistent with the material properties and wave types specified in the setup structure parameters. We will use a MATLAB function that is passed the setup structure and returns the appropriate transmission coefficient. The MATLAB function fluid_solid, (see Code Listing 12.3) for example, calculates the plane wave transmission coefficient for a fluid-solid interface using the explicit expressions given in Appendix D (Eq. (D.59)). For a refracted S-wave, this transmission coefficient will be complex if the first critical angle is exceeded. The function fluid_solid calculates this complex transmission coefficient *for positive frequencies only*. Thus, if one wants to synthesize a pulse with these calculations, one will need to follow the steps discussed in Appendix A in performing the necessary FFT.

Code Listing 12.3. A MATLAB function for calculating the plane wave transmission coefficient for a fluid-solid interface.

```
function  T12 = fluid_solid(setup)
% fluid_solid(setup) computes the P-P (tpp)
% and P-S (tps) transmission coefficients based on velocity ratios
% for a plane fluid-solid interface. It obtains the necessary input
% parameters from the setup structure and then returns the
% appropriate transmission coefficient


% get setup parameters
type1 =setup.type1;
type2 =setup.type2;
inc= setup.geom.i_ang;
d1 = setup.matl.d1;
d2 =setup.matl.d2;
cp1 = setup.matl.cp1;
cs1 =setup.matl.cs1;
cp2 =setup.matl.cp2;
cs2 =setup.matl.cs2;

% consistency check (if material one is not a fluid
% then can't use this fluid-solid trans. coefficient)

if strcmp(type1, 's') | cs1 ~=0
    error('wrong wave type or wave speed for medium 1')
end


% calculate transmission coefficients
```

```
iang = (inc.*pi)./180;
sinp = (cp2/cp1)*sin(iang);
sins =(cs2/cp1)*sin(iang);
len = length(sinp);
for j=1:len
if sinp(j) >= 1
        cosp(j) = i*sqrt(sinp(j)^2 - 1);
        else
        cosp(j) = sqrt(1 - sinp(j)^2);
        end
end
for j=1:len
if sins(j) >= 1
        coss(j) = i*sqrt(sins(j)^2 - 1);
        else
        coss(j) =sqrt(1 - sins(j)^2);
        end
end
denom = cosp + (d2/d1)*(cp2/cp1)*sqrt(1-sin(iang).^2).*(4.*((cs2/cp2)^2)...
.*(sins.*coss.*sinp.*cosp) + 1 - 4.*(sins.^2).*(coss.^2));
tpp = (2*sqrt(1 - sin(iang).^2).*(1 - 2*(sins.^2)))./denom;
tps = -(4*cosp.*sins.*sqrt(1 - sin(iang).^2))./denom;

%select appropriate coefficient
if strcmp(type2, 'p')
   T12 = tpp;
elseif strcmp(type2, 's')
   T12 = tps;
else
   error('wrong wave type specification')
end
```

Having the setup structure, the multi-Gaussian beam coefficients, and the plane wave transmission coefficient, we now are in a position to develop the complete multi-Gaussian beam model. The MATLAB function MGbeam extracts the setup parameters it needs from the setup structure (which is the only input to MGbeam); calls the function c_gauss15 to obtain the Gaussian beam coefficients; updates the setup.wave parameters c1 and c2 to be consistent with the wave types; calls the fluid_solid function to compute the plane wave transmission coefficient (and then updates the setup structure with that coefficient); computes some of the additional parameters appearing explicitly in the beam model, and then computes the ideal velocity field in Eq. (12.20). A function init_z is called to generate an empty array of velocity values before the beam model calculations are performed. That function is given in Code Listing 12.4. This function decides what the largest size of matrix is present for the parameters f, z1, z2, x2, and y2, and pre-allocates an array of zeros of the same size for the velocity field, v, to be calculated. This pre-allocation is for efficiency only. One could have instead simply initialized v with v = 0. MGbeam is coded to allow f, z1, z2, x2, and y2 to be either scalars, vectors, or 2 by 2 arrays so that one can perform a number of different studies and plot various combinations of parameters, as will be shown shortly. MGbeam is not coded to allow the incident angle with the interface to be other than a single scalar value. However, multiple calls to MGbeam with different values of setup.geom.i_ang could be used to perform those types of studies.

Code Listing 12.4. A MATLAB function for pre-allocating memory for the velocity calculations of the same size as the largest array present in the input parameters f, z1, z2, x2, y2.

```
function v =init_z(setup)
% get parameters that may not be scalars
f =setup.f;
z1 = setup.geom.z1;
z2=setup.geom.z2;
x2 =setup.geom.x2;
y2 = setup.geom.y2;
%get dimensions, put in rows
A = [size(f); size(z1);size(z2);size(x2); size(y2)];
%get product of dimensions for each varaible
prod =A(:,1).*A(:,2); % this is a column vector
%find which row (or rows) have largest dimension
ind = find( prod = = max(prod));
%pick first row with largest dimension
val = ind(1);
% initialize v with  zeros of same size
% as the parameter(s) with largest dimensions
v = zeros(A(val,:));
```

For a spherically focused probe the Gaussian beam coefficients B_r are simply changed by letting $B_r \rightarrow B_r + iD_R/F$, where D_R is the Rayleigh length and F is the focal length, as discussed in Chapter 9. The propagation term $\exp(ik_{p1}z_1 + ik_{\gamma2}z_2)$ is not included in the calculations since this term only generates a time delay $t_0 = z_1/c_{p1} + z_2/c_{\gamma2}$ in going from the transducer to the point in the solid and this delay can easily be added in separately, if needed, by simply shifting the time axis appropriately. Thus, for pulses calculated using MGbeam the time $t = 0$ corresponds to the time when the incident quasi-plane wave is at the "center" of the flaw. MGbeam returns the ideal velocity field, V_i^γ/v_0, and the updated setup structure. As can be seen from Code Listing 12.5, the multi-Gaussian beam model is calculated in only the last fourteen lines of that Code. All the other parts of MGBeam simply prepare the necessary input parameters. Thus, except in very special cases there are no alternative beam models as simple and fast as a multi-Gaussian beam model.

Code Listing 12.5. A MATLAB function MGbeam for calculating the wave field of circular piston transducer (planar or focused) radiating through a fluid-solid interface into a solid. The function uses a multi-Gaussian beam model.

```
function [v,setup ]=MGbeam(setup)

% get setup parameters
f = setup.f;                    %frequency or frequencies (MHz)
type1 = setup.type1;            % wave type in medium one
type2 = setup.type2;            % wave type in medium two

a = setup.trans.d/2;            % transducer radius (mm)
Fl = setup.trans.fl;            % transducer focal length (mm)

z1 = setup.geom.z1;             % water path length (mm)
z2 = setup.geom.z2;             % path length in solid (mm)
x2 =setup.geom.x2;              % distance (mm) from ray axis in POI
y2 = setup.geom.y2;             % distance (mm) perpendicular to the POI
Rx = setup.geom.R1;             % interface radius of curvature (mm) in POI
Ry =setup.geom.R2;          % interface radius of curvature (mm) out of POI
iang = setup.geom.i_ang;        % incident angle (deg)

d1 = setup.matl.d1;             % density (fluid)
d2 =setup.matl.d2;              % density (solid)
cp1 = setup.matl.cp1;           % compressional wave speed -fluid  (m/sec)
```

```
cp2 = setup.matl.cp2;              % compressional wave speed -solid (m/sec)
cs2 = setup.matl.cs2;              % shear wave speed -solid (m/sec)

[A, B] = gauss_c15;                % Wen and Breazeale coefficients (15)

% update setup.wave wave speeds
if strcmp(type1, 'p')
   setup.wave.c1 =cp1;
elseif strcmp(type1, 's')
   setup.wave.c1 = cs1;
else
   error('wrong wave type (must be p or s) ')
end

if strcmp(type2, 'p')
   setup.wave.c2 =cp2;
elseif strcmp(type2, 's')
   setup.wave.c2 = cs2;
else
   error('wrong wave type (must be p or s)')
end
% calculate transmission coefficient, update setup
setup.wave.T12 = fluid_solid(setup);

% wave speeds and transmission coefficient for the beam model
c1 =setup.wave.c1;
c2 =setup.wave.c2;                  % wave speed for wave type2
T = setup.wave.T12;                % transmission coefficient

% parameters appearing in beam model

cosi = cos(pi*iang/180);           % cosine of incident angle
sinr = (c2/c1)*sin(pi*iang/180);   % sine of refracted angle from Snell's law
if sinr >= 1
   error('Beyond the Critical angle')   % no transmitted wave of given wave type
else
   cosr = sqrt( 1 - sinr^2);
end

   h11 = 1/Rx;  %curvature
   h22 = 1/Ry;  %curvature
zr = eps*(f == 0) + 1000*pi*(a^2)*f./c1; % "Rayleigh" distance
k1 = 2*pi*1000*f./c1;              % wave number in fluid

%initialize predicted velocity with zeros of a size
% compatible with largest array in f, z1, z2, x2, y2 parameters
```

```
v = init_z(setup);

%multi-Gaussian beam model

for j = 1:15                          % form up multi-Gaussian beam model

  b =B(j) + i*zr./Fl;                 % modify coefficients for focused probe
                                      % Fl = inf for planar probe

q = z1 - i*zr./b;
K = q.*(cosi -(c1/c2)*cosr);
M1 = (cosi^2 +K.*h11)./cosr^2;
M2 =1 + K.*h22;
ZR1 = q./M1;
ZR2 =q./M2;
m11 = 1./(ZR1 +(c2/c1).*z2);
m22 = 1./(ZR2 +(c2/c1).*z2);
   t1 = A(j)./(1 + (i.*b./zr).*z1);
   t2 = t1.*T.*sqrt(ZR1).*sqrt(ZR2).*sqrt(m11).*sqrt(m22);
   v = v + t2.*exp(i.*(k1./2).*(m11.*(x2.^2) + m22.*(y2.^2)));

end
```

As a simple test of this multi-Gaussian beam model we can use the default setup structure to simulate the on-axis wave field of a 5MHz, 12.7 mm diameter circular piston transducer radiating into water. The following MATLAB commands will generate the plot shown in Fig. 12.2:

```
>> setup = setup_maker;
>> [v, setup] = MGbeam(setup);
>> z2 =setup.geom.z2;
>> plot(z2, abs(v))
>> xlabel('z-distance (mm)')
>> ylabel('|v/v_0|')
```

As seen in Fig. 12.2 the beam model accurately predicts the near-field of the transducer down to a distance of approximately a transducer diameter, as discussed in Chapter 9.

Other plots also easy to simulate. From Fig. 12.2 we see that there is an on-axis null near $z2 = 70$ mm, so we can examine the cross-axis behavior at that distance through the commands:

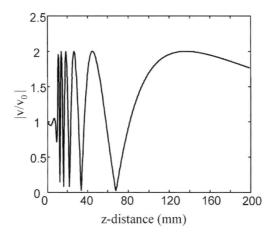

Fig. 12.2. The on-axis field of a 5 MHz, 12.7 mm diameter circular piston transducer radiating into water as calculated with a multi-Gaussian beam model.

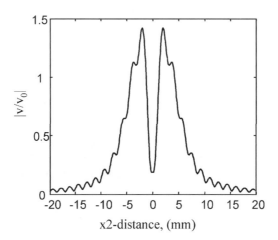

Fig. 12.3. The wave field in a plane perpendicular to the axis of a 5 MHz, 12.7 mm diameter planar piston transducer radiating into water at a distance approximately equal to one-half a near field distance along the axis.

```
>> setup.geom.z2 =70;
>> x2 = linspace(-20,20, 512);
>> setup.geom.x2 = x2;
>> [v, setup] = MGbeam(setup);
>> plot(x2, abs(v))
>> xlabel('x2-distance, (mm)')
>> ylabel(' | v/v_0 |')
```

The results are shown in Fig. 12.3. In a similar fashion we can see a 2-D cross-section of the entire wave field with the commands:

```
>> %  recall, we already had set x2 = linspace(-20,20, 512);
>> z2 = linspace(0, 200, 512);
>> [zz, xx] =meshgrid(z2, x2);
>> setup.geom.z2 = zz;
>> setup.geom.x2 = xx;
>> [v, setup] = MGbeam(setup);
>> image(z2, x2,abs(v)*50)          % scale the result to get a good
                                    % color map

>> xlabel('z2-distance (mm)')
>> ylabel('x2-distance (mm)')
```

The results are shown in Fig. 12.4.

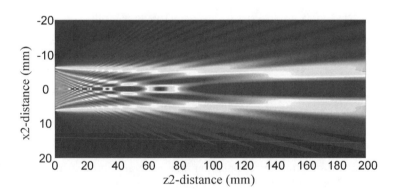

Fig. 12.4. A 2-D image of the near-field beam profile for a 5 MHz, 12.7 mm diameter planar piston transducer radiating into water. Note the scales on the two axes are very different.

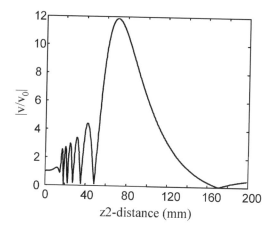

Fig. 12.5. The on-axis wave field of a 10 MHz, 12.7mm diameter, 76.2mm focal length focused transducer radiating into water as calculated with a multi-Gaussian beam model.

To simulate a spherically focused probe and examine the on-axis response, consider a 10 MHz, 12.7 mm diameter, 76.2 mm focal length transducer radiating into water. This can be simulated via the commands:

```
>> setup.f =10;
>> setup.geom.x2 =0.;
>> setup.geom.z2 =z2;  % put a vector set of values back into setup
>> setup.trans.fl = 76.2;
>> [v, setup] = MGbeam(setup);
>> plot(z2, abs(v))
>> xlabel('z2-distance (mm)')
```

The results are shown in Fig. 12.5. Note that we changed the frequency of the calculation by changing the setup.f parameter, not the setup.system.fc (center frequency) parameter. The center frequency parameter refers to a parameter of the frequency profile of the system function which is needed to synthesize a time domain waveform. This center frequency parameter will not affect beam calculations performed at a single frequency. To synthesize a transducer pulse, however, we would have to let setup.f be an array of frequencies and multiply the output of MGbeam function by a system function to simulate the spectral behavior of the system. We will show simulation examples of this type later.

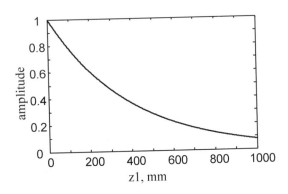

Fig. 12.6. The attenuated amplitude versus distance for propagation in water at room temperature and at a frequency of 10 MHz.

12.5 Ultrasonic Attenuation in the Measurement Model

Ultrasonic material attenuation is a part of the measurement model which must be determined experimentally. The linear attenuation terms appearing in the attenuation expression $\exp\left[-\alpha_{p1}(\omega)z_1 - \alpha_{\gamma2}(\omega)z_2\right]$ are frequency dependent so that normally one fits the measured values of these linear attenuation terms to functions with a simple frequency dependency (linear, quadratic, etc.) that best match the experimental results over the bandwidth of the measurement system. The MATLAB function attenuate in Code Listing 12.6 defines each of the linear attenuation coefficients for the appropriate wave types traveling in medium one and two in terms of five fitting coefficients for a polynomial of up to fourth order in frequency, i.e. we use a fitting expression for an attenuation coefficient α in the form $\alpha = a_1 + a_2 f + a_3 f^2 + a_4 f^3 + a_5 f^4$. Those fitting coefficients must be placed in setup.matlp1, setup.matls1, setup.matlp2, and setup.matls2 .

Code Listing 12.6. A MATLAB function for calculating attenuation terms for propagation in two adjacent media.

```
function y = attenuate(setup)
% atten(setup) generates a frequency dependent attenuation factor
% as a function of the frequency, f, and the distances z1, z2 in (mm)
% traveled in two media
% For water at room temp for the first medium , take p1(1) = p1(2) = p1(4)
% =p1(5)=0,
% and p1(3) = 25.3E-06 if f is in MHz, distances are in mm

f=setup.f;
type1=setup.type1;
type2=setup.type2;
z1 =setup.geom.z1;
z2 =setup.geom.z2;
p1 =setup.matl.p1;
s1 =setup.matl.s1;
p2=setup.matl.p2;
s2=setup.matl.s2;
if strcmp(type1, 'p')
        a1 =p1;
elseif  strcmp(type1, 's')
        a1 =s1;
else
error('wrong wave type')
end

if strcmp(type2, 'p')
        a2 =p2;
elseif  strcmp(type1, 's')
        a2 =s2;
else
error('wrong wave type')
end

d1 = a1(1) + a1(2)*f + a1(3)*f.^2  + a1(4)*f.^3  + a1(5)*f.^4;
d2 = a2(1) + a2(2)*f + a2(3)*f.^2  + a2(4)*f.^3  + a2(5)*f.^4;

y = exp(-d1.*z1).*exp(-d2.*z2);
```

To illustrate this function, consider the attenuated amplitude versus distance in water at room temperature for a frequency of 10 MHz where the attenuation coefficient is $\alpha = 25.3 \times 10^{-6} f^2$ with f the frequency in MHz. Using the default setup structure and the MATLAB commands:

```
>> setup.f =10.;
>> z1 =linspace(0,1000,512);
>> setup.geom.z1  = z1;
>> setup.geom.z2 =0.0;
>> setup.matl.p1 = [ 0  0  25.3E-06  0  0];
>> y=attenuate(setup);
>> plot(z1, y)
>> xlabel('z1, mm')
>> ylabel('amplitude')
```

we obtain the plot show in Fig. 12.6 (the default type1 ='p' here and the other attenuation fitting coefficients are all zero).

12.6 The System Function

The system function, $s(\omega)$, is found in practice by either performing a measurement of the received voltage in a calibration setup or by measuring all the ultrasonic system components in the sound generation and reception processes and combining them to form up the $s(\omega)$, as described in previous Chapters. However, we can also simulate this function directly to model its effects on the measurement process.

To model the system function we will use a simple Gaussian function of the type discussed in Appendix A given by

$$F(f) = A \exp\left[-4\pi^2 a^2 (f - f_c)^2\right] = A \exp\left[-a^2 (\omega - \omega_c)^2\right], \quad (12.22)$$

where A is the amplitude, $f = 2\pi\omega$ is the frequency and f_c is the center frequency, both measured in MHz. The inverse Fourier transform of this function can be obtained analytically as

$$f(t) = \frac{A}{2a\sqrt{\pi}} \exp(-2\pi i f_c t) \exp(-t^2 / 4a^2), \quad (12.23)$$

which is complex since we have not included any negative frequency components in $F(f)$. As shown in Appendix A we can recover a real time domain signal, $v(t)$, from only the positive frequency components if we take twice the real part of Eq. (12.23) which gives

$$v(t) = \frac{A}{a\sqrt{\pi}} \cos(2\pi f_c t) \exp(-t^2 / 4a^2).$$

(12.24)

In all the model terms in our measurement models, we will likewise only model those terms for positive frequencies and then take twice the real part of the result to recover real time domain functions.

It is convenient to rewrite $F(f)$ in a form which is parameterized not in terms of a but instead in terms of the bandwidth, bw, where bw is the width of the Gaussian, in MHz, where its amplitude is one-half of its maximum value (see Fig. A.5). This gives

$$\exp\left[-4\pi^2 a^2 (f_0 - f_c)^2\right] = \exp\left[-\pi^2 a^2 (bw)^2\right] = \frac{1}{2}$$

(12.25)

so solving for a in terms of bw we find

$$a = \frac{\sqrt{\ln 2}}{\pi\, bw}.$$

(12.26)

For small center frequencies and large bandwidths, the simple Gaussian function in Eq. (12.22) will have a non-zero D.C. (zero frequency) component. Most transducers band limit the measured ultrasonic response so that the response should be very small at low frequencies. To model this behavior we therefore modify the Gaussian slightly through a sine function that tapers the response to zero at zero frequency. Thus, the simulated system transfer factor, $s(f)$, we will model is given by

$$s(f) = \begin{cases} F(f)\sin\left[\dfrac{\pi f}{2 f_c}\right] & f < f_c \\ F(f) & f \geq f_c \end{cases}.$$

(12.27)

This modification means that the corresponding time domain waveform will not be given exactly by Eq. (12.24) but in many cases the difference is small. The MATLAB function in Code Listing 12.7 returns the system function given in Eq. (12.27):

Code Listing 12.7. A MATLAB function for simulating the system function.

```
function y = systf (setup)
% systf(setup) models the system function by a Gaussian window function
% of amplitude amp centered at frequency fc and with a bandwidth bw defined to
% be the spread in frequency at the half amplitude point in the Gaussian.
% The Gaussian is tapered to zero at frequencies below fc with a sine function to
% guarantee the dc value is always zero.
% For small fc and large bw, this tapering will distort the Gaussian
%
f =setup.f;
amp = setup.trans.amp;
fc = setup.trans.fc;
bw = setup.trans.bw;
a = sqrt(log(2))/(pi*bw);
s1 = exp(-(2*a*pi*(f - fc)).^2).*(f > fc);
s2 = exp(-(2*a*pi*(f - fc)).^2).*sin(pi*f/(2*fc)).*(f <= fc);
y = amp*(s1 + s2);
```

To illustrate this function we can use the default setup structure where amp = .05 volts/MHz, fc = 5 MHz, and bw = 3 MHz with the commands:

```
>> f = linspace(0, 20, 512);
>> setup.f = f;
>> y=systf(setup);
>> plot(f, y)
>> xlabel( ' f, MHz')
>> ylabel('volts/MHz')
```

to obtain the system function shown in Fig. 12.7. Note that the system function modeled here is a purely real function. A measured system function, however, will generally be a complex-valued function.

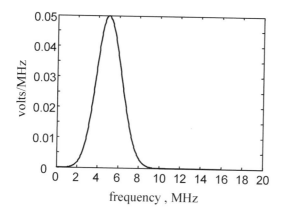

Fig. 12.7. A simulated system function.

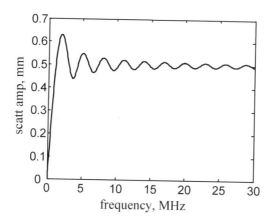

Fig. 12.8. The pulse-echo far-field scattering amplitude versus frequency for a 1 mm radius spherical void in steel, calculated using the Kirchhoff approximation.

12.7 Flaw Scattering Models

As shown in Chapter 10, the Kirchhoff approximation is a very useful approximation for obtaining the flaw scattering properties of a number of flaws. We will develop MATLAB functions that will use the Kirchhoff approximation for modeling the pulse-echo far-field scattering amplitude of a spherical void and a circular crack. The explicit expressions for these scattering amplitudes were given in Chapter 10. For the spherical void of radius b we found (Eq. (10.14):

$$A\left(\mathbf{e}_i^\beta;-\mathbf{e}_i^\beta\right)=\frac{-b}{2}\exp\left(-ik_\beta b\right)\left[\exp\left(-ik_\beta b\right)-\frac{\sin\left(k_\beta b\right)}{k_\beta b}\right],\tag{12.28}$$

while for wave incident on a circular crack of radius b at an angle, θ, with respect to the crack normal we found (Eq. (10.36)):

$$A\left(\mathbf{e}_i^\beta;-\mathbf{e}_i^\beta\right)=\frac{ib\cos\theta}{2\sin\theta}J_1\left(2k_\beta b\sin\theta\right).\tag{12.29}$$

Code Listing 12.8 describes the function A_void that uses Eq. (12.28) and returns the pulse-echo scattering amplitude of the spherical void.

Code Listing 12.8. A MATLAB function for modeling the pulse-echo far-field scattering amplitude of a spherical void.

```
function A = A_void(setup)
% A_VOID calculates the pulse-echo far-field scattering amplitude
% of a spherical void in the Kirchhoff approximation, using
% the frequency f in setup.f, the radius b in  setup.flaw.b,
% and the wave speed for the wave type in setup.wave.c2.
% The calling sequence is A = A_void(setup). The scattering
% amplitude, A, (in mm) is returned.

%get the parameters
f =setup.f;
c = setup.wave.c2;
b = setup.flaw.b;

%calculate the wave number kb (f in MHz, b in mm, c in m/sec)
kb = (2000*pi*b*f)./c;

%calculate the pulse-echo scattering amplitude
kb = kb + eps*(kb == 0);   % prevent division by zero
A =(-b/2)*exp(-i*kb).*(exp(-i*kb)-sin(kb)./(kb));
```

Similarly, Code Listing 12.9 gives the MATLAB function A_crack that uses Eq. (12.29) and returns the pulse-echo scattering amplitude of the circular crack.

Code Listing 12.9. A MATLAB function for modeling the pulse-echo far-field scattering amplitude of a circular crack.

```
function A = A_crack(setup)
% A_CRACK calculates the pulse-echo far-field scattering amplitude
% of a circular crack in the Kirchhoff approximation, using the
% frequency f in setup.f, the radius b in setup.flaw.b, the acute
% angle between the incident wave direction and the crack normal in
% setup.flaw.f_ang, and the wave speed for the wave type in
% setup.wave.c2.
% The calling sequence is A = A_crack(setup). The
% scattering amplitude,A, (in mm) is returned.

%get the parameters
f = setup.f;
c = setup.wave.c2;
b = setup.flaw.b;
ang = setup.flaw.f_ang;

% put the angle in radians, calculate the wave number
iang = ang.*pi./180;
kb = (2000*pi*b*f)./c;

% calculate the pulse-echo scattering amplitude
arg = 2*sin(iang).*kb;      % argument of bessel function
arg = arg + eps*(arg == 0); % prevent division by zero
A = i*kb.*b.*cos(iang).*(besselj(1, arg)./arg);
```

We can use these functions to verify some of the results presented in Chapter 10. First, consider the pulse-echo frequency domain response of a 1 mm radius spherical void in steel (c_{p2} = 5900 m/sec). Using the commands:

```
>> clear
>> setup=setup_maker;
>> setup.f =linspace(0,30,512);
```

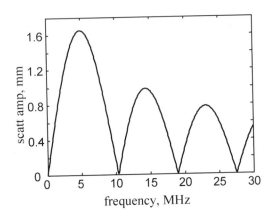

Fig. 12.9. The pulse-echo far-field scattering amplitude versus frequency for a 1 mm radius circular crack in steel, calculated using the Kirchhoff approximation. The incident angle $\theta = 10°$ with respect to the crack normal.

```
>> setup.wave.c2 =5900;
>> setup.flaw.b =1.;
>> setup.flaw.Afunc ='A_void';
>> f = setup.f;
>> A = feval(setup.flaw.Afunc, setup);
>> plot(f, abs(A))
>> xlabel('frequency, MHz')
>> ylabel('scatt amp, mm')
```

generates the plot shown in Fig. 12.8 which is identical to Fig. 10.6. Notice that we put the frequencies and wave speed into the appropriate parameters in setup and we have placed the name of the flaw function in the setup structure and then retrieved it to evaluate it with the function feval. This process was done simply to illustrate how in a measurement model the setup structure will be used to obtain the flaw response. In this case we could have just called the function A_void directly with setup as its argument.

The same type of pulse-echo response for a 1 mm radius crack in steel where the incident direction is at $10°$ from the crack normal can be found using the same setup parameters just defined plus the commands

```
>> setup.flaw.f_ang = 10;
>> setup.flaw.Afunc ='A_crack';
>> Ac =feval(setup.flaw.Afunc, setup);
```

```
>> plot(f, abs(Ac))
>> xlabel('frequency, MHz')
>> ylabel('scatt amp, mm')
```

These commands generate the plot shown in Fig. 12.9 which is identical to the same plot shown in Fig. 10.17.

12.8 The Thompson-Gray Measurement Model

We now have all the MATLAB functions defined that will allow us to construct a complete ultrasonic measurement model of the type given in Eq.(12.6) where the flaw is assumed to be small enough so that we can neglect the beam variations over the flaw surface. Thompson and Gray first developed this type of measurement model in 1983 [11.2]. The MATLAB function TG_PE_MM (Code Listing 12.10), like all our other functions uses only the setup structure as its input. TG_PE_MM returns an updated setup structure and the measured voltage, V_R, in the frequency domain obtained from a flaw in the solid using the Thompson-Gray measurement model for a pulse-echo immersion setup of the type shown in Fig 12.1. The multi-Gaussian beam model function MGbeam is used to predict the transducer velocity field at the flaw and the far-field scattering amplitude is obtained by the MATLAB function whose name is specified in the setup parameter setup.flaw.Afunc. The system function is modeled by the MATLAB function systf if the setup.sysf contains the string 'systf' (the default) or this function is obtained experimentally by use of the function whose name is contained in setup.sysf. The attenuation of the materials in the measurement model is accounted for by the MATLAB function attenuate.

Code Listing 12.10. The MATLAB function TG_PE_MM for modeling the response of a flaw using the Thompson-Gray ultrasonic measurement model.

```
function [Vf, setup] =TG_PE_MM(setup)
% TG_PE_MM generates the frequency components of the
% output voltage, Vf, of an ultrasonic pulse-echo immersion
% measurement system generated by a flaw.
% The function returns Vf as well as an updated setup structure
% The calling sequence is [Vf, setup] =TG_mm(setup);

% First, compute the incident beam velocity and update
```

```
% the setup structure
[v, setup] = MGbeam(setup);

%get the setup parameters  needed for the constant term
%in the measurement model
f = setup.f;
r= setup.trans.d/2;    % transducer radius
d1 =setup.matl.d1;
d2 =setup.matl.d2;
c1 = setup.wave.c1;
c2 = setup.wave.c2;

%compute wave number in medium two and
%the constant term in the measurement model

k2 = (2000.*pi.*f)./c2;
k2 =k2 + eps*( k2 == 0);  % prevent division by zero
K= (4.*d2.*c2)./(-i.*k2.*r^2.*d1.*c1);

% check to see if a model-based or experimentally determined system
% function is to be used
if strcmp(setup.sysf, 'systf')
    sys = systf(setup);
else
    sys =feval(setup.sysf, setup);
end

% find flaw type to be used
if strcmp( setup.flaw.Afunc, 'empty')
    error('flaw function not specified in setup')
else
    A = feval(setup.flaw.Afunc, setup);
end

%compute output voltage, Vf, (volts/MHz)
Vf = sys.*(v.*attenuate(setup)).^2.*A.*K;
```

To illustrate an application of the MATLAB function TG_PE_MM we will describe a MATLAB calculation that uses the setup shown in Fig. 12.10 (b), where a planar, 5 MHz transducer is being used in pulse-echo to examine a spherical 0.6921 mm diameter void in a glass block at normal incidence through a water-solid interface. These parameters are

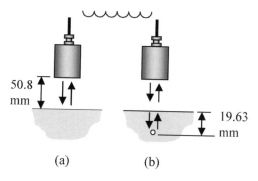

50.8 mm

19.63 mm

(a)   (b)

Fig. 12.10. (a) A reference scattering configuration where a planar 12.7 mm diameter, 5 MHz transducer receives the P-waves reflected from a water-glass interface. **(b)** A pulse-echo flaw measurement setup where the transducer in (a) receives the P-waves scattered from a 0.6921 mm diameter spherical void in glass located on the central axis of the transducer. The water path length is the same (50.8 mm) in both measurements.

similar to those of a experimental setup that we will discuss next. We will simulate the received voltage time-domain waveform from the void. If we call the function setup_maker then we need to change only those parameters that are different from the default setup structure that is generated by this function. In this case we will set up a range of frequencies from 0 to 20 MHz to do our calculations and define the measured wave speed of the water (the water density was taken as the default value of 1.0) and also the density and wave speed of the glass:

```
>> setup = setup_maker;
>> f = s_space(0, 20, 200);
>> cp1 = 1484;
>> d2 = 2.2;
>> cp2 = 5969.4;
>> cs2 = 3774.1;
```

The MATLAB function s_space (xmin, xmax, num) used here (the MATLAB code listing is given in Appendix G) is similar to the MATLAB function linspace. The s-space function gives a set of num evenly spaced sampled values from xmin to xmax - dx, where dx = (xmax - xmin)/num is the sample spacing, whereas the MATLAB function linspace(xmin, xmax, num) gives set of num evenly sampled values from xmin to xmax with sample spacing $dx = (xmax - xmin)/(num-1)$. As discussed in Appendix A the function s_space generates precisely the sampled values needed in both the time and frequency domains to perform Fourier analysis with FFTs, but the built-in MATLAB function linspace does not.

We will also specify the water path length from the transducer to the interface and distance from the interface to center of the spherical void in the solid (see Fig. 12.10 (b)):

```
>> z1 = 50.8;
>> z2 = 19.62725;
```

The default system function center frequency of 5 MHz can be left unchanged but the system function amplitude and bandwidth will be chosen to be similar to the experimental example we will discuss shortly:

```
>> amp = 0.08;
>> bw = 4;
```

Although in this example the parameters amp and bw are the only values needed to predict the system function, when we determine this function experimentally we will also need to specify the water path length to be used in a reference experiment so that anticipating the need for that variable, we will also set it appropriately here:

```
>> z1r = 50.8;
```

The transducer diameter (12.7 mm) and focal length (infinity) are compatible with the default values generated by setup_maker. The attenuation of the glass block is very small so that it will be neglected. The P-wave attenuation of the water is included as a quadratic function of frequency:

```
>> p1 = [ 0 0 .02479E-03 0 0];
```

Finally, the flaw radius is specified and the name of the function that calculates the pulse-echo far-field scattering amplitude of a spherical void in the Kirchhoff approximation is given:

```
>> b = .34605;
>> flaw_name = 'A_void';
```

All of the other default setup parameters can be used unchanged so it is only necessary to update these parameters:

```
>> setup.f = f;
>> setup.trans.amp = amp;
>> setup.trans.bw = bw;
>> setup.z1r = z1r;
>> setup.geom.z1 =z1;
>> setup.geom.z2 =z2;
>> setup.matl.cp1 = cp1;
>> setup.matl.d2 = d2;
>> setup.matl.cp2 = cp2;
>> setup.matl.cs2 =cs2;
>> setup.matl.p1 = p1;
>> setup.flaw.b =b;
>>setup.flaw.Afunc =flaw_name;
```

With these changes then the output voltage in the frequency domain, Vf, and an updated setup structure can be calculated:

```
>> [Vf, setup] = TG_PE_MM(setup);
```

If we want to examine the time-domain waveform from the void, we must extend the maximum frequency beyond the 20 MHz value used in the calculations and zero pad the Vf values. Here we have extended the maximum frequency to 100 MHz, using the same frequency spacing, df, used in calculating Vf. The sampling time interval, dt, is then the reciprocal of this max frequency, and we can use this time interval to generate a time window, t. Since we are only going to use the positive frequency components of the response to calculate the wave form, we have also divided the zero frequency value of Vf by two:

```
>> df = f(2) - f (1);
>> dt = 1/(1000*df);
>> t= s_space(0,1000*dt , 1000);
>> Vfe = [ Vf  zeros(1, 800)];
>> Vfe(1) = Vfe(1)/2;
```

We are now able to calculate the time domain void response with an inverse FFT of these positive frequency components:

```
>> vt = 2*real(IFourierT(Vfe, dt));
```

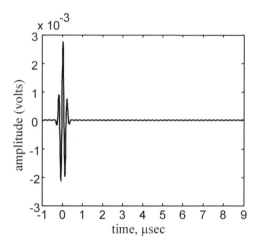

Fig. 12.11. The simulated response pulse-echo P-wave response of a spherical void for the setup shown in Fig. 12.10 (b).

and we can plot the result. Since we have omitted all the time delay terms in these calculations, t = 0 corresponds to when the waves reach the center of the flaw so that we need to use the t_shift and c-shift functions to obtain a result where the responses before t = 0 are not in the upper part of the window:

```
>> plot(t_shift (t, 100), c_shift(vt,100))
```

The simulated wave form (in volts) is shown in Fig. 12.11. All of the above MATLAB commands are contained in the MATLAB script TG_sphere_example1(Code Listing 12.11). This simple example shows how one can use the MATLAB functions to model a flaw response where the system function was taken to be the simple Gaussian function described previously.

Code Listing 12.11. A MATLAB script for calculating the pulse-echo response of an on-axis pore at normal incidence through a fluid-solid interface.

```
% TG_sphere_example1 script
% This script calculates the pulse-echo P-wave response of an on-axis
% spherical pore interrogated by a 5 MHz planar probe through a
% fluid-solid interface at normal incidence
clear
setup = setup_maker;
```

```
% setup parameters that need to be specified for this example
f =s_space(0, 20, 200);
cp1 = 1484.;
d2 = 2.2;
cp2 = 5969.4;
cs2 = 3774.1;
z1 = 50.8;
z2 = 19.62725;
amp =0.08;
bw = 4.;
z1r =50.8;
p1 = [ 0 0 0.02479E-03  0 0];
b =0.34605;
flaw_name = 'A_void';
setup.f =f;
setup.system.amp = amp;
setup.system.bw = bw;
setup.system.z1r =z1r;
setup.geom.z1 = z1;
setup.geom.z2 = z2;
setup.matl.cp1 = cp1;
setup.matl.d2 = d2;
setup.matl.cp2 = cp2;
setup.matl.cs2 = cs2;
setup.matl.p1 = p1;
setup.flaw.b = b;
setup.flaw.Afunc = flaw_name;
% calculate received voltage
[Vf, setup] = TG_PE_MM(setup);
% extend frequency components to permit
% taking FFT
df = f(2)-f(1);
dt = 1/(1000*df);
t = s_space(0, 1000*dt, 1000);
Vfe = [Vf zeros(1,800)];
Vfe(1) = Vfe(1)/2;
vt =2*real(IFourierT(Vfe, dt));
plot(t_shift(t,100), c_shift(vt,100))
```

As shown in Chapter 7, it is relatively easy to calculate the system function experimentally in a reference experiment, and this function then truly represents the effects of all the electrical and electromechanical compo nents of the system (pulser/receiver, cabling, transducers) at a specific

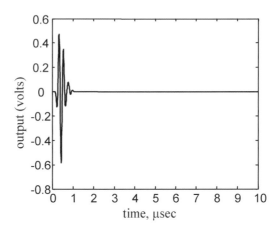

Fig. 12.12. The voltage received from the fluid-solid interface for the reference scattering configuration shown in Fig. 12.10 (a).

set of instrument settings. It is also easy to incorporate such a measured system function into our measurement model. All that is needed is to replace the output of the systf function in the previous example with a compatible set of measured values of the system function. This can be done for the example just discussed by measuring the waves received from the front surface of the glass block, as shown in Fig. 12.10(a). Since the acoustic/elastic transfer function is known for this configuration, deconvolution (with the aid of a Wiener filter) of the frequency components of the measured response by the transfer function, as shown in Chapter 7, will give us the measured system function. Figure 12.12 shows the experimental wave form received by a 5 MHz, 12.7 mm diameter planar transducer from the interface as shown in Fig. 12.10 (a). The 1000 point wave form and its corresponding time axis are stored as MATLAB variables ref and t_ref, respectively in the MATLAB MAT-file sphere_ref.mat. The MATLAB function exp_systf is used in place of the model-based systf function to calculate the system function experimentally. The function exp_systf loads the ref and t_ref variables into MATLAB (assuming that the sphere_ref file is contained in the current MATLAB directory), computes the frequency components of this measured response and then deconvolves those components with the acoustic/elastic transfer function for this configuration, using the MATLAB function Wiener_filter defined in Appendix C with a noise constant defined by the parameter setup.system.en. The function then returns the measured system function. The listing of exp_systf is given in Code Listing 12.12.

Code Listing 12.12. A function for calculating the system function from an experimentally measured wave form in the reference scattering configuration of Fig. 12.10 (a).

```
function s = exp_systf(setup)
% EXP_SYSTF generates the system function from the
% measured voltage received by a circular, planar or focused
% transducer from the planar front surface of a
% solid. It is assumed that the solid is the same as the one
% in the flaw measurement where this system function is to be used
% as is the rest of the measurement setup except that the fluid
% path length can be different from the one used in a flaw measurement.
% This function assumes that there are 1000 sampled
% values in the reference wave form and time axis
% and the sampling frequency is 100MHz
filename =setup.system.ref_file;
load(filename) % load reference wave form (in the variable ref)
% and the time axis values (in the variable t_ref) from a MAT-file

dt = t_ref(2)-t_ref(1);
% calculate Fourier Transform
V =FourierT(ref, dt);
%generate frequency axis
fs = s_space(0, 1/dt, 1000);


% get setup frequency values and check for consistency
f = setup.f;
df= f(2) - f(1);
dfs =fs(2) - fs(1);
fsize=size(f);
numf = fsize(2);
if df > (dfs + .001) | df < (dfs - .001)
    error('frequency spacing mismatch of setup and exp values')
end
if f(end) > (fs(end) +dfs)/2
    error('max frequency in setup exceeds Nyquist')
end
% keep number of measured voltage frequency components
% compatible with that in setup
Vc=V(1:numf);
% get remaining setup parameters
```

```
z1r =setup.system.z1r;
en =setup.system.en;
d1 =setup.matl.d1;
cp1 = setup.matl.cp1;
d2 = setup.matl.d2;
cp2 = setup.matl.cp2;
cs2 = setup.matl.cs2;
a = setup.trans.d/2;
p1 =setup.matl.p1;
alphac =p1(3); % frequency squared attenuation coefficient
fl = setup.trans.fl;

% if transducer is focused, z1r must be the same as the focal length
if fl ~= inf
   if z1r > fl +.01 | z1r < fl - .01
       warning(' reference water path is not the focal length, using focal length')
       z1r = fl;
   end
end

% calculate wave number , reflection coefficient of fluid-solid interface
% and argument for acoustic/elastic transfer function
ka =2000.*pi.*f.*a./cp1;
R12 = (cp2*d2 - cp1*d1)/(cp2*d2 + cp1*d1);
arg = (a/z1r)*ka;
alpha = alphac*f.^2;
% calculate acoustic-elastic transfer function, leave out propagation phase

ta = 2*R12*exp(-2*alpha.*z1r).*(1 -exp(i*arg/2).*(BesselJ(0, arg/2)...
    -i*BesselJ(1, arg/2)));
if fl ~= inf
   ta = -conj(ta);
end

% deconvolve measured voltage frequency components with transfer function
% to get system function
s = Wiener_filter(Vc, ta, en);
```

To use exp_systf for our spherical void example in place of the function systf which generates a model-based system function, we need only have the appropriate setup parameters, which can be obtained by first running the script TG_sphere_example1, and then updating setup.sysf to indicate we now are going to use an experimentally determined system function.

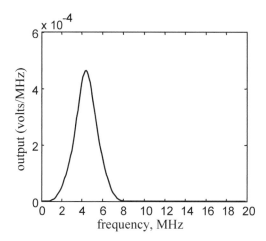

Fig. 12.13. The magnitude of the frequency components of the voltage received from an on-axis spherical void for the configuration shown in Fig. 12.10 (b) as predicted by the Thompson-Gray measurement model using an experimentally determined system function and the Kirchhoff approximation for the far-field scattering of the void.

The Wiener filter constant, en, is set at a default value of 0.01 in the setup parameters but it can be changed, if necessary. We also need to specify the MAT-file that contains the reference wave form obtained from the configuration in Fig. 12.10 (a). Note that the distance z1r has already been defined appropriately.

```
>> clear
>> TG_sphere_example1
>> setup.system.sysf = 'exp_systf';
>> setup.system.ref_file = 'sphere.ref';
```

Then we can run the measurement model and plot the output:

```
>> [Vout, setup] = TG_PE_MM(setup);
>> plot(f, abs(Vout))
```

The results are shown in Fig. 12.13. If we now pad these frequency domain values with zeros to extend the frequency range to 100 MHz and do an inverse FFT, the time domain received wave form can be plotted:

```
>> Ve =[Vout, zeros(1, 800)];
```

```
>> Ve(1) = Ve(1)/2 ;
>> vt = 2*real(IFourierT(Ve, dt));
>> plot(t, vt)
```

The results are shown in Fig. 12.14. All of the MATLAB commands needed to generate this waveform are in the MATLAB script TG_sphere_example2 (see Code Listing 12.13). The intermediate frequency plot of Fig. 12.13, however, is omitted in that script.

Code Listing 12.13. A MATLAB script for calculating the A-scan wave form for a spherical void using an experimentally determined system function.

```
% script TG_sphere_example2
% calculates the waveform for a spherical void
% using an experimentally determined system function
clear
% run TG_sphere_example1 script to get system parameters
TG_sphere_example1
%specify use of experimentally determined system function
%and reference waveform
setup.system.sysf='exp_systf';
setup.system.ref_file ='sphere_ref';
%run measurement model
[Vout, setup] = TG_PE_MM(setup);
% plot(f, abs(Vout))  intermediate plot omitted
% pad frequency domain amplitude with zeros
Ve= [ Vout, zeros(1,800)];
Ve(1) = Ve(1)/2;  % Now, compute wave form and plot
vt =2*real(IFourierT(Ve, dt));
plot(t, vt)
```

For comparison, the actual measured wave form from the flaw can also be plotted. This wave form, vexp, and its corresponding time axis, t_exp, are contained in the file sphere_flaw.mat. We can load that file and display that flaw signal on the same plot as the one just obtained:

```
>> hold on
>> load 'sphere_flaw'
>> plot(t, vexp, '--')
>> hold off
```

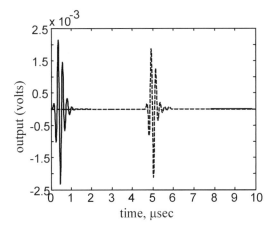

Fig. 12.14. The voltage received from an on-axis spherical void for the configuration shown in Fig. 12.10 (b) as predicted by the Thompson-Gray measurement model using an experimentally determined system function and the Kirchhoff approximation for the far-field scattering of the void (solid line) and the experimentally measured flaw signal (dashed line).

We can see in Fig. 12.14 that the two waveforms are close in amplitude and general shape. No attempt was made to match the time of arrivals of the two signals. In fact, in the calculation of these signals the phase terms that represent the time delays present due to propagation in the fluid and solid media were omitted. The measurement model predicts a slightly larger response than the measured response and there are some very small late time differences between the two signals. Fig. 12.14 shows that the Thompson-Gray measurement model coupled with the Kirchhoff approximation does a remarkably good job of predicting the flaw signal in this example even though the non-dimensional wave number, $k_{p2}b$, of the flaw for P-waves based on the transducer center frequency of 5 MHz is only $k_{p2}b = 1.8$. Formally the Kirchhoff approximation is a high frequency approximation where we must have $k_{p2}b \gg 1$ but we see this approximation still works well at much lower frequencies (or smaller flaw sizes) where $k_{p2}b$ is not large. This is consistent with our discussion of that approximation in Chapter 10. However, as shown in Chapter 10, if $k_{p2}b < 1$ then the Kirchhoff approximation generally will not be accurate. Also note that even the completely modeled signal of Fig. 12.11, has

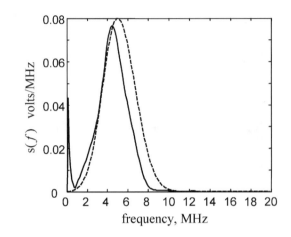

Fig. 12.15. The magnitude of the measured system function for the configuration of Fig 12.10 (a) (solid line) and the magnitude of the system function synthesized using the function systf (dashed line).

approximately the same amplitude as the experimental signal although the waveform details are different. Those differences in waveform shape come primarily from the fact that a purely real model-based system function was used in calculating the response in Fig. 12.11 while the complex-valued measured system function was used in Fig. 12.14. There are also some differences in the amplitudes and widths of the two different system functions used in Figs. 12.11 and 12.14. Figure 12.15 compares the magnitudes of these two system functions versus frequency. It can be seen that although the transducer being used is listed as a 5 MHz transducer, the system function determined experimentally peaks at a slightly lower value. For the modeled system function, we centered the Gaussian function at the 5 MHz value. Likely we could improve our predictions of the wave form obtained using a model-based system function by making the amplitude and bandwidth of that function agree more closely with the experimentally determined system function.

In Chapter 10 we gave the separations of variables solution for the pulse echo P-wave response of a spherical void. Those expressions have been encoded in the MATLAB function A_void_Psep (see Appendix G for a code listing). We can simply replace the Kirchhoff-based function A_void in the setup structure by this function:

```
>> setup.flaw.Afunc ='A_void_Psep';
```

and then rerun the measurement model and compare with the experimentally measured sphere response:

```
>> [Vout, setup] = TG_PE_MM(setup);
>> Ve = [Vout, zeros(1,800)];
>> Ve(1) = Ve(1)/2 ;
>> vt = 2*real(IFourierT(Ve, dt));
>> plot(t, vt)
>> hold on
>> load 'sphere_flaw'
>> plot(t, vexp, '--')
>> hold off
```

The results are shown in Fig. 12.16. From that figure we see that the amplitude of the modeled flaw signal is now very close to that of the experimental signal.

We can also examine the sphere with a spherically focused probe. The script TG_sphere_example3 given in Code Listing 12.14 again uses the TG_sphere_example1 script to set up most of the parameters. The transducer used is a 12.46 mm diameter, 172.9 mm focal length probe, so those parameters in setup are changed. These transducer parameters are both measured effective values, found by the methods discussed in Chapter 7. In this case the water path length for the flaw measurement is again 50.8 mm so that value need not be changed but the reference experiment to determine the system function must be carried out with the spherically focused transducer at a water path equal to the focal length to use the transfer function found in Chapter 8. Thus, the setup.system.z1r must also be changed. The function exp_systf again can calculate the system function for this focused probe. In this case the reference waveform is contained in the MAT-file 'sphere_ref_foc'. For this example we will also use the Kirchhoff approximation to determine the scattering amplitude of the void, so that we set setup.flaw.Afunc = 'A_void'. With these updates made to setup, the measurement model can be run and the waveform synthesized as before. The experimentally measured response of the void to this focused probe is contained in the .mat file 'sphere_flaw_foc' in the variable vexp so if we load this file and then plot it alongside our modeled response we obtain the results shown in Fig. 12.17. It can be seen from that figure that the Kirchhoff approximation does a very good job of reproducing the measured flaw signal.

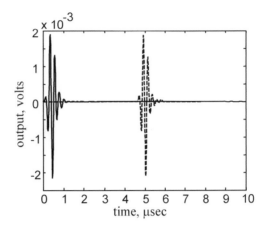

Fig. 12.16. The voltage received from an on-axis spherical void for the configuration shown in Fig. 12.10 (b) as predicted by the Thompson-Gray measurement model using an experimentally determined system function and the method of separation of variables for the far-field scattering of the void (solid line). The experimentally measured flaw signal is shown for comparison (dashed line).

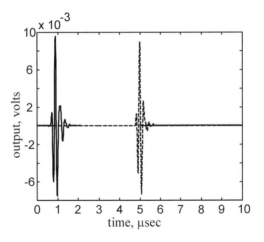

Fig. 12.17. The voltage received from an on-axis spherical void for the configuration shown in Fig. 12.10 (b) using a spherically focused probe. The wave form was predicted by the Thompson-Gray measurement model using an experimentally determined system function and the Kirchhoff approximation for the far-field scattering of the void (solid line). The experimentally measured flaw signal is shown for comparison (dashed line).

Code Listing 12.14. A script for calculating the response of a spherical void in the configuration shown in Fig. 12.10 (b) where a spherically focused probe is used. The predicted response uses an experimentally determined system function and a flaw response given by the Kirchhoff approximation which is then plotted and compared to an experimentally measured signal.

```
% script TG_sphere_example3
% calculates the waveform for a spherical void
% using an experimentally determined system function; focused probe case
clear
% run TG_sphere_example1 script to get most system parameters
TG_sphere_example1
%update setup
setup.trans.d = 12.46;
setup.trans.fl =172.9;
setup.system.z1r =172.9;
setup.flaw.Afunc = 'A_void';
%specify use of experimentally determined system function
%and reference waveform
setup.system.sysf='exp_systf';
setup.system.ref_file ='sphere_ref_foc';
%run measurement model
[Vout, setup] = TG_PE_MM(setup);
% plot(f, abs(Vout))  intermediate plot omitted
% pad frequency domain amplitude with zeros
Ve= [ Vout, zeros(1,800)];
Ve(1) = Ve(1)/2 ;   %Now, compute wave form and plot
vt =2*real(IFourierT(Ve, dt));
plot(t, c_shift(vt, 600))
load 'sphere_flaw_foc'
hold on
plot(t, vexp,'--')
hold off
```

12.9 A Large Flaw Measurement Model

We could also use the Thompson-Gray measurement model to predict the response of other scatterers in the configuration of Fig. 12.10 (b) such as

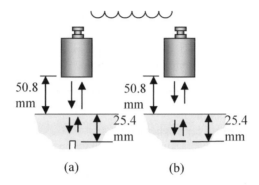

Fig. 12.18. A scattering configuration where **(a)** a flat-bottom hole or **(b)** a flat circular crack is interrogated by a planar transducer at normal incidence through a fluid-solid interface. In both cases the center of the scatterer is located on the central axis of the transducer.

the flat-bottom hole shown in Fig. 12.18 (a) or the flat circular crack shown in Fig. 12.18 (b). However, both of these scatterers are very "specular", i.e. they reflect much of the incident waves directly back to the transducer from their flat surfaces. As a consequence, the assumption of the Thompson-Gray measurement model that the wave field of the transducer beam is nearly constant over the flaw surface leads to significant errors if the sizes of the flat-bottom hole or crack being considered are not very small. In contrast, it has been found that the spherical void is much more tolerant to the small flaw assumption and the Thompson-Gray measurement model works well even for large spherical flaws. To account for beam variations we will use the more general measurement model of Eq. (12.1) coupled with a Kirchhoff approximation model for the scattering of a crack. In the Kirchhoff approximation this same flaw scattering model is appropriate also for the flat-bottom hole since the sides of the hole do not contribute anything in that approximation when the incident waves are at normal incidence to the circular, flat end of the hole. Since we are considering a pulse-echo setup for P-waves we have $\hat{V}^{(1)} = \hat{V}^{(2)} = \hat{V}(\mathbf{x}, \omega)$ in Eq. (12.1) and from the Kirchhoff approximation and the fact that we have a stress-free surface, we find (see Eq. (10.12))

$$A(\mathbf{x}, \omega) = -\frac{ik_{p2}}{2\pi} \left(\mathbf{d}_i^p \cdot \mathbf{n} \right) \exp\left[ik_{p2} \left(\mathbf{d}_i^p \cdot \mathbf{x} \right) \right]$$
$$= \frac{ik_{p2}}{2\pi},$$

(12.30)

where we have used the fact that on the flat surface S $\mathbf{d}_i^p \cdot \mathbf{n} = -1$ and $\mathbf{d}_i^p \cdot \mathbf{x} = 0$ .Then Eq. (12.1) becomes

$$V_R(\omega) = s(\omega) \left[\frac{4\pi\rho_2 c_{p2}}{-ik_{p2}Z_r^{T;a}} \right] \frac{ik_{p2}}{2\pi} \int_{S_f} \left[\hat{V}(\mathbf{x},\omega) \right]^2 dS. \tag{12.31}$$

Note that because of the symmetry of the incident field in the configuration of Fig. 12.18 we have $\hat{V}(\mathbf{x},\omega) = \hat{V}(r,\omega)$, where r is the radial distance from the center of the scatterer and the transducer axis. Thus, in this case we have

$$V_R(\omega) = s(\omega) \left[\frac{4\pi\rho_2 c_{p2}}{-ik_{p2}Z_r^{T;a}} \right] ik_{p2} \int_{r=0}^{r=b} \left[\hat{V}(r,\omega) \right]^2 r dr. \tag{12.32}$$

If we break the total integration into a series segments from $r = r_m$ to $r = r_{m+1}$, with $r_m = (m-1)b/(M-1)$ $(m = 1,2,...M-1)$ then we can approximate the velocity field as constants over the centroids of those segments given by $\hat{V}(\bar{r}_m,\omega)$, where $\bar{r}_m = (r_{m+1} + r_m)/2$ is an average radius. Each of these segments represent a circular ring except the first one which is a complete circular area of radius $b/(M-1)$ since $r_1 = 0$. For that circular segment we let $\bar{r}_1 = 0$ so that fields over that segment are calculated on the transducer axis, which is consistent with what we would do normally for a very small on-axis crack or flat-bottom hole. Equation (12.32) becomes

$$V_R(\omega) = \sum_{m=1}^{M-1} s(\omega) \left[\frac{4\pi\rho_2 c_{p2}}{-ik_{p2}Z_r^{T;a}} \right] \left[\hat{V}(\bar{r}_m,\omega) \right]^2 \frac{ik_{p2}\left(r_{m+1}^2 - r_m^2\right)}{2}. \tag{12.33}$$

In the Kirchhoff approximation the normal incidence pulse-echo P-wave far-field scattering amplitude of a flat crack of radius r_m is just (see Eq. (10.38)):

$$A_m\left(\mathbf{e}_i^p; -\mathbf{e}_i^p\right) = \frac{ik_{p2}r_m^2}{2} \tag{12.34}$$

so that we can write Eq. (12.33) as:

$$V_R(\omega) = \sum_{m=1}^{M-1} s(\omega) \left[\frac{4\pi\rho_2 c_{p2}}{-ik_{p2}Z_r^{T;a}} \right] \left[\hat{V}(\bar{r}_m, \omega) \right]^2$$
$$\cdot \left[A_{m+1}(\mathbf{e}_i^p; -\mathbf{e}_i^p) - A_m(\mathbf{e}_i^p; -\mathbf{e}_i^p) \right].$$

(12.35)

Comparing Eq. (12.35) and Eq. (12.6), we see that we can obtain the voltage by merely combining appropriately a number of Thompson-Gray measurement model terms for the scattering of a circular crack. Thus, we can use the TG_PE_MM function in conjunction with A_crack to model this case.

The MATLAB script FBH_example1 (Code Listing 12.15) implements Eq. (12.35) for a #8 flat-bottom hole in a steel block. The reference wave form for calculating the system function resides in the file FBH_ref.mat and the experimental flaw response is in the file FBH_flaw_n8.mat. The script calculates the FBH response and then plots both it and the experimental signal. The results are shown in Fig. 12.19.

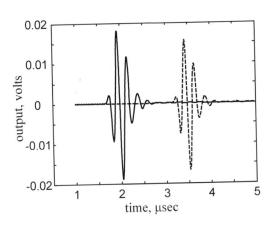

Fig. 12.19. The voltage received from an on-axis #8 flat-bottom hole for the configuration shown in Fig. 12.18 (a) as predicted by a measurement model that accounts for field variations over the end of the flat-bottom hole and uses an experimentally determined system function and the Kirchhoff approximation for the far-field scattering of the hole (solid line). The experimentally measured flat-bottom hole signal is shown for comparison (dashed line).

Code Listing 12.15. A script for calculating the response of a #8 flat-bottom hole, taking into account the variations of the incident transducer beam over the bottom of the hole.

```
% FBH_example1 script
% This script calculates the pulse-echo P-wave response of an on-axis
% #8 flat-bottom hole interrogated by a 5 MHz planar probe through a
% fluid-solid interface at normal incidence

clear
setup = setup_maker;
% setup parameters that need to be specified
% for this example
f =s_space(0, 20, 200);
cp1 = 1484.;
d2 = 7.86;
cp2 = 5940.;
cs2 = 3230.;
z1 = 50.8;
z2 = 25.4;
amp =0.12;
bw = 3.;
z1r =50.8;
p1 = [ 0 0 0.02479E-03  0 0];
b =1.5875;   % number eight FBH
flaw_name = 'A_crack';
sysfunc ='exp_systf';
reffile='FBH_ref';
setup.f =f;
setup.system.amp = amp;
setup.system.bw = bw;
setup.system.z1r =z1r;
setup.system.sysf = sysfunc;
setup.system.ref_file = reffile;
setup.geom.z1 = z1;
setup.geom.z2 = z2;
setup.matl.cp1 = cp1;
setup.matl.d2 = d2;
setup.matl.cp2 = cp2;
setup.matl.cs2 = cs2;
setup.matl.p1 = p1;
setup.flaw.b = b;
setup.flaw.Afunc = flaw_name;

% break up hole end into rings
```

```
nR= 10;  % use 9 rings (10 points)
rm = linspace(0, b, nR);    %ring edges
rmu = rm(2:nR); %upper edges
rml =rm(1:nR-1); %lower edges
rc =(rmu-rml)/2 + rml; %ring centroids
rc(1) = 0; %make first centroid at origin

Vf = zeros(size(f));

% calculate received voltage

for nd = 1:nR-1
   setup.geom.x2 = rc(nd);
   setup.flaw.b =rm(nd);
   [Vf1, setup] = TG_PE_MM(setup);
   setup.flaw.b = rm(nd+1);
   [Vf2, setup] = TG_PE_MM(setup);
   Vf = (Vf2-Vf1) +Vf;
end

% extend frequency components to permit
% taking FFT

df = f(2)-f(1);
dt = 1/(1000*df);
t = s_space(0, 1000*dt, 1000);
Vfe = [Vf zeros(1,800)];
Vfe(1) = Vfe(1)/2;
vt =2*real(IFourierT(Vfe, dt));
vs =c_shift(vt, 700);
plot(t(100:500), vs(100:500))
%plot(t_shift(t,700), c_shift(vt,700))
hold on
load 'FBH_flaw';
plot(t(100:500), vexp(250:650), '--')
hold off
```

12.10 A Measurement Model for Cylindrical Reflectors

The third measurement model discussed previously was for treating the pulse-echo response of cylindrical reflectors such as a side-drilled hole (SDH) where the beam variations can be neglected over the cross-section of the scatterer. In terms of the geometry parameters defined in Fig. 12.1, this measurement model (see Eq. (12.5)) is:

$$V_R(\omega) = s(\omega) \left[\int_L \left(\hat{V}_0^{(1)}(y_2, \omega) \right)^2 dy_2 \right] \left[\frac{A(\omega)}{L} \right] \left[\frac{4\pi\rho_2 c_{\alpha 2}}{-ik_{\alpha 2} Z_r^{T;a}} \right]. \qquad (12.36)$$

This measurement model is similar to the Thompson-Gray measurement model (Eq. (12.6)) but now we must replace the square of the incident velocity field in that model (for pulse-echo) by the integrated velocity squared term in Eq. (12.36) and the 3-D scattering amplitude in the

Code Listing 12.16. A MATLAB function for calculating the normalized far-field scattering amplitude of a side-drilled hole in pulse-echo using the Kirchhoff approximation.

```
function A =A_SDH(setup)
% A_SDH calculates the pulse-echo 3-D normalized far-field scattering
% amplitude,A/L, of a side-drilled hole in the Kirchhoff approximation
% using the frequency f in setup.f, the radius b in setup.flaw.b,
% and the wave speed for the wave type in setup.wave.c2.
% The calling sequence is A = A_SDH(setup). The scattering
% amplitude, A, (in mm) is returned. In the calculation of the
% Struve function, an integration routine is used. Thus, the
% frequency, f, must be at most a vector to use this function
% effectively. It is not vectorized for f being a matrix.

f =setup.f;
b =setup.flaw.b;
c=setup.wave.c2;
kb =2000*pi*b.*f./c;
A =(kb./2).*(besselj(1, 2*kb)-i*struve(2*kb)) +i*kb./pi;

function y = struve(z)
num = length(z);
y=zeros(1,num);
for k = 1:num
y(k) = quadl(@struve_arg, 0, 1, [ ],[ ], z(k));
end

function y = struve_arg(x, z)
y = (4./pi).*z.*x.^2.*sin(z.*(1-x.^2)).*sqrt(2-x.^2);
```

Thompson-Gray model is now replaced by the normalized 3-D scattering amplitude, A/L, of the cylindrical scatterer, where L is the scatterer length. In the Kirchhoff approximation this normalized scattering amplitude was previously given by Eq. (10.53) for a SDH and has been coded in the MATLAB function A_SDH (Code Listing 12.16).

The multi-Gaussian beam model defined by the MATLAB function MGbeam has been modified so that it returns the integral of the square of the velocity field at the center of the SDH as well as an updated setup structure. The new MATLAB function is called I_MGbeam (Code-Listing 12.17). It is assumed that the y_2-coordinate of the flaw is now given in setup.geom.y2 as a vector of values and the integral in Eq. (12.36) is calculated approximately in I_MGbeam as a simple sum:

$$\int_L \left(\hat{V}(y_2,\omega)\right)^2 dy_2 = \sum_{i=1}^{N}\left(\hat{V}(y_i,\omega)\right)^2 \Delta y, \tag{12.37}$$

where $\hat{V}$ is the ideal velocity field (no attenuation) calculated by the multi-Gaussian beam model. In most cases the length of the hole extends the full width of a test block so that the hole length may be larger than the width of the incident beam. In that case, we can treat the SDH as infinitely long and simply sum over y_2-values where the fields are significant.

Code Listing 12.17. A MATLAB function for returning the integrated square of the velocity field for use in a measurement model for cylindrical reflectors where beam variations along the length of the reflector must be considered.

```
function [vi,setup ]=I_MGbeam(setup)

% get setup parameters
fin = setup.f;                  %frequency or frequencies (MHz)
type1 = setup.type1;            % wave type in medium one
type2 = setup.type2;            % wave type in medium two

a = setup.trans.d/2;            % transducer radius (mm)
Fl = setup.trans.fl;            % transducer focal length (mm)

z1 = setup.geom.z1;             % water path length (mm)
z2 = setup.geom.z2;             % path length in solid (mm)
x2 =setup.geom.x2;              % distance (mm) from ray axis in POI
yin = setup.geom.y2;            % distance (mm) perpendicular to the POI
Rx = setup.geom.R1;             % interface radius of curvature (mm) in POI
Ry =setup.geom.R2;              % interface radius of curvature (mm) out of POI
```

```
iang = setup.geom.i_ang;          % incident angle (deg)

d1 = setup.matl.d1;               % density (fluid)
d2 =setup.matl.d2;                % density (solid)
cp1 = setup.matl.cp1;             % compressional wave speed -fluid  (m/sec)
cp2 = setup.matl.cp2;             % compressional wave speed -solid (m/sec)
cs2 = setup.matl.cs2;             % shear wave speed -solid (m/sec)

% form frequency, y2-values needed for integration into arrays
[f,y2]=meshgrid(fin, yin);
% update setup with these values temporarily (need for init_z)
% setup values will be returned to fin, yin values later
setup.f =f;
setup.geom.y2 = y2;

% define y -increment
dy = yin(2) - yin(1);


[A, B] = gauss_c15;       % Wen and Breazeale coefficients (15)

% update setup.wave wave speeds
if strcmp(type1, 'p')
   setup.wave.c1 =cp1;
elseif strcmp(type1, 's')
   setup.wave.c1 = cs1;
else
   error('wrong wave type (must be p or s) ')
end

if strcmp(type2, 'p')
   setup.wave.c2 =cp2;
elseif strcmp(type2, 's')
   setup.wave.c2 = cs2;
else
   error('wrong wave type (must be p or s)')
end
% calculate transmission coefficient, update setup
setup.wave.T12 = fluid_solid(setup);

% wave speeds and transmission coefficient for the beam model
c1 =setup.wave.c1;
c2 =setup.wave.c2;        % wave speed for wave type2
T = setup.wave.T12;       % transmission coefficient

% parameters appearing in beam model
```

```
cosi = cos(pi*iang/180);              % cosine of incident angle
sinr = (c2/c1)*sin(pi*iang/180);      % sine of refracted angle from Snell's law
if sinr >= 1
   error('Beyond the Critical angle') % no transmitted wave of given wave type
else
  cosr = sqrt( 1 - sinr^2);
end


  h11 = 1/Rx;  %curvature
  h22 = 1/Ry;  %curvature
zr = eps*(f == 0) + 1000*pi*(a^2)*f./c1;        % "Rayleigh" distance
k1 = 2*pi*1000*f./c1;                           % wave number in fluid

%initialize predicted velocity with zeros of a size
% compatible with largest array in f, z1, z2, x2, y2 setup parameters
v = init_z(setup);
% return to original frequency, fin, and distance, yin, values in setup
setup.f = fin;
setup.geom.y2 =yin;


%multi-Gaussian beam model

for j = 1:15                           % form up multi-Gaussian beam model

  b =B(j) + i*zr./Fl;                  % modify coefficients for focused probe
                                       % Fl = inf for planar probe

q = z1 - i*zr./b;
K = q.*(cosi -(c1/c2)*cosr);
M1 = (cosi^2 +K.*h11)./cosr^2;
M2 =1 + K.*h22;
ZR1 = q./M1;
ZR2 =q./M2;
m11 = 1./(ZR1 +(c2/c1).*z2);
m22 = 1./(ZR2 +(c2/c1).*z2);
  t1 = A(j)./(1 + (i.*b./zr).*z1);
  t2 = t1.*T.*sqrt(ZR1).*sqrt(ZR2).*sqrt(m11).*sqrt(m22);
  v = v + t2.*exp(i.*(k1./2).*(m11.*(x2.^2) + m22.*(y2.^2)));

end
% sum over y-values squared times dy to integrate
vs =v.^2;
vi=sum(vs.*dy, 1);
```

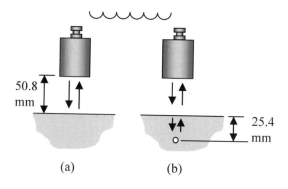

50.8 mm

25.4 mm

(a)   (b)

Fig. 12.20. (a) The reference scattering configuration for determining the system function and (b) the setup for measuring the pulse-echo response of a side-drilled hole.

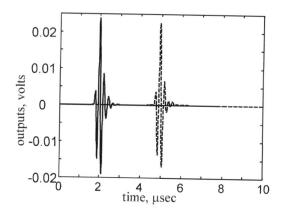

Fig. 12.21. The output voltage simulated for the pulse-echo P-wave response of a 1 mm side-drilled hole in the configuration shown in Fig. 12.20 (b) (solid line) and the corresponding experimentally measured response (dashed line).

The MATLAB function SDH_PE_MM (Code Lisitng 12.18) uses I_MGbeam and A_SDH to generate the system output voltage. The MATLAB script SDH_example1(Code Listing 12.19) uses SDH_PE_MM to simulate the response of a one mm diameter SDH in an aluminum sample in a configuration shown in Fig. 12.20 (b). Again, the system function is determined experimentally from a measured front-surface reflection as shown in Fig. 12.20 (a). The integration over the length of the hole here is taken from −50 mm to +50 mm based on an evaluation of the incident fields on the SDH for this problem (that evaluation is not shown here explicitly but can be easily done with the MGbeam function). For other SDH problems the limits of integration will have to be determined in this same way on a case by case basis. The predicted voltage using the MATLAB script SDH_example1 is shown in Fig. 12.21 along with the corresponding experimentally observed signal. Again, the Kirchhoff approximation does a very good job of representing the measured signal.

Code Listing 12.18. A MATLAB function that computes the output voltage for a cylindrical reflector using the measurement model of Eq. (12.21).

```
function [Vf, setup] =SDH_PE_MM(setup)
% SDH_PE_MM generates the frequency components of the
% output voltage, Vf, of an ultrasonic pulse-echo immersion
% measurement system generated by a side-drilled hole.
% The function returns Vf as well as an updated setup structure
% The calling sequence is [Vf, setup] =SDH_PE_MM(setup);

% First, compute the integrated beam velocity squared term
% and update the setup structure. This does not include
% attenuation
[vs, setup] = I_MGbeam(setup);

%get the setup parameters  needed for the constant term
%in the measurement model
f = setup.f;
r= setup.trans.d/2;   % transducer radius
d1 =setup.matl.d1;
d2 =setup.matl.d2;
c1 = setup.wave.c1;
c2 = setup.wave.c2;

%compute wave number in medium two and
%the constant term in the measurement model

k2 = (2000.*pi.*f)./c2;
k2 =k2 + eps*( k2 == 0); % prevent division by zero
K= (4.*d2.*c2)./(-i.*k2.*r^2.*d1.*c1);

% check to see if a model-based or experimentally determined system
% function is to be used
if strcmp(setup.system.sysf, 'systf')
    sys = systf(setup);
```

```
else
    sys =feval(setup.system.sysf, setup);
end

% find flaw type to be used
if strcmp( setup.flaw.Afunc, 'empty')
    error('flaw function not specified in setup')
else
    A = feval(setup.flaw.Afunc, setup);
end

%compute output voltage, Vf, (volts/MHz)
Vf = sys.*(vs).*(attenuate(setup)).^2.*A.*K;
```

Code Listing 12.19. A MATLAB script for calculating the pulse-echo P-wave response of a 1 mm diameter side-drilled hole in the configuration of Fig. 12.20 (b) using the Kirchhoff approximation to calculate the scattering of the side-drilled hole and an experimentally determined system function found from the reference configuration of Fig. 12.20 (a). The predicted response is compared to the experimentally observed signal.

```
%SDH_example1 script
% This script calculates the pulse-echo P-wave response of an on-axis
% 1 mm diam side-drilled hole interrogated by a 5 MHz planar probe through a
% fluid-solid interface at normal incidence
clear
setup = setup_maker;
% setup parameters that need to be specified
% for this example
f =s_space(0, 20, 200);
y2 =linspace(-50, 50, 500);
cp1 = 1484.;
d2 = 2.75;
cp2 = 6416.;
cs2 = 3163.;
z1 = 50.8;
z2 = 25.4;
amp =0.12;
bw = 3.;
z1r =50.8;
p1 = [ 0 0 0.02479E-03  0 0];
```

```
b =0.5;  % 0.5 mm radius
flaw_name = 'A_SDH';
sysfunc ='exp_systf';
reffile='SDH_ref';

% put parameters in setup

setup.f =f;
setup.system.amp = amp;
setup.system.bw = bw;
setup.system.z1r =z1r;
setup.system.sysf = sysfunc;
setup.system.ref_file = reffile;
setup.geom.z1 = z1;
setup.geom.z2 = z2;
setup.geom.y2 = y2;
setup.matl.cp1 = cp1;
setup.matl.d2 = d2;
setup.matl.cp2 = cp2;
setup.matl.cs2 = cs2;
setup.matl.p1 = p1;
setup.flaw.b = b;
setup.flaw.Afunc = flaw_name;

[Vf, setup] = SDH_PE_MM(setup);

% extend frequency components to permit
% taking FFT
df = f(2)-f(1);
dt = 1/(1000*df);
t = s_space(0, 1000*dt, 1000);
Vfe = [Vf zeros(1,800)];
Vfe(1) = Vfe(1)/2;
vt =2*real(IFourierT(Vfe, dt));
vs =c_shift(vt, 700);
plot(t, vs)
hold on
load 'SDH_flaw_1';
plot(t, vexp, '--')
hold off
```

12.11 References

12.1 Lopez-Sanchez A, Kim HJ, Schmerr LW, Gray TA (2006) Modeling the response of ultrasonic reference reflectors. Research in NDE 17: 49-70

12.2 Song SJ, Schmerr LW, Thompson RB (2006) Ultrasonic benchmarking study: overview up to year 2005. In: Thompson DO, Chimenti DE (eds) Review of progress in quantitative nondestructive evaluation. American Institute of Physics, Melville, NY, pp 1844-1853

13 Applications of Ultrasonic Modeling

In this Chapter we will describe several applications that use the ultrasonic models developed in previous Chapters and consider some extensions of those models to cases not previously discussed such as angle beam inspections. In [Fundamentals] one can see additional examples of model-based applications such as equivalent flaw sizing.

13.1 Obtaining Flaw Scattering Amplitudes Experimentally

In the paper where the Thompson-Gray measurement model was developed [11.2], the authors also showed that because a component of the far-field plane wave scattering amplitude of the flaw appeared explicitly in their model they could obtain an experimental measure of this component through the deconvolution of the measured voltage with the other measurement model terms, as discussed in Chapter 11 (see Eq. (11.38)). They gave examples of such experimentally determined scattering amplitude components for several spherical inclusions and showed that they agreed with the corresponding theoretical scattering amplitudes. Here, we will demonstrate the deconvolution process to extract the P-wave pulse-echo scattering amplitude component for the 1 mm diameter side-drilled hole (SDH) considered in Chapter 12 (see script SDH_example1 and Fig. 12.20). In that example the 1 mm diameter SDH was interrogated by a 5 MHz, 12.7 mm diameter planar transducer radiating at normal incidence through the fluid-solid interface. The center of the hole was located at a depth of 25.4 mm in an aluminum block and the water path length was fixed in this configuration to be 50.8 mm. The received A-scan voltage versus time signal for this case was shown previously in Fig. 12.21. Recall the system factor, $s(\omega)$, for this experiment was obtained from the front surface normal incidence reflection from the block. The

term $E(\omega) = \left[\int_L \hat{V}_0^{(1)}(z,\omega) \hat{V}_0^{(2)}(z,\omega) dz \right] \left[\dfrac{4\pi \rho_2 c_{\alpha 2}}{-ik_{\alpha 2} Z_r^{T;a}} \right]$ which also appears in the small flaw measurement model of Eq. (12.5) is completely known once we specify the material constants and geometry parameters and use our multi-Gaussian beam model in conjunction with appropriate attenuation terms for the water and aluminum. Determining the frequency components the measured voltage, $V_R(\omega)$, by taking the FFT of the waveform shown in Fig. 13.2, we then have formally

$$V_R(\omega) = G(\omega)\left[\frac{A(\omega)}{L}\right], \tag{13.1}$$

where $G(\omega) = s(\omega)E(\omega)$ is known. Then by deconvolution with the aid of a Wiener filter we have

$$\frac{A(\omega)}{L} = \frac{V_R(\omega)G^*(\omega)}{|G(\omega)|^2 + \varepsilon^2 \max\left\{|G(\omega)|^2\right\}}. \tag{13.2}$$

Note that the $G(\omega)$ factor is given by the side-drilled hole pulse-echo measurement model by simply setting the scattering amplitude term, A/L, to be unity in Eq. (12.36). Thus the MATLAB function SDH_PE_MM can be used to generate $G(\omega)$ by simply placing a modified scattering amplitude function, A_unity, in setup that returns a value of one for all frequencies. This function is given in Code Listing 13.1:

Code Listing 13.1. A MATLAB function that simply returns unity for the far field scattering amplitude.

```
function A = A_unity(setup)
f = setup.f;
A = ones(size(f));
```

The MATLAB script SDH_deconvolve1 (Code Listing 13.2), which uses the same setup parameters as given in the SDH_example1 script (see Code Listing 12.19) combines A_unity and SDH_PE_MM in this manner to calculate $G(\omega)$, loads the file containing the experimentally measured A-scan SDH response and computes the FFT of that response to obtain $V_R(\omega)$. The MATLAB function Wiener_filter then is used to calculate the

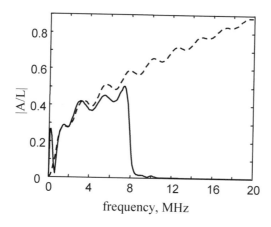

Fig. 13.1. The pulse echo P-wave far field scattering amplitude, A_{3D}/L, of a 1 mm diameter side-drilled hole obtained experimentally (solid line) compared to the exact separation of variables solution for the hole (dotted line).

far field scattering amplitude according to Eq. (13.2), where in this case a noise factor of $\varepsilon = 0.03$ was specified. Figure 13.1 shows the output of the script SDH_deconvolve1 which plots the magnitude of the non-dimensional 3-D scattering amplitude, A_{3D}/L, of the 1 mm diameter SDH obtained experimentally by this procedure and compares it to the theoretical scattering amplitude as calculated by the method of separation of variables with the MATLAB function A_SDH_Psep. It can be seen that there is relatively good agreement between the model-based and experimental results over the bandwidth of the system, with the oscillations present in the theoretical scattering amplitude (which are due to interference of the frequency components of the early time hole response with that of a creeping wave) are clearly visible in the experimentally obtained result. This example also shows the value of having as wide a bandwidth possible in a flaw experiment so that the features in the response can be used for quantitative purposes, such as flaw classification or sizing.

Code Listing 13.2. A MATLAB script for obtaining the measured pulse-echo P-wave far field scattering amplitude of a 1 mm diameter side-drilled hole by deconvolution and comparing it to the theoretical separation of variables solution.

```
%SDH_deconvolve1 script
% This script uses the measured pulse-echo P-wave response of an on-axis
% 1 mm diam side-drilled hole interrogated by a 5 MHz planar probe through a
% fluid-solid interface at normal incidence and the side-drilled hole
% measurement model for this case to obtain an experimental far field scattering
% amplitude by deconvolution. the experimental result is plotted versus
% frequency and compared to the theoretical scattering amplitude calculated
% by the method of separation of variables.
clear
setup = setup_maker;
% setup parameters that need to be specified
% for this example
f =s_space(0, 20, 200);
y2 =linspace(-50, 50, 500);
cp1 = 1484.;
d2 = 2.75;
cp2 = 6416.;
cs2 = 3163.;
z1 = 50.8;
z2 = 25.4;
amp =0.12;
bw = 3.;
z1r =50.8;
en =0.03;   % noise parameter for this example
p1 = [ 0 0 0.02479E-03  0 0];
b =0.5;   % 0.5 mm radius
flaw_name = 'A_unity';  % scattering amplitude set to unity
sysfunc ='exp_systf';
reffile='SDH_ref';
setup.f =f;
setup.system.amp = amp;
setup.system.bw = bw;
setup.system.z1r =z1r;
setup.system.sysf = sysfunc;
setup.system.ref_file = reffile;
setup.system.en = en;
setup.geom.z1 = z1;
setup.geom.z2 = z2;
setup.geom.y2 = y2;
setup.matl.cp1 = cp1;
setup.matl.d2 = d2;
```

```
setup.matl.cp2 = cp2;
setup.matl.cs2 = cs2;
setup.matl.p1 = p1;
setup.flaw.b = b;
setup.flaw.Afunc = flaw_name;
% calculate all terms in measurement model except scattering amplitude
[G, setup] = SDH_PE_MM(setup);
% get experimental wave form and calculate FFT
load 'SDH_flaw_1';  % loads time and voltage variables into texp and vexp
dt = texp(2) - texp(1);
Vf = FourierT(vexp, dt);
% consider only same number of frequency components as modeled
Vfl = Vf(1:200);
% compute Wiener filter with en value in setup
Aexp =Wiener_filter(Vfl, G, en);
% plot experimental scattering amplitude
plot(f,abs(Aexp))
hold on
% compare with theoretical scattering amplitude calculated
% by separation of variables
Asep = A_SDH_Psep(setup);
plot(f, abs(Asep), ':');
hold off
```

13.2 Distance-Amplitude-Correction Transfer Curves

In ultrasonic NDE testing, the location of a flaw in the beam of an ultrasonic transducer greatly affects the amplitude of the signal received since the beam amplitude itself can vary considerably. For example, for a transducer radiating into a fluid, the on-axis pressure varies like $1/z$ in the far field, where z is the distance from the transducer. This behavior occurs since in the far field the beam is spreading out from the transducer in a spherical wave fashion. Thus, for every doubling of the distance in the far field the pressure drops by 6 dB due to transducer beam spread alone. Material attenuation causes additional amplitude changes. Such pressure changes produce amplitude changes of a received flaw signal that are not associated with the flaw itself, so it is desirable to compensate for these non-flaw related changes. Often this is done with a series of calibration blocks containing reference reflectors of a given size and type (such as flat-bottom holes) located at different depths. Plotting the measured   peak-to-peak

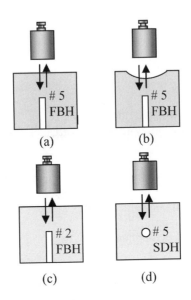

Fig. 13.2. (a) Examining a # 5 flat-bottom hole through a planar interface, **(b)** examining a #5 flat-bottom hole through a curved interface, **(c)** examining a #2 flat-bottom hole through a plane interface, and **(d)** examining a #5 side-drilled hole through a planar interface.

voltage received from these reference samples versus depth then gives what is called a distance-amplitude-correction (DAC) curve. Making such DAC test samples is rather expensive and for every change of testing condition a new set of samples must be manufactured. With the use of ultrasonic measurement models, however, there is a much more effective approach that can be taken. If a calibration experiment is done on a simple, inexpensive test sample with a reference reflector such as a #5 flat-bottom hole (where #n = n/64 inch diameter), as shown in Fig. 13.2 (a), then models can be used to predict the amplitude changes due to sample geometry changes such as surface curvature (Fig. 13.2 (b)), changes in size of the reference reflector (Fig. 13.2 (c)) or even a change in the type of the reflector being used such as a change from a flat-bottom hole to a side-drilled hole (Fig. 13.2 (d)). Following the approach outlined in [13.1], it will be shown here that all of these types of changes (or any combination of them) can be easily accounted for in a model-based approach where we use the ultrasonic measurement models to develop *DAC transfer curves* that relate the DAC curves to one another in different testing setups.

Since DAC curves define the changes in peak-to-peak measured voltages versus scatterer depth for that scatterer in a given reference setup,

DAC transfer curves involve the ratios of such peak-to-peak voltages. In order to simulate peak-to-peak voltage responses with our measurement models, it is necessary to specify the system function. This could be done with an experimentally measured system function, but here we will use a purely model-based approach that follows the specifications made in a previous set of ultrasonic benchmark studies [13.2]. Specifically, we model the voltage received by a circular planar transducer due to the waves reflected from the plane surface of a test block immersed in water as

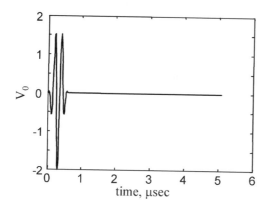

Fig. 13.3. The voltage specified as the response received from the planar front surface of a calibration block.

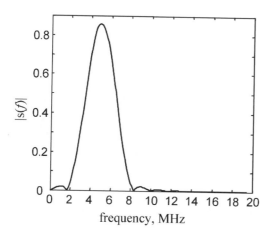

Fig. 13.4. Magnitude of the system function used in the generation of DAC transfer curves.

$$V_0(t) = \begin{cases} A\big[1 - \cos(2\pi Ft/N)\big]\cos(2\pi Ft) & (0 < t < N/F) \\ 0 & otherwise \end{cases} \tag{13.3}$$

where A is the amplitude, F is the center frequency of the system response and N controls the amount of ringing, and hence the bandwidth of the response. In this case we chose $A = 1$, $F = 5$ MHz and $N = 3$, which gives the time domain response shown in Fig. 13.3. To obtain the system function from this voltage we also need to specify the size of the trans-

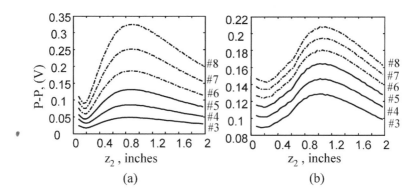

(a)   (b)

Fig. 13.5. Model-based DAC curves generated for **(a)** flat-bottom holes, and **(b)** side-drilled holes for a planar transducer immersed in water and the holes drilled in a planar aluminum block (see Figs. 13.2 (a) and 13.2 (d)). The depth of the reflectors in the aluminum was varied from 2 mm (0.07 in.) to 50.8 mm (2 in.). Each curve is for a reflector of the size indicated.

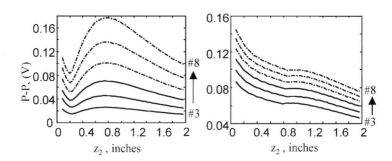

Fig. 13.6. Model-based DAC curves generated for a defocusing cylindrical interface with a radius of curvature 4 in. (101.6 mm) using **(a)** flat-bottom hole reflectors, and **(b)** side-drilled hole reflectors. Each curve is for a reflector of a given size, where the sizes range over the values indicated.

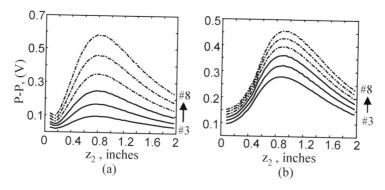

Fig. 13.7. Model-based DAC curves generated for a focusing cylindrical interface with a radius of curvature -4 in. (-101.6 mm) using **(a)** flat-bottom hole reflectors, and **(b)** side-drilled hole reflectors of the sizes indicated.

ducer, the water path length, and the material properties of the water and block . In the case considered below the transducer diameter is 6.35 mm, the water path length is 50.8 mm, the density and wave speed of the water are $\rho_1 = 1$ gm/cm^3, $c_{p1} = 1470$ m/sec, and for the aluminum block $\rho_2 = 2.71$ gm/cm^3, $c_{p2} = 6374$ m/sec. Then with the acoustic/elastic transfer, t_A, function given by Eq. (5.18), we have via a Wiener filter:

$$s(\omega) = \frac{V_0(\omega)t_A^*(\omega)}{\left|t_A(\omega)\right|^2 + \varepsilon^2 \max\left\{\left|t_A(\omega)\right|^2\right\}}. \tag{13.4}$$

The system function obtained in this manner is shown in Fig. 13.4 where we have used a noise factor $\varepsilon = 0.03$.

With a simulated system function given in this manner it is then possible to generate DAC curves for any of the cases shown in Fig. 13.2. Here we will consider DAC curves for flat-bottom holes and side-drilled holes, but other reference reflectors such as a spherical pore also could be modeled. The range of reference reflector diameters was varied from #3 to #8 (1.19 mm to 3.175 mm) in #1 steps. The radius of curvature of the cylindrical fluid-solid interface was varied from $R = -8$ in. (−203.2 mm) to $R = +8$ in. (+203.2 mm) in one inch steps. The curved interfaces with negative R values are focusing interfaces while those with positive values are defocusing interfaces (see Fig. 8.30). The measurement models used here were the large flaw measurement model for the flat-bottom hole (Eq. (12.33)) and the small side-drilled hole model (Eq. (12.36)) since for the range of sizes considered here studies have shown that beam variations over the surface of the side-drilled hole do not affect the peak-to-peak voltage responses [13.3]. The Kirchhoff approximation was used for both the flat-bottom holes and side-drilled holes to calculate their scattering amplitudes.

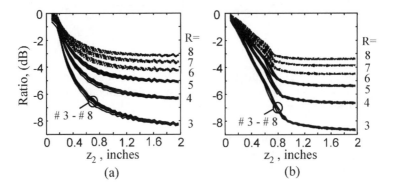

Fig. 13.8. DAC interface curvature transfer curves defined as the ratio of a curved interface response divided by the plane interface response for defocusing cylindrical interfaces (R = 3 to 8 inches) using **(a)** flat-bottom holes of sizes #3 - #8, and **(b)** side-drilled holes of sizes #3 - #8.

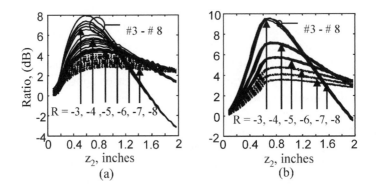

Fig. 13.9. DAC interface curvature transfer curves defined as the ratio of a curved interface response divided by the plane interface response for focusing cylindrical interfaces using **(a)** flat-bottom holes of sizes #3 - #8, and **(b)** side-drilled holes of sizes #3 - #8.

Figures 13.5 (a), (b) show model-based DAC curves generated for both flat-bottom holes and side-drilled holes of various sizes in a block with a planar interface. These are idealized DAC curves in that material attenuation has not been included in the calculations. Thus, the changes of amplitude shown are only due to beam diffraction effects. It can be seen from those figures that both the shape and the amplitude of the DAC curves are dependent on the type of reflector present. Figures 13.6 (a), (b) show the corresponding DAC curves for a cylindrical interface having a radius of curvature of 4 inches. This type of curved interface defocuses the ultrasonic beam, so that the amplitudes seen in Fig. 13.6 are generally smaller than those seen in Fig. 13.5 for the planar interface. Similarly, Figs. 13.7 (a), (b) give the DAC curves for a cylindrical interface having a curvature of −4 inches, which is a focusing interface that generally increases the amplitudes from the planar interface case.

Having DAC curves such as shown for these various cases, it then is possible to generate DAC transfer curves by taking the ratios of the DAC responses. Note that unlike the DAC curves themselves the DAC transfer curves are indeed independent of attenuation effects so that attenuation need not be considered in their generation. All of these DAC transfer curves will be given in terms of response ratios as measured in decibels (dB). Figures 13.8 (a), (b) show the DAC transfer curves for both flat-bottom holes and side-drilled holes corresponding to the ratio of the curved interface response to the planar interface response of these reflectors for a series of defocusing interfaces ranging in curvatures from

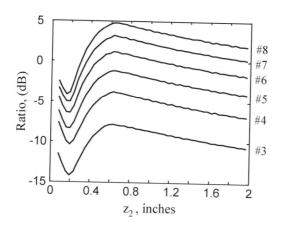

Fig. 13.10. DAC transfer curves for change of flaw type, defined as the ratio of the flat-bottom hole response to the side-drilled hole response for the planar interface case.

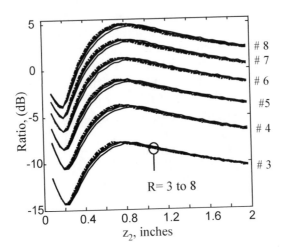

Fig. 13.11. DAC transfer curves for change of flaw type, defined as the ratio of the flat-bottom hole response to the side-drilled hole response for the cylindrical defocusing interfaces.

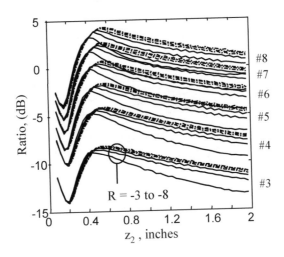

Fig. 13.12. DAC transfer curves for change of flaw type, defined as the ratio of the flat-bottom hole response to the side-drilled hole response for cylindrical focusing interfaces.

$R = 3$ in. to $R = 8$ in. It can be seen that for a given curvature, holes for all the sizes considered fall on essentially the same curve. In contrast for focusing interfaces side-drilled holes of the different sizes again fall on essentially the same curve for a given radius of curvature (Fig. 13.9 (b)), but for the flat-bottom holes there is more variability seen in the curves as a function of the flaw size at the smaller radii of curvature interfaces considered of -3 and -4 inches (Fig. 13.9 (a)). This behavior might be expected since the focusing curved interfaces concentrate the sound in the solid and a flat-bottom hole is a much more specular reflector than a side-drilled hole and so is more sensitive to beam variations caused by the focusing effect of the curved interface.

It is also possible to generate DAC transfer curves defined as the ratio of the flat-bottom hole response to the side-drilled hole response. Figs. 13.10-13.12 give these DAC flaw type transfer curves for planar, cylindrical defocusing, and cylindrical focusing interfaces, respectively. It can be seen that for the defocusing interface (Fig. 13.11), there is relatively little change in the DAC curves at a given flaw size with changes of the radius of curvature, but for the focusing interface (Fig. 13.12) there are more significant changes with radius of curvature, especially for the more tightly curved interfaces. Finally, we also have developed DAC transfer curves for changes in size of the flat-bottom holes considered. As a reference, we arbitrarily chose the #5 flat-bottom hole response, so that the ratios are defined here as that of a #n flat-bottom hole response divided by the #5 flat-bottom hole response. Figs. 13.13-13.15 give these flat-bottom

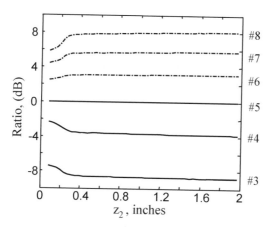

Fig. 13.13. DAC transfer curves for change in size of a flat-bottom hole, defined as the ratio of a #n flat-bottom hole response to the same response of a #5 flat-bottom hole as measured through a planar interface.

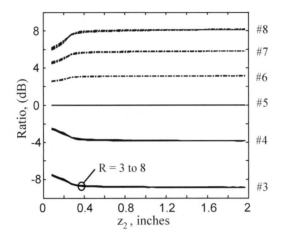

Fig. 13.14. DAC transfer curves for change in size of a flat-bottom hole, defined as the ratio of a #n flat-bottom hole response to the same response of a #5 flat-bottom hole as measured through a cylindrical defocusing interface.

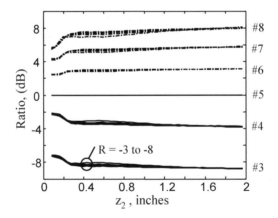

Fig. 13.15. DAC transfer curves for change in size of a flat-bottom hole, defined as the ratio of a #n flat-bottom hole response to the same response of a #5 flat-bottom hole as measured through a cylindrical focusing interface.

hole size DAC transfer curves for a planar interface, cylindrical defocusing interface, and focusing interface, respectively. All of these flat-bottom hole size DAC transfer curves are very similar for the range of radii of curvatures considered here but with some differences seen as a function of the radius of curvature for the largest flat-bottom holes and a cylindrical focusing interface (Fig. 13.15).

With these DAC transfer curves, it then is easy to go from one calibration configuration response to another. For example, consider the case where we have the DAC curve for a #4 side-drilled hole as measured through a planar interface and we want to obtain the corresponding DAC curve for a #6 flat-bottom hole when measured through a defocusing cylindrical interface having a radius of curvature $R = 5$ in. Figures 13.16 (a)-(d) show the steps in using the DAC transfer curves to make that change. Figure 13.16 (a) gives the original #4 side-drilled hole DAC curve. Only an idealized curve is shown here but it could also be a curve that included attenuation explicitly or it could be an experimentally obtained DAC curve. In either of those latter two cases the resulting flat-bottom

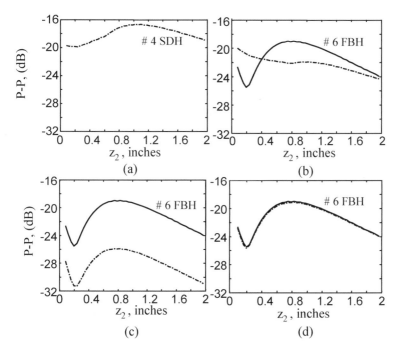

Fig. 13.16. Example of using the DAC transfer curves to change from the DAC curve for a #4 side-drilled hole (SDH), as measured through a planar interface, to the DAC curve for a #6 flat-bottom hole (FBH), as measured through a defocusing cylindrical interface with radius of curvature of 5 inches. **(a)** Original #4 SDH, **(b)** DAC curve for a #4 SDH through the curved interface after compensation for curvature effects, **(c)** the DAC curve for a #4 FBH through the curved interface, and **(d)** the DAC curve for a #6 FBH through the curved interface. In cases (b)-(d), the directly computed DAC curve for the target case (response of a #6 FBH measured through the curved interface) is labeled as #6 FBH and shown for comparison.

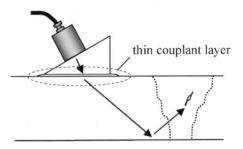

thin couplant layer

Fig. 13.17. An example angle beam shear wave test configuration for the inspection of a welded specimen.

hole DAC curve obtained would then also include the losses present as well. After multiplication with the DAC transfer curve for interface curvature effects, the new DAC curve is shown in Fig. 13.16 (b). This would be the DAC curve for a #4 side-drilled hole when measured through the curved interface. Also shown in Fig. 13.16 (b) is the idealized (no attenuation) DAC curve for the case we want to obtain, i.e. the curve for the #6 flat-bottom hole though this same curved interface. Figure 13.16 (c) shows the result of applying the flaw type DAC transfer curve to generate the DAC curve corresponding to a #4 flat-bottom hole as measured through the curved interface. Again the "target" #6 flat-bottom hole response is shown in Fig. 13.16 (c). Finally, in Fig. 13.16 (d) we have applied the DAC transfer curve for change in flat-bottom hole size to obtain the predicted response of the #6 flat-bottom hole, which agrees very closely with the directly calculated DAC curve for this target case, as shown in Fig. 13.16 (d).

The use of the DAC transfer curves in this manner is a very powerful tool and one that is easily implemented with the MATLAB measurement models discussed in the last chapter for the cases considered here as well as others.

13.3 Angle Beam Inspection Models and Applications

An angle beam pulse-echo setup is an ultrasonic testing configuration that is widely adopted in practice for the nondestructive evaluation of welded joints, as shown in Fig. 13.17. In this case a contact P-wave transducer radiates a bounded ultrasonic beam into a solid wedge. This beam crosses the interface between the wedge and the welded specimen and propagates obliquely into that specimen. In many weld tests the angle of the wedge is

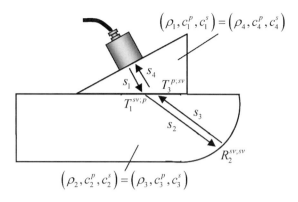

$$\left(\rho_1,c_1^p,c_1^s\right)=\left(\rho_4,c_4^p,c_4^s\right)$$

$$\left(\rho_2,c_2^p,c_2^s\right)=\left(\rho_3,c_3^p,c_3^s\right)$$

Fig. 13.18. A calibration setup where the reflection of the angle beam shear wave from the cylindrical surface of a test block is used to determine the system function.

chosen beyond the first critical angle so that primarily an SV-wave is generated in the solid. The SV-wave can be used to interrogate a weld directly, or it may be reflected off the back surface of the specimen as shown in Fig. 13.17. In this type of angle beam shear wave test, therefore, the transducer beam can be transmitted or reflected multiple times, making it an ideal candidate for the use of a multi-Gaussian beam model and the **ABCD** matrices described in Chapter 9. In this section we will give a number of examples of models that use this approach to simulate angle beam shear wave tests.

To use the ultrasonic measurement model concepts described in the previous Chapters, it is also necessary to have a reference calibration setup where one can experimentally determine the system function. For angle beam testing a convenient choice is to reflect the beam from the cylindrical interface of a IIW Type I standard block (or, equivalently STB A-1 block), as shown in Fig. 13.18, since this block is widely used for the calibration of angle beam transducers in practical field inspections.

We can see from Figs. 13.17 and 13.18 that in the weld test configuration and the calibration setup there are a series of parallel planar or cylindrical interfaces involved where the plane of incidence defined by the central ray of the transducer beam and the interface normal remains the same after the beam undergoes multiple transmissions and/or reflections. Also, the cylindrical interface of Fig. 13.18 has a principal axis that is aligned with the plane of incidence of the incident beam. Thus, in all these cases a Gaussian beam that is initially described by a diagonal **M**-matrix on the face of the transducer will continue to have an **M**-matrix that is diagonal after propagation and transmission/reflection and there is also no need to perform coordinate rotations to obtain the correct components of the transmitted waves, as discussed in Chapter 9. This means that the general multi-Gaussian beam model of Eq. (9.133) reduces to the simpler form

$$
\mathbf{v}_{M+1}^{\alpha;(y)} = \sum_{r=1}^{10} \left[V_1^p(0) \right]_r \frac{\sqrt{\det\left[\mathbf{M}_{M+1}^{\alpha}\left(s_{M+1} \right) \right]_r}}{\sqrt{\det\left[\mathbf{M}_{M+1}^{\alpha}(0) \right]_r}} \cdot \mathbf{d}_{M+1}^{\alpha}
$$

$$
\cdot \left[\prod_{m=1}^{M} T_m^{\gamma_{m+1};\gamma_m} \frac{\sqrt{\det\left[\mathbf{M}_m^{\gamma_m}\left(s_m \right) \right]_r}}{\sqrt{\det\left[\mathbf{M}_m^{\gamma_m}(0) \right]_r}} \right]
$$
(13.5)

$$
\cdot \exp\left[i\omega \sum_{m=1}^{M+1} \frac{s_m}{c_m^{\gamma_m}} + i\frac{\omega}{2} \mathbf{y}^T \left[\hat{\mathbf{M}}_{M+1}^{\alpha}\left(s_{M+1} \right) \right]_r \mathbf{y} \right],
$$

where $\mathbf{d}_{M+1}^{\alpha}$ is the polarization of the transducer beam in medium $M+1$ of type α, and $T_m^{\gamma_{M+1};\gamma_m}$ is the ordinary scalar plane wave transmission or reflection coefficient at the mth interface. Since the waves in the wedge are taken as P-waves we have set the wave type $\gamma_1 = p$. As discussed in Chapter 9 and Appendix F the starting values $\left[V_1^p(0) \right]_r$ and $\left[\mathbf{M}_1^p(0) \right]_r$ can be defined in terms of the Wen and Breazeale multi-Gaussian beam coefficients $\left(A_r, B_r \right)$. Since all of the **M**-matrices in Eq. (13.5) are diagonal the square roots appearing in that equation can also be obtained directly without any need for the transformation described in Chapter 9. At the interface between the solid wedge and the solid specimen being inspected a thin layer of liquid couplant is present as shown in Fig. 13.17. However, the thickness of this layer is taken to be negligible and the wedge/specimen interface is assumed to act like two solids directly in smooth contact. Thus, in Eq. (13.5) the plane wave transmission coefficients for this type of interface, which were explicitly given in Appendix D, can be used. We should also note that although Eq. (13.5) was developed in Chapter 9 based on an immersion testing model, it also can be used for the contact transducer case of Fig. 13.17 since, as discussed in Section 8.11, only the P-waves in the wedge are significant so that the wedge can be treated as an equivalent "fluid" [Fundamentals].

As a first example of the use of Eq. (13.5) for an angle beam shear wave problem, consider the use of the calibration setup of Fig. 13.18 to determine the system function by measuring the reflection from the cylindrical surface of a IIW Type I (or STB A-1) standard block. As discussed in Chapter 9 we cannot expect a beam model like the multi-Gaussian beam model to be accurate if the curvature of a surface is too small or varies significantly over the "footprint" of the beam on the surface. In this case, however, the radius of curvature of the curved interface in Fig. 13.18 is constant and has a radius of curvature of 100 mm, which is approximately 11 times larger than the transducer diameter, so that for a well collimated beam incident on this surface we expect the conditions required for the multi-Gaussian beam model to be applied will be very well satisfied.

There does not exist an analytical expression for the acoustic/elastic transfer function, t_A, for the configuration of Fig. 13.18 but the multi-Gaussian beam model can be used to numerically determine that transfer function. To see this first note that from Eq. (13.5) we have

$$
\frac{v_n(\mathbf{y}_s, \omega)}{v_0(\omega)} = \sum_{r=1}^{10} \frac{\left[V_1^p(0)\right]_r}{v_0} T_1^{sv:p} R_2^{sv:sv} T_3^{p:sv} \frac{\sqrt{\det\left[\mathbf{M}_1^p(0)\right]}}{\sqrt{\det\left[\mathbf{M}_1^p(s_1)\right]}}
$$

$$
\cdot \frac{\sqrt{\det\left[\mathbf{M}_2^{sv}(0)\right]}}{\sqrt{\det\left[\mathbf{M}_2^{sv}(s_2)\right]}} \frac{\sqrt{\det\left[\mathbf{M}_3^{sv}(0)\right]}}{\sqrt{\det\left[\mathbf{M}_3^{sv}(s_3)\right]}} \frac{\sqrt{\det\left[\mathbf{M}_4^p(0)\right]}}{\sqrt{\det\left[\mathbf{M}_4^p(s_4)\right]}} \qquad (13.6)
$$

$$
\cdot \exp\left[i\omega\left(\frac{s_1 + s_4}{c_1^p} + \frac{s_2 + s_3}{c_2^s} + i\frac{\omega}{2}\mathbf{y}_s^T\left[\mathbf{M}_4^p(s_4)\right]_r \mathbf{y}_s \right) \right],
$$

where v_n/v_0 is the normalized velocity of the received P-wave at the transducer face in a direction opposite to the outward normal to the transducer (i.e. along the direction of the incoming P-wave) and $v_0(\omega)$ is the velocity on the face of the transducer when it acts as a transmitter. This normalized velocity is evaluated at a point, $\mathbf{y}_s$, which is an arbitrary point on the surface of the transducer. The distances s_1 (and $s_4 = s_1$) and s_2 (and $s_3 = s_2$) are the propagating distances in the wedge and the standard block, respectively, and $T_1^{sv:p}$ and $T_3^{p:sv}$ are the plane wave transmission coefficients of plane wave going from/to the wedge, respectively, and $R_2^{sv:sv}$ is the plane wave reflection coefficient from the cylindrical surface of the block.

The specific terms appearing in Eq. (13.6) arise from Eq. (13.5) since in this setup the bounded beam radiated from the angle beam transducer propagates into the solid wedge ("medium 1") as a longitudinal wave $(\gamma_1 = p)$, and is transmitted into the STB-A1 standard block ("medium 2") as a mode-converted shear wave $(\gamma_2 = sv)$. Upon reaching the cylindrical surface of the block the beam is reflected back to the block ("medium 3") as a shear wave $(\gamma_3 = sv)$ and propagates as a mode-converted longitudinal wave $(\gamma_4 = p)$ back into the wedge ("medium 4") after crossing the interface between the block and the wedge, and then finally reaches the transducer.

The frequency components of the received voltage, $V_R(\omega)$, in this setup are given in terms of the system function, $s(\omega)$, and the acoustic transfer function, $t_A(\omega)$, which is the ratio of the received blocked force, $F_B(\omega)$, to the transmitted force, $F_t(\omega)$, generated by the transducer when it is acting as a transmitter. Thus, we have (see Eq. (7.7))

$$V_R(\omega) = s(\omega)t_A(\omega) = s(\omega)\frac{F_B(\omega)}{F_t(\omega)}$$
$$= s(\omega)\frac{2p_{ave}(\omega)}{\rho_1 c_1^p v_0},$$
(13.7)

where p_{ave} is the average pressure in the waves incident on the transducer. Let this incident pressure be given as p_{inc}. Then we can relate this pressure to the velocity, v_n, as simulated by the multi-Gaussian beam model, through the plane wave relationship $p_{inc} = \rho_1 c_1^p v_n$ (see the discussion in Appendix F) and we have

$$p_{ave}(\omega) = \frac{1}{S}\int_S p_{inc}(\mathbf{y}_s, \omega)dS(\mathbf{y}_s)$$
$$= \frac{\rho_1 c_1^p}{S}\int_S v_n(\mathbf{y}_s, \omega)dS(\mathbf{y}_s),$$
(13.8)

where S is the area of the transducer. Equation (13.7) then becomes

$$V_R(\omega) = 2s(\omega)\frac{1}{S}\int_S \frac{v_n(\mathbf{y}_s, \omega)}{v_0(\omega)}dS(\mathbf{y}_s).$$
(13.9)

By measuring the received voltage and numerically computing the integral in Eq. (13.9) with the use of Eq. (13.6) we can then determine the system function by deconvolution, as discussed in Chapter 7. We see the

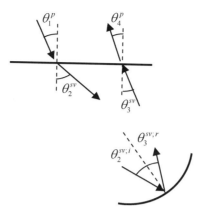

Fig. 13.19. The definitions of the angles that the waves make at the plane interface between the wedge and the test block and at the cylindrical surface of the block.

multi-Gaussian beam model has allowed us to determine the acoustic/elastic transfer function, $t_A(\omega)$, for this setup in the form

$$t_A(\omega) = \frac{2}{S} \int_S \frac{v_n(\mathbf{y}_s, \omega)}{v_0(\omega)} dS(\mathbf{y}_s).$$
(13.10)

Although Eq. (13.10) is applied here to the calibration block geometry of Fig. 13.18, it is an important general result that can also be used to determine the acoustic/elastic transfer function for any contact or immersion setup where the transfer function cannot be obtained in a simple form. All that is needed is a beam model capable of predicting the received velocity $v_n(\mathbf{y}_s, \omega)$ in the given setup. As discussed in Chapter 7, the use of Eq. (13.10) for contact problems must be done carefully as changes in the thin couplant layer between the transducer and the component being inspected or changes in surface condition may introduce considerable variability in the measurements of $s(\omega)$ that are not accounted for in Eq. (13.9).

All of the 2x2 M-matrices appearing in Eq. (13.6) can be conveniently calculated by the use of **ABCD** matrices as discussed in Chapter 9 and Appendix F since each of these matrices can be expressed in terms of the starting values, $\mathbf{M}_1^p(0)$, through the **ABCD** relationship

$$\mathbf{M}_m^{\gamma_m}\left(s_m\right)=\frac{\mathbf{C}'+\mathbf{D}'\mathbf{M}_1^p\left(0\right)}{\mathbf{A}'+\mathbf{B}'\mathbf{M}_1^p\left(0\right)},\tag{13.11}$$

where $M_m^{\gamma_m}\left(s_m\right)$ can be any of the matrices in Eq. (13.6) and the $\left(\mathbf{A}',\mathbf{B}',\mathbf{C}',\mathbf{D}'\right)$ matrices are obtained by an appropriate multiplication of the propagation and transmission/reflection matrices that describe the propagation and transmission/reflection occurring between the transducer and the place at which $M_m^{\gamma_m}\left(s_m\right)$ is being evaluated, as described in Chapter 9. Using the definitions of the angles at the wedge/block boundary and the cylindrical interface as defined in Fig. 13.19 the specific propagation and transmission/reflection matrices needed for Eq. (13.6) are:

propagation in medium1 and medium 4 (which are both the wedge):

$$\mathbf{A}_1^d=\mathbf{A}_4^d=\begin{bmatrix}1&0\\0&1\end{bmatrix}\qquad\qquad\mathbf{B}_1^d=\mathbf{B}_4^d=c_1^p s_1\begin{bmatrix}1&0\\0&1\end{bmatrix}$$

$$\mathbf{C}_1^d=\mathbf{C}_4^d=\begin{bmatrix}0&0\\0&0\end{bmatrix}\qquad\qquad\mathbf{D}_1^d=\mathbf{D}_4^d=\begin{bmatrix}1&0\\0&1\end{bmatrix}\tag{13.12}$$

transmission across the wedge/block interface from medium 1 (the wedge) to medium 2 (the block):

$$\mathbf{A}_1^t=\begin{bmatrix}\dfrac{\cos\theta_2^{sv}}{\cos\theta_1^p}&0\\0&1\end{bmatrix}\qquad\qquad\mathbf{B}_1^t=\begin{bmatrix}0&0\\0&0\end{bmatrix}$$

$$\mathbf{C}_1^t=\begin{bmatrix}0&0\\0&0\end{bmatrix}\qquad\qquad\mathbf{D}_1^t=\begin{bmatrix}\dfrac{\cos\theta_1^p}{\cos\theta_2^{sv}}&0\\0&1\end{bmatrix}\tag{13.13}$$

propagation in medium 2 and medium 3 (which are both the block):

$$\mathbf{A}_2^d = \mathbf{A}_3^d = \begin{bmatrix} 1 & 0 \\ 0 & 1 \end{bmatrix} \qquad\qquad \mathbf{B}_2^d = \mathbf{B}_3^d = c_2^s s_2 \begin{bmatrix} 1 & 0 \\ 0 & 1 \end{bmatrix}$$

$$\mathbf{C}_2^d = \mathbf{C}_3^d = \begin{bmatrix} 0 & 0 \\ 0 & 0 \end{bmatrix} \qquad\qquad \mathbf{D}_2^d = \mathbf{D}_3^d = \begin{bmatrix} 1 & 0 \\ 0 & 1 \end{bmatrix} \qquad (13.14)$$

reflection from the cylindrical interface between medium 2 and medium 3 (which are both the block):

$$\mathbf{A}_2^r = \begin{bmatrix} -\dfrac{\cos\theta_2^{sv;r}}{\cos\theta_2^{sv;i}} & 0 \\ 0 & 1 \end{bmatrix} \qquad\qquad \mathbf{B}_2^r = \begin{bmatrix} 0 & 0 \\ 0 & 0 \end{bmatrix}$$

$$\mathbf{C}_2^r = \left(\dfrac{\cos\theta_2^{sv;i}}{c_2^s} + \dfrac{\cos\theta_2^{sv;r}}{c_3^s} \right)$$
$$\cdot \begin{bmatrix} \dfrac{-h_{11}}{\cos\theta_2^{sv;r}\cos\theta_2^{sv;i}} & 0 \\ 0 & 0 \end{bmatrix} \qquad \mathbf{D}_2^r = \begin{bmatrix} -\dfrac{\cos\theta_2^{sv;r}}{\cos\theta_2^{sv;i}} & 0 \\ 0 & 1 \end{bmatrix} \qquad (13.15)$$

transmission across the block/wedge interface from medium 3 to 4:

$$\mathbf{A}_3^t = \begin{bmatrix} \dfrac{\cos\theta_4^p}{\cos\theta_2^{sv}} & 0 \\ 0 & 1 \end{bmatrix} \qquad\qquad \mathbf{B}_3^t = \begin{bmatrix} 0 & 0 \\ 0 & 0 \end{bmatrix}$$

$$\mathbf{C}_3^t = \begin{bmatrix} 0 & 0 \\ 0 & 0 \end{bmatrix} \qquad\qquad \mathbf{D}_3^\gamma = \begin{bmatrix} \dfrac{\cos\theta_3^{sv}}{\cos\theta_4^p} & 0 \\ 0 & 1 \end{bmatrix} \qquad (13.16)$$

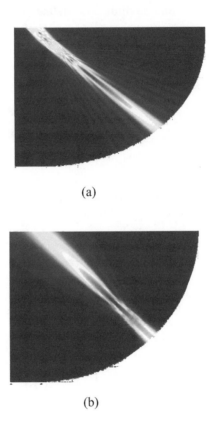

(a)

(b)

Fig. 13.20. Examples of calculated wave fields produced in a STB A-1 standard block from a circular piston transducer mounted on the solid wedge; **(a)** the radiation beam (of SV wave type) through the wedge/specimen interface, and **(b)** the reflected beam (of SV wave type) from the circular part of the STB A-1 block.

The specific example we will consider is a 45 degree angle beam shear wave transducer where the wedge is located at a position on the top surface of a STB A-1 standard block so that the waves are reflected from the cylindrical surface of the block at normal incidence. In this case the wave speeds and angles appearing in Eqs. (13.12)-(13.16) are given by: $c_1^p = c_4^p = 2680$ m/s , $c_2^s = c_3^s = 3240$ m/s , $\theta_1^p = \theta_4^p = 35.8°$, $\theta_2^{sv;p} = \theta_3^{p;sv} = 45.0°$, $\theta_2^{sv;i} = \theta_2^{sv;r} = 0.0°$. Using these values we obtained a multi-Gaussian beam model of the incident SV-wave velocity field produced in the STB A-1 standard block as shown in Fig. 13.20 (a). Figure 13.20 (b) similarly shows a model of the reflected SV wave from the cylindrical surface of

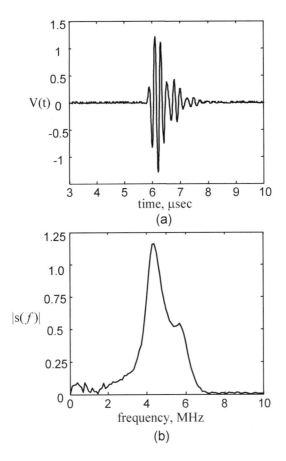

Fig. 13.21. (a) The experimental reference reflected signal, and **(b)** the system function for a 5 MHz center frequency, 0.375 inch diameter transducer with the refracting angle of 45 degrees in the STB-A1 block.

that block. The calculations of these wave fields were both obtained from Eq. (13.21) but in the calculation of the incident field (Fig. 13.20 (a)) only the terms in Eq. (13.6) corresponding to the first two beam paths $(s_1$ and $s_2)$ were considered, while in the case of the reflected beam field (Fig. 13.20 (b)) only the terms corresponding to first three beam paths (s_1 through s_3) were used. In Fig. 13.20 (b) one can see the focusing behavior that the curved cylindrical interface has on the reflected beam.

Figure 13.21 shows an example of using the complete set of terms in Eq. (13.6) to determine the system function, by the procedure described previously. Here, an angle beam pulse-echo setup was considered where a

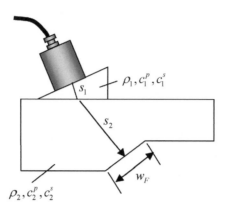

Fig. 13.22. Geometry of the counter bore reflection.

circular planar transducer (Panametrics A551S-SM) having a center frequency of 5 MHz and a diameter of 0.375 inch was mounted on an acrylic wedge so that a shear wave having a refracted angle of 45 degrees in steel was obtained. The reflected signal was captured from the cylindrical surface of the STB-A1 block and the system function was determined by deconvolution. Fig. 13.21 (a) shows the time domain experimental voltage signal, $V_R(\omega)$, obtained in this reference setup and Fig. 13.21 (b) shows the system function obtained.

Consider now the use of the system function obtained in this manner in a modeling application. Specifically, we will develop an ultrasonic measurement model for the angle beam shear wave response of a counter bore, as shown in Fig. 13.22. [13.4]. A counter bore is an incline (reduction in thickness) fabricated intentionally on the inner surface of a pipe to relax the stresses in welded joints. It produces significant ultrasonic reflections when an angle beam transducer is used to interrogate defects in such a welded joint so it is important to be able to characterize this type of reflector response and to discriminate its ultrasonic signals from those of defects. In the next section we will discuss the discrimination problem. In most cases the size, w_F, of the counter bore (the length of incline - see Fig. 13.22) is smaller than the interrogating beam width but the circumferential length is normally larger than the beam width. Thus, it is necessary to consider the counter bore as a large but finite scatterer. The appropriate measurement model for this reflector is given by Eq. (11.31) for a pulse-echo setup, i.e.

$$V_R(\omega) = s(\omega) \left[\frac{4\pi\rho_2 c_2^s}{-ik_2^s Z_r^{T;a}} \right]$$

$$\cdot \int_{S_{cb}} \left[\hat{V}^{(1)}(\mathbf{x},\omega) \right]^2 A(\mathbf{x},\omega) \exp\left[ik_2^s \mathbf{e}_2^{sv} \cdot \mathbf{x} \right] dS(\mathbf{x}), \tag{13.17}$$

where $V_R(\omega)$ are the frequency components of the received voltage, $s(\omega)$ is the system function, $Z_r^{T;a} = \rho_1 c_1^p S_T$ is the acoustic impedance of the transducer whose area is S_T. The unit vector, $\mathbf{e}_{sv}$, is in the direction of propagation of the incident SV-wave and S_{cb} is the area of the counter bore. Since the counter bore is a stress-free surface if we use the Kirchhoff approximation to model its ultrasonic response we have from Chapter 10:

$$A(\mathbf{x},\omega) = \frac{-ik_2^s \left(\mathbf{e}_2^{sv} \cdot \mathbf{n} \right)}{2\pi} \exp\left(ik_2^s \mathbf{e}_2^{sv} \cdot \mathbf{x} \right), \tag{13.18}$$

where $\mathbf{n}$ is the a unit normal to the counter bore surface. Using a multi-Gaussian beam model, the velocity field, $\hat{V}^{(1)}(\mathbf{x},\omega)$, is just the incident field part of Eq. (13.6) as computed for the cylindrical surface response, i.e.

$$\hat{V}^{(1)}(\mathbf{x},\omega) = \sum_{r=1}^{10} \frac{\left[V_1^p(0) \right]_r}{V_0} T_1^{sv:p} \frac{\sqrt{\det\left[\mathbf{M}_1^p(0) \right]}}{\sqrt{\det\left[\mathbf{M}_1^p(s_1) \right]}} \frac{\sqrt{\det\left[\mathbf{M}_2^{sv}(0) \right]}}{\sqrt{\det\left[\mathbf{M}_2^{sv}(s_2) \right]}}$$

$$\cdot \exp\left[i\omega\left(\frac{s_1}{c_1^p} + \frac{s_2}{c_2^s} + i\frac{\omega}{2}\mathbf{y}^T \left[\mathbf{M}_2^{sv}(s_2) \right]_r \mathbf{y} \right) \right] \tag{13.19}$$

where $\mathbf{x}$ is a point on the surface of the counter bore. In Eq. (13.19) $\mathbf{y} = (y_1, y_2)$ are coordinates normal to the central ray of the refracted beam taken from an origin at a distance s_2 along the central ray. The **ABCD** matrices defined in Eqs. (13.12)-(13.14) can also be used to compute all the terms in Eq. (13.19). In the calculation of the integral in Eq. (13.17), it is convenient to divide the surface of the counter bore into a large number of small planar elements and calculate the individual responses from each element first. Then the total response from the entire counter bore can be obtained by simply summing up all the element contributions. The system function for the angle beam transducer can be obtained using the reflection from the cylindrical portion of a reference block as described previously. An inverse FFT of $V_R(\omega)$ then yields a model of the time domain signal

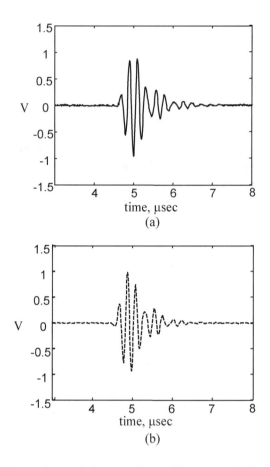

Fig. 13.23. Comparison of the experimentally measured signal to the predicted time domain signal from a 4 mm wide counter bore: **(a)** the experimental signal, and **(b)** the signal calculated using the multi-Gaussian beam model.

from the counter bore. Figure 13.23 (a) shows the experimentally measured signal obtained with a 45 degree angle beam shear wave transducer from a small counter bore at an inclined angle of $\theta = 45°$. Here the width, w_F, of the counter bore was 4 mm in a steel specimen whose thickness was 10 mm. Figure 13.23 (b) shows the corresponding signal synthesized from Eq. (13.17). It can be seen that there is very good agreement between the experimental and model-based results.

Another weld scattering case that will be considered is the angle beam shear wave inspection of a surface-breaking thin, rectangular-shaped vertical slot of height, h_c, and width, w_c, as shown in Figs. 13.24 (a), (b). This configuration is useful for simulating the angle beam inspection of a

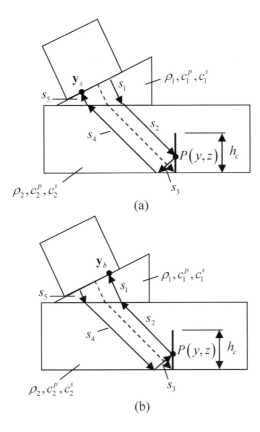

Fig. 13.24. Angle beam shear wave inspection of a vertical surface-breaking slot to determine the corner trap signal where **(a)** the beam first strikes the slot in returning to the transducer and **(b)** where the beam strikes the bottom surface of the specimen first.

vertical surface crack on the opposite face of a welded plate or pipe. One could model this configuration with the use of the Kirchhoff approximation in much the same manner as done for the counter bore. However, when the plane of incidence of the interrogating beam is perpendicular to the surface of the slot and the central ray of the transducer beam intersects the corner between the slot and the bottom surface, as shown in Fig. 13.24, it is well known that a large "corner trap" signal is generated in such a pulse-echo setup. This corner trap signal results from a series of direct multiple reflections from the slot surface and the bottom of the specimen so that it is possible to develop a model of this signal using an approach that is similar to what was done for the reflection of the beam from the cylindrical surface of the STB A-1 reference block where Eq. (13.9) was used to simulate the received signal. The basis for this model is to note that a wave leaving the transducer can follow a path shown in Fig. 13.24 (a) where the beam leaves the transducer surface, passes through a point on the interface, reflects at a point on the surface of the vertical slot, reflects once more at a point on the bottom of the specimen, propagates through another point on the interface, and finally reaches to a point on the transducer face. However, the beam could also travel this same path but in a completely reversed order, as shown in Fig. 13.24 (b), i.e. where the beam reflects at the bottom of the specimen first. The total response from the surface-breaking vertical slot obviously comes from both of these reflected wave contributions. It is easy to show that the time of arrival of both of these multiply reflected waves is identical to a wave traveling down to the corner and back along the central ray path shown as the dashed line in Figs. 13.24 (a), (b). Therefore, both of these wave contributions to the total response could be treated as if they were Gaussian beams that traveled along this central ray path but with their amplitudes and phases appropriately modified to characterize the individual ray paths just described. For example, consider the velocity generated at a point $\mathbf{y}_s$ on the face of the transducer from the wave that reflected first from the side of the slot, $v_{side}(\mathbf{y}_s,\omega)$. Using Eq. (13.5) this velocity (normalized by the velocity on the face of the transducer) is given by

$$
\frac{v_{side}(\mathbf{y}_s,\omega)}{v_0(\omega)} = \sum_{r=1}^{10} \left[\frac{V_1^P(0)}{v_0(\omega)}\right]_r T_1^{sv;P} R_2^{sv;sv} R_3^{sv;sv} T_4^{P;sv} \frac{\sqrt{\det\left[\mathbf{M}_1^P(0)\right]_r}}{\sqrt{\det\left[\mathbf{M}_1^P(s_1)\right]_r}}
$$

$$
\cdot \frac{\sqrt{\det\left[\mathbf{M}_2^{sv}(0)\right]_r}}{\sqrt{\det\left[\mathbf{M}_2^{sv}(s_2)\right]_r}} \frac{\sqrt{\det\left[\mathbf{M}_3^{sv}(0)\right]_r}}{\sqrt{\det\left[\mathbf{M}_3^{sv}(s_3)\right]_r}}
$$

$$
\cdot \frac{\sqrt{\det\left[\mathbf{M}_4^{sv}(0)\right]_r}}{\sqrt{\det\left[\mathbf{M}_4^{sv}(s_4)\right]_r}} \frac{\sqrt{\det\left[\mathbf{M}_5^{P}(0)\right]_r}}{\sqrt{\det\left[\mathbf{M}_5^{P}(s_5)\right]_r}}
$$

$$
\cdot \exp\left[i\omega\left(\frac{s_1+s_5}{c_1^P} + \frac{s_2+s_3+s_4}{c_2^s} + \frac{1}{2}\mathbf{y}_s^T\left[\hat{\mathbf{M}}_5^P(s_5)\right]_r \mathbf{y}_s\right)\right].
$$

(13.20)

All of the terms appearing in Eq. (13.20) can be calculated with the following **ABCD** matrices:

propagation in medium 1 (the wedge):

$$\mathbf{A}_1^d = \mathbf{A}_5^d = \begin{bmatrix} 1 & 0 \\ 0 & 1 \end{bmatrix} \qquad \mathbf{B}_1^d = \mathbf{B}_5^d = c_1^P s_1 \begin{bmatrix} 1 & 0 \\ 0 & 1 \end{bmatrix}$$

$$\mathbf{C}_1^d = \mathbf{C}_5^d = \begin{bmatrix} 0 & 0 \\ 0 & 0 \end{bmatrix} \qquad \mathbf{D}_1^d = \mathbf{D}_5^d = \begin{bmatrix} 1 & 0 \\ 0 & 1 \end{bmatrix} \qquad (13.21)$$

transmission across the wedge/specimen interface from medium 1 (the wedge) to medium 2 (the specimen containing the slot):

$$\mathbf{A}_1^t = \begin{bmatrix} \dfrac{\cos\theta_2^{sv}}{\cos\theta_1^{P}} & 0 \\ 0 & 1 \end{bmatrix} \qquad \mathbf{B}_1^t = \begin{bmatrix} 0 & 0 \\ 0 & 0 \end{bmatrix}$$

$$\mathbf{C}_1^t = \begin{bmatrix} 0 & 0 \\ 0 & 0 \end{bmatrix} \qquad \mathbf{D}_1^t = \begin{bmatrix} \dfrac{\cos\theta_1^{P}}{\cos\theta_2^{sv}} & 0 \\ 0 & 1 \end{bmatrix} \qquad (13.22)$$

propagation in medium 2 from the wedge to the slot:

$$\mathbf{A}_2^d = \begin{bmatrix} 1 & 0 \\ 0 & 1 \end{bmatrix} \qquad \mathbf{B}_2^d = c_2^s s_2 \begin{bmatrix} 1 & 0 \\ 0 & 1 \end{bmatrix}$$

$$\mathbf{C}_2^d = \begin{bmatrix} 0 & 0 \\ 0 & 0 \end{bmatrix} \qquad \mathbf{D}_2^d = \begin{bmatrix} 1 & 0 \\ 0 & 1 \end{bmatrix} \qquad (13.23)$$

reflection from the surface of the slot (incident and reflected SV-modes):

$$\mathbf{A}_2^r = \begin{bmatrix} -1 & 0 \\ 0 & 1 \end{bmatrix} \qquad\qquad \mathbf{B}_2^r = \begin{bmatrix} 0 & 0 \\ 0 & 0 \end{bmatrix}$$

$$\mathbf{C}_2^r = \begin{bmatrix} 0 & 0 \\ 0 & 0 \end{bmatrix} \qquad\qquad \mathbf{D}_2^r = \begin{bmatrix} -1 & 0 \\ 0 & 1 \end{bmatrix} \qquad (13.24)$$

propagation in medium 3 (same as medium 2) from the slot to the bottom surface of the specimen:

$$\mathbf{A}_3^d = \begin{bmatrix} 1 & 0 \\ 0 & 1 \end{bmatrix} \qquad\qquad \mathbf{B}_3^d = c_2^s s_3 \begin{bmatrix} 1 & 0 \\ 0 & 1 \end{bmatrix}$$

$$\mathbf{C}_3^d = \begin{bmatrix} 0 & 0 \\ 0 & 0 \end{bmatrix} \qquad\qquad \mathbf{D}_3^d = \begin{bmatrix} 1 & 0 \\ 0 & 1 \end{bmatrix} \qquad (13.25)$$

reflection from the bottom of the specimen (incident and reflected SV-modes):

$$\mathbf{A}_3^r = \begin{bmatrix} -1 & 0 \\ 0 & 1 \end{bmatrix} \qquad\qquad \mathbf{B}_3^r = \begin{bmatrix} 0 & 0 \\ 0 & 0 \end{bmatrix}$$

$$\mathbf{C}_3^r = \begin{bmatrix} 0 & 0 \\ 0 & 0 \end{bmatrix} \qquad\qquad \mathbf{D}_3^r = \begin{bmatrix} -1 & 0 \\ 0 & 1 \end{bmatrix} \qquad (13.26)$$

propagation in medium 4 (same as medium 2) from the bottom of the specimen to the wedge:

$$\mathbf{A}_4^d = \begin{bmatrix} 1 & 0 \\ 0 & 1 \end{bmatrix} \qquad \mathbf{B}_4^d = c_2^s s_4 \begin{bmatrix} 1 & 0 \\ 0 & 1 \end{bmatrix}$$

$$\mathbf{C}_4^d = \begin{bmatrix} 0 & 0 \\ 0 & 0 \end{bmatrix} \qquad \mathbf{D}_4^d = \begin{bmatrix} 1 & 0 \\ 0 & 1 \end{bmatrix} \qquad (13.27)$$

transmission across the wedge/specimen interface from medium 4 to medium 5:

$$\mathbf{A}_4^t = \begin{bmatrix} \dfrac{\cos\theta_1^p}{\cos\theta_2^{sv}} & 0 \\ 0 & 1 \end{bmatrix} \qquad \mathbf{B}_4^t = \begin{bmatrix} 0 & 0 \\ 0 & 0 \end{bmatrix}$$

$$\mathbf{C}_4^t = \begin{bmatrix} 0 & 0 \\ 0 & 0 \end{bmatrix} \qquad \mathbf{D}_4^y = \begin{bmatrix} \dfrac{\cos\theta_2^{sv}}{\cos\theta_1^p} & 0 \\ 0 & 1 \end{bmatrix} \qquad (13.28)$$

propagation in medium 5 (same as medium 1-the wedge):

$$\mathbf{A}_5^d = \begin{bmatrix} 1 & 0 \\ 0 & 1 \end{bmatrix} \qquad \mathbf{B}_5^d = c_1^p s_5 \begin{bmatrix} 1 & 0 \\ 0 & 1 \end{bmatrix}$$

$$\mathbf{C}_5^d = \begin{bmatrix} 0 & 0 \\ 0 & 0 \end{bmatrix} \qquad \mathbf{D}_5^d = \begin{bmatrix} 1 & 0 \\ 0 & 1 \end{bmatrix} \qquad (13.29)$$

In the same manner we can consider the velocity at a point $\mathbf{y}_b$ on the face of the transducer generated by the wave that has first reflected from the bottom of the specimen with the slot, $v_{bottom}(\mathbf{y}_b, \omega)$. In this case we can write this velocity in normalized form as:

$$
\frac{v_{bottom}(\mathbf{y}_b, \omega)}{v_0(\omega)} = \sum_{r=1}^{10} \frac{\left[V_1^p(0)\right]_r}{v_0(\omega)} T_1^{sv;p} R_2^{sv;sv} R_3^{sv;sv} T_4^{p;sv}
$$

$$
\cdot \frac{\sqrt{\det\left[\mathbf{M}_1^p(0)\right]_r} \ \sqrt{\det\left[\mathbf{M}_2^{sv}(0)\right]_r}}{\sqrt{\det\left[\mathbf{M}_1^p(s_5)\right]_r} \ \sqrt{\det\left[\mathbf{M}_2^{sv}(s_4)\right]_r}}
$$

$$
\cdot \frac{\sqrt{\det\left[\mathbf{M}_3^{sv}(0)\right]_r} \ \sqrt{\det\left[\mathbf{M}_4^{sv}(0)\right]_r} \ \sqrt{\det\left[\mathbf{M}_5^p(0)\right]_r}}{\sqrt{\det\left[\mathbf{M}_3^{sv}(s_3)\right]_r} \ \sqrt{\det\left[\mathbf{M}_4^{sv}(s_2)\right]_r} \ \sqrt{\det\left[\mathbf{M}_5^p(s_1)\right]_r}} \tag{13.30}
$$

$$
\cdot \exp\left[i\omega\left(\frac{s_1 + s_5}{c_1^p} + \frac{s_2 + s_3 + s_4}{c_2^s} + \frac{1}{2}\mathbf{y}_s^T\left[\hat{\mathbf{M}}_5^p(s_1)\right]_r \mathbf{y}_s\right)\right]
$$

and all the terms in Eq. (13.30) can also be generated by appropriate multiplication of the **ABCD** matrices appearing in Eqs. (13.21)- (13.29).

If the slot was larger than the beam width we could simply add the contributions of Eq. (13.20) and Eq. (13.30) to get the total velocity field on the face of the transducer and then calculate the response of the slot directly with Eq. (13.9). However, if the slot is smaller than the beam width we must first suitably truncate the velocity fields in Eqs. (13.20) and (13.30) to account for the fact that there are points on the transducer surface that do not receive the corner trap signals of the type given by Eqs. (13.20) and (13.30). We can perform such truncation by noting that every point $\mathbf{y}_s$ and $\mathbf{y}_b$ on the transducer surface has a corresponding point P on the surface of the slot where the incident or reflected beam strikes the slot. If we let (y, z) be the coordinates of point P (see Fig. 13.24) then $\mathbf{y}_s = \mathbf{y}_s(y, z)$ and $\mathbf{y}_b = \mathbf{y}_b(y, z)$. If the slot occupies the region $(0 < z < h_c, -z_c/2 < y < +z_c/2)$ therefore we can define a total truncated received velocity as:

$$
v_{trunc} = \begin{cases} v_{side}(\mathbf{y}_s, \omega) + v_{bottom}(\mathbf{y}_b, \omega) & 0 \le z \le h_c, -w_c/2 \le y \le w_c/2 \\ 0 & \text{elsewhere} \end{cases} \tag{13.31}
$$

and calculate the received voltage from the slot in the same form as Eq.(13.10), namely

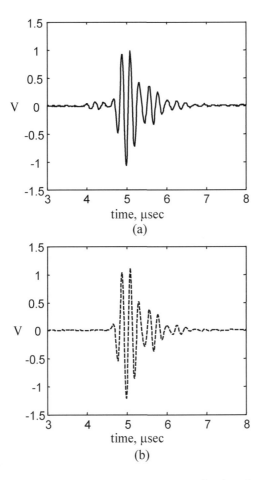

Fig. 13.25. Comparison of the predicted corner reflection from a 2 mm vertical surface breaking slot with the experiments: **(a)** the experimental signal, **(b)** the signal calculated using a multi-Gaussian beam model.

$$V_R(\omega) = 2s(\omega)\frac{1}{S}\int_S \frac{v_{trunc}}{v_0} dS, \qquad (13.32)$$

where S is the surface of the transducer. Figure 13.25 shows a comparison between experimentally measured signal and a simulated corner trap signal from a slot obtained using Eq. (13.32) and performing an inverse FFT to obtain a time domain signal. In this case the interrogating transducer was a

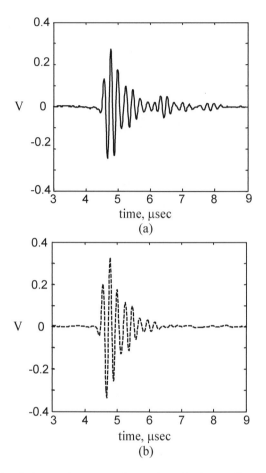

Fig. 13.26. Angle beam shear wave inspection of a 2 mm diameter side drilled hole: **(a)** the experimental signal, and **(b)** the signal calculated using a multi-Gaussian beam model.

45 degree angle beam shear wave transducer and the surface breaking vertical slot had a height $h_c = 2$ mm and a width $w_c = 10$ mm which extended across the entire thickness of a steel specimen. As can be seen from Fig. 13.25 there is very good agreement between the model-based result and the experiment.

As a final example of modeling angle beam transducer responses we will consider the case where the beam from an angle beam shear wave transducer is scattered from a side drilled hole (SDH). This is a reference reflector that is often used in angle beam testing and was considered in an immersion testing example in Chapter 12. The ultrasonic measurement model appropriate for the SDH was given in Eq. (12.36). For our angle beam shear wave problem this measurement model can be expressed as:

$$V_R(\omega) = s(\omega)\left[\int_L \left(\hat{V}^{(1)}(\mathbf{x},\omega)\right)^2 dy_2\right]\left[\frac{A(\omega)}{L}\right]\left[\frac{4\pi\rho_2 c_2^s}{-ik_2^s Z_r^{T;a}}\right], \qquad (13.33)$$

where $\hat{V}^{(1)}(\mathbf{x},\omega) = \hat{V}^{(1)}(y_2,\omega)$ is the normalized velocity field incident on the SDH evaluated along its central axis, which is taken here as the y_2-axis. This velocity field for the angle beam transducer can again be calculated with the multi-Gaussian beam model expression given by Eq. (13.19) and the **ABCD** matrices defined earlier. The normalized pulse-echo scattering amplitude $A(\omega)/L$ of the SDH can be obtained using the Kirchhoff approximation (see Eq. (10.53)) where, for a SDH of radius b:

$$\frac{A(\omega)}{L} = \frac{(k_2^s b)}{2}\left[J_1(2k_2^s b) - iS_1(2k_2^s b)\right] + \frac{i(k_2^s b)}{\pi}. \qquad (13.34)$$

Figure 13.26 shows a comparison between the model prediction and the experimental measurement for the response of a 2 mm diameter SDH to a 45 degree shear wave generated by an angle beam transducer located on the top surface of a steel specimen. The response from the SDH was predicted by using Eqs. (13.33) and (13.34). The early part of the theoretical prediction agrees very well with the experimental response. However, there is a small later arriving creeping wave in the experimental signal that is not predicted by the model. This is to be expected since the SDH model is based on the Kirchhoff approximation which neglects such creeping waves. Such a creeping wave could be included in the model-based results by instead calculating the scattering amplitude of the SDH with the method of separation of variables, as discussed in Chapter 12.

13.4 Model-Assisted Flaw Identification

In conducting angle beam inspection of welded joints, the signals acquired often consist of flaw signals together with non-relevant signals caused by geometrical reflectors such as corners, counter bores and weld roots [13.5]. In some cases these non-relevant signals can be discriminated from the flaw signals based on prior information on the testing location and geometry of the welded joint. Echo-dynamic patterns obtained by scanning the transducer around the indication can also be used in order to identify

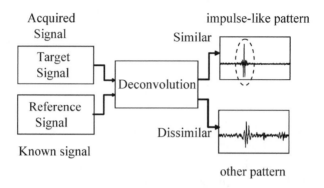

Fig. 13.27. A schematic representation of the TIFD.

the flaw signals. However, these traditional methods have considerable ambiguity since they are based on the subjective experience of inspectors. B- and C-Scans obtained by an automated ultrasonic testing system can also provide information on the identification of flaw signals. However, those usually require expensive equipment and time-consuming (and expensive) scanning.

To overcome these difficulties a technique called the TIFD (Technique for Identification of Flaw signals using Deconvolution), was proposed [13.6]. The concept of the TIFD was quite simple. Let $f(t)$ and $g(t)$ be a reference and target signal, respectively. Then, a deconvolution pattern, $h(t)$, called the similarity function was defined as

$$h(t) = f(t) \otimes^{-1} g(t), \tag{13.35}$$

where, $\otimes^{-1}$ symbolically denotes deconvolution. The basic idea behind the use of this similarity function is illustrated in Fig. 13.27. When the reference and target signals are similar, the deconvolution pattern of the similarity function becomes a sharp impulse-like shape. Otherwise the deconvolution pattern will be quite different in its characteristics. Thus, by comparing a measured signal with a set of reference signals from non-relevant reflectors it was possible to discriminate a flaw signal from those non-relevant reflectors. The major limitation of the TIFD, when implemented in this manner, was that the various kinds of reference signals had to be obtained experimentally. This approach was costly, time-consuming, and in some cases impractical.

Ultrasonic modeling, using the angle beam ultrasonic testing models discussed in Section 13.3, however, can make the TIFD approach

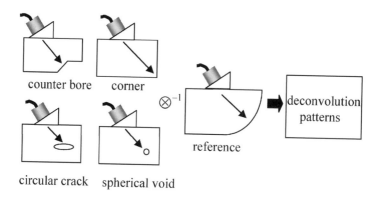

counter bore   corner

circular crack   spherical void

reference

deconvolution patterns

Fig. 13.28. A schematic representation of the model-based TIFD.

more viable [13.7]. This new approach, called model-based TIFD, also relaxes the requirement of acquiring many different kinds of reference signals. Instead, it adopts only one reference signal which is the specular reflection from the cylindrical surface of the STB-A1 standard block that was discussed in the previous section. The idea is that geometrical reflectors such as counter bores and corners should look similar to the cylindrical surface reflection, which is also a simple geometrical reflector, but flaws should not. Models, instead of experimental studies can be used to validate this approach.

As an example of the application of angle beam ultrasonic testing models for flaw identification consider a model-based TIFD approach for discriminating a crack or void flaw signal from the signals generated by reflectors such as counter bores or corners. As shown in Fig. 13.28, deconvolution of these four target signals by the reference signal produces four deconvolution patterns. The characteristics of the counter bore and corner patterns can be predicted by use of the ultrasonic testing models discussed in the previous section. We will discuss how the spherical void and crack responses were simulated in this section.

Figure 13.29 shows an example of a reflection signal captured experimentally from the cylindrical surface of the STB-A1 standard block with a 45 degree angle beam shear wave transducer (the center frequency was 5 MHz and the diameter of the transducer on the wedge was 0.375 inch). This wave form is the reference signal used in the model-based TIFD. It is also the signal that can be used to determine the system function, $s(\omega)$. Using this system function and Eq. (13.17) the received signal from the counter bore can be simulated. The result is shown in

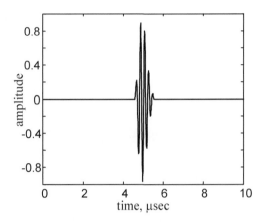

Fig. 13.29. An experimental reference signal obtained by the reflection of the beam of a 45 degree angle beam shear wave transducer (5 MHz center frequency, 0.375 inch diameter) from the cylindrical surface of an STB-A1 standard block.

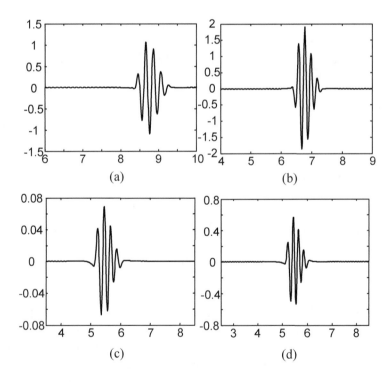

Fig. 13.30. Model-based predictions of time-domain waveforms for (a) a counter bore, (b) a corner, (c) a crack tip, and (d) a spherical void.

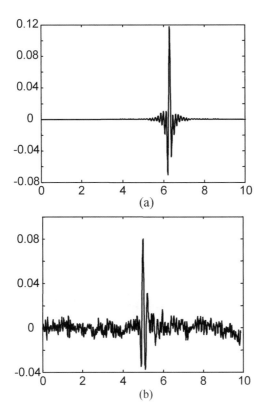

Fig. 13.31. The deconvolution patterns obtained using the model-based TIFD for the counter bore signals by **(a)** theoretical prediction and **(b)** the experiments.

Fig. 13.30 (a). The corner signal can be obtained from Eq. (13.32) but since the corner is larger than the incident beam, the (y,z) truncation limits are taken as the width of the specimen and the beam width, respectively. The corner signal simulated in this manner is shown in Fig. 13.30 (b). For both the spherical void and the crack, the Thompson-Gray measurement model of Eq. (12.6) can be used. Writing that measurement model in the notation used in this Chapter we obtain

$$V_R(\omega) = s(\omega)\left[\hat{V}_0^{(1)}(\omega)\right]^2 A(\omega)\left[\frac{4\pi\rho_2 c_2^s}{-ik_2^s Z_r^{T;a}}\right], \qquad (13.36)$$

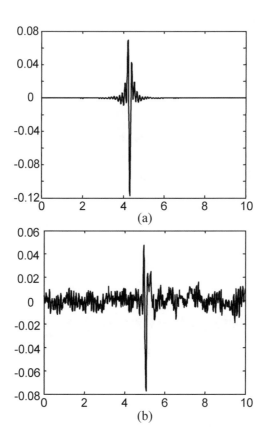

Fig. 13.32. The deconvolution patterns obtained using the model-based TIFD for the corner signals by **(a)** theoretical prediction and **(b)** the experiments.

where the normalized velocity $\hat{V}_0^{(1)}(\omega) = \hat{V}^{(1)}(\mathbf{x}_c, \omega)$ can be calculated from Eq. (13.19) at the point, $\mathbf{x}_c$, which is taken as the center of the void. The scattering amplitude component, $A(\omega)$, for the void can be calculated, in the Kirchhoff approximation, by Eq. (10.14). Figure 13.30 (d) shows the simulated angle beam response of this flaw. If the crack is large enough so that beam variations across its surface are significant, then we must use a "large flaw" measurement model like Eq. (13.17) to simulate the crack response. However, we can take a simpler approach by noting that the first-arriving crack tip flash point signal is generated by the interaction of the incident waves with a small part of the entire crack surface so that even if the crack is large we can still calculate this surface-breaking crack tip signal by placing the Kirchhoff approximation for the response of an

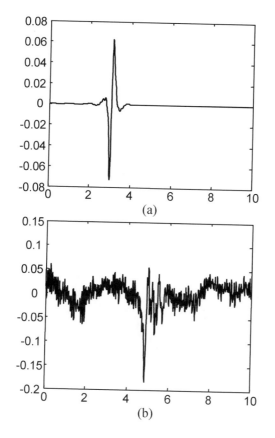

Fig. 13.33. The deconvolution patterns obtained using the model-based TIFD for the crack signals by **(a)** theoretical prediction and **(b)** the experiments.

isolated crack (Eq. (10.33)) into the "small flaw" measurement model of Eq. (13.36), inverting the results into the time domain and then keeping only the first-arriving flash point response. A time domain crack tip signal simulated in this fashion is given in Fig. 13.30 (c).

Comparing Figs. 13.30 (a)-(d) one sees it is very difficult to distinguish these various reflectors/flaws from their scattering waveforms. Certainly there are some amplitude differences in these signals but such amplitude changes are not unique to any particular flaw characteristic. Thus, it would be very difficult for an NDE inspector, seeing these four responses, to separate the flaw responses (the crack and pore) from the geometrical reflectors (counter bore and corner). Now, consider the results of obtaining the similarity function by deconvolving these four model-based patterns with the reference signal, as shown in Figs. 13.31 (a), and

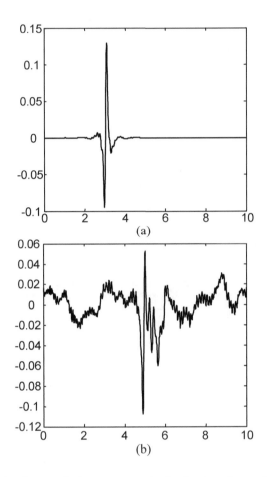

Fig. 13.34. The deconvolution patterns obtained using the model-based TIFD for the spherical void signals by **(a)** theoretical prediction and **(b)** the experiments.

Figs. 13.32 (a), 13.33 (a) and 13.34 (a). One sees impulse-like patterns for the counter bore and corner signals, indicating that these reflectors are very similar to the cylindrical surface reference signal, but bi-polar like responses are obtained instead for both the crack and spherical void cases. Figures 13.31 (b), 13.32 (b), 13.33 (b) and 13.34 (b) show the deconvolved patterns of experimental signals obtained from these same types of scatterers. In the experiments, the "crack" geometry was not that of a circular crack but instead a thin, through-thickness slot in a block was used as the scatterer. However, simulations of the tip signals from such a slot show the same type of flash point response as for the circular crack so we expect similar results upon deconvolution. For the experimental signals

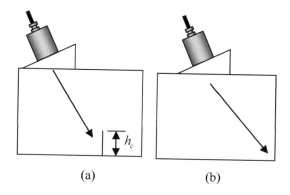

(a)   (b)

Fig. 13.35. A schematic representation of angle beam ultrasonic testing of **(a)** a vertical crack (corner signal) and **(b)** a specimen corner.

again one sees the same impulse-like or bi-polar patterns present, in agreement with the model-based predictions. These model-based TIFD results suggest that such similarity functions should be useful in developing an automated procedure for discriminating flaws from other non-relevant responses such as the counter bore and corner signals considered here.

13.5 Model-Assisted Flaw Sizing

Flaw sizing is one of the fundamental issues in the ultrasonic NDE of various materials and structures since the estimation of structural integrity using methods such as fracture mechanics requires flaw size information. Unfortunately, current flaw sizing approaches are often dependent on the skill of the NDE inspector, and in many cases do not perform well in practice. Models, therefore, have an important role to play in improving flaw sizing methods. This section describes one example of the use of models to help make flaw size estimates [13.8]. The specific case that will be considered here is the sizing of surface breaking vertical cracks using an angle beam shear wave inspection of the type discussed in Section 13.3.

In section 13.3, we developed an angle beam ultrasonic testing model that can predict the corner trap signal received from a vertical slot. That model can be used to consider the response of a surface-breaking crack, as shown in Fig. 13.35 (a). In fact, a crack corner trap signal can be acquired very easily during the inspection of surface breaking vertical cracks in many practical situations. As a reference for the sizing of this

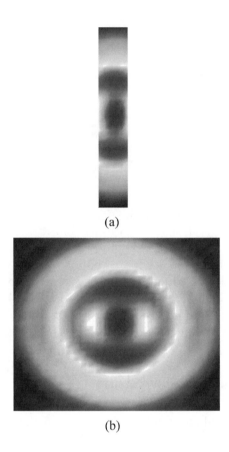

(a)

(b)

Fig. 13.36. Ultrasonic beam footprints for **(a)** a vertical crack corner, and **(b)** the specimen corner: Transducer: 5 MHz center frequency, 0.375 inch diameter, and 45 degrees diffraction angle.

crack let us consider the signal reflected from a specimen corner as shown in Fig. 13.35 (b). The difference between the corner trap signal received from the crack and the one received from the specimen corner is primarily due to different "footprints" of the incident ultrasonic beam that reaches the receiver. Since the size of this footprint is dependent on the size of the crack, dividing the crack response by the specimen corner response should give a quantity that is strongly dependent on that size. Figure 13.36 (a) shows the footprint of the incident beam that is reflected by the crack while Fig. 13.36 (b) shows the same footprint that is reflected from the specimen corner.

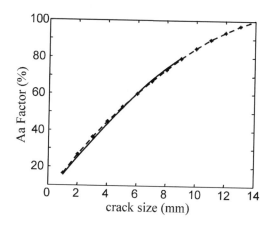

Fig. 13.37. SACs constructed for two specimens with different heights. Transducer: 5 MHz center frequency, 0.375 inch diameter, and 45 degrees diffraction angle. Solid line: D = 10 mm, dashed line with markers: D = 15 mm.

The corner trap signal of a surface-breaking vertical crack with crack height, h_c, and crack width, w_c, can be calculated by use of Eq. (13.32), as discussed in Section 13.3. The specimen corner signal can also be evaluated by Eq. (13.32), as discussed in Section 13.4, by simply not truncating the reflected signal.

We will define the amplitude-area factor (A_a) as the peak-to-peak amplitude ratio (in the time-domain) of the crack corner trap signal to that of the specimen corner signal, i.e.

$$A_a = \frac{V_{crack}^{p-p}}{V_{corner}^{p-p}} \times 100 \ (\%) \tag{13.37}$$

where V_{crack}^{p-p} is the peak-to-peak amplitude of the time domain vertical crack corner trap signal and V_{corner}^{p-p} is the peak-to-peak amplitude of the time domain signal received from the specimen corner. Both of these quantities can be obtained by applying an inverse fast Fourier transform to Eq. (13.32). With this definition of the amplitude-area factor, it is possible to construct a plot that shows the variation of the A_a factor according to the size of a surface breaking vertical crack. This plot we have called a *size-amplitude-curve* (SAC). Figure 13.37 shows examples of the SACs constructed for two specimens having heights of 10 mm and 15 mm. As shown in Fig. 13.37, the SAC is not sensitive to the specimen height so

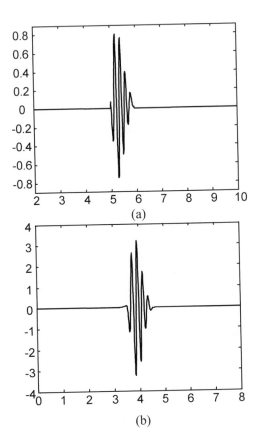

Fig. 13.38. (a) An experimentally measured corner trap signal from a 2 mm vertical crack, **(b)** the predicted corner reflection signal from a specimen with the height of 15 mm.

long as that height is larger than the beam width at the specimen corner. Figures 13.38 and 13.39 show an example of sizing of a through-thickness 2 mm high surface-breaking crack in a specimen having the height of 15 mm. Figure 13.38 (a) shows the corner trap signal captured from this surface breaking crack with a 45 degree angle beam shear wave transducer. The peak-to-peak voltage of this crack signal was measured to be 1.55 mV. The system function for this flaw measurement setup was obtained using the method discussed in Section 13.3 and placed into Eq. (13.32) (used without truncation, as described previously) to obtain the predicted specimen corner signal shown in Figure 13.38 (b). The peak-to-peak voltage of this specimen corner signal was estimated to be 6.5 mV. Based

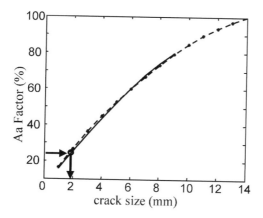

Fig. 13.39. An estimation of the vertical crack size using the SAC: Transducer: 5 MHz center frequency, 0.375 inch diameter, and 45 degrees diffraction angle.

on these two values, the A_a factor was calculated to be 23.85 %. Using this factor the size of the crack can then be estimated from the SAC, as shown in Fig. 13.39. In this case, the estimated crack size was 1.95 mm, which is very close to the actual size of 2.0 mm. Although this is a simple flaw sizing example, it is a nice illustration of the value that models can have in making flaw size estimates.

13.6 References

13.1 Kim HJ, Schmerr LW, Song SJ, Sedov A (2003) Transferring distance-amplitude correction curves – a model-based approach. Journ. Korean Soc. NDT 23: 605-615
13.2 Schmerr LW (2002) Ultrasonic modeling of benchmark problems. In: Thompson DO, Chimenti DE (eds) Review of progress in quantitative nondestructive evaluation 21. American Institute of Physics, Melville, NY, pp 1933-1940
13.3 Lopez AL, Kim HJ, Schmerr LW, Sedov A (2005) Measurement models and scattering models for predicting the ultrasonic pulse-echo response of side-drilled holes. Journ. Nondestr. Eval. 24: 83-96
13.4 Kim HJ, Park JS, Song SJ, Schmerr LW (2004) Modeling angle beam ultrasonic testing using multi-Gaussian beams. J. Nondestr. Eval. 23: 81-93
13.5 Kräutkramer J, Kräutkramer H (1990) Ultrasonic testing of materials, 4th ed. Springer-Verlag, Berlin, Germany, pp 431-465

13.6 Song SJ, Kim JY, Kim YH (2002) Identification of flaw Signals in the angle beam ultrasonic testing of welded joints with geometric reflectors. In: Thompson DO, Chimenti DE (eds) Review of progress in quantitative nondestructive evaluation 21, American Institute of Physics, Melville, NY, pp 691-698

13.7 Jung HJ, Kim HJ, Song SJ, Kim YH (2003) Model-based enhancement of the TIFD for flaw signal identification in ultrasonic testing of welded joints. In: Thompson DO, Chimenti DE (eds) Review of progress in quantitative nondestructive evaluation 22. American Institute of Physics, Melville, NY, pp 628-634

13.8 Kim HJ, Song SJ, Kim YH (2003) Quantitative approaches to flaw sizing using ultrasonic testing models. In: Thompson DO, Chimenti DE (eds) Review of progress in quantitative nondestructive evaluation 22. American Institute of Physics, Melville, NY, pp 703-710

A Fourier Transforms and the Delta Function

Ultrasonic NDE involves the propagation of short, transient pulses. A pulser, for example, generates voltage pulses that drive an ultrasonic transducer. That transducer transforms the electrical pulses into mechanical pulses that travel as waves and are converted back into electrical pulses at the receiving transducer. The received electrical pulses are then often displayed on an oscilloscope as a voltage versus time signal. In order to model ultrasonic systems ultimately we must be able to describe such transient behavior. If we directly simulate these time varying signals this is referred to as modeling in the *time domain*. However, it is often more convenient to describe the ultrasonic system and its components in terms of their *frequency domain* response, which is obtained by applying the Fourier transform to the time domain signals. It is always possible to recover the time domain signal from its Fourier transform through an inverse Fourier transform, so that working in the frequency domain does not imply a loss of information. In this Appendix we will describe some of the basic properties of Fourier and inverse Fourier transforms and we will show how these transforms can be implemented numerically with Fast Fourier Transform (FFT) algorithms. We will also introduce the delta function and its Fourier transform since that function plays a key role in modeling linear systems, as shown in Appendix C.

A.1 The Fourier Transform and Its Inverse

Consider a pulse, $v(t)$, a signal that is a function of the time, t. The Fourier transform of $v(t)$, $V(f)$, is given by [A.1], [A.3],

$$V(f) = \int_{-\infty}^{+\infty} v(t)\exp(2\pi i f t)\,dt \,. \tag{A.1}$$

The variable f is the frequency. Typically in ultrasonic NDE problems f is given in MHz (millions of cycles/sec), where 1 cycle/sec = 1 Hz, and the corresponding time, t, is given in microseconds. Although we integrate over all times in Eq. (A.1), most ultrasonic pulses are non-zero only over a finite time interval. The inverse Fourier transform, which allows us to obtain $v(t)$, is

$$v(t) = \int_{-\infty}^{+\infty} V(f)\exp(-2\pi i f t)df .$$ (A.2)

Equation (A.2) shows that in order to recover $v(t)$ one must integrate over both negative and positive frequencies. If the time domain function is real, however, it can be shown that its Fourier transform satisfies $V(-f) = V^*(f)$, where $(\ )^*$ denotes the complex conjugate. Thus, the negative frequency components can be obtained from the positive components and in this sense they are redundant. Later, we will show how to recover the time domain signal, $v(t)$, from only the positive frequency components.

There are other definitions of the Fourier transform and its inverse that are used in the literature so that one must be careful when comparing results, using Eqs. (A.1) and (A.2), to similar results from other authors. We will use the definitions of Eqs. (A.1) and (A.2) exclusively in this work, or their equivalent definitions given by

$$V(\omega) = \int_{-\infty}^{+\infty} v(t)\exp(i\omega t)dt$$

$$v(t) = \frac{1}{2\pi} \int_{-\infty}^{+\infty} V(\omega)\exp(-i\omega t)d\omega$$ (A.3)

in terms of the *circular frequency*, ω, as measured in rad/sec, where $\omega = 2\pi f$. Since $v(t)$ and $V(f)$ can be obtained from each other, we can write this relationship as $v(t) \leftrightarrow V(f)$. In a similar fashion we write the corresponding relationships for a time shifted or differentiated signal as:

$$v(t - t_0) \leftrightarrow \exp(2\pi i f t_0)V(f)$$

$$\frac{dv}{dt} \leftrightarrow -2\pi i f V(f).$$ (A.4)

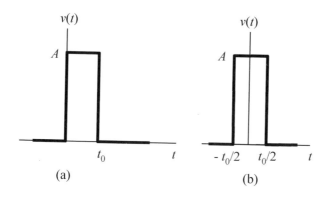

Fig. A.1. (a) An example of a simple "box" time domain function and (b) the same function shifted in time.

For many other relationships see [A.3] or [Fundamentals].

As an example of a Fourier transform, consider the simple "box" function shown in Fig. A.1 (a). The Fourier transform of this function can be obtained analytically as

$$
\begin{aligned}
V(f) &= \int_{0}^{t_0} A \exp(2\pi i f t)\, dt \\
&= \frac{A t_0 \exp(i\pi f t_o)\sin(\pi f t_o)}{\pi f t_0}.
\end{aligned}
\tag{A.5}
$$

The magnitude of this complex Fourier transform is shown in Fig. A.2 (a) and its phase is given in Fig. A.2 (b). The phase plot shows periodic jumps of π radians corresponding to the sign changes at these points of the $\sin(\pi f t_0)$ function. Otherwise the phase of the Fourier transform is a linearly increasing function of frequency due to the $\exp(i\pi f t_0)$ term in Eq. (A.5). If we shift this box function to the left by $t_0/2$, the shifting property of the Fourier transform given in Eq. (A.4) shows that for this symmetrical function (see Fig. A.1 (b)) we instead obtain a purely real Fourier transform given by

$$
V(f) = \frac{A t_0 \sin(\pi f t_o)}{\pi f t_0}
$$

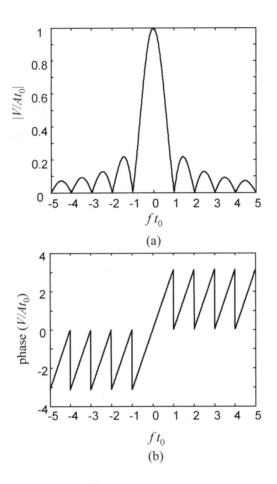

Fig. A.2. **(a)** The magnitude of the Fourier transform of the "box" function of Fig. A.1 (a) and **(b)** the phase of this transform.

As another example of the use of the Fourier transform, consider the following time function and its Fourier transform (also called the *Fourier spectrum* of the signal):

$$v(t) = \cos(2\pi f_c t)\exp\left[-t^2/4A^2\right]$$
$$V(f) = A\sqrt{\pi}\left\{\exp\left[-4\pi^2 A^2(f-f_c)^2\right] + \exp\left[-4\pi^2 A^2(f+f_c)^2\right]\right\}. \tag{A.6}$$

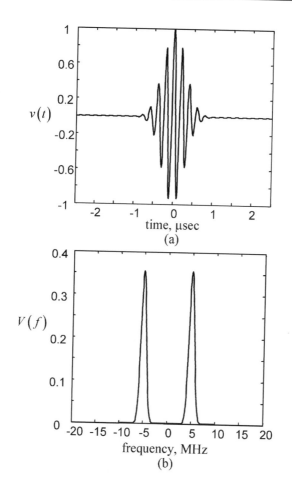

Fig. A.3. (a) The time domain function in Eq. (A.6) and (b) the Fourier transform of Eq. (A.6) for $A = 0.2$, $f_c = 5$ MHz.

These functions are shown in Fig. A.3 for $A = 0.2$ and $f_c = 5$ MHz. The time domain function is a transient that has a shape typical of many ultrasonic signals. The frequency domain spectrum is a pair of Gaussians whose maxima are located at the frequencies f_c and $- f_c$. The constant A controls both the amplitude of these Gaussians and their widths. Normally, one specifies the characteristics of a frequency domain spectrum as shown in Fig. A.4 by a *center frequency*, f_c, defined as the frequency at which the maximum frequency domain response occurs, and a width of the spectrum. One measure of the width that is commonly used in ultrasonic NDE is the

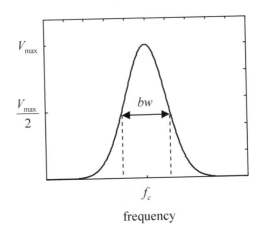

Fig. A.4. Definition of the center frequency, f_c, and −6 dB bandwidth, bw, of a spectrum.

−6 *dB bandwidth*, bw, which is defined in this case as the width of the Gaussian when the amplitude drops 6 decibels (*dB*) below its maximum at $f = f_c$, as shown in Fig. A.4. Note that an amplitude ratio V/V_r is defined in terms of decibels (*dB*) as

$$\frac{V}{V_r}(dB) = 20\log_{10}\left(\frac{V}{V_r}\right). \tag{A.7}$$

Since $20\log_{10}(1/2) = -6.02\ dB$, at −6 *dB* the amplitude has been reduced to a value of approximately one half that of the reference value, where in our case the reference value is the maximum at $f = f_c$. The parameter, A, can be shown to be related to the −6 *dB* bandwidth, bw, by the relation $A = \sqrt{\ln 2}/(\pi bw)$. Figure A.5 shows the functions in Eq. (A.6) for a center frequency $f_c = 10$ MHz and for various choices of the −6 *dB* bandwidth, bw. It can be seen that a narrow bandwidth results in a wide pulse with significant ringing in the time domain signal while a wide bandwidth generates a very short pulse with little ringing. Ultrasonic transducers with a large bandwidth are called wide-band, high resolution transducers because the short time domain signals they generate allow one to resolve signals that are near to one another. Ultrasonic transducers with

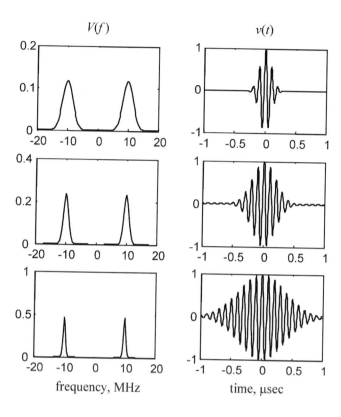

Fig. A.5. Wave forms and spectra obtained from Eq. (A.6) for f_c = 10 MHz and (from top to bottom) bw = 4, 2, 1 MHz, respectively. Note that the amplitudes of the spectra are getting larger and their widths narrower as the bandwidth decreases although the amplitudes appear to be decreasing with decreasing bandwidth because of the changing vertical scale.

small bandwidths are called narrow-band, high sensitivity transducers. Although they cannot as easily resolve closely separated signals, the longer duration pulses they generate normally have more energy than wide-band signals and they can therefore penetrate deeper into materials.

As a final example of a Fourier transform pair, consider a 1-D wave pressure pulse traveling along the positive x-direction in a fluid.

Such a pulse has the form $p(t - x/c)$ where c is the wave speed. If we take the Fourier transform of this traveling wave we find

$$
\int_{-\infty}^{+\infty} p(t - x/c) \exp(2\pi i f t) dt
$$

$$
= \exp(2\pi i f x/c) \int_{-\infty}^{+\infty} p(u) \exp(2\pi i f u) du \tag{A.8}
$$

$$
= P(f) \exp(2\pi i f x/c),
$$

where $P(f)$ is the Fourier transform of $p(t)$. Note that we could also have obtained this result directly by using the shifting property relationship of Eq. (A.4). If we put this Fourier transform back into the inverse Fourier transform expression, we obtain

$$
p(t - x/c) = \int_{-\infty}^{+\infty} P(f) \exp(2\pi i f x/c - 2\pi i f t) df
$$

$$
= \frac{1}{2\pi} \int_{-\infty}^{+\infty} P(\omega) \exp(i k x - i \omega t) d\omega, \tag{A.9}
$$

where $k = \omega/c = 2\pi f/c$ is called the *wave number*. Equation (A.9) shows that we can consider a 1-D pulse as a superposition of terms of the form $p = A \exp(i k x - i \omega t)$ which is a harmonic plane wave traveling in the positive x-direction. The amplitude, A, of the plane wave is just proportional to the spectrum of the pressure wave, i.e. $A = P(\omega)/2\pi$. The wave number is related to the *wavelength, λ*, of the wave through the relation $k = 2\pi/\lambda$. To see the meaning of the terms in the exponential factor of this harmonic wave, first fix x (i.e. sit at a fixed location in space) and watch the wave go by as a function of the time, t. The pressure will go through a complete cycle (the exponential will change by 2π radians) in a time, T_p, (in seconds) called the period of the wave. Thus, over one cycle $2\pi f T_p = 2\pi$. This shows that the frequency, f, (in Hz or cycles/sec) is just $f = 1/T_p$. Since $\omega = (2\pi \, rad/cycle)(f \, cycles/\sec)$ we see that ω is just the rate at which the argument of the exponential term of the pressure is changing in time at a fixed location in units of rad/sec. Now, instead fix the time t and consider the pressure changes as a function of x. Physically, this would correspond to taking a "snapshot" of the wave variations at a fixed time as a function of the distance, x. Again the pressure will go through a complete cycle over a distance, D, when $kD = 2\pi D / \lambda = 2\pi$ so that the wave length, λ, is just that distance, measured in units of length/cycle. But $k = \left(2\pi \, rad \, / \, cycle \right) / \left(\lambda \, length \, / \, cycle \right)$ so the wave number is just the rate at which the argument of the exponential term of the pressure is changing in distance at a fixed time in terms of units of rad/length.

Equation (A.9) shows that in solving wave propagation and interaction problems, we can consider the behavior of harmonic waves and then obtain the solution for a pulse by Fourier superposition. We have shown this fact here only for 1-D plane waves, but it is also true for other types of waves as well.

A.2 The Discrete Fourier Transform

In practice, experimental ultrasonic NDE signals are manipulated digitally, i.e. the analog (continuous) time domain signals are first sampled and then these sampled values are stored digitally for later processing. Thus, it is also important to be able to deal with Fourier transforms and their inverses in terms of discrete, sampled signal values. This can be done using forms similar to Eqs. (A.1) and (A.2) given by the discrete Fourier transform pair [A.2]:

$$V\left(f_n\right) = \Delta t \sum_{j=1}^{N} v\left(t_j\right) \exp\left[2\pi i (j-1)(n-1) / N \right] \quad (n = 1, 2, ..., N)$$

$$v\left(t_k\right) = \frac{1}{N \Delta t} \sum_{n=1}^{N} V\left(f_n\right) \exp\left[-2\pi i (n-1)(k-1) / N \right] \quad (k = 1, 2, ..., N),$$

(A.10)

where $v\left(t_k\right), V\left(f_k\right)$ are values of a time domain function and its Fourier transform at discrete frequency and time values, respectively, $\Delta t = t_{k+1} - t_k$ is the sampling time interval, and N is the total number of sampled points. As with the Fourier transform and its inverse, the discrete Fourier transform pair of Eq. (A.10) may appear in different forms in the literature. Here, Eq. (A.10) is the discrete transform pairs corresponding directly to Eqs. (A.1) and (A.2).

While the Fourier transform and its inverse are usually applied to non-periodic functions their discrete Fourier transform counterparts in Eq. (A.10) are always periodic functions. For example, the first sampled

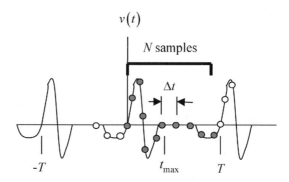

Fig. A.6. A sampled periodic time domain function showing the N sampled values used in the discrete Fourier transform (dark circles) and other sampled values (light circles). The sampling time interval is Δt, the time t_{max} is the time at which the transient signal ends, and the time $T = N\Delta t$ is the period.

value in the time domain, $v(t_1)$, is normally taken to be the value sampled at $t = 0$. There are then N sampled values from $t = 0$ to $t = (N-1)\Delta t$. The sampled value $v(t_{N+1}) = v(N\Delta t)$, however, is the same as the value at time $t = 0$ and subsequent samples $v(t_{N+2}), v(t_{N+3})$ etc. also repeat previous values. As Fig. A.6 shows the sampled time function is periodic with period $T = N\Delta t$. Similarly, in the frequency domain the first sampled value, $V(f_1)$, is the frequency component for $f = 0$ (the d.c. value) and there are then N sampled frequency components from $f = 0$ to $f = (N-1)\Delta f$, where $\Delta f = 1/T$. The sampled value $V(f_{N+1}) = V(N\Delta f)$, is again the d.c. value. The frequency domain function is therefore also periodic with period $f_s = 1/\Delta t$, where f_s is the sampling frequency (see Fig. A.7). In Appendix G a MATLAB function s_space is given that generates a sampled time or frequency axis with precisely these values needed for application of the discrete Fourier transform or its inverse. For example, s_space(0, T, N) generates a set of N evenly spaced sampled values going from 0 to T - Δt, where $\Delta t = T/N$ is the sample spacing. Note that if we used the built-in MATLAB function linspace(0, T, N) we would obtain instead a set of N evenly sampled values going from 0 to T with sample spacing $\Delta t = T/(N-1)$.

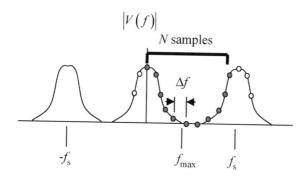

Fig. A.7. The magnitude of a sampled periodic frequency domain function showing the N sampled values use in the discrete inverse Fourier transform (dark circles) and other sampled values (light circles). The frequency sampling interval is $\Delta f = 1/T$, the frequency f_{max} is the maximum frequency contained in the signal and the sampling frequency $f_s = 1/\Delta t$ is the period.

As long as the original time domain signal is shorter than the sampling period, T, and the sampling frequency is sufficiently high to capture all the significant "wiggles" in the signal it can be seen from Fig. A.6 that the N sampled time domain components contained in the discrete Fourier transform will capture the entire signal adequately. Note that if the time signal has non-zero values before $t = 0$, those values will appear in the upper half of the N time domain samples as shown in Fig. A.6. The MATLAB function c_shift (see section G.8 in Appendix G for a code listing) can be used to shift the entire time domain function and place these negative time values back into their proper position so that the function does not appear to be split. Similarly, a MATLAB function t_shift given in Appendix G changes the time-axis appropriately.

In the frequency domain the negative frequency components in the discrete Fourier transform are also contained in the upper half of the N frequency domain samples as can be seen from Fig. A.7. Since inherently the frequency spectrum of a real time domain function must contain both negative as well as positive frequency components, Fig. A.7 shows that unless the sampling frequency, f_s, is at least double f_{max}, the highest frequency contained in the signal, these periodically repeated functions will overlap and we will not recover the original spectrum of the signal. To prevent this phenomenon, which is called *aliasing,* we must therefore always choose a high enough sampling frequency. This requirement is embodied in what is called the Nyquist criterion (or the sampling theorem) which is [A.1]:

The sampling frequency, $f_s = 1/\Delta t$, must be at least twice the maximum significant frequency, f_{max}, contained in the waveform being sampled.

In ultrasonic NDE, the transducers commonly used do not produce significant frequencies above about 20 MHz and inspected materials (steel,

Code Listing A.1. The FFT corresponding to Eq.(A.10).

```
function y = FourierT(x, dt)
% FourierT(x, dt) computes forward FFT of x, with sampling time interval dt
% FourierT assumes the Fourier transform  is in terms of exp(2*pi*i*f*t)
% For NDE, frequency components are normally in MHz, dt in microseconds
% If x is a matrix, the transform is performed on the columns of x
[nr, nc] = size(x);
if nr == 1
N = nc;
else
    N = nr;
end
y = N*dt*ifft(x);
```

Code Listing A.2. The Inverse FFT corresponding to Eq.(A.10).

```
function y = IFourierT(x, dt)
% IFourierT(x, dt) computes the inverse FFT of x, for a sampling time interval dt
% IFourierT assumes the inverse transform is in terms of exp(-2*pi*i*f*t)
% For NDE, frequency components are normally in MHz, dt in microseconds
% If x is a matrix, the inverse transform is performed on the columns of x
[nr,nc] = size(x);
if nr == 1
    N = nc;
else
    N = nr;
end
y =(1/(N*dt))*fft(x);
```

aluminum, etc.) also attenuate ultrasound severely above such frequencies. Thus, a frequency of 100 MHz is normally a conservative choice for the sampling frequency that will satisfy the Nyquist criterion. This corresponds to a sampling time interval $\Delta t = 10$ nanoseconds. The number of sampling points, N, then determines the total length of the time record, $T = N\Delta t$, being digitized which in turn determines the sampling interval in the frequency domain, Δf, since $\Delta f = f_s / N = 1 / N\Delta t = 1/T$.

To implement the discrete Fourier transform pair of Eq. (A.10) efficiently, one uses Fast Fourier Transform (FFT) algorithms, which are widely available [A.4]. Numerous books have been written on FFTs if one is interested in the details of those algorithms. To perform these discrete transforms in MATLAB, we must be aware that the built-in MATLAB FFT functions fft and ifft are defined such that the signs are exchanged in the exponentials appearing in Eq. (A.10) and the MATLAB functions do not include the sampling time constant Δt or N appearing in the coefficients of Eq. (A.10). Thus, we have defined a new set of MATLAB functions, FourierT and IFourierT, to implement the discrete transforms in Eq. (A.10). Those functions are defined in Code Listings A.1 and A.2. The functions FourierT and IFourierT, like fft and ifft, will perform Fast Fourier Transforms and their inverse on either vectors or matrices. If the input data is a vector, it can be a column or row vector. If the input data of these functions is a matrix, then they will perform the FFTs or inverse FFTs on the columns of the matrix. Fast Fourier Transform algorithms are often implemented with the number of samples $N = 2^m$ for some integer m. In fact some FFT algorithms require the number of samples be a power of two. The MATLAB functions fft and ifft do not have this restriction so that neither do the functions FourierT and IFourierT. However, these functions are also more efficient when the number of samples is a power of two.

If we give the function IFourierT only the positive frequency components of a real time domain function, then to recover that real time function we need to compute twice the real part of the output of IFourierT (which will be complex), This is necessary since if we compute the inverse Fourier transform of a function $V(f)$ using only the positive frequency components we do not obtain the function $v(t)$ but instead find the function $v_+(t)$ where [Fundamentals]:

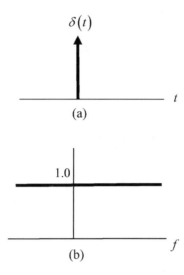

Fig. A.8. (a) A delta function, and **(b)** its Fourier transform.

$$v_+(t) = \int_0^\infty V(f)\exp(-2\pi i f t)\,df$$

$$= \frac{1}{2}v(t) + \frac{i}{2\pi}\int_{-\infty}^{+\infty}\frac{v(\tau)}{\tau - t}\,d\tau \tag{A.11}$$

so that one-half of the original function v is in the real part of $v_+(t)$ and one half of the *Hilbert transform* [A.3] of v shows up in the imaginary part. [Note that if $V(f_1) = V(f)\big|_{f=0}$ is non-zero, which can happen with time functions whose average (dc) value is not zero, then one half of that value must be associated with the positive frequencies and one half with the negative frequencies, i.e. we should compute IFourierT on the positive frequencies only after first making the replacement $V(f_1) \to V(f_1)/2$].

A.3 The Delta Function

One function that plays a key role in analyzing linear systems is the *delta function* [A.1]. We can define a delta function from a limit of the "box"

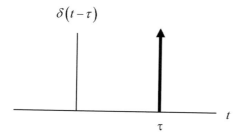

$\delta(t-\tau)$

Fig. A.9. A shifted delta function.

function shown in Fig. A.1 (a). If we let $t_0 \to 0$ but keep the product $At_0 = 1$ so that the function always contains unit area, then the box function becomes an infinite spike at $t = 0$ as shown in Fig. A.8 (a), which we will denote symbolically by the delta function $\delta(t)$. In the same limit the Fourier transform of the box function becomes simply unity at all frequencies, as shown in Fig. A.8 (b) so that we have the Fourier transform pair $\delta(t) \leftrightarrow 1$. Thus, a delta function generates all frequencies equally. It is this property that makes a delta function an ideal function to serve as a system input since the output of a system with such an input will then reflect how the system modifies this uniform input at all frequencies. As discussed in Appendix C, this allows us to obtain the transfer function of a linear system. The shifted delta function $\delta(t - \tau)$ is an infinite spike at time $t = \tau$, as shown in Fig. A.9. Some important properties of $\delta(t - \tau)$ are:

$$\delta(t-\tau) = 0 \quad t \neq \tau$$

$$\int_a^b g(t)\delta(t-\tau)\,dt = \begin{cases} 0 & \tau < a \text{ or } \tau > b \\ g(\tau) & a < \tau < b \\ g(\tau)/2 & \tau = a \text{ or } \tau = b \end{cases} \tag{A.12}$$

and

$$\int_{u=-\infty}^{u=t} \delta(u-\tau)\,du = H(t-\tau) = \begin{cases} 0 & t < \tau \\ 1/2 & t = \tau, \\ 1 & t > \tau \end{cases} \tag{A.13}$$

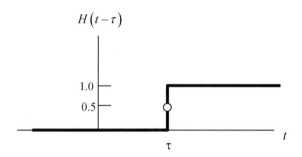

Fig. A.10. A unit step function at t = τ.

where $g(t)$ is an arbitrary function and $H(t)$ is the unit step function. Equation (A.12) shows the sampling properties of the delta function while Eq. (A.13) shows that the integral of the delta function $\delta(t-\tau)$ is the step function $H(t-\tau)$, as shown in Fig. A.10. If we examine the several Fourier transform relations of Eq. (A.4) for the delta function we find

$$\delta(t-t_0) \leftrightarrow \exp(2i\pi f t_0)$$
$$\frac{d\delta}{dt} \leftrightarrow -2\pi i f. \tag{A.14}$$

A.4 References

A.1 Bracewell RN (2000) The Fourier transform and its applications, 3rd ed. McGraw-Hill, New York, NY
A.2 Burrus CS, Parks TW (1985) DFT/FFT and convolution algorithms. John Wiley and Sons, New York, NY
A.3 Sneddon IN (1951) Fourier transforms. McGraw-Hill, New York, NY
A.4 Walker JS (1996) Fast Fourier transforms, 2nd ed. CRC Press, New York, NY

A.5 Exercises

1 (a). Write a MATLAB function, spectrum1, which computes the positive frequency components of a signal given by

$$V(f) = A\sqrt{\pi} \exp\left[-4\pi^2 A^2 (f - f_c)^2\right],$$

where A is given in terms of the -6 dB bandwidth, bw, as $A = \sqrt{\ln 2}/(\pi bw)$. Spectrum1 should return sampled values of V for a set of sampled frequencies, f (in MHz), and a specified center frequency, fc , (in MHz), and bandwidth, bw, (in MHz), i.e. we should have for the MATLAB function call:

>> V = spectrum1(f, fc, bw)

Show that your function is working by evaluating the spectrum for 512 frequencies ranging from zero to 100 MHz with fc = 5 MHz, bw = 1 MHz. Generate the frequencies with the function s_space (see Appendix G), i.e. evaluate

>> f = s_space(0, 100, 512);

Plot V over a range of frequencies 0 – 10 MHz (approximately).

1 (b). Use the MATLAB function IFourierT to obtain a sampled time domain function, $v(t)$, from the sampled spectrum computed in problem 1 (a). Plot $v(t)$ by first generating a set of 512 time domain values with the function s_space, i.e.

>> t = s_space(0, 512*dt, 512);

where dt is the time interval between samples (which in this case is dt = 1/100). Then use the t_shift and c_shift functions given in Appendix G so that the sampled values of $v(t)$ shown are not split between the first and last half of the window. Show that your results agree with the analytical result:

$$v(t) = \cos(2\pi f_c t)\exp\left[-t^2/4A^2\right]$$

if we take into account the fact that we only used the positive frequency components.

1 (c). Now apply the FourierT function to the sampled $v(t)$ obtained from part 1 (b) and plot the magnitude of the resulting spectrum $V(f)$. How is this $V(f)$ different from the one you started with?

1 (d). If you look carefully at your plot of $V(f)$ in part 1 (a) you should see that the sampling interval, Δf, in the frequency domain is not quite small enough to give an accurate representation of the Gaussian function. Take the $v(t)$ signal obtained in 1 (b), shift it so that all the values in the second half of the time domain window are zero, and then append 512 zeros to that signal. Apply FourierT to this longer signal to obtain the spectrum. Show that this process, which is called *zero padding*, improves our resolution in the frequency domain. Note that zero padding does not affect the sampling frequency.

2. The Hilbert transform, $H\left[f(t)\right]$, of a function $f(t)$ is defined as

$$H\left[f(t)\right] = \frac{1}{\pi}\int_{-\infty}^{+\infty}\frac{f(\tau)d\tau}{\tau - t}.$$

When a plane traveling wave of the form $f(t - x/c)$ is reflected from a surface beyond a critical angle, as discussed in Appendix D, the Hilbert transform of the function $f(t)$ appears in the reflected wave causing the reflected waveform to be distorted from the incident wave. It can be shown [A.2] that if $F(\omega)$ is the Fourier transform of $f(t)$ then the Fourier transform of the Hilbert transform of $f(t)$, $\tilde{H}(\omega)$, is given by $\tilde{H}(\omega) = -i\,\mathrm{sgn}(\omega)F(\omega)$ where

$$\mathrm{sgn}(\omega) = \begin{cases} +1 & \omega > 0 \\ -1 & \omega < 0 \end{cases}.$$

We can use this fact in conjunction with the Fast Fourier transform as a convenient way to compute the Hilbert transform of a function. To see this consider the follow example:

In MATLAB, define a sampled time axis consisting of 1024 points going from $t = 0$ to $t = 10$ μsec. Over this time interval define a sampled function $f(t)$ that has unit amplitude for $4.5 < t < 5.5$ μsec and is zero otherwise. Use FourierT to calculate the Fourier transform, $F(\omega)$, of $f(t)$. Then use the relationship $\tilde{H}(\omega) = -i\operatorname{sgn}(\omega)F(\omega)$ to find the Fourier transform of the Hilbert transform of $f(t)$ and compute the Hilbert transform itself by performing an inverse Fourier transform on this result with IFourierT. [Note that if you use only the positive frequency values to compute the inverse, then we only need to multiply those values by $-i$]. Plot your results versus time. In this case the Hilbert transform of $f(t)$ can be obtained analytically [Fundamentals]. It is

$$\mathbb{H}[f(t)] = \frac{1}{\pi}\ln\left|\frac{t - 5.5}{t - 4.5}\right|.$$

Using a different plotting symbol, plot this function also on the same graph. How do your results compare?

B Impedance Concepts and Equivalent Circuits

Impedance is very important concept for ultrasonic systems since it appears in a variety of contexts [B.1]. Thus, in this Appendix we will discuss briefly impedance as it appears in both electrical and acoustical components. We will also examine the concept of equivalent circuits and the use of Thévenin's theorem to represent active electrical systems such as an ultrasonic pulser.

B.1 Impedance

Impedance is a quantity that is most often associated with electrical circuits. Consider, for example, the electrical elements shown in Fig. B.1. The time varying voltage and current for these elements are related to one another through the following relations:

Resistor–
$$V(t) = R\,I(t) \tag{B.1a}$$

Capacitor–
$$C\frac{dV(t)}{dt} = I(t) \tag{B.1b}$$

Inductor–
$$V(t) = L\frac{dI(t)}{dt}. \tag{B.1c}$$

If we assume these voltages and currents are harmonic, i.e. $V = V_0 \exp(-i\omega t)$, $I = I_0 \exp(-i\omega t)$, then for these elements we have

Resistor–
$$V_0 = R\,I_0 \tag{B.2a}$$

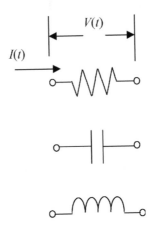

Fig. B.1. The voltage and current flowing (top to bottom) in a resistor, capacitor, and inductor, respectively.

Capacitor–
$$V_0 = \frac{I_0}{-i\omega C} \qquad (B.2b)$$

Inductor–
$$V_0 = -i\omega L\, I_0. \qquad (B.2c)$$

We could instead take the Fourier transform of all the relations in Eqs. (B.1a-c) and view Eqs. (B.2a-c) as the relations between the Fourier transform of the voltage, $V_0(\omega)$, and the Fourier transform of the current, $I_0(\omega)$, for these elements. In general, we see we can write for all these elements $V_0(\omega) = Z^e(\omega) I_0(\omega)$, where $Z^e(\omega)$ is the *complex electrical impedance*. The impedance has the dimensions of volts/amps = ohms (Fourier transforms of voltage and current will have dimensions such as volts/Hz and amps/Hz but their ratio is still ohms). We see that $V_0(\omega) = Z^e(\omega) I_0(\omega)$ is true for these simple individual circuit elements. However, we can also take a complex circuit composed of many of these elements and also replace them by an equivalent complex impedance in the same fashion.

Impedance also is associated with mechanical systems [B.1]. Consider, for example, the mechanical elements shown in Fig. B.2. The

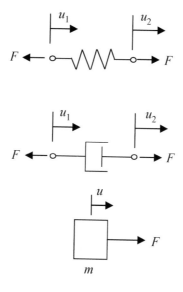

Fig. B.2. The forces and displacements acting (top to bottom) in a spring, dashpot, and mass, respectively.

time varying forces and displacements for these elements are related to one another through the following relations:

Spring–
$$F = k\left(u_2 - u_1\right)$$
(B.3a)

Dashpot–
$$F = c_d\left(\frac{du_2}{dt} - \frac{du_1}{dt}\right)$$
(B.3b)

Mass–
$$F = m\frac{d^2u}{dt^2}.$$
(B.3c)

Again, if we assume these forces and displacements are harmonic so that $F = F_0 \exp(-i\omega t)$, $u = U \exp(-i\omega t)$ or take the Fourier transforms of Eq. (B.3a-c) we find

Spring–
$$F_0 = k(U_2 - U_1) \qquad \text{(B.4a)}$$

Dashpot–
$$F_0 = -i\omega c_d (U_2 - U_1) \qquad \text{(B.4b)}$$

Mass–
$$F_0 = -m\omega^2 U, \qquad \text{(B.4c)}$$

which also can all be expressed in terms of a *complex mechanical impedance*, $Z^m(\omega)$, where $F_0(\omega) = Z^m(\omega)\Delta U(\omega)$. In this case the dimensions of the mechanical impedance are that of stiffness, i.e. force/displacement.

Ultrasonics inherently involves the propagation of waves and the concept impedance also is an important one for wave motion [Fundamentals]. Consider, for example, a 1-D plane pressure wave in a fluid propagating in the positive x-direction. The pressure and x-component of the velocity in the wave can be expressed in the forms

$$p = P f(t - x/c)$$
$$v_x = V f(t - x/c), \qquad \text{(B.5)}$$

where P, V are pressure and velocity amplitudes of the waves (the function f is dimensionless) and c is the wave speed. However, the pressure and velocity in the fluid are related to one another though the equation of motion of the fluid, which is (see Appendix D)

$$-\frac{\partial p}{\partial x} = \rho \frac{\partial v_x}{\partial t}, \qquad \text{(B.6)}$$

where ρ is the density of the fluid. Placing the pressure and velocity expressions of Eq. (B.5) into Eq. (B.6) then gives

$$\frac{P}{c} f'(t - x/c) = \rho V f'(t - x/c), \qquad \text{(B.7)}$$

where $f' = df(u)/du$. It then follows that

$$P = \rho c V. \qquad \text{(B.8)}$$

The quantity $z^a = \rho c$ is called the *specific acoustic impedance* of a plane wave. If we consider the force in the wave, F, generated by the pressure

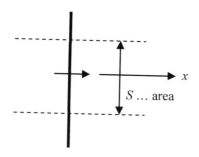

Fig. B.3. A plane wave traveling in the x-direction and a cross-sectional area, S, of the wave front.

acting on a cross-sectional area, S, of the wave front, as shown in Fig. B.3, then we have

$$F = PS = \rho c S V, \tag{B.9}$$

where $Z^a = \rho c S$ is called the *acoustic impedance* of the plane wave. This acoustic impedance has the dimensions in the SI system of Newtons-second/meter (N-s/m) and the specific acoustic impedance has the dimensions N-s/m^3. For more general wave types the acoustic impedance or specific acoustic impedance is in general a complex quantity.

An ultrasonic system inherently contains electrical and electromechanical elements as well as propagating acoustic and elastic waves. Thus, the system will be described by a variety of different impedances and we need to distinguish between them. In this book we will use the symbol "Z" for impedances and denote electrical impedance by an "e" superscript and acoustical impedance by an "a" superscript, a notation also followed in this section. For example, in Chapter 4 the electrical input impedance of a transmitting transducer A is given as $Z_{in}^{A;e}$ while the same transducer's acoustic radiation impedance is given as $Z_r^{A;a}$.

B.2 Thévenin's Theorem

An ultrasonic system contains both active and passive electrical and electromechanical elements. The pulser, for example is an active electrical network since it contains the driving elements of the ultrasonic system.

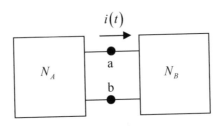

Fig. B.4. An electrical network with sources connected to a passive network.

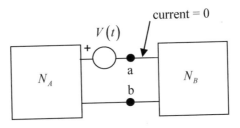

Fig. B.5. Introduction of a voltage source that makes the current flow between the two networks zero.

Cables and transducers are passive networks since they merely transfer and/or transform energy but do not generate it. To model in detail an active electrical network like a pulser would be a very challenging task since a pulser is a very complex set of circuits. If we assume the pulser acts as a linear device then Thévenin's theorem allows us to avoid this complexity and replace a pulser with a simple equivalent circuit consisting of a voltage source and electrical impedance in series. Here, we will outline briefly the proof of this important theorem [B.2].

Consider an electrical network, N_A, that contains both passive elements and sources and connect it at its terminals to a network, N_B, that is passive (no sources) as shown in Fig. B.4. Let $i(t)$ be the current flowing between two networks at terminals a-b. It is assumed here that both N_A and N_B are linear networks. Now introduce an opposing voltage source, $V(t)$, in front of the terminals a-b such that the current is driven to zero, as shown in Fig. B.5. Since now there is no current flowing between N_A and

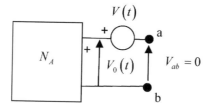

Fig. B.6. The network N_A detached from the passive network N_B.

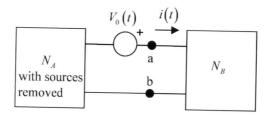

Fig. B.7. Re-attaching the network A with all the sources replaced by $V_0(t)$.

N_B, the voltage across terminals a-b, V_{ab}, is zero and we can break the circuit at a-b , as shown in Fig. B.6, without disturbing any voltages or currents. Because there is no current flowing out of the network N_A in Fig. B.6 the voltage $V_0(t)$ in that figure is the output voltage of the network N_A under open-circuit conditions and it follows that

$$V_0(t) - V(t) = V_{ab} = 0, \tag{B.10}$$

which shows that $V(t)$ is just the open-circuit voltage, $V_0(t)$, of network N_A. If we now reverse the polarity of $V(t)$ and remove all the sources in N_A (by replacing them with short circuits), when we reattach N_A to N_B the original current will be set up between N_A and N_B, as shown in Fig. B.7. Since N_B is a passive network, the voltage across terminals a-b will also be the same as in the original setup. Thus, we can say that the

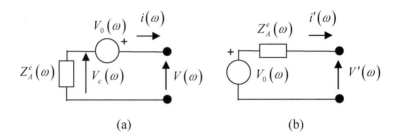

Fig. B.8. (a) The Thévenin equivalent network of Fig B.7 represented in the frequency domain by a complex source and impedance, and **(b)** the same equivalent circuit with the source and impedance exchanged.

circuit shown in Fig. B.7 is equivalent to the original circuit of Fig. B.4. This is the essence of Thévenin's theorem which states that:

A two terminal network containing sources and passive elements is equivalent (as far as its external effects are concerned) to a voltage source in series with the network with all the sources removed; the voltage of the equivalent source has the same magnitude and polarity as those of the voltage appearing across the terminals of the original network under open-circuit conditions.

If we Fourier transform all the voltages and currents appearing in these networks and work in the frequency domain, then the original network with it sources removed is equivalent to a complex electrical impedance, $Z_A^e(\omega)$, and we can replace our original network with the equivalent circuit of Fig. B.8 (a). However, from that figure we have the two equations

$$V_c(\omega) = Z_A^e(\omega)i(\omega)$$
$$V(\omega) - V_c(\omega) = V_0(\omega),$$

(B.11)

which give

$$V(\omega) - V_0(\omega) = Z_A^e(\omega)i(\omega).$$

(B.12)

It is often customary to exchange the positions of the source and impedance so that the source "drives" that impedance, as shown in Fig. B.8 (b). If we let $V' = -V, i' = -i$ and $V_0 = -V_0$ in Eq. (B.12) then that equation still holds for the equivalent circuit shown in Fig. B.8 (b). Thus,

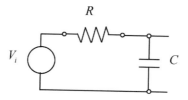

Fig. B.9. An RC-circuit and voltage source.

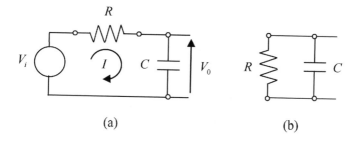

(a)   (b)

Fig.B.10. (a) The RC-circuit showing the open-circuit voltage, V_0, and the current, I, flowing in the circuit, and **(b)** the source-free circuit that must be placed in series with the open-circuit voltage to obtain the Thévenin equivalent circuit.

the complex source and impedance of Fig. B.8 (b) is the Thévenin equivalent circuit corresponding to our original network.

Example: Consider the simple circuit shown in Fig. B.9 where a voltage source with frequency components $V_i(\omega)$ is connected to a resistance, R, and a capacitance, C. Determine the Thévenin equivalent source and impedance that replaces this circuit.

Consider the open-circuit voltage, $V_0(\omega)$, and the current, $I(\omega)$, in the circuit as shown in Fig. B.10 (a). We have the relations

$$V_i - V_0 = IR$$

$$V_0 = \frac{I}{-i\omega C}. \tag{B.13}$$

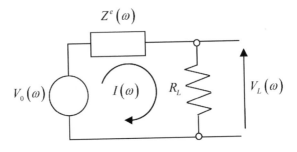

Fig. B.11. The Thévenin equivalent circuit for a pulser attached to a known external resistance, R_L, for measuring the impedance, $Z^e(\omega)$.

If we eliminate I from these two equations we find the Thévenin equivalent source is

$$V_0(\omega) = \frac{V_i(\omega)}{1 - i\omega RC}. \tag{B.14}$$

This circuit in series with the original circuit with the sources removed (short-circuited) is our Thévenin equivalent circuit. The source-free circuit is shown in Fig. B.10 (b) where we see that we just have the resistor and capacitor in parallel. Thus, the equivalent impedance, Z^e, of this source-free circuit is just

$$\frac{1}{Z^e(\omega)} = \frac{1}{R} + \frac{1}{(1/-i\omega C)}, \tag{B.15}$$

which gives

$$Z^e(\omega) = \frac{R}{1 - i\omega RC}. \tag{B.16}$$

B.3 Measurement of Equivalent Sources and Impedances

A pulser in an ultrasonic measurement system is an example of an electrical network containing sources. As shown in Chapter 2 if we assume

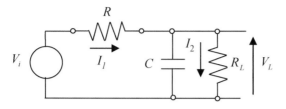

Fig. B.12. A setup for determining the impedance of the RC-circuit.

the pulser acts as a linear system we can find the Thévenin equivalent source for the pulser, $V_0(\omega)$, by measuring the open-circuit voltage, $V_0(t)$, at the output terminals of the pulser and Fourier transforming this measured voltage to obtain $V_0(\omega)$. But how do we find the equivalent impedance, $Z^e(\omega)$, of a real instrument such as a pulser since we cannot go into the instrument and physically short circuit the sources, as we did with the known circuit in the previous example? Instead, as shown in Chapter 2, we can place a known load resistance, R_L, at the output terminals of the pulser and measure the voltage, $V_L(t)$, across this load. Fourier transforming this voltage then gives $V_L(\omega)$ (see Fig. B.11). But from the Thévenin equivalent circuit of the pulser shown in Fig. B.11 (see also Chapter 2), we find that

$$V_0 - V_L = Z^e I$$
$$V_L = R_L I.$$
(B.17)

So eliminating the current, I, we find

$$Z^e(\omega) = R_L\left(\frac{V_0(\omega)}{V_L(\omega)} - 1\right).$$
(B.18)

Equation (B.18) shows that with measurements of both $V_0(\omega)$ and $V_L(\omega)$, with the resistance, R_L, known, we can determine the Thévenin equivalent impedance, $Z^e(\omega)$. Note that this impedance does not depend on the value of the known resistance. It is only a function of the properties of the pulser itself. We can demonstrate this method for determining the impedance for our RC-circuit example again. Figure B.12 shows that circuit and the resistance, R_L, at its output terminals. From Fig. B.12, we have

$$V_i - V_L = I_1 R$$
$$V_L = I_2 R_L$$
$$V_L = \frac{(I_1 - I_2)}{-i\omega C}. \tag{B.19}$$

Eliminating the currents I_1, I_2 from these equations gives

$$V_L = \frac{V_i}{(1 - i\omega RC) + R/R_L}. \tag{B.20}$$

Using Eq. (B.20) and Eq. (B.14) for the Thévenin equivalent source, we find

$$Z^e(\omega) = R_L \left(\frac{V_0(\omega)}{V_L(\omega)} - 1 \right) = R_L \left\{ \frac{(1 - i\omega RC) + R/R_L}{(1 - i\omega RC)} - 1 \right\}$$
$$= R_L \left\{ \frac{R/R_L}{(1 - i\omega RC)} \right\} = \frac{R}{(1 - i\omega RC)}, \tag{B.21}$$

which is the same value for the impedance obtained earlier in Eq. (B.16) and is indeed independent of R_L.

B.4 References

B.1 Cremer L, Heckl M, Ungar EE (1973) Structure-borne sound. Springer-Verlag, Berlin, Germany
B.2 Cheng DK (1959) Analysis of linear systems. Addison Wesley, Reading, PA

B.5 Exercises

1. A propagating harmonic spherical pressure wave from a point source in a fluid is given by $p = P \exp(ikr - i\omega t)/r$, where r is the radial distance from the source (see Fig. B.13). The radial velocity can also similarly be

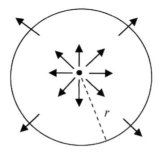

Fig. B.13. A spherical wave arising from a point source in a fluid.

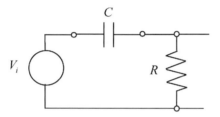

Fig. B.14. An example circuit.

written as $v_r = V \exp(ikr - i\omega t)/r$. If the equation of motion of the fluid in spherical coordinates is given by

$$-\frac{\partial p}{\partial r} = \rho \frac{\partial v_r}{\partial t},$$

determine the specific acoustic impedance P/V of this spherical wave. What happens to this impedance when the frequency, ω, is very large?

2. For the circuit shown in Fig. B.14 obtain the Thévenin equivalent source and impedance in terms of the given circuit elements.

C Linear System Fundamentals

In this book an ultrasonic system is modeled as a series of interconnected linear systems. Thus, linear system theory will be a fundamental part of all our discussions. This Appendix will outline a number of key linear systems concepts such as two port systems and linear time-shift invariant systems. We will also discuss the role that the convolution theorem plays in linear systems as well as related quantities such as impulse response functions and transfer functions.

C.1 Two Port Systems

The pulser in an ultrasonic system is an active circuit (a circuit with sources) that drives the rest of the ultrasonic system through the pulser output port. The cabling and transducer(s) in an ultrasonic system normally are passive elements (no sources) and they contain both input and output ports, as shown in Fig. C.1. In the case of a cable, it is purely an electrical system so the inputs and outputs are both of the same type (voltage, current). An ultrasonic transducer transforms voltage, V, and current, I, at its electrical port into a mechanical force, F, (arising from a pressure distribution on the face of a piezoelectric crystal as shown in Fig. C.1 (b)) and a velocity, v, (which represents the average velocity of motion of the crystal) at its acoustic port. The underlying velocity distribution is shown in Fig. C.1 (b) as being uniform at the acoustic port. A transducer with this type of velocity profile is called a *piston transducer*. [Note: piston transducer models have been shown to often be very effective for modeling real commercial ultrasonic transducers but one should be aware that this idealized model may not be suitable for all transducers. In this book we will generally assume a piston transducer model is valid].

We can represent a purely electrical two port system such as a cable schematically as shown in Fig. C.2 [C.1]. Note that it is customary to assign the currents so that they flow into the two port system on the input side and flow out on the output side, a convention that we will also follow here.

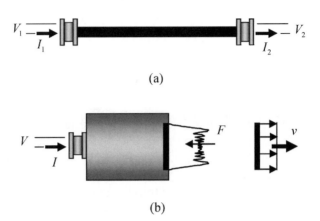

(a)

(b)

Fig. C.1. **(a)** Cabling as a two port electrical system, and **(b)** an ultrasonic transducer as two port system with voltage, V, and current, I, at the electrical port and force, F, and velocity, v, at the acoustic port. The force F is the net compressive force generated by the pressure distribution acting across the face of the transducer and v is the average velocity due to the velocity distribution of the transducer face. The pressure distribution is non-uniform, as shown, but the velocity distribution is taken to be the uniform velocity profile of a piston transducer.

Since we will assume this is a linear system the inputs and outputs are proportional to each other through a 2x2 transfer matrix, $[\mathbf{T}]$, where

$$\begin{Bmatrix} V_1(\omega) \\ I_1(\omega) \end{Bmatrix} = \begin{bmatrix} T_{11} & T_{12} \\ T_{21} & T_{22} \end{bmatrix} \begin{Bmatrix} V_2(\omega) \\ I_2(\omega) \end{Bmatrix}. \tag{C.1}$$

The dimensions of the elements of the transfer matrix are: T_{11}, T_{22}: dimensionless, T_{12} :ohms, T_{21} :1/ohms. Note that the voltages and currents in Eq. (C.1) are all in the frequency domain, i.e. they are the Fourier transforms of the time varying voltages and currents present at the input and output ports. Thus, the transfer matrix is also in the frequency domain. Another common way to represent a two port system is in terms of a 2x2 impedance matrix, $[\mathbf{Z}^e]$. In this case it is usual to represent the currents on both sides of the two port system as flowing into the system, as shown in Fig. C.3, and write:

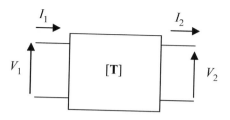

Fig. C.2. An electrical two port system represented by a transfer matrix. $[\mathbf{T}]$.

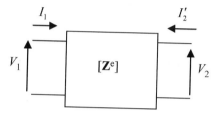

Fig. C.3. An electrical two port system represented by an impedance matrix, $[\mathbf{Z}^e]$.

$$\begin{Bmatrix} V_1(\omega) \\ V_2(\omega) \end{Bmatrix} = \begin{bmatrix} Z_{11}^e & Z_{12}^e \\ Z_{21}^e & Z_{22}^e \end{bmatrix} \begin{Bmatrix} I_1(\omega) \\ I_2'(\omega) \end{Bmatrix}, \tag{C.2}$$

where $I_2' = -I_2$. In this case the dimensions of the elements of the impedance matrix are all ohms. In addition to linearity, we will assume that a two port system is reciprocal. The meaning of reciprocity is as follows. Consider a two port system, characterized by its transfer matrix $[\mathbf{T}]$ (or, equivalently, by its impedance matrix, $[\mathbf{Z}^e]$). Let us attach this two port system to electrical networks A and B at its input and output terminals, respectively, as shown in Fig. C.4. We will call this connected set of systems state (1). Under these conditions the voltage and current at the input port are $V_1^{(1)}, I_1^{(1)}$ and the voltage and current at the output port

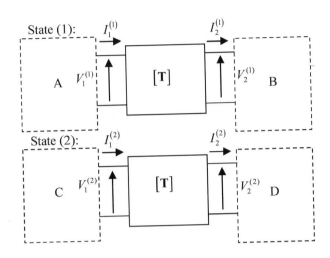

Fig. C.4. A two port system with its ports terminated differently in two states, labeled states (1) and (2).

are $V_2^{(1)}, I_2^{(1)}$. Now, attach the same two port system to two other networks C and D, as shown in Fig. C.4. Call this connected set of systems state (2). Then in this state we have $V_1^{(2)}, I_1^{(2)}$ and $V_2^{(2)}, I_2^{(2)}$ for the voltages and currents at the input and output port, respectively. Our two port system is said to be reciprocal if for any two states (1) and (2) the inputs and outputs satisfy the reciprocity relation given by

$$V_1^{(1)} I_1^{(2)} - V_1^{(2)} I_1^{(1)} = V_2^{(1)} I_2^{(2)} - V_2^{(2)} I_2^{(1)}. \tag{C.3}$$

Equation (C.3) is a rather "opaque" equation in that it is difficult to see what it really means. However, when it is applied to our two port system written in terms of its impedance matrix, one can show that reciprocity simply implies that the impedance matrix is symmetric, i.e. $Z_{21}^e = Z_{12}^e$ [C.1]. Similarly, Eq. (C.3) implies that determinant of the transfer matrix of the two port system equals one, i.e. $\det[\mathbf{T}] = T_{11}T_{22} - T_{12}T_{21} = 1$ [C.1].

For a linear, reciprocal two port system the components of the transfer matrix and the impedance matrix are obviously related. It is not difficult to show that the transfer matrix can be expressed in terms of the impedance matrix components as:

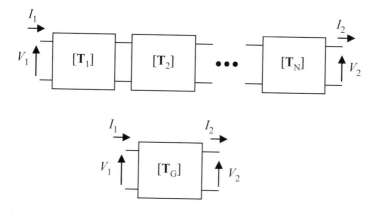

Fig. C.5. A cascade of linear, reciprocal two port systems and their replacement by a single "global" two port system.

$$[\mathbf{T}]=\begin{bmatrix} Z_{11}^e / Z_{12}^e & \left(Z_{11}^e Z_{22}^e - \left(Z_{12}^e\right)^2\right)/ Z_{12}^e \\ 1/ Z_{12}^e & Z_{22}^e / Z_{12}^e \end{bmatrix}. \tag{C.4}$$

From Eq. (C.4) it follows directly that $\det[\mathbf{T}]=1$, as it should be. Similarly, the impedance matrix can be written in terms of the transfer matrix components as:

$$\left[\mathbf{Z}^e\right]=\begin{bmatrix} T_{11}/T_{21} & 1/T_{21} \\ 1/T_{21} & T_{22}/T_{21} \end{bmatrix}, \tag{C.5}$$

which shows that $Z_{21}^e = Z_{12}^e$ is automatically satisfied.

One can express impedance components in terms of transfer matrix components and vice versa so in principle it does not matter which of these representations we use for a two port system. However, when one is dealing with a series of connected two port systems, as is the case for an ultrasonic system (e.g. the cabling is attached to the transducer, both of which are two port systems) then the transfer matrix is more convenient to use since one can replace a series of connected two port systems, each characterized by their own transfer matrices $[\mathbf{T}_1],[\mathbf{T}_2],....,[\mathbf{T}_N]$ as shown in Fig. C.5, by a single global 2x2 transfer matrix , $[\mathbf{T}_G]$, where the global

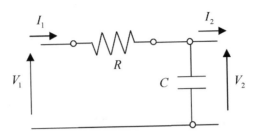

Fig. C.6. An RC-circuit modeled as a two port system.

matrix is obtained by matrix multiplication of each of the individual transfer matrices, i.e.

$$[\mathbf{T}_G] = [\mathbf{T}_1][\mathbf{T}_2]\dots[\mathbf{T}_N].$$ (C.6)

This global transfer matrix is also reciprocal if the individual transfer matrices are reciprocal since

$$\det[\mathbf{T}_G] = \det[\mathbf{T}_1]\det[\mathbf{T}_2]\dots\det[\mathbf{T}_N] = 1.$$ (C.7)

As a simple example of a linear, reciprocal two port system, consider the RC-circuit example used in Appendix B with the voltage source removed to form the two port system shown in Fig. C.6. To determine the transfer matrix for this circuit, consider first the voltage across the resistance and the current flowing through it. We have

$$V_1 - V_2 = RI_1.$$ (C.8)

Also, considering the voltage across the capacitor and the current flowing through it (which is $I_1 - I_2$ flowing downwards) we find

$$V_2 = \frac{I_1 - I_2}{-i\omega C}.$$ (C.9)

Equation (C.9) can be written directly in transfer matrix form (inputs in terms of outputs) as

$$I_1 = -i\omega C V_2 + I_2.$$ (C.10)

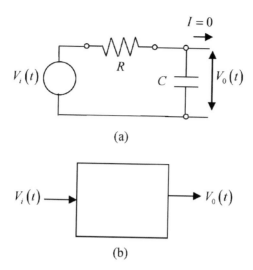

(a)

(b)

Fig. C.7. (a) An RC-circuit with a voltage source at the input and open-circuit conditions at the output, and (b) the representation of this terminated system as a single input-single output system.

If we now place Eq. (C.10) into Eq. (C.8), the resulting equation can also be placed in transfer matrix form as

$$V_1 = (1 - i\omega RC)V_2 + RI_2. \tag{C.11}$$

From Eqs. (C.10) and (C.11) the transfer matrix follows directly, giving

$$\begin{Bmatrix} V_1 \\ I_1 \end{Bmatrix} = \begin{bmatrix} (1 - i\omega RC) & R \\ -i\omega C & 1 \end{bmatrix} \begin{Bmatrix} V_2 \\ I_2 \end{Bmatrix}. \tag{C.12}$$

Equation (C.12) shows that $\det[\mathbf{T}] = 1$ is indeed satisfied for this system so that it is reciprocal. Using Eq. (C.5) we can also obtain the impedance matrix directly for this two port system, where

$$\begin{Bmatrix} V_1 \\ V_2 \end{Bmatrix} = \begin{bmatrix} (i\omega RC - 1)/i\omega C & -1/i\omega C \\ -1/i\omega C & -1/i\omega C \end{bmatrix} \begin{Bmatrix} I_1 \\ -I_2 \end{Bmatrix}, \tag{C.13}$$

and obviously we also have $Z_{21}^e = Z_{12}^e$.

Fig. C.8. A general linear time-shift invariant (LTI) system.

C.2 Linear Time-Shift Invariant (LTI) Systems

If we have a two port linear system that is terminated in some fashion at both its ports, then this two port system reduces to a system where single inputs and outputs can be linearly related to each other [C.2], [C.3], [C.4]. As an example, consider again the RC-circuit two port system of Fig. C.6. If we attach a voltage source $V_i(t)$ at its input port and leave the output port open-circuited (Fig. C.7 (a)), we have a linear system where we can relate the open-circuit voltage, $V_0(t)$, to the input voltage, $V_i(t)$. This type of single input-single output system can be represented schematically as shown in Fig. C.7 (b). For this simple system it is easy to see that

$$V_i(t) - V_0(t) = i(t)R$$
$$i(t) = C\frac{dV_0}{dt}. \tag{C.14}$$

Eliminating the current between the two equations in Eq. (C.14), we see that V_0 is related implicitly to V_i through the solution of the differential equation given by

$$\frac{dV_0(t)}{dt} + \frac{V_0(t)}{RC} = \frac{V_i(t)}{RC}. \tag{C.15}$$

We can write this relation symbolically as

$$V_0(t) = L[V_i(t)], \tag{C.16}$$

where $L[\ ]$ is a linear operator since the underlying RC-circuit is linear.

An important class of linear single input, single output systems is called a linear time-shift invariant (LTI) system, as shown schematically in

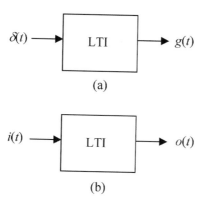

(a)

(b)

Fig. C.9. (a) An LTI system driven by a delta function input, and **(b)** driven by a general input.

Fig. C.8. for a general input, $i(t)$, and output, $o(t)$. An LTI system is defined as a linear system where a time shift of the input signal produces exactly the same time shift of the output signal. These properties can be stated mathematically as follows:

Linearity:

If
$$o_1(t) = L\big[i_1(t)\big]$$
$$o_2(t) = L\big[i_2(t)\big]$$
then
$$o(t) = L\big[a_1 i_1(t) + a_2 i_2(t)\big]$$
$$= a_1 L\big[i_1(t)\big] + a_2 L\big[i_2(t)\big]$$
(C.17)

Time-Shift Invariance:

If
$$o(t) = L\big[i(t)\big]$$
then
$$o(t - t_0) = L\big[i(t - t_0)\big]$$
(C.18)

It is clear that The RC-circuit example just considered is an LTI system. We expect that elements of an ultrasonic NDE system in general may also be modeled as LTI systems. LTI systems have the important property that they can be characterized completely by their response to a delta function input, $\delta(t)$. This delta function response is called the *impulse response function*, $g(t)$, of the system, and the Fourier transform of this impulse response, $G(f)$, we will call the *transfer function* of the LTI system. Figure C.9 (a) shows an LTI system being driven by a delta function input,

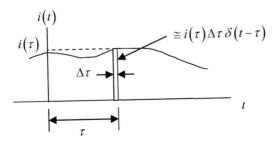

Fig. C.10. Representing a general input function as a superposition of delta function inputs.

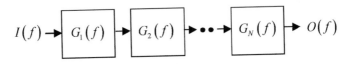

Fig. C.11. A series of LTI systems.

while Fig. C.9 (b) shows the same system under a general input. It can be shown that the output, $o(t)$, of an LTI system to a general input, $i(t)$, is given in terms of a convolution integral of that input with the impulse response function, $g(t)$, i.e.

$$o(t) = \int_{-\infty}^{+\infty} i(\tau)g(t-\tau)d\tau$$

$$= \int_{-\infty}^{+\infty} g(\tau)i(t-\tau)d\tau.$$

(C.19)

Equation (C.19) follows directly from the properties of an LTI system since we can take a general input function and consider it as a super-position of small rectangular elements as shown in Fig. C.10. A general rectangular element at time τ of width $\Delta\tau$ and amplitude $i(\tau)$ is shown in that figure. This element, however, acts like a shifted delta function (located at $t = \tau$) a with strength (area) $i(\tau)\Delta\tau$. Thus, from the linearity and time shift invariance properties of the system, we have that the output, Δo, from this rectangular element is given by $\Delta o(t) \cong i(\tau)\Delta\tau g(t-\tau)$

and so by superposition over all elements, we have the total output, $o(t)$, due to the total input given by

$$o(t) \cong \sum i(\tau) \Delta \tau g(t - \tau)$$

$$= \int_{-\infty}^{+\infty} i(\tau) g(t - \tau) d\tau. \tag{C.20}$$

The convolution integral of Eq. (C.19) is a fundamental relationship for LTI systems. If we take the Fourier transform of this relationship we obtain an even simpler result since, if we define the following Fourier transforms of the input, output, and impulse response functions, respectively:

$$I(f) = \int_{-\infty}^{+\infty} i(t) \exp(2\pi i f t) dt$$

$$O(f) = \int_{-\infty}^{+\infty} o(t) \exp(2\pi i f t) dt \tag{C.21}$$

$$G(f) = \int_{-\infty}^{+\infty} g(t) \exp(2\pi i f t) dt$$

and if the output and input are related through the convolution integral of Eq. (C.19), then it is easy to show that their Fourier transforms are related through [Fundamentals]

$$O(f) = G(f) I(f), \tag{C.22}$$

i.e. convolution in the frequency domain is just obtained by complex-valued multiplication. In a similar fashion, *deconvolution* in the frequency domain is in principle accomplished by complex-valued division. For example, we can write

$$G(f) = \frac{O(f)}{I(f)}. \tag{C.23}$$

In practice, however, such division must be done with care since noise may contaminate both the numerator and denominator and make the ratio unreliable. Often filters are used to desensitize the deconvolution process to such errors. A *Wiener filter* is a particular filter commonly used for deconvolution purposes in ultrasonic NDE. With that filter, Eq. (C.23) is replaced by

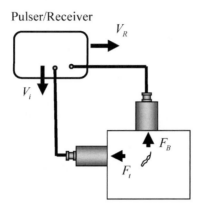

Fig. C.12. An ultrasonic flaw measurement system.

$$G(f) = \frac{O(f)I^*(f)}{\left|I(f)\right|^2 + \varepsilon^2 \max\left\{\left|I(f)\right|^2\right\}},$$
(C.24)

where $(\ )^*$ denotes the complex conjugate and ε is a small constant that is used to represent the noise level present. The quantity $\max\left\{\left|I(f)\right|^2\right\}$ is a constant. It is the maximum value of the magnitude squared of the values of $I(f)$ present. In this form ε gives a measure of the noise as a fraction of the size of the signals present. Generally, small values such as $\varepsilon = 0.01$ to 0.05 work well for many ultrasonic problems. When $\varepsilon \to 0$, Eq. (C.24) reduces to Eq. (C.23). Code listing C.1 gives a MATLAB function for implementing the Wiener filter of Eq. (C.24).The use of transfer functions such as $G(f)$ is very convenient, particularly when we have a series of connected LTI systems as shown in Fig. C.11 since the input and output of the entire system can be related through simply a product of the transfer functions of each subsystem, i.e.

$$O(f) = G_1(f)G_2(f)\cdots G_N(f)I(f).$$
(C.25)

In the time domain, the relationship equivalent to Eq. (C.25) would be a series of nested convolution integrals. By working in the frequency domain we can avoid having to deal with multiple integrations and instead we need

Code Listing C.1. A MATLAB function for the generation of a Wiener filter.

```
function Y = Wiener_filter( O, I, e)
% WIENER_FILTER provides a 1-D filter for desensitizing
% division in the frequency domain (deconvolution) to noise.
% The filter takes a sampled output spectrum ,O, and an
% input spectrum, I, and computes Y = O*conj(I)/(|I|^2 + e^2*M^2)
% where M is the maximum value of |I| and conj(I) is the
% complex conjugate of I. The constant e is generally taken as
% a constant to represent the noise level. Small values of e
% such as e = 0.01 often work well for ultrasonic systems.
%The calling sequence is:
%Y = Wiener_filter(O,I,e);
%
M = max(abs(I));
Y = O.*conj(I)./((abs(I)).^2 + e^2*M^2);
```

only a series of complex multiplications to obtain $O(f)$ from the input. The time domain output, $o(t)$, can then be obtained by an inverse Fourier transform.

As an example of such a cascade of LTI systems, consider an ultrasonic pitch-catch flaw measurement system, as shown in Fig. C.12. Let $V_i(f)$ be the frequency components of the Thévenin equivalent input voltage of the pulser. This input then travels through the sending cable and drives the sending transducer which outputs a mechanical force, $F_t(f)$ at its acoustic port. This force launches a wave into the specimen which then interacts with a flaw and in turn produces a driving force, $F_B(f)$, on the receiving transducer. This driving force is converted into electrical energy which is transmitted by the receiving cable back to the receiver, where it is amplified and output as the received flaw signal, $V_R(f)$. If we treat this entire measurement system as a series of LTI systems, then we can write:

$$V_R(f) = \frac{V_R(f)}{F_B(f)} \frac{F_B(f)}{F_t(f)} \frac{F_t(f)}{V_i(f)} V_i(f)$$
$$= t_R(f) t_A(f) t_G(f) V_i(f),$$

(C.26)

where $t_G(f)$ is the transfer function for the sound generation process (containing properties of the pulser, cabling, and sending transducer), $t_R(f)$ is the transfer function for the sound reception process (containing properties of the receiving transducer, cabling, and receiver), and $t_A(f)$ is the transfer function describing the acoustic/elastic processes (wave propagation to the flaw, scattering from the flaw, and propagation from the flaw to the receiving transducer). We will see that it is possible to model and/or measure all of these transfer functions so that through Eq. (C.26) we have an *ultrasonic measurement model* of our entire ultrasonic system. The challenge, of course, is to obtain explicit expressions for the transfer functions in Eq. (C.26). Much of this book is devoted to just that task.

C.3 References

C.1 Pozar DM (1998) Microwave engineering, 2nd ed. John Wiley and Sons, New York, NY

C.2 Cheng DK (1959) Analysis of linear systems. Addison Wesley, Reading, PA

C.3 Gaskill JD (1978) Linear systems, transforms, and optics. McGraw-Hill, New York, NY

C.4 Papoulis A (1968) Systems and transforms with applications in optics. McGraw-Hill, New York, NY

C.4 Exercises

1. Consider a two port electrical system where

$$\begin{Bmatrix} V_1(\omega) \\ I_1(\omega) \end{Bmatrix} = \begin{bmatrix} T_{11} & T_{12} \\ T_{21} & T_{22} \end{bmatrix} \begin{Bmatrix} V_2(\omega) \\ I_2(\omega) \end{Bmatrix}.$$

We wish to measure the transfer matrix components (as a function of frequency). This is easy to do if we first measure the inputs and outputs under open-circuit conditions at the output port since $I_2 = 0$. Thus, if we let the voltages and currents be $\left(V_1^{oc}, I_1^{oc} \right)$ and $\left(V_2^{oc}, I_2^{oc} = 0 \right)$ we have

$$T_{11} = \frac{V_1^{oc}}{V_2^{oc}}, T_{21} = \frac{I_1^{oc}}{V_2^{oc}}.$$

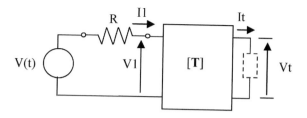

Fig. C.13. A measurement setup for obtaining the transfer matrix components of a two port system using different output terminations.

Similarly, if under short- circuit conditions at the output port ($V_2 = 0$) we measure the voltages and currents $\left(V_1^{sc}, I_1^{sc}\right)$, $\left(V_2^{sc} = 0, I_2^{sc}\right)$ we have

$$T_{12} = \frac{V_1^{sc}}{I_2^{sc}}, T_{22} = \frac{I_1^{sc}}{I_2^{sc}}.$$

Now, perform these "measurements" in MATLAB for an unknown two port system, two_portX, which is written in terms of a MATLAB function which has the calling sequence:

```
>> [ v1, i1, vt, it] = two_portX( V, dt, R, 'term');
```

The input arguments of two_portX are as follows. V is a sampled voltage source versus time, where the sampling interval is dt. R is an external resistance (in ohms). This source and resistance are connected in series to one end of the two port system as shown in Fig. C.13. The other end of the system can be either open-circuited or short-circuited. The string 'term' specifies the termination conditions. It can be either 'oc' for open-circuit or 'sc' for short-circuit. The function two_portX then returns the "measured" sampled voltages and currents versus time: v1, i1, vt, it (Note: for open-circuit conditions the function returns it = 0 and for short-circuit conditions vt = 0).

As a voltage source to supply the V input to two_portX use the MATLAB function pulserVT. For a set of sampled times this function returns a sampled voltage output that is typical of a "spike" pulser. Make a vector, t, of 512 sampled times ranging from 0 to 5 µsec, and call the pulserVT function with the following call sequence:

```
>> V = pulserVT(200, 0.05, 0.2, 12, t);
```

For the resistance, take R = 200 ohms. Using Eq. (C.14), determine the four transfer matrix components and plot their magnitude and phase from zero to approximately 30 MHz. Note that the outputs of two_portX are all time domain signals but the quantities we wish to measure are all in the frequency domain.

2. It is not physically possible to generate a delta function as the input of an LTI system to obtain its impulse response. However, it is possible to obtain the transfer function of an LTI system by deconvolution of a measured output with a known input as shown in Eq. (C.23). Consider a MATLAB function LTI_X that represents a "black box" LTI system. It can be evaluated in the form

```
>> O =LTI_X(I, dt)
```

Where I is a sampled time domain input (with sampling interval dt) and O is the time domain output. Use as an input for this LTI system the voltage output of the pulserVT function of problem 1 and obtain the transfer function of this system as a function of frequency by deconvolution. Plot the magnitude and phase (in degrees) of this transfer function from zero to approximately 30 MHz. To obtain the impulse response function from this transfer function we would have to compute its inverse Fourier transform. Is that possible with this function?

3. Consider an LTI system which has as its transfer function

$$G(f) = \begin{cases} \cos(\pi f / 40) & 0 < f < 20 MHz \\ 0 & otherwise \end{cases}.$$

Also, consider an input spectrum to this system given by

$$I(f) = \begin{cases} 1 - f / 20 & 0 < f < 20 MHz \\ 0 & otherwise \end{cases}.$$

We expect the output of this system will then have the spectrum

$$O(f) = G(f)I(f).$$

However, if we add noise to these functions then given O and I it may not be possible to reliably obtain G by simple division and we must use some filter instead such as the Wiener filter. The MATLAB function noisy will generate noisy sampled versions of both the O and I given previously over a range 0-40 MHz. The function call is:

```
>> [O, I] = noisy( ) ;
```

Plot both O and I from 0 to 40 MHz to verify those functions are correct (the noise you will see is very small) and then attempt to obtain G by direct division, i.e. compute

$$G(f) = \frac{O(f)}{I(f)}$$

and plot your results 0-40 MHz. Then use a Wiener filter instead to find G and plot your results. Take $\varepsilon = 0.01$. Are the results sensitive to ε? Is there any other way (besides using the Wiener filter) that you can get the "right" answer?

D Wave Propagation Fundamentals

D.1 Waves in a Fluid

In immersion testing the waves are generated by a transducer radiating sound into a fluid. Sound propagation in the fluid can be modeled by considering the fluid to be an ideal (viscous-free) compressible fluid. In this case, an element of the fluid only has pressures (compressive normal stresses) acting on its surfaces. If a wave in the fluid generates pressure changes in the x_1-direction, as shown in Fig. D.1, then we can relate those changes to the motion of the fluid by simply applying Newton's third law to a small element as shown in Fig. D.1 [Fundamentals], [D.1]. We find from

$$\sum F_x = ma_x \qquad (D.1a)$$

that

$$p dx_2 dx_3 - \left(p + \frac{\partial p}{\partial x_1} dx_1 \right) dx_2 dx_3 + f_1 dx_1 dx_2 dx_3 = \rho dx_1 dx_2 dx_3 \frac{\partial^2 u_1}{\partial t^2}, \qquad (D.1b)$$

which gives the equation of motion of the fluid in the x_1-direction as

$$-\frac{\partial p}{\partial x_1} + f_1 = \rho \frac{\partial^2 u_1}{\partial t^2}, \qquad (D.1c)$$

where p is the pressure, ρ is the density of the fluid, u_1 is the displacement in the x_1-direction and f_1 is the body force (force/unit volume) acting on the fluid. Similarly, if we consider the pressure changes in the x_2, x_3 directions we find the equations of motion:

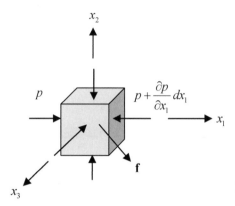

Fig. D.1. The pressures and body force acting on an element of an ideal, compressible fluid. Only the pressure changes in the x_1-direction are shown explicitly.

$$-\frac{\partial p}{\partial x_2} + f_2 = \rho \frac{\partial^2 u_2}{\partial t^2}$$

$$-\frac{\partial p}{\partial x_3} + f_3 = \rho \frac{\partial^2 u_3}{\partial t^2}. \tag{D.2}$$

These three equations of motion of the fluid can also be written in vector form as

$$-\nabla p + \mathbf{f} = \rho \frac{\partial^2 \mathbf{u}}{\partial t^2} = \rho \frac{\partial \mathbf{v}}{\partial t}, \tag{D.3}$$

where $\mathbf{v} = \partial \mathbf{u}/\partial t$ is the velocity of the fluid. If we assume that the fluid is an ideal compressible fluid, then the pressure is related to the relative volume changes, dV/V, occurring in the fluid through the constitutive equation

$$p = -\lambda \frac{dV}{V}, \tag{D.4}$$

where λ is the *bulk modulus* of the fluid. For water, for example, the bulk modulus is approximately 2 GPa. These relative volume changes can be written in terms of the displacements, so we also have

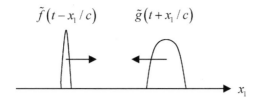

$$\tilde{f}(t - x_1/c) \qquad \tilde{g}(t + x_1/c)$$

Fig. D.2. 1-D waves traveling in a fluid.

$$p = -\lambda \nabla \cdot \mathbf{u} = -\lambda \left(\frac{\partial u_1}{\partial x_1} + \frac{\partial u_2}{\partial x_2} + \frac{\partial u_3}{\partial x_3} \right). \tag{D.5}$$

To place this constitutive equation in the equations of motion, we first take the divergence ($\nabla \cdot$) of Eq. (D.3) which gives

$$-\nabla^2 p - f_b = \rho \frac{\partial^2 (\nabla \cdot \mathbf{u})}{\partial t^2}, \tag{D.6}$$

where $f_b = -\nabla \cdot \mathbf{f}$ and $\nabla^2 = \partial^2/\partial x_1^2 + \partial^2/\partial x_2^2 + \partial^2/\partial x_3^2$ is the Laplacian operator. We then can place Eq. (D.5) into Eq. (D.6) to obtain the inhomogeneous wave equation for the pressure given by

$$\nabla^2 p - \frac{1}{c^2} \frac{\partial^2 p}{\partial t^2} = -f_b, \tag{D.7}$$

where $c = \sqrt{\lambda/\rho}$ is the wave speed of compressional waves (also called P-waves) in the fluid. For water, for example, $c = 1480$ m/sec, approximately. In NDE tests the ultrasonic waves that are generated are freely traveling so that they must satisfy the homogeneous wave equation, i.e. where $f_b = 0$.

D.2 Plane Waves in a Fluid

If we consider 1-D disturbances of the fluid where $p = p(x_1, t)$, then these disturbances must satisfy the 1-D homogeneous wave equation:

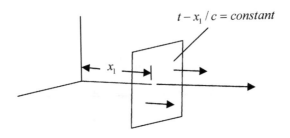

Fig. D.3. A plane wave traveling along the x_1-direction.

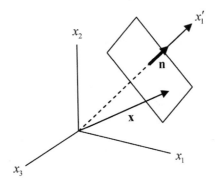

Fig. D.4. A plane wave traveling in a general direction, **n**, in three dimensions.

$$\frac{\partial^2 p}{\partial x_1^2} - \frac{1}{c^2}\frac{\partial^2 p}{\partial t^2} = 0,$$
(D.8)

which has general solutions of the form $p = \tilde{f}(t - x_1/c) + \tilde{g}(t + x_1/c)$. The $\tilde{f}$ function represents a wave traveling in the plus x_1-direction while the $\tilde{g}$ function represents a wave traveling in the negative x_1-direction, as shown in Fig. D.2. Consider the pressure wave $p = \tilde{f}(t - x_1/c)$. The pressure in this 1-D wave is constant on the moving plane $t - x_1/c = constant$, so this is a plane wave traveling in the fluid (see Fig. D.3). Now, consider this plane wave traveling along an x_1'-axis which is oriented along the **n** direction as shown in Fig. D.4 (**n** is a unit vector). Then this plane wave can be written as

$$p = \tilde{f}\left(t - x_1'/c\right)$$
$$= \tilde{f}\left(t - \mathbf{x} \cdot \mathbf{n}/c\right),$$

(D.9)

which is the general expression for a plane wave traveling in the **n**-direction in three dimensions, where here $\mathbf{x} = (x_1, x_2, x_3)$. It can be easily verified that this 3-D plane wave satisfies the full 3-D homogeneous wave equation for the fluid. Plane wave solutions are important types of waves since they can be used to model many of the wave propagation and wave interaction problems we encounter in ultrasonic NDE. An important special type of plane wave solution is a harmonic plane wave. For example, we can write a 1-D harmonic wave of frequency f (measured in Hz = cycles/sec) traveling in the x-direction as

$$p = F(f)\exp\left[2\pi i f\left(x/c - t\right)\right].$$

(D.10)

As discussed in Appendix A, such harmonic wave solutions can be used to synthesize an arbitrary traveling plane wave since we have the Fourier transform relationship

$$\tilde{f}\left(t - x/c\right) = \int_{-\infty}^{+\infty} F(f)\exp\left[2\pi i f\left(x/c - t\right)\right] df,$$

(D.11)

where $F(f)$ is the Fourier transform of the function $\tilde{f}(t)$. Thus, there is no loss in generality in considering harmonic plane wave solutions. We can write such 1-D harmonic plane waves in a number of forms. For example, we have

$$F(f)\exp\left[2\pi i f\left(\pm x/c - t\right)\right]$$
$$= F(f)\exp\left[ik\left(\pm x - ct\right)\right]$$
$$= F(f)\exp\left[\frac{2\pi i}{\lambda}\left(\pm x - ct\right)\right]$$
$$= F(\omega)\exp\left[i\omega\left(\pm x/c - t\right)\right],$$

(D.12)

where $\omega = 2\pi f$ is the circular frequency (in rad/sec), $k = \omega/c$ is the wave number (in rad/length) and $\lambda = 2\pi/k = c/f$ is the wave length (in length/cycle). The plus sign is for plane waves traveling in the positive x-direction and the minus sign is for waves traveling in the negative x-direction. A harmonic plane wave traveling in the plus $\mathbf{n}$-direction in three dimensions can also be written in a number of forms. The most commonly used forms seen in the literature are

$$F(\omega)\exp(ik\mathbf{n}\cdot\mathbf{x} - i\omega t)$$
$$F(\omega)\exp(i\mathbf{k}\cdot\mathbf{x} - i\omega t), \tag{D.13}$$

where $\mathbf{k} = k\,\mathbf{n}$ is a vector wave number.

Note, however, in all our forms we have used the time dependent factor $\exp(-i\omega t)$. Other authors may assume a factor $\exp(+i\omega t)$ or $\exp(+j\omega t)$ instead ($i = j = \sqrt{-1}$). In that case, we must also change the signs appropriately on the spatial terms as well. For example, $F(f)\exp[2\pi i f(-x/c + t)]$ represents a plane wave traveling in the positive x-direction.

Also note that the last form in Eq. (D.13) can alternatively be written as

$$F(\omega)\exp\left[i\left(k_x x + k_y y + k_z z - \omega t\right)\right]. \tag{D.14}$$

But we must have $k_x^2 + k_y^2 + k_z^2 = k^2 = \omega^2/c^2$ for Eq. (D.14) to represent a plane wave solution of the wave equation so we must have $k_z = \pm\sqrt{k^2 - k_x^2 - k_y^2}$ where the plus sign would represent a plane wave traveling in three dimensions in the positive z-direction while the minus sign would give a wave traveling in the negative z-direction. In Chapter 8 these forms arise when we discuss the use of plane waves to synthesize the wave field of an ultrasonic transducer.

D.3 Waves in an Isotropic Elastic Solid

The equations of motion for waves in an isotropic elastic solid can be obtained in the same manner as for the fluid. They are [Fundamentals], [D.1-D.3]:

$$\sum_{j=1}^{3} \frac{\partial \tau_{ji}}{\partial x_j} = \rho \frac{\partial^2 u_i}{\partial t^2} \quad (i=1,2,3),$$ (D.15)

where τ_{ij} are the stresses, u_i the displacement components in the x_i-direct-ions, and ρ is the density of the solid. The constitutive equations for an isotropic elastic solid are more complicated than that of a fluid. They are given by generalized Hooke's law:

$$
\left.
\begin{aligned}
\tau_{11} &= \lambda\Delta + 2\mu\frac{\partial u_1}{\partial x_1} & \tau_{12} &= \mu\left(\frac{\partial u_1}{\partial x_2} + \frac{\partial u_2}{\partial x_1}\right) \\
\tau_{22} &= \lambda\Delta + 2\mu\frac{\partial u_2}{\partial x_2} & \tau_{13} &= \mu\left(\frac{\partial u_1}{\partial x_3} + \frac{\partial u_3}{\partial x_1}\right) \\
\tau_{33} &= \lambda\Delta + 2\mu\frac{\partial u_3}{\partial x_3} & \tau_{23} &= \mu\left(\frac{\partial u_2}{\partial x_3} + \frac{\partial u_3}{\partial x_2}\right)
\end{aligned}
\right\}
$$ (D.16)

where λ, μ are the Lame' constants. For an isotropic elastic solid there are only two independent material constants. Some authors instead may use as the independent constants E, v, where E is Young's modulus and v is Poisson's ratio. In terms of these constants the Lame' constants are given by

$$
\begin{aligned}
\lambda &= \frac{Ev}{(1+v)(1-2v)} \\
\mu &= \frac{E}{2(1+v)}.
\end{aligned}
$$ (D.17)

If one places the constitutive equations into the equations of motion we obtain *Navier's equations* for the displacements. In vector form we have

$$\mu\nabla^2 \mathbf{u} + (\lambda+\mu)\nabla(\nabla\cdot\mathbf{u}) - \rho\frac{\partial^2 \mathbf{u}}{\partial t^2} = 0,$$ (D.18)

where $\mathbf{u}$ is the displacement vector. This vector will be written in terms of its scalar components as $\mathbf{u} = (u_1, u_2, u_3) \equiv (u_x, u_y, u_z)$.

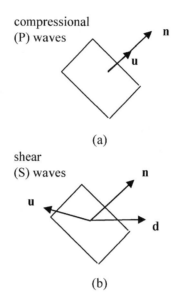

Fig. D.5. The displacements for P-waves and S-waves traveling in the **n**-direction in an isotropic elastic solid.

D.4 Plane Waves in an Isotropic Elastic Solid

Navier's equations are not wave equations, but they do have plane wave solutions. However, unlike the fluid case, there are actually two types of plane waves possible in an isotropic, elastic solid. They are called plane P-waves and plane S-waves. The P-waves are also referred to as pressure, compressional, primary, longitudinal (L), dilatational, or irrotational waves. Similarly, S-waves are also called shear, secondary, transverse (T), distortional, equivoluminal, or rotational waves. Both of these waves are *bulk waves* since they travel throughout the volume of a solid. A bulk P-wave is by far the most commonly used type of wave in NDE testing. If one places a plane wave solution of the form $\mathbf{u} = U\mathbf{n}\, f\left(t - \mathbf{x} \cdot \mathbf{n}/c_p\right)$ in Navier's equation (see Fig. D.5), that equation will be satisfied if

$$c_p = \sqrt{\frac{\lambda + 2\mu}{\rho}} = \sqrt{\frac{E(1-v)}{(1+v)(1-2v)\rho}}. \tag{D.19}$$

If we instead assume a plane wave solution that has the form $\mathbf{u} = U\mathbf{d} \times \mathbf{n} \, g(t - \mathbf{x} \cdot \mathbf{n}/c_s)$, where $\mathbf{d}$ is an arbitrary unit vector (see Fig. D.5(b)), in order to satisfy Navier's equations we find

$$c_s = \sqrt{\frac{\mu}{\rho}} = \sqrt{\frac{E}{2(1+v)\rho}}. \tag{D.20}$$

The quantities c_p, c_s are just the wave speeds for plane bulk P-waves and S-waves, respectively. For a structural material such as steel, for example, the P-wave speed is approximately 5900 m/sec while the S-wave speed is about 3200 m/sec, both of which are considerably larger than the wave speed for water (see Table D.1 for wave speeds and other properties of some selected materials). From Eqs. (D.19) and (D.20) we can see that the ratio of these wave speeds is just a function of Poisson's ratio, i.e.

$$\frac{c_p}{c_s} = \sqrt{\frac{2(1-v)}{(1-2v)}}, \tag{D.21}$$

which for many structural materials gives a ratio of about two to one.

Table D.1. Acoustic properties of some common materials.

Material	P-wave speed [m/s x 10^3]	S-wave speed [m/sx10^3]	Density [kgm/m^{3}x10^3]	Impedance (P-wave) [kgm/(m^2-s) x 10^6]
Air	0.33	--	0.0012	0.0004
Aluminum	6.42	3.04	2.70	17.33
Brass	4.70	2.10	8.64	40.60
Copper	5.01	2.27	8.93	44.60
Glass	5.64	3.28	2.24	13.10
Lucite	2.70	1.10	1.15	3.10
Nickel	5.60	3.00	8.84	49.50
Steel, mild	5.90	3.20	7.90	46.00
Titanium	6.10	3.10	4.48	27.30
Tungsten	5.20	2.90	19.40	101.00
Water	1.48	--	1.00	1.48

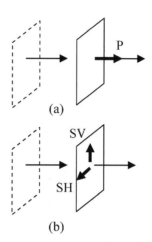

(a)

(b)

Fig. D.6. The polarizations of **(a)** P-waves, and **(b)** vertically polarized (SV) shear waves and horizontally polarized (SH) waves.

As shown in Fig. D.5 (a), the direction of the displacement in the P-wave is along the direction of propagation, **n**, while for an S-wave (Fig. D.5 (b)), the displacement lies in the plane of the wave front, i.e. perpendicular to **n**. Thus, P-waves are said to have longitudinal polarizations while S-wave are said to have transverse polarizations. Figure D.6 shows the polarizations for P-waves and S-waves and also shows that if the polarization (direction of motion) of the plane shear wave lies in a vertical plane, it is called an SV-wave (vertically-polarized shear), while if the polarization lies in a horizontal plane it is called an SH-wave (horizontally-polarized shear). In general, an S-wave may have both vertical and horizontal polarizations combined.

To solve wave propagation problems in elastic solids, many authors represent the displacement in terms of potential functions in the form

$$\mathbf{u} = \nabla \phi + \nabla \times \boldsymbol{\psi}, \tag{D.22}$$

where ϕ is a scalar potential and $\boldsymbol{\psi} = (\psi_1, \psi_2, \psi_3) \equiv (\psi_x, \psi_y, \psi_z)$ is a vector potential. The advantage of using such potentials is that in order to satisfy Navier's equations the potentials must satisfy the ordinary wave equations:

$$\nabla^2 \phi - \frac{1}{c_p^2} \frac{\partial^2 \phi}{\partial t^2} = 0$$

$$\nabla^2 \boldsymbol{\psi} - \frac{1}{c_s^2} \frac{\partial^2 \boldsymbol{\psi}}{\partial t^2} = 0. \tag{D.23}$$

Equation (D.23) shows that the scalar potential, ϕ, represents P-waves while the vector potential, $\boldsymbol{\psi}$, represents S-waves. In solving wave problems with potentials if the disturbance is two-dimensional where the displacements (u_x, u_y) are the only non-zero displacements and they only depend on the x- and y-coordinates, only two potentials are needed:

$$\phi = \phi(x, y, t)$$

$$\psi_z = \psi(x, y, t), \psi_x = \psi_y = 0 \tag{D.24}$$

and the displacements are given by

$$u_x = \frac{\partial \phi}{\partial x} + \frac{\partial \psi}{\partial y}$$

$$u_y = \frac{\partial \phi}{\partial y} - \frac{\partial \psi}{\partial x} \tag{D.25}$$

$$u_z = 0.$$

In this case the stresses are also given by

$$\tau_{xx} = \mu \left[\kappa^2 \nabla^2 \phi + 2 \left(\frac{\partial^2 \psi}{\partial x \partial y} - \frac{\partial^2 \phi}{\partial y^2} \right) \right]$$

$$\tau_{yy} = \mu \left[\kappa^2 \nabla^2 \phi - 2 \left(\frac{\partial^2 \psi}{\partial x \partial y} + \frac{\partial^2 \phi}{\partial x^2} \right) \right]$$

$$\tau_{xy} = \mu \left[2 \frac{\partial^2 \phi}{\partial x \partial y} + \frac{\partial^2 \psi}{\partial y^2} - \frac{\partial^2 \psi}{\partial x^2} \right] \tag{D.26}$$

$$\tau_{zz} = \nu \left[\tau_{xx} + \tau_{yy} \right], \tau_{xz} = \tau_{yz} = 0$$

where $\kappa = c_p / c_s$.

For a harmonic plane P-wave traveling in the positive x-direction, as shown in Fig. D.7 (a), we can express the wave either in terms of its potential, ϕ, displacement, u_x, velocity, v_x or stress, τ_{xx}:

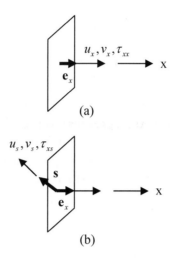

Fig. D.7. The displacement, velocity and stress **(a)** for a plane P-wave, and **(b)** for a plane S-wave, both traveling in the x-direction.

$$\phi = \Phi \exp\left(ik_p x - i\omega t\right)$$
$$u_x = U_x \exp\left(ik_p x - i\omega t\right)$$
$$v_x = V_x \exp\left(ik_p x - i\omega t\right) \tag{D.27}$$
$$\tau_{xx} = T_{xx} \exp\left(ik_p x - i\omega t\right)$$

where $k_p = \omega / c_p$ is the wave number for P-waves and the amplitudes are all related:

$$U_x = ik_p \Phi$$
$$V_x = -i\omega U_x \tag{D.28}$$
$$T_{xx} = -\rho c_p V_x$$

For a harmonic plane S-wave we can also use the potential, $\boldsymbol{\psi}$, displacement, u_s, velocity, v_s, or stress, τ_{xs}, to describe the wave and we have instead (Fig. D.7 (b)):

$$\boldsymbol{\psi} = \Psi \mathbf{t} \exp\left(ik_s x - i\omega t\right)$$
$$\mathbf{u} = U_s \mathbf{s} \exp\left(ik_s x - i\omega t\right)$$
$$\mathbf{v} = V_s \mathbf{s} \exp\left(ik_s x - i\omega t\right)$$
$$\tau_{xs} = T_{xs} \exp\left(ik_s x - i\omega t\right)$$

$$(\text{D.29})$$

where $\mathbf{t}$ is an arbitrary unit vector, $\mathbf{e}_x$ is a unit vector in the x-direction, $\mathbf{s} = \left(\mathbf{e}_x \times \mathbf{t}\right)/\left|\mathbf{e}_x \times \mathbf{t}\right|$ is a unit vector in the plane of the wave front, and $k_s = \omega/c_s$ is the wave number for shear waves. In this case the amplitude relations are

$$U_s = ik_s \Psi$$
$$V_s = -i\omega U_s$$
$$T_{xs} = -\rho c_s V_s$$

$$(\text{D.30})$$

To obtain these relations in the P-wave case we have used the fact that u_x is the only non-zero displacement component in the wave and the only corresponding velocity component is $v_x = \partial u_x / \partial t$. In this case the constitutive equation for the solid (generalized Hooke's law) gives

$$\tau_{xx} = \left(\lambda + 2\mu\right)\frac{\partial u_x}{\partial x} = \frac{E\left(1 - v\right)}{\left(1 + v\right)\left(1 - 2v\right)}\frac{\partial u_x}{\partial x}$$

$$= \rho c_p^2 \frac{\partial u_x}{\partial x}.$$

$$(\text{D.31})$$

Similarly, in the S-wave case we have used the fact that the only non-zero displacement component in the wave is u_s, the displacement in the s-direction, and so the only corresponding velocity component is also $v_s = \partial u_s / \partial t$. In this case the constitutive equation gives

$$\tau_{xs} = \mu \frac{\partial u_s}{\partial x}$$

$$= \rho c_s^2 \frac{\partial u_s}{\partial x}.$$

$$(\text{D.32})$$

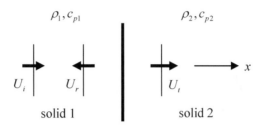

Fig. D.8. A plane P-wave incident on a planar interface between two solids.

D.5 Reflection/Refraction of Plane Waves – Normal Incidence

As a simple but important example of the use of these plane wave relations, consider the reflection and transmission of a plane harmonic P-wave that strikes a planar interface between two solids at normal incidence as shown in Fig. D.8. The density and compressional wave speed in solids one and two are (ρ_1, c_{p1}), and (ρ_2, c_{p2}), respectively. The displacements of the incident, reflected, and transmitted plane waves are given by

$$
\begin{aligned}
u_x &= U_i \exp(ik_1 x - i\omega t) \\
u_x &= -U_r \exp(-ik_1 x - i\omega t) \\
u_x &= U_t \exp(ik_2 x - i\omega t)
\end{aligned}
\tag{D.33}
$$

where $k_1 = \omega/c_{p1}, k_2 = \omega/c_{p2}$. We have taken the reflected wave expression with a minus sign so that U_r represents the amplitude of a plane wave traveling in the $-x$-direction with polarization vector in the direction of propagation, i.e. the vector displacement of the reflected wave would be given by $\mathbf{u} = U_r \mathbf{e}_r \exp(ik_1 \mathbf{e}_r \cdot \mathbf{x} - i\omega t)$ where $\mathbf{e}_r = -\mathbf{e}_x$, $\mathbf{x} = x\mathbf{e}_x$. At the interface $x = 0$ the displacement u_x and the stress $\tau_{xx} = \rho c_p^2 \, \partial u_x / \partial x$ must be continuous so we find

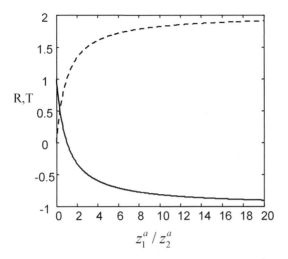

Fig. D.9. The reflection coefficient (solid line) and transmission coefficient (dashed line) versus the impedance ratio z_1^a / z_2^a.

Continuity of displacement:
$$U_i - U_r = U_t \qquad (D.34)$$

Continuity of stress:
$$\rho_1 c_{p1} U_i + \rho_1 c_{p1} U_r = \rho_2 c_{p2} U_t \qquad (D.35)$$

Solving Eqs. (D.34) and (D.35) simultaneously we find

$$T_u = \frac{U_t}{U_i} = \frac{2\rho_1 c_{p1}}{\rho_1 c_{p1} + \rho_2 c_{p2}}$$

$$R_u = \frac{U_r}{U_i} = \frac{\rho_2 c_{p2} - \rho_1 c_{p1}}{\rho_1 c_{p1} + \rho_2 c_{p2}} \qquad (D.36)$$

where (T_u, R_u) are the plane wave transmission and reflection coefficients (based on ratios of displacements). From Eq. (D.36) it follows that these reflection and transmission coefficients are dependent only on the specific acoustic impedances $z_1^a = \rho_1 c_{p1}, z_2^a = \rho_2 c_{p2}$. Figure D.9 plots the behavior of these coefficients versus z_1^a / z_2^a.

The limit $z_1^a / z_2^a \to \infty$ would correspond to the reflection from a free surface. In that case we see $R \to -1$ so that the total displacement $U_i - U_r$ at the interface in the first medium would be double that of the incident wave. The other limit where $z_1^a / z_2^a \to 0$ would correspond to the wave incident on a very rigid boundary. In that case $R \to 1$ so the total displacement at the interface in the first medium would be zero. For the special case where the acoustic impedances of the two solids are matched ($z_1^a / z_2^a = 1$), we see that $R = 0, T = 1$ so that there is no reflected wave and the incident wave passes through the interface with its amplitude unchanged.

These same reflection and transmission coefficients could be used for the reflection of a plane S-wave at normal incidence to a solid-solid interface if we simply replace the compressional wave speeds by the corresponding shear wave speeds, i.e.

$$T_u = \frac{U_t}{U_i} = \frac{2\rho_1 c_{s1}}{\rho_1 c_{s1} + \rho_2 c_{s2}}$$

$$R_u = \frac{U_r}{U_i} = \frac{\rho_2 c_{s2} - \rho_1 c_{s1}}{\rho_1 c_{s1} + \rho_2 c_{s2}}. \tag{D.37}$$

The coefficients could also be used for a fluid-fluid or fluid-solid interface (as encountered in immersion testing) by appropriately replacing the densities and wave speeds in Eq. (D.36) or Eq. (D.37). However, note that these coefficients are based on displacement ratios and if we want to use the ratios of other quantities we may have to make appropriate adjustments. To use velocity ratios, for example, we do not need to make any changes since

$$R_v = \frac{V_r}{V_i} = \frac{-i\omega U_r}{-i\omega U_i} = \frac{U_r}{U_i} = R_u$$

$$T_v = \frac{V_t}{V_i} = \frac{-i\omega U_t}{-i\omega U_i} = \frac{U_t}{U_i} = T_u. \tag{D.38}$$

Equations (D.31) and (D.32) show that $\tau = \rho c^2 \, \partial u / \partial x$ is valid for either plane P-waves or S-waves in a solid if we use the appropriate τ, c and u in this relationship. Similarly, for a fluid we have $p = -\rho c^2 \partial u / \partial x$. As mentioned previously, we combined these relations with the relationship between displacement and velocity, $v = \partial u / \partial t$ to obtain the various plane wave amplitude relationships given by Eqs. (D.28) and (D.29). For a plane wave traveling in the $+ x$-direction with a stress amplitude, T, and velocity amplitude, V, we found $T = -\rho c V$. For a pressure amplitude, P, we found $P = \rho c V$. Obviously, we can use these relations for the incident and transmitted waves since they are both traveling in the $+ x$-direction. However, because we placed the minus sign on the reflected wave in Eq. (D.33), we can also use these same relations for the reflected wave as well. Thus, if we define, for example, reflection and transmission coefficients based on stress ratios we would find (also using Eq. (D.38))

$$R_\tau = \frac{-\rho_1 c_{p1} V_r}{-\rho_1 c_{p1} V_i} = R_u = \frac{\rho_2 c_{p2} - \rho_1 c_{p1}}{\rho_1 c_{p1} + \rho_2 c_{p2}}$$

$$T_\tau = \frac{-\rho_2 c_{p2} V_t}{-\rho_1 c_{p1} V_i} = \frac{\rho_2 c_{p2}}{\rho_1 c_{p1}} T_u = \frac{2\rho_2 c_{p2}}{\rho_1 c_{p1} + \rho_2 c_{p2}}$$

(D.39)

which are also valid for reflection and transmission coefficients based on pressure ratios since there are then changes of signs in the coefficients shown in Eq. (D.39) in both the numerator and denominator that cancel. In the SI system the units of specific acoustic impedance are $kgm/(m^2\text{-sec})$. This set of units is also called a *Rayl*, i.e. 1 Rayl = 1 $kgm/(m^2\text{-sec})$.

For water we have $z_w^a = 1.5 \times 10^6$ $kgm/(m^2 -sec) = 1.5$ MRayls and for steel $z_s^a = 46.0 \times 10^6$ $kgm/(m^2 -sec) = 46$ MRayls, so that for a plane wave traveling in water at normal incidence to a water-steel interface $R_u = -0.937$, $T_u = 0.06$. Because of this high impedance mismatch, we see that in immersion testing most of an ultrasonic beam will be reflected back into the water at normal incidence.

D.6 Reflection/Refraction of Plane Waves – Oblique Incidence

When plane waves are incident on a plane interface at oblique incidence, there are additional aspects of the interactions that one does not see with the normal incidence case. Consider, for example, the simple problem of a plane wave at oblique incidence to a plane interface between two fluids, as shown in Fig. D.10, where ρ_1, c_{p1} are the density and compressional wave speed of medium 1 and ρ_2, c_{p2} are the corresponding density and wave speed for medium 2. Although this problem does not correspond to one we would likely see in NDE testing, most of the physics involved in more

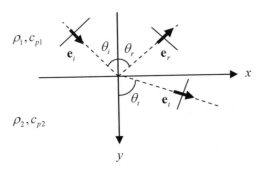

Fig. D.10. A plane wave incident on an interface between two fluids.

complicated plane wave interactions at fluid-solid and solid-solid interfaces are the same as in this problem [Fundamentals]. Here, the total pressure, p_1, due the incident and reflected waves in medium 1 and the total pressure, p_2, due to the transmitted waves in medium 2 are given by

$$
\begin{aligned}
p_1 &= P_i \exp\left[ik_{p1}\left(x\sin\theta_i + y\cos\theta_i\right) - i\omega t\right] \\
&+ P_r \exp\left[ik_{p1}\left(x\sin\theta_r - y\cos\theta_r\right) - i\omega t\right] \\
p_2 &= P_t \exp\left[ik_{p2}\left(x\sin\theta_t + y\cos\theta_t\right) - i\omega t\right].
\end{aligned}
\tag{D.40}
$$

From the equations of motion we have

$$
-\frac{\partial p}{\partial y} = -i\omega\rho\, v_y
\tag{D.41}
$$

so that the total velocity, v_y, in each medium is

$$
\begin{aligned}
\left(v_y\right)_1 &= \frac{P_i\cos\theta_i}{\rho_1 c_{p1}}\exp\left[ik_{p1}\left(x\sin\theta_i + y\cos\theta_i\right) - i\omega t\right] \\
&- \frac{P_r\cos\theta_r}{\rho_1 c_{p1}}\exp\left[ik_{p1}\left(x\sin\theta_r - y\cos\theta_r\right) - i\omega t\right] \\
\left(v_y\right)_2 &= \frac{P_t\cos\theta_t}{\rho_2 c_{p2}}\exp\left[ik_{p2}\left(x\sin\theta_t + y\cos\theta_t\right) - i\omega t\right].
\end{aligned}
\tag{D.42}
$$

At the interface ($y = 0$), the boundary conditions are

$$P_1 = P_2$$
$$\left(v_y\right)_1 = \left(v_y\right)_2 \tag{D.43}$$

so that we find (dividing out all common terms)

$$P_i \exp\left(ik_{p1} x \sin\theta_i\right) + P_r \exp\left(ik_{p1} x \sin\theta_r\right)$$
$$= P_t \exp\left(ik_{p2} x \sin\theta_t\right)$$

$$\frac{P_i \cos\theta_i}{\rho_1 c_{p1}} \exp\left(ik_{p1} x \sin\theta_i\right) - \frac{P_r \cos\theta_r}{\rho_1 c_{p1}} \exp\left(ik_{p1} x \sin\theta_r\right) \tag{D.44}$$
$$= \frac{P_t \cos\theta_t}{\rho_2 c_{p2}} \exp\left(ik_{p2} x \sin\theta_t\right).$$

For these boundary conditions to be satisfied for all x along the boundary we must have the phase terms in Eq. (D.44) all match, which gives

$$k_{p1} \sin\theta_i = k_{p1} \sin\theta_r = k_{p2} \sin\theta_t. \tag{D.45}$$

The first pair of these equations gives

$$\theta_i = \theta_r \tag{D.46a}$$

while the second pair gives

$$\frac{\sin\theta_i}{c_{p1}} = \frac{\sin\theta_t}{c_{p2}}. \tag{D.46b}$$

Equation (D.46a) shows that the angle of incidence equals the angle of reflection while Eq. (D.46b) is a statement of *Snell's law* for the transmitted angle in terms of the incident angle. Applying these phase matching conditions to Eq. (D.44), we obtain

$$P_i + P_r = P_t$$
$$\frac{P_i \cos\theta_i}{\rho_1 c_{p1}} - \frac{P_r \cos\theta_i}{\rho_1 c_{p1}} = \frac{P_t \cos\theta_t}{\rho_2 c_{p2}}. \tag{D.47}$$

These equations can be solved for the transmission and reflection coefficient (in terms of pressure) given by:

$$T_p = \frac{P_t}{P_i} = \frac{2\rho_2 c_{p2} \cos\theta_i}{\rho_1 c_{p1} \cos\theta_t + \rho_2 c_{p2} \cos\theta_i}$$

$$R_p = \frac{P_r}{P_i} = \frac{\rho_2 c_{p2} \cos\theta_i - \rho_1 c_{p1} \cos\theta_t}{\rho_1 c_{p1} \cos\theta_t + \rho_2 c_{p2} \cos\theta_i} \qquad (D.48)$$

or, equivalently, in terms of velocity ratios (using $P = \rho c V$)

$$T_v = \frac{V_t}{V_i} = \frac{2\rho_1 c_{p1} \cos\theta_i}{\rho_1 c_{p1} \cos\theta_t + \rho_2 c_{p2} \cos\theta_i}$$

$$R_v = \frac{V_r}{V_i} = \frac{\rho_2 c_{p2} \cos\theta_i - \rho_1 c_{p1} \cos\theta_t}{\rho_1 c_{p1} \cos\theta_t + \rho_2 c_{p2} \cos\theta_i}. \qquad (D.49)$$

These coefficients are functions of the acoustic impedances of the two media and the incident angle, θ_i, since by Snell's law

$$\cos\theta_t = \sqrt{1 - \sin^2\theta_t} = \sqrt{1 - \frac{c_{p2}^2}{c_{p1}^2}\sin^2\theta_i}. \qquad (D.50)$$

At normal incidence we see that these results simply reduce to those found previously for that special case.

Equation (D.50) shows that when $\sin\theta_i < c_{p1}/c_{p2}$ the $\cos\theta_t$ term is real and both the reflection and transmission coefficients are merely real numbers. This condition is always true when the wave speed for the second medium is slower than that of the first medium since in that case $c_{p1}/c_{p2} > 1$. For the case when the second medium has a faster wave speed, however, the cosine term will only be real for a range of incident angles $0 \leq \theta \leq \theta_{cr}$, where

$$\theta_{cr} = \sin^{-1}\left(c_{p1}/c_{p2}\right) \qquad (D.51)$$

is called the *critical angle*. For incident angles exceeding this critical angle, the reflection and transmission coefficients will become complex. In fact these coefficients will also become frequency dependent. To see this, consider Eq. (D.40). From that equation we see that the only exponential term affected by the critical angle is in the transmitted wave pressure term where $\cos\theta_t$ appears. That term is

$$p_2 = P_t \exp\left[i\omega(x\sin\theta_t + y\cos\theta_t)/c_{p2} - i\omega t\right]. \qquad (D.52)$$

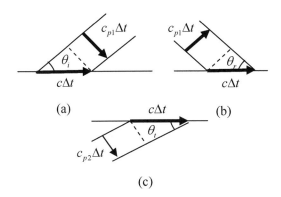

(a)

(b)

(c)

Fig. D.11. (a) The incident wave front, showing its propagation in a time Δt, and its apparent wave speed $c = c_{p1} / \sin \theta_i$ along the interface ; **(b)** the corresponding reflected wave front and its apparent wave speed $c = c_{p1} / \sin \theta_r$ along the interface; **(c)** the transmitted wave front and its apparent wave speed $c = c_{p2} / \sin \theta_t$ along the interface. By Snell's law, the wave speed, c, along the interface is the same for all three waves.

Beyond the critical angle we can let

$$\cos \theta_t = \pm i \sqrt{\sin^2 \theta_t - 1} = \pm i \sqrt{\frac{c_{p2}^2}{c_{p1}^2} \sin^2 \theta_i - 1} \qquad (D.53)$$

and Eq. (D.52) becomes

$$p_2 = P_t \exp \left[\pm \omega y \sqrt{\frac{c_{p2}^2}{c_{p1}^2} \sin^2 \theta_i - 1} \right] \exp \left[i \omega x \sin \theta_t / c_{p2} - i \omega t \right] \qquad (D.54)$$

$$= P_t \exp \left[\pm \omega y \gamma \right] \exp \left[i \omega x / c - i \omega t \right].$$

where $\gamma = \sqrt{\dfrac{c_{p2}^2}{c_{p1}^2} \sin^2 \theta_i - 1}$ is a real constant and $c = c_{p2} / \sin \theta_t = c_{p1} / \sin \theta_i$

is the apparent wave speed along the interface of all the waves (incident, reflected, and transmitted) as shown in Fig. D.11. From Eq. (D.54) the transmitted pressure will be a wave traveling along the interface with wave speed c and an amplitude that varies exponentially in y. However, this amplitude physically must decay to zero as y becomes infinitely large for all frequencies, ω, both positive and negative, so we must choose the

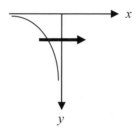

Fig. D.12. An inhomogeneous wave traveling along the interface.

positive sign in Eq. (D.54) for $\omega > 0$ and the negative sign in Eq. (D.54) for $\omega < 0$, i.e. we must let

$$\cos\theta_t = i\,\mathrm{sgn}\,\omega\sqrt{\sin^2\theta_t - 1} = i\,\mathrm{sgn}\,\omega\sqrt{\frac{c_{p2}^2}{c_{p1}^2}\sin^2\theta_i - 1} \qquad (D.55)$$

where

$$\mathrm{sgn}\,\omega = \begin{cases} +1 & \omega > 0 \\ -1 & \omega < 0 \end{cases}. \qquad (D.56)$$

With this choice then the transmitted pressure is given by

$$\begin{aligned} p_2 &= P_t \exp\left[-|\omega|y\gamma\right]\exp\left[i\omega x/c - i\omega t\right] \\ &= P_t \exp\left[-|\omega|y\sqrt{\frac{c_{p2}^2}{c_{p1}^2}\sin^2\theta_i - 1}\right]\exp\left[i\omega x\sin\theta_t/c_{p2} - i\omega t\right], \end{aligned} \qquad (D.57)$$

which represents an *inhomogeneous wave* traveling along the interface and decaying exponentially into the second medium as shown in Fig. D.12. Beyond the critical angle, both the transmission and reflection coefficients are complex and frequency dependent because the $\cos\theta_t$ appearing in those coefficients is given by Eq. (D.55). When we consider an incident plane pulse and use these reflection and transmission coefficients and Fourier transforms to obtain the reflected and transmitted pulses at the interface, the frequency dependency in these coefficients will lead to reflected and transmitted waves that do not have the same shape as the incident waves, a phenomenon called *pulse distortion*. Note that below the critical angle, a reflected or transmitted wave pulse will have different

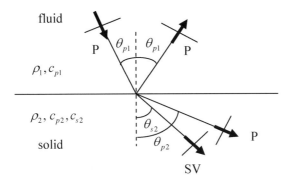

Fig. D.13. A plane wave in a fluid incident on a fluid-solid interface at oblique incidence.

amplitudes from the incident wave but will have exactly the same shape waveform as that of the incident wave.

Although we have only considered a fluid problem here, the behavior of plane waves at solid interfaces is very similar. Consider, for example, the reflection and transmission of a plane wave in a fluid at a fluid-solid interface, as would be encountered in immersion testing (Fig. D.13).The main difference between this case and the fluid-fluid case just considered is that the plane P-wave in the fluid generates both plane P- and SV-waves in the solid. The generation of a wave type by a different wave type is called *mode conversion*. The angles of each of the waves are given here by *generalized Snell's law* so that we have the angle of the reflected P-wave in the fluid equal to the incident P-wave angle, as shown in Fig. D.13, and we have

$$\frac{\sin\theta_{p1}}{c_{p1}} = \frac{\sin\theta_{p2}}{c_{p2}} = \frac{\sin\theta_{s2}}{c_{s2}}, \tag{D.58}$$

where c_{p1} is the compressional wave speed of the fluid and c_{p2}, c_{s2} are the compressional and shear wave speeds of the solid, respectively. Another difference from the fluid-fluid problem is that in this case there can be two critical angles. Above the first critical angle $\theta_{p1} = (\theta_{cr})_1 = \sin^{-1}(c_{p1}/c_{p2})$ the transmitted P-wave becomes an inhomogeneous P-wave traveling along the interface and there is only a transmitted SV-wave, as shown in Fig. D.14 (a). Such a critical angle will exist as long as the compressional

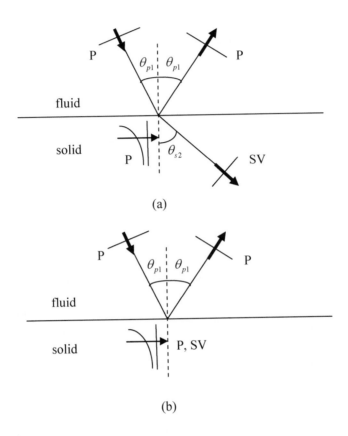

Fig. D.14. (a) The case when the incident angle is greater than the first critical angle and **(b)** the case when the incident angle is greater than the second critical angle.

wave speed of the solid is larger than the compressional wave speed of the fluid ($c_{p2} > c_{p1}$), which is satisfied for water and most structural materials. Above the second critical angle $\theta_{p1} = (\theta_{cr})_2 = \sin^{-1}(c_{p1}/c_{s2})$ the SV-wave also becomes an inhomogeneous wave as shown in Fig. D.14 (b). This critical angle will exist if the shear wave speed in the solid is larger than the compressional wave speed of the fluid ($c_{s2} > c_{p1}$), which again is normally satisfied for water and most common structural materials.

The fluid-solid interface problem can be solved in manner similar to the fluid-fluid problem to obtain the plane wave reflection and transmission

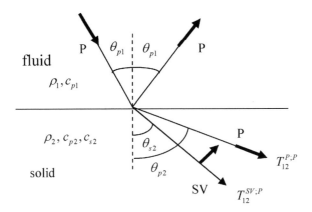

Fig. D.15. The polarization directions chosen for the reflected and transmitted waves.

coefficients. The transmission coefficients, for example, (based on velocity ratios) are given by:

$$T_{12}^{P;P} = \frac{2\cos\theta_{p1}\left[1 - 2\left(\sin\theta_{s2}\right)^2\right]}{\cos\theta_{p2} + \dfrac{\rho_2 c_{p2}}{\rho_1 c_{p1}}\cos\theta_{p1}\Delta}$$

$$T_{12}^{SV;P} = \frac{-4\cos\theta_{p1}\cos\theta_{p2}\sin\theta_{s2}}{\cos\theta_{p2} + \dfrac{\rho_2 c_{p2}}{\rho_1 c_{p1}}\cos\theta_{p1}\Delta}$$

(D.59a)

with

$$\Delta = \left[4\left(\frac{c_{s2}}{c_{p2}}\right)^2 \sin\theta_{s2}\cos\theta_{s2}\sin\theta_{p2}\cos\theta_{p2}\right.$$

$$\left. +1 - 4\left(\sin\theta_{s2}\cos\theta_{s2}\right)^2\right].$$

(D59.b)

Both transmission coefficients are given in the form $T_{12}^{\alpha;\beta}$, which denotes a transmission from medium 1 to medium 2 of a plane wave of type α $(\alpha = P, SV)$ due to an incident plane wave of type β $(\beta = P)$. The signs of these coefficients depend on the specific choice made for the polarization

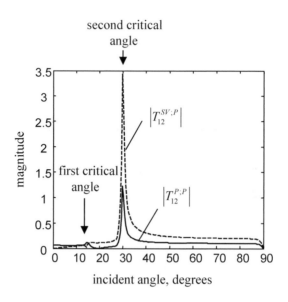

Fig. D.16. The magnitude of the plane wave transmission coefficients at a water-steel interface.

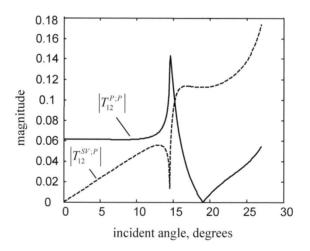

Fig. D.17. The magnitude of the plane wave transmission coefficients at a water-steel interface for incident angles below the second critical angle.

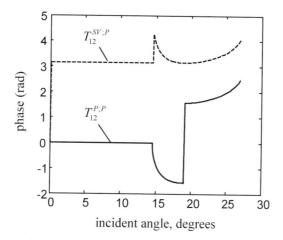

Fig. D.18. The phase (in radians) versus incident angle of the transmission coefficients at a water-steel interface.

directions of the transmitted P- and SV-waves. Here, the polarization directions are chosen as shown in Fig. D.15. When the shear wave speed $c_{s2} \to 0$, $T_{12}^{SV;P} \to 0$ and $T_{12}^{P;P}$ is the same transmission coefficient found previously for the fluid-fluid problem (see Eq. (D.49)). Figure D.16 shows a plot of the magnitude of these transmission coefficients versus the incident angle for a water-steel interface. Because of the relatively large values of these coefficients near the second critical angle, it is useful to consider only angles below that second critical angle, which is the range of most interest anyway since beyond the second critical angle there are no waves transmitted into the solid. Figure D.17 shows this expanded plot. The transmitted shear wave transmission coefficient is zero at normal incidence (incident angle = 0) where there is no mode conversion and increases almost linearly until near the first critical angle. The transmitted P-wave coefficient is small at normal incidence because of the large impedance mismatch between the water and steel and is almost constant until near the first critical angle. For angles near the first critical angle, both coefficients change rapidly in their magnitudes. Figure D.18 shows the corresponding behavior of the phase of the transmission coefficients for angles below the second critical angle. The phase of the transmitted P-wave is zero below the first critical angle because the coefficient is real. There is a phase jump of π radians at an incident angle of about 18.0 degrees (Fig. D.18) where the transmission coefficient changes sign.

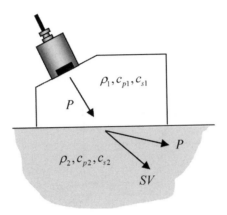

Fig. D.19. An angle beam transducer setup.

The phase of the transmitted SV-wave is π radians below the critical angle because the transmission coefficient is real but negative, i.e. the velocity of the transmitted wave is opposite to the assumed polarization direction shown in Fig. D.15.

As discussed in Appendix E, in angle beam testing a P-wave transducer is placed on a solid wedge which in turn is in contact with a solid that is to be inspected (see Fig. D.19). In this case a thin fluid couplant layer exists between the wedge and the underlying solid to guarantee that there is a good acoustic coupling between the wedge and the solid. If we neglect the thickness of the couplant layer then we can model this setup as two elastic solids in "smooth" and direct contact with each other where the shear stress must vanish at the wedge-solid boundary. In this case the transmission coefficients are given by [Fundamentals]

$$T_{12}^{P;P} = \frac{2\cos\theta_{p1}\left(1-2\sin^2\theta_{s2}\right)\left(1-2\sin^2\theta_{s1}\right)}{\Delta_1+\Delta_2}$$

$$T_{12}^{SV;P} = \frac{-4\sin\theta_{s2}\cos\theta_{p1}\cos\theta_{p2}\left(1-2\sin^2\theta_{s1}\right)}{\Delta_1+\Delta_2}$$

(D.60)

where

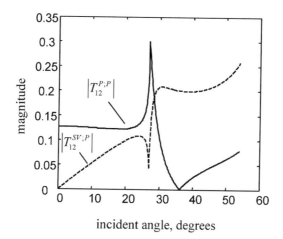

incident angle, degrees

Fig. D.20. The plane wave transmission coefficients for a Lucite-steel interface where $\rho_1 = 1.18\ gm/cm^3$, $\rho_2 = 7.9\ gm/cm^3$, $c_{p1} = 2670\ m/\sec$, $c_{s1} = 1120\ m/\sec$, $c_{p2} = 5900\ m/\sec$, $c_{s2} = 3200\ m/\sec$.

$$\Delta_1 = \cos\theta_{p2}\left[1 - 4\sin^2\theta_{s1}\cos^2\theta_{s1}\right.$$
$$\left. + 4\frac{c_{s1}^2}{c_{p1}^2}\sin\theta_{s1}\cos\theta_{s1}\sin\theta_{p1}\cos\theta_{p1}\right] \tag{D.61a}$$

and

$$\Delta_2 = \frac{\rho_2 c_{p2}}{\rho_1 c_{p1}}\cos\theta_{p1}\left[1 - 4\sin^2\theta_{s2}\cos^2\theta_{s2}\right.$$
$$\left. + 4\frac{c_{s2}^2}{c_{p2}^2}\sin\theta_{s2}\cos\theta_{s2}\sin\theta_{p2}\cos\theta_{p2}\right]. \tag{D.61b}$$

Again, these transmission coefficients are based on velocity ratios and the polarization directions are the same as shown in Fig. D.15. If we let the shear wave speed in the wedge (c_{s1}) go to zero in these expressions, then these transmission coefficients simply reduce to those for fluid-solid case. The magnitudes of these coefficients are plotted versus angle of incidence in Fig. D.20 for a Lucite (plexiglass) wedge in smooth contact with steel.

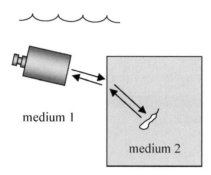

Fig. D.21. Pulse-echo immersion testing showing the transmission and reception of sound from a flaw along a completely reversed path through an interface.

Comparing Figs. D.17 and D.20 we see that the absolute magnitudes and critical angles are different in the two cases because of the wave speed differences but the overall behavior of the curves are very similar.

All the transmission and reflection coefficients discussed so far have been based on amplitude ratios. It is also possible to define similar coefficients that use energy intensity ratios instead. It can be shown that the *wave intensity*, *I*, which is defined as the average power flux (power/unit area) in a harmonic pressure wave in a fluid (where the average is carried out over one complete cycle of the wave) is given by [Fundamentals]

$$I = \frac{P^2}{2\rho c_p} = \frac{\rho c_p V^2}{2},$$
(D.62)

where ρ is the density of the fluid, c_p is the compressional wave speed and *P*, *V* are the pressure and velocity amplitudes, respectively. Similarly, for harmonic plane waves in a solid we have for P-waves

$$I = \frac{T_{nn}^2}{2\rho c_p} = \frac{\rho c_p V_n^2}{2},$$
(D.63)

where T_{nn}, V_n are the normal stress and velocity amplitudes, respectively, and c_p is the P-wave speed. For shear waves

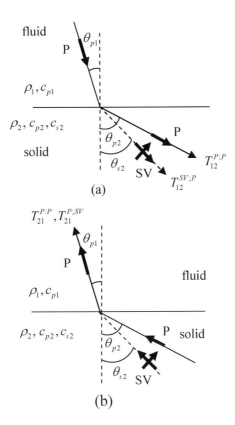

Fig. D.22. (a) Transmission coefficients when going from a fluid to a solid, and **(b)** the corresponding transmission coefficients for a completely reversed path going from the solid to the fluid.

$$I = \frac{T_{ns}^2}{2\rho c_s} = \frac{\rho c_s V_s^2}{2},$$
(D.64)

where T_{ns}, V_s are shear stress and velocity amplitudes, respectively, and c_s is the shear wave speed.

In pulse-echo NDE immersion testing the same transducer is used as both a transmitter of sound and a receiver, as shown in Fig. D.21. In an ultrasonic flaw measurement, for example, if the waves transmitted to a flaw involve a transmission coefficient $T_{12}^{\alpha;P}\ (\alpha = P, SV)$ going from medium 1 (the fluid) to medium 2 (the flawed solid), the received waves

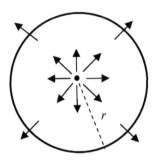

Fig. D.23. A spherical wave arising from a symmetrical point source in a fluid.

from the flaw will involve a transmission coefficient $T_{21}^{P;\alpha}$ going from medium 2 back to medium 1 along a completely reversed path, as shown. These transmission coefficients, however, are related to each other through *Stokes' relations* [Fundamentals], which are (see Fig. D.22):

$$T_{21}^{P;P} = \frac{\rho_1 c_{p1} \cos\theta_{p2}}{\rho_2 c_{p2} \cos\theta_{p1}} T_{12}^{P;P}$$

$$T_{21}^{P;SV} = \frac{\rho_1 c_{p1} \cos\theta_{s2}}{\rho_2 c_{s2} \cos\theta_{p1}} T_{12}^{SV;P}.$$

(D.65)

D.7 Spherical Waves

A spherical wave, like a plane wave, is a special wave type that is very useful for describing the scattering properties of flaws and for constructing more general waves, including the waves generated from ultrasonic transducers [Fundamentals]. First, examine a spherical wave in a fluid. If we consider harmonic waves where the pressure, p, and velocity $\mathbf{v}$, are given by

$$p(\mathbf{x},t) = p(\mathbf{x},\omega)\exp(-i\omega t)$$

$$\mathbf{v}(\mathbf{x},t) = v(\mathbf{x},\omega)\exp(-i\omega t)$$

(D.66)

then the equation of motion for the fluid (recall Eq. (D.3)) is

$$\nabla p(\mathbf{x},\omega) = i\omega \mathbf{v}(\mathbf{x},\omega)$$

(D.67)

and the wave equation for the pressure becomes the *Helmholtz equation*

$$\nabla^2 p(\mathbf{x}, \omega) + k_p^2 p(\mathbf{x}, \omega) = 0, \tag{D.68}$$

where $k_p = \omega / c_p$.

Consider a spherical wave in a fluid arising from a symmetrical point source as shown in Fig. D.23. Because of the symmetry, the equations of motion and the Helmholtz equation in spherical coordinates that describe this spherical wave are given by

$$\frac{\partial p}{\partial r} = i \omega \rho \, v_r \tag{D.69}$$

and

$$\frac{\partial^2 p}{\partial r^2} + \frac{2}{r} \frac{\partial p}{\partial r} + \frac{\omega^2}{c^2} p = 0, \tag{D.70}$$

where v_r is the radial velocity. There are two solutions of Eq. (D.70) given by

$$p = P_1 \frac{r_0}{r} \exp(i k_p r) + P_2 \frac{r_0}{r} \exp(-i k_p r), \tag{D.71}$$

where P_1, P_2 are pressure amplitudes and r_0 is a constant reference radius.

The first of these solutions represents a wave moving outwards from the source while the second moves toward the source. Since the source only generates outward-going waves we must set $P_2 = 0$. Letting $P_1 = P$ we find the pressure and velocity (using Eq. (D.69)) are

$$p = P \frac{r_0}{r} \exp(i k_p r)$$

$$v_r = \frac{P r_0}{\rho c_p} \left[1 - \frac{1}{i k_p r} \right] \frac{\exp(i k_p r)}{r} = V \frac{r_0}{r} \exp(i k_p r). \tag{D.72}$$

Unlike plane waves we see from Eq. (D.72) that the pressure amplitude, P, and velocity amplitude, V, of spherical waves are not just proportional to each other. However, at high frequencies (i.e. $k_p r \gg 1$), we have approximately

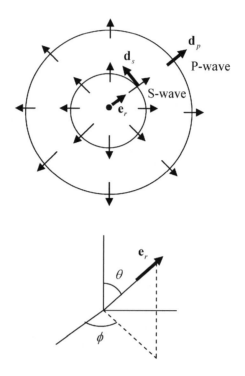

Fig. D.24. Spherical P- and S-waves from a point source in an elastic solid.

$$p = Pr_0 \frac{\exp\left(ik_p r\right)}{r}$$

$$v_r = \frac{Pr_0}{\rho c_p} \frac{\exp\left(ik_p r\right)}{r} = Vr_0 \frac{\exp\left(ik_p r\right)}{r} \tag{D.73}$$

so that P and V do just satisfy the plane wave relation $P = \rho c_p V$. In many ultrasonic NDE problems the frequencies and distances are large enough so that this high frequency approximation is valid.

In elastic solids one can also look for point source solutions to Navier's equations. The details are much more complicated than the fluid case because the sources of interest are usually not symmetric and there can be spherically spreading P-waves and S-waves that are coupled [Fundamentals]. However, at high frequencies, one can treat the waves from a source in a solid as separate traveling spherical waves, as shown in

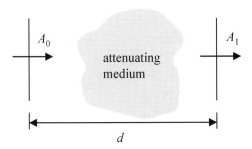

Fig. D.25. Propagation of a plane wave in an attenuating medium.

Fig. D.24 where the polarization of the P-wave, $\mathbf{d}_p$, is in the direction of propagation of the wave and the polarization, $\mathbf{d}_s$, of the shear wave is in a plane perpendicular to the propagation direction.

The displacement $\mathbf{u}(\mathbf{x},t) = \mathbf{u}(\mathbf{x},\omega)\exp(-i\omega t)$ in the solid is then given by

$$\mathbf{u}(\mathbf{x},\omega) = U_p \mathbf{d}_p r_0 \frac{\exp(ik_p r)}{r} + U_s \mathbf{d}_s r_0 \frac{\exp(ik_s r)}{r}, \tag{D.74}$$

where in general $U_p = U_p\left(\mathbf{e}_r(\theta,\phi),\omega\right), U_s = U_s\left(\mathbf{e}_r(\theta,\phi),\omega\right)$, i.e. the amplitudes are angular dependent, where θ,ϕ are the spherical coordinates defining a radial unit vector, $\mathbf{e}_r$, pointing in the direction of propagation as shown in Fig. D.24.

D.8 Ultrasonic Attenuation

All of the wave propagation models discussed in this Appendix have been for ideal, lossless media. At ultrasonic frequencies, however, there are material dependent losses that cause waves to attenuate as they propagate. Generally, the sources of the attenuation can be very complex. In metals, for example, attenuation can be due to scattering of the wave from the grain structure of the solid. One can use models to describe in some detail those scattering processes, but in most cases one can characterize the attenuation losses in a simpler, ad hoc fashion [Fundamentals]. Consider, for example, a plane wave traveling through an attenuating medium as shown in Fig. D.25. The amplitude of this wave will change as it propagates

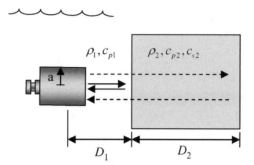

Fig. D.26. A measurement setup for determining the attenuation of P-waves in the solid block.

agates. We will model the effects of attenuation by an exponential factor that contains a frequency dependent attenuation coefficient, $\alpha(f)$, and express the amplitude changes in the form

$$\frac{A_1}{A_0} = \exp\left[-\alpha(f)d\right],$$
(D.75)

where d is the distance traveled in the medium. This attenuation coefficient is measured in Nepers/unit length (Np/l), where a Neper (Np) is a dimensionless quantity. It is also common to express the attenuation in terms of decibels/unit length (dB/l). To convert from Np/l to dB/l we have the relationship

$$\alpha_{dB/l} = 8.686\,\alpha_{Np/l}.$$
(D.76)

The attenuation of water as a function of temperature has been measured. At room temperature, the attenuation of water is [Fundamentals]

$$\alpha_w(f) = 25.3 \times 10^{-3} f^2 \quad Np/m,$$
(D.77)

where f is the frequency in MHz. Equation (D.77) is convenient to use to characterize the attenuation in the water tank of immersion studies. However, for other materials, such as metals, the attenuation is highly dependent on the material processing the metal has undergone so that tabulated values are generally not available and the attenuation must be measured. A convenient setup for making attenuation measurements is one of the calibration setups discussed in Chapter 5 and shown in Fig. D.26. A planar transducer is used in a pulse-echo immersion arrangement to measure the waves reflected at normal incidence from both the front surface and back surface of a solid block whose P-wave attenuation is to be determined. As shown in Chapter 7 the frequency components of the received voltage from the front surface, $V_{fs}(\omega)$, and the frequency components of the voltage received from the back surface, $V_{bs}(\omega)$, can be expressed in the form

$$V_{fs}(\omega) = s(\omega)t_A^{fs}(\omega)$$
$$V_{bs}(\omega) = s(\omega)t_A^{bs}(\omega)$$

(D.78)

where $s(\omega)$ is the system function of the measurement system that accounts for all the electrical and electromechanical components (pulser/receiver, cabling, transducer) and $t_A^{fs}(\omega), t_A^{bs}(\omega)$ are the acoustic/elastic transfer functions that account for all the wave processes, including attenuation, between the sending and receiving transducer. If both front and back surface measurements are done with the same components and at the same system settings, the system function is the same for both measurements, as indicated in Eq. (D.78).

The front surface transfer function is given in Chapter 5. We write this transfer function as

$$t_A^{fs}(\omega) = \tilde{t}_A^{fs}(\omega)\exp\left[-2\alpha_w(\omega)D_1\right],$$

(D.79)

where $\alpha_w(\omega)$ is the attenuation of the water and $\tilde{t}_A^{fs}(\omega)$ is the acoustic/elastic transfer function for the waves in the water without attenuation, given by (see Eq. (5.20)):

$$\tilde{t}_A^{fs}(\omega) = \tilde{D}_p\left(k_{p1}a^2/2D_1\right)R_{12}\exp\left(2ik_{p1}D_1\right),$$

(D.80)

where R_{12} is the plane wave reflection coefficient (Eq. (5.17)) for the fluid-solid interface. The $\tilde{D}_p$ coefficient is a diffraction correction (Eq. (5.20)) that accounts for the deviation of the waves in this setup from plane waves. In a similar manner, we can write the acoustic/elastic transfer function for the waves reflected from the back surface of the block as a transfer function for ideal materials, $\tilde{t}_A^{bs}(\omega)$, multiplied by an attenuation term to account for the attenuation of the waves in both the water and the solid [Fundamentals]:

$$t_A^{bs}(\omega) = \tilde{t}_A^{bs}(\omega)\exp\left[-2\alpha_w(\omega)D_1 - 2\alpha_{p2}(\omega)D_2\right] \tag{D.81}$$

with a loss free transfer function given by

$$\tilde{t}_A^{bs}(\omega) = \tilde{D}_p\left(k_{p1}a^2/2\bar{D}\right)T_{12}R_{21}T_{21}\exp\left(2ik_{p1}D_1 + 2ik_{p2}D_2\right), \tag{D.82}$$

where $\alpha_{p2}(\omega)$ is the attenuation coefficient for P-waves in the solid, R_{21} is the reflection coefficient from the back face (solid-fluid interface) of the block, T_{12} is the plane wave transmission coefficient at normal incidence (based on a pressure ratio) in going from the fluid to the solid, and T_{21} is the corresponding transmission coefficient in going from the solid to the fluid. The distance, $\bar{D}$, is given by

$$\bar{D} = D_1 + \frac{c_{p2}}{c_{p1}}D_2 \tag{D.83}$$

and c_{p1}, c_{p2} are the P-wave speeds of the fluid and the solid. It follows from Eqs. (D.78), (D.79) and (D.81) that

$$\left|\frac{V_{bs}(\omega)}{V_{fs}(\omega)}\right| = \left|\frac{\tilde{t}_A^{bs}(\omega)}{\tilde{t}_A^{fs}(\omega)}\right|\exp\left[2\alpha_{p2}(\omega)D_2\right] \tag{D.84}$$

or, equivalently

$$\exp\left[2\alpha_{p2}(\omega)D_2\right] = \left|\frac{V_{bs}(\omega)}{V_{fs}(\omega)}\right|\left|\frac{\tilde{t}_A^{fs}(\omega)}{\tilde{t}_A^{bs}(\omega)}\right|. \tag{D.85}$$

By measuring the received voltages for the front and back surface reflections from the block and using the known acoustic/elastic transfer functions in Eq. (D.85), one can solve for the attenuation coefficient as a function of frequency. Notice that the attenuation of the water is not needed as the water attenuation term is the same for both the front and back surface responses. So it cancels out in Eq. (D.85). Similarly, one does not need the system function since it also cancels out. Normally, the attenuation coefficient is fitted to a simple polynomial function (in frequency) over the bandwidth of the measurement. If the attenuation is needed over a wider range of frequencies, measurements with other transducers are needed. The setup of Fig. D.26 is suitable for measuring the attenuation of P-waves in the solid. However, in order to determine the

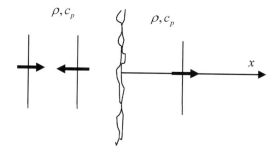

Fig. D.27. A plane P-wave incident on a rough interface.

attenuation of S-waves in the solid would require a different setup involving shear-wave transducers.

Introducing attenuation in this ad-hoc manner works well in describing attenuation effects when the attenuation is not too severe. For highly attenuating materials, however, the wave speed as well as the amplitude of the waves is affected by the attenuation, leading to *material dispersion* effects where this simple method of characterizing the attenuation is inadequate.

D.9 References

D.1 Cheeke JDN (2002) Fundamentals and applications of ultrasonic waves. CRC Press, Boca Raton, FL

D.2 Rose JL (1999) Ultrasonic waves in solid media. Cambridge University Press, Cambridge, UK

D.3 Harker AH (1988) Elastic waves in solids. Adam Hilger, Bristol, UK

D.10 Exercises

1. Consider the case where a solid is split along a rough planar interface which lies in the plane $x = 0$ as shown in Fig. D.27. The two parts of the solid are in contact over some places on the interface and are not in contact at other places. Where the two sides touch the stress τ_{xx} is continuous, and where the sides do not touch $\tau_{xx} = 0$ (on both sides) so again the stress is continuous. Where the sides touch the displacement u_x is continuous but where the sides do not touch there can be a displacement of one side

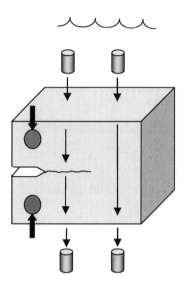

Fig. D.28. A measurement setup for ultrasonically examining a partially closed crack. Note that the crack extends across the entire width of the block.

relative to another. It is reasonable to expect that the amount of this relative displacement is proportional to the stress at the interface. Thus, under these conditions, we could expect that we might specify boundary conditions on the interface as:

continuity of stress:

$$\tau_{xx}\left(x=0^{-},t\right)=\tau_{xx}\left(x=0^{+},t\right)$$

stress proportional to the relative displacement:

$$\tau_{xx}\left(x=0,t\right)=\kappa_{s}\left[u_{x}\left(x=0^{+},t\right)-u_{x}\left(x=0^{-},t\right)\right].$$

The constant κ_{s} determines the relative "springiness" of the interface. The case $\kappa_{s}=0$ corresponds to a stress free interface (no transmission) while $\kappa_{s}\to\infty$ means that the displacement is also continuous so that we have perfect contact and complete transmission (no reflection).

(a) Determine the reflection and transmission coefficients (based on stress ratios) for a P-wave at normal incidence to this rough interface and plot the magnitude and phase of these coefficients versus frequency from 0-20 MHz for steel with $\kappa_s = 150$ MPa/μm.

(b) The magnitude of transmission coefficient, T, you obtain in part (a) should be of the form, $|T(f)| = 1/\sqrt{1 + C^2 f^2}$, where f is the frequency and the constant C is related to κ_s and ρc_p. We could use this transmission coefficient to try to estimate the effects of crack closure of a rough crack as follows. Figure D.28 shows a compact tension specimen which is used to grow a through-thickness crack from a starter notch. If the compact tension specimen is then loaded, the sides of this rough crack will touch at some points and not at others, so the crack surface will look like two rough surfaces in partial contact, the same problem as shown in Fig. D.27. Suppose we now examine this crack with a through-transmission immersion ultrasonic experiment, as shown in Fig. D.28, and also do a reference experiment where we move the transducers laterally so that they are not over the crack. Let $V_c(f)$ be the frequency components of the measured voltage for the case when we are over the crack, and let $V_r(f)$ be the frequency components of the measured voltage for the reference experiment, Since the only difference between two setups is the transmission coefficient at the crack, we expect that the voltages to satisfy $V_c(f) = T(f)V_r(f)$. The MATLAB function rough_crack gives the sampled received voltage versus time, vc, for the case when the transducers are placed over the crack, the sampled voltage versus time, vr, for the reference setup, and the sampled time values, t. The function call is:

```
>> [ vc, vr, t] =rough_crack
```

Using this function, obtain the measured transmission coefficient, $T(f)$, versus frequency [Note: if you want to use a Wiener filter here, the numerical round off "noise" is extremely small so choose a small constant value such as $\varepsilon = 0.001$ or smaller]. Using this transmission coefficient, determine a best fit value of C, and the corresponding value of κ_s (in MPa/μm), which is a measure of how closed the crack is. Assume the compact tension specimen is made of steel whose specific plane wave impedance is $z^a = 46 \times 10^6$ kg/(m²-sec).

2. A 6 mm thick aluminum plate is immersed in water and an immersion transducer is placed in the water at normal incidence to this plate. The

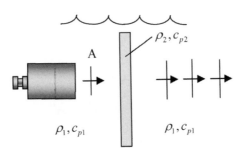

Fig. D.29. A transducer sending multiple pulses through an aluminum plate.

pulse generated from the transducer will pass through the plate and also be multiply reflected within the plate many times, causing a series of transmitted pulses to appear on the other side of the plate (see Fig. D.29). If we assume that the transducer beam incident on the plate acts as if it were a plane wave of pressure amplitude A, what would be the pressure amplitudes of the first three transmitted pulses (in terms of A) and what would be the time separation between them? For the water take $\rho_1 = 1$ gm/cm^3 , $c_{p1} = 1480$ m/sec and for the aluminum take $\rho_2 = 2.7$ gm/cm^3, $c_{p2} = 6420$ m/sec.

3. The stress-strain (constitutive) relations for an isotropic elastic solid can be written as:

$$\tau_{xx} = \frac{E}{(1+v)(1-2v)}\left[(1-v)e_{xx} + v\left(e_{yy} + e_{zz}\right)\right]$$

$$\tau_{yy} = \frac{E}{(1+v)(1-2v)}\left[(1-v)e_{yy} + v\left(e_{xx} + e_{zz}\right)\right]$$

$$\tau_{zz} = \frac{E}{(1+v)(1-2v)}\left[(1-v)e_{zz} + v\left(e_{xx} + e_{yy}\right)\right]$$

$$\tau_{xy} = \mu\gamma_{xy}$$

$$\tau_{xz} = \mu\gamma_{xz}$$

$$\tau_{yz} = \mu\gamma_{yz}$$

where $\mu = E/2(1+v)$ and the strains are given by

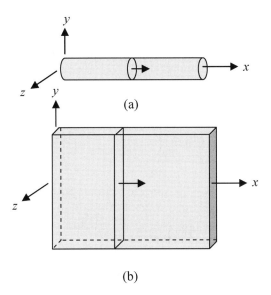

Fig. D.30. A compressional wave traveling in **(a)** a rod and **(b)** a plate.

$$e_{xx} = \frac{\partial u_x}{\partial x}, \quad e_{yy} = \frac{\partial u_y}{\partial y}, \quad e_{zz} = \frac{\partial u_z}{\partial z}$$

$$\gamma_{xy} = \frac{\partial u_x}{\partial y} + \frac{\partial u_y}{\partial x}, \quad \gamma_{xz} = \frac{\partial u_x}{\partial z} + \frac{\partial u_z}{\partial x}, \quad \gamma_{yz} = \frac{\partial u_y}{\partial z} + \frac{\partial u_z}{\partial y}.$$

(a) If we consider a compressional wave propagating along a long slender rod as shown in Fig. D.30 (a), it is reasonable to assume that $u_x = u_x(x,t)$ and that the only non-zero stress is τ_{xx} ($\tau_{yy} = \tau_{zz} = \tau_{xy} = \tau_{xz} = \tau_{yz} = 0$). The equations of motion in this case become simply

$$\frac{\partial \tau_{xx}}{\partial x} = \rho \frac{\partial^2 u_x}{\partial t^2}.$$

Use the stress-strain relations and the conditions $\tau_{yy} = \tau_{zz} = 0$ to determine the relationship between τ_{xx} and $e_{xx} = \partial u_x / \partial x$ for this case. What is the wave speed for compressional waves in the rod?

(b) Now consider a compressional wave traveling in a plate as shown in Fig. D.30 (b). In this case it is reasonable to assume $\tau_{zz} = \tau_{xy} = \tau_{xz} = \tau_{yz} = 0$ and $e_{yy} = 0$. If again we assume $u_x = u_x(x,t)$ the equation of motion for the plate is the same as for part (a). Use the stress-strain relations and the conditions $\tau_{zz} = e_{yy} = 0$ to again determine the relationship between τ_{xx} and $e_{xx} = \partial u_x / \partial x$ for this case. What is the wave speed for compressional waves in the plate? For steel (take E = 210 GPa, $v = 0.3$, $\rho = 7.9 \times 10^3$ kgm/m^3) how does the compressional rod wave speed and plate wave speed compare to the wave speed for bulk compressional waves?

4. A plane wave travels 100 mm in a material to a point where its amplitude is P_1. After the wave travels through an additional 100 mm of material its amplitude is reduced to $P_2 = 0.45 P_1$. What is the average attenuation of this material in dB/m?

5. A transducer beam spreads as it propagates. In the far-field of the transducer this spreading is just like that of a spherical wave, i.e. the amplitude varies as $1/r$ where r is the distance from the transducer. At a distance of 100 mm from the transducer the amplitude of the pressure is P_1. After the beam has propagated another 100 mm the amplitude is reduced to $P_2 = 0.45 P_1$. What is the average attenuation of the material in dB/m, assuming that at both of these distances we are in the transducer far-field?

6. Consider a harmonic plane P-wave traveling in water at room temperature. Determine an expression for the distance (as a function of the frequency, f) that this wave must travel (in meters) to reduce its amplitude by 10% due to attenuation. Plot this function from $f = 1$ MHz to $f = 20$ MHz.

7. The reflection and transmission coefficients (based on stress ratios) for a plane P-wave wave at normal incidence to a plane interface between two elastic solids were given by Eq. (D.39). Determine the corresponding transmission and reflection coefficients based on ratios of the energy intensities (see Eq. (D.63)). Plot these intensity-based coefficients versus the impedance ratio of the two solids, as done in Fig. D.9. What is the sum of these intensity-based reflection and transmission coefficients?

E Waves Used in Nondestructive Evaluation

Bulk P-waves and S- waves are the types of waves most frequently used in NDE testing. Thus, the wave propagation models developed in Appendix D and in Chapters 8-12 are all bulk wave models. In this Appendix we discuss some of the issues associated with the generation of bulk S-waves in solids and briefly describe surface (Rayleigh) and plate waves since these wave types also have important NDE applications.

E.1 Shear Waves

Many ultrasonic nondestructive evaluation inspections are performed with P-wave transducers operating either in a contact mode or in immersion testing. It is also possible to have a piezoelectric crystal generate a shearing motion when it is excited by a voltage pulse and use that shearing motion to launch a shear wave into a solid component in a contact setup as shown in Fig. E.1. In order to couple the motion of the crystal to the solid, however, the shear wave transducer must be attached to the solid in a permanent or semi-permanent fashion. Highly viscous shear wave couplants or glues can be used for this purpose, but the transducer is then not able to be scanned along the surface which greatly limits the usefulness of such a shear wave setup. As a consequence, most shear waves are instead generated through the process of mode conversion from a P-wave to an SV-wave at oblique incidence to an interface. This is the basic mechanism used in an *angle beam shear wave transducer*, as shown in Fig. E.2. An ordinary P-wave type of crystal and backing is placed on a plastic wedge. The P-wave this crystal generates strikes the interface between the wedge and the solid to be inspected at oblique incidence. If the incident angle in the wedge is chosen so that the first critical angle in the solid is exceeded, then only a transmitted SV-wave propagates into the solid as shown in Fig. E.2. The angle of propagation of the shear wave in the solid is determined by generalized Snell's law so that

$$\theta_{s2} = \sin^{-1}\left[c_{s2} \sin \theta_{p1} / c_{p1} \right] \tag{E.1}$$

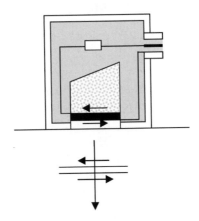

Fig. E.1. A contact shear wave transducer on the free surface of a solid showing the shear motion of the piezoelectric and the corresponding shear wave that is generated.

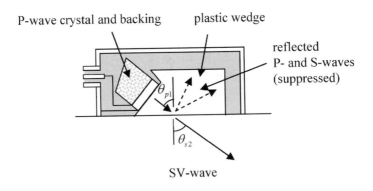

Fig. E.2. An angle beam shear wave transducer on the free surface of an elastic solid.

where c_{p1} is the P-wave speed in the wedge, c_{s2} is the shear wave speed in the solid, and θ_{p1} is the angle that the P-wave in the wedge makes with the normal to the surface. Some refracted shear wave angles that are commonly used in practice are: $\theta_{s2} = 45°, 60°, 70°$. The angle beam shear wave transducer, like an ordinary P-wave contact transducer, can be coupled to the solid by a thin fluid layer so that it can be scanned along the surface. Angle beam shear wave transducers are often used for weld inspection

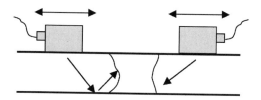

Fig. E.3. An angle beam shear wave inspection of a welded plate geometry using a directly generated SV-wave as shown on the right side of the weld or a SV-wave reflected from a back surface as shown on the left side.

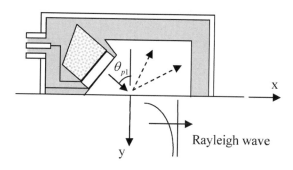

Fig. E.4. A Rayleigh wave transducer on the free surface of an elastic solid.

problems as shown in Fig. E.3, where the entire weld zone can be probed by scanning the transducer along the surface and using either the SV-wave directly or a wave reflected from a back surface. Models of angle beam shear wave inspections are discussed in Chapter 13.

E.2 Rayleigh Waves

A transducer arrangement very similar to the angle beam shear wave case can also be used to generate Rayleigh surface waves, as shown in Fig. E.4 [E.1]. In this case the angle of the incident P-waves in the wedge must be slightly larger than the second critical angle. Specifically,

$$\theta_{p1} = \sin^{-1}\left[c_{p1}/c_{r2}\right], \tag{E.2}$$

where c_{r2} is the wave speed for Rayleigh waves in the solid. Typically the Rayleigh wave speed is about 90 per cent the shear wave speed. At this angle, there are only inhomogeneous P- and SV-waves generated in the solid which combine to form the Rayleigh wave mode. The Rayleigh wave travels along the stress free surface of the solid and decays in depth from the surface. Lord Rayleigh first discovered these waves by choosing P- and SV-wave potentials given by [Fundamentals]

$$\phi = A\exp\left[-\alpha y\right]\exp\left[ik\left(x-ct\right)\right]$$
$$\psi = B\exp\left[-\beta y\right]\exp\left[ik\left(x-ct\right)\right] \tag{E.3}$$

which represent inhomogeneous waves propagating along the surface with the common wave speed , c. These waves must satisfy the wave equations

$$\nabla^2\phi - \frac{1}{c_{p2}^2}\frac{\partial^2\phi}{\partial t^2} = 0$$
$$\nabla^2\psi - \frac{1}{c_{s2}^2}\frac{\partial^2\psi}{\partial t^2} = 0, \tag{E.4}$$

where c_{p2} and c_{s2} are the wave speeds for compressional and shear waves in the solid, respectively. Also, these waves must satisfy the free surface (zero stress) boundary conditions which are $\tau_{yy} = \tau_{xy} = 0$ on $y = 0$. Rayleigh showed that potentials of the form given in Eq. (E.3) could be found that satisfy both the wave equations and the boundary conditions if

$$\alpha = |\omega/c|\sqrt{1-c^2/c_{p2}^2}$$
$$\beta = |\omega/c|\sqrt{1-c^2/c_{s2}^2} \tag{E.5}$$

and the wave speed , c, is a root of the equation

$$\left(2-c^2/c_{s2}^2\right)^2 - 4\sqrt{1-c^2/c_{p2}^2}\sqrt{1-c^2/c_{s2}^2} = 0. \tag{E.6}$$

Equation (E.6) is called the *Rayleigh equation*. It can be shown [E.3] that for an isotropic elastic solid there is always one real root of Eq. (E.5) $c = c_{r2}$, where $c_{r2} < c_{s2}$, called the *Rayleigh wave speed*. Eq. (E.5) then shows that α and β are both real so that the Rayleigh wave potentials

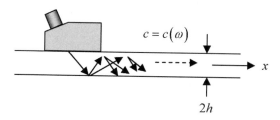

Fig. E.5. An angle beam transducer on a thin plate generating a series of reflected and mode-converted waves that combine to form a dispersive (frequency dependent) plate wave traveling with the wave speed $c = c(\omega)$.

have an exponential decay in distance from the surface. A simple approximate expression for the Rayleigh wave speed is given by

$$c_{r2} \cong \frac{0.862 + 1.14\nu}{1 + \nu} c_{s2}, \tag{E.7}$$

where ν is Poisson's ratio.

If one examines the displacements and stresses in the Rayleigh wave, one finds that like the potentials they also decay in depth from the interface but the decay is not a simple exponential behavior as given in Eq. (E.3). However, at high frequencies, these quantities are all confined near the surface while at lower frequencies they have deeper penetration. Thus, in an inspection with Rayleigh waves one can adjust the depth of the region one is interrogating by adjusting the frequency. Since they are confined to the surface Rayleigh waves are very useful for inspecting for surface flaws or near-surface flaws. Also, since Rayleigh waves travel and spread out in two-dimensions on the surface whereas bulk waves spread out in three-dimensions as they propagate through the volume of a material the amplitudes of Rayleigh waves do not decay as fast as bulk waves and they can travel long distances.

E.3 Plate (Lamb) Waves

If an angle beam shear wave transducer is placed on a thin plate, as shown in Fig. E.5, a series of reflected and mode converted waves are generated in the plate and these combine to form a new wave mode traveling with a wave speed, c, in the x-direction called a plate (or Lamb) wave [E.2].

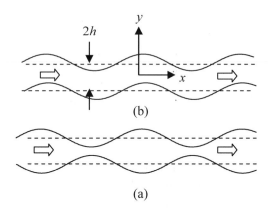

Fig. E.6. **(a)** An extensional plate wave traveling in the x-direction in a thin plate and **(b)** a flexural plate wave traveling in the x-direction. The type of deformation present in each of these wave types is shown.

Unlike bulk waves or Rayleigh waves whose wave speeds are just a function of material constants, $c = c(\omega)$ i.e. the wave speed of a plate wave is generally frequency dependent, a phenomenon called *geometric dispersion*. Actually as we will see there are many different plate waves that can be generated, each with a different frequency dependency.

Plate waves are solutions of the wave equations for the potentials where we assume

$$\phi = f(y)\exp\left[ik(x-ct)\right]$$
$$\psi = g(y)\exp\left[ik(x-ct)\right].$$

$$(\text{E.8})$$

In this case, we find that

$$f = A\cosh(\alpha y)$$
$$g = B\sinh(\beta y)$$

$$(\text{E.9})$$

or

$$f = A'\sinh(\alpha y)$$
$$g = B'\cosh(\beta y),$$

$$(\text{E.10})$$

where α and β are again given by Eq. (E.5). Solutions of the form given by Eq. (E.9) are *extensional plate waves*, while those given by Eq. (E.10)

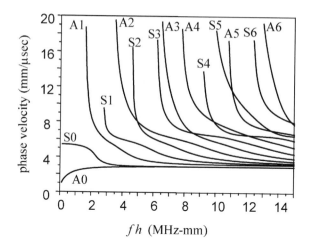

Fig. E.7. The phase velocity versus non-dimensional frequency (dispersion) curves. Extensional wave modes are labeled An and flexural modes are labeled Sn (n = 0, 1, 2...).

are *flexural plate waves*. In general, we may have both types of waves generated. The extensional plate waves are waves with the symmetric thickness variations shown in Fig. E.6 (a) while the flexural plate waves generate a bending deformation of the plate as they propagate, as shown in Fig. E.6 (b).

Both types of these plate waves must satisfy the boundary conditions $\tau_{yy} = \tau_{xy} = 0$ on $y = \pm h$ which yields the Rayleigh-Lamb equations

$$\frac{\tanh(\beta h)}{\tanh(\alpha h)} = \left[\frac{4\omega^2 \alpha \beta}{c^2 \left(\omega^2 / c^2 + \beta^2\right)^2} \right]^{\pm 1}, \tag{E.11}$$

where the plus sign is for extensional waves and the negative sign for flexural waves. Solutions of Eq. (E.11) for the frequency dependent wave speed $c = c(\omega)$ are rather complex and generally must be determined numerically. Plots of the wave speed versus frequency are shown in Fig. E.7 for a number of extensional and flexural wave modes.

We can obtain some information on the behavior of these waves by noting that for sufficiently high frequencies we have $\tanh(\alpha h) = 1$,

$\tanh(\beta h) = 1$ and the Rayleigh-Lamb equations for both extensional and flexural waves reduce to just our equation for Rayleigh waves, Eq. (E.6). This is a reasonable result since at very high frequencies the two sides of the plate appear very far away from each other and the waves can propagate independently on each side (as Rayleigh waves) as if the other side did not exist. Thus, all the curves in Fig. E.7 asymptote to the Rayleigh wave speed at sufficiently high frequencies.

At low frequencies, one can also extract some explicit results from the Rayleigh-Lamb equation. For the lowest order extensional wave mode (see Fig. E.7), one finds [Fundamentals]

$$c \cong \sqrt{\frac{E}{\rho(1-v^2)}} \tag{E.12}$$

which is the non-dispersive wave speed for extensional plate waves found from elementary plate theory. In contrast, the lowest order flexural mode at low frequencies produces a wave speed [Fundamentals]

$$c \cong \left(\frac{D_p}{2\rho h}\right)\sqrt{\omega}, \tag{E.13}$$

where $D_p = \dfrac{8\mu(\lambda+\mu)h^3}{3(\lambda+2\mu)}$ is the flexural rigidity of the plate. In this case we see the flexural waves remain dispersive even at low frequencies.

Plate waves are good candidates for inspecting thin plates and pipes and are frequently used in those applications [E.2]. The inherent dispersive nature of plate waves and the fact that one often simultaneously generates many different modes often makes inspections with these waves challenging from a data interpretation standpoint.

E.4 References

E.1 Uberall H (1973) Surface waves in acoustics. In: Mason WP, Thurston, RN (eds) Physical acoustics, vol. X. Academic Press, New York, NY

E.2 Rose JL (1999) Ultrasonic waves in solid media. Cambridge Univ. Press, Cambridge, UK

E.3 Achenbach JD (1973) Wave propagation in elastic solids. Am. Elsevier Publishing Co., New York, NY

F Gaussian Beam Fundamentals

A Gaussian beam is a very important type of propagating wave since it is an elementary wave that can be used as an efficient building block for constructing the more complex wave fields present in NDE inspections. Chapter 9 develops in detail the propagation and transmission/reflection laws for Gaussian beams in fluid and solid media. That Chapter also describes how a multi-Gaussian beam model of circular and rectangular piston NDE transducers can be constructed by superimposing only 10-15 Gaussian beams. The Gaussian beam discussions and derivations given in Chapter 9, however, are inherently rather complex since they involve the types of Gaussian beams and beam interactions that are needed to model general NDE testing situations. In this Appendix we will discuss Gaussian beams in a much more restricted context in order to illustrate some of the important properties of this type of wave in as simple a manner as possible. Specifically, we will examine the propagation of a circularly symmetrical Gaussian beam in a fluid medium along a single coordinate direction and describe the interactions of that beam with spherical or planar interfaces that are normal to the propagation direction. Special cases of this type are commonly encountered when using Gaussian beams to represent the fields present in lasers, so we will also use a notation that is consistent with many references found in the laser science literature.

F.1 Gaussian Beams and the Paraxial Wave Equation

Let the pressure, p, of a propagating harmonic wave (of $\exp(-i\omega t)$ time dependency) be written in cylindrical coordinates (ρ, z) in the form of a *quasi-plane wave* propagating in the z-direction, i.e.

$$p = P(\rho, z, \omega)\exp(ikz - i\omega t). \tag{F.1}$$

This is called a quasi-plane wave because the amplitude, P, has variations in (ρ, z) while in a true plane wave P would be constant. Placing Eq. (F.1) into the wave equation (see Eq. (D.8)) then shows that P must satisfy

$$\frac{1}{\rho}\frac{\partial}{\partial\rho}\left(\rho\frac{\partial P}{\partial\rho}\right) + \frac{\partial^2 P}{\partial z^2} + 2ik_p\frac{\partial P}{\partial z} = 0, \tag{F.2}$$

where $k_p = \omega/c_p$ is the wave number. Note that this is an axially symmetrical wave since there are no angular variations in the plane perpendicular to the z-axis. As discussed in Chapter 9 if Eq. (F.1) represents a wave disturbance propagating primarily in the z-direction we can assume that the $\partial^2 P/\partial z^2$ term in Eq. (F.2) will be smaller than the other terms in that equation, leading to the *paraxial wave equation* for P given by [F.1], [F.2]

$$\frac{1}{\rho}\frac{\partial}{\partial\rho}\left(\rho\frac{\partial P}{\partial\rho}\right) + 2ik_p\frac{\partial P}{\partial z} = 0. \tag{F.3}$$

The paraxial wave equation assumes that

$$\left|\frac{\partial^2 P}{\partial z^2}\right| << \left|2ik_p\frac{\partial P}{\partial z}\right|, \left|\frac{1}{\rho}\frac{\partial}{\partial\rho}\left(\rho\frac{\partial P}{\partial\rho}\right)\right|, \tag{F.4}$$

but since the magnitudes of both terms on the right side of Eq. (F.4) are always equal by virtue of the paraxial wave equation, it is sufficient to require only that

$$\left|\frac{\partial^2 P}{\partial z^2}\right| << \left|2ik_p\frac{\partial P}{\partial z}\right|. \tag{F.5}$$

We will discuss the consequences of this inequality shortly. Now consider a Gaussian beam solution of Eq. (F.3) in the form

$$P = \tilde{P}\exp\left(ik_p\rho^2/2q\right), \tag{F.6}$$

where $\tilde{P} = \tilde{P}(z)$ and $q = q(z)$ can both be complex-valued functions. Placing Eq. (F.6) into Eq. (F.3) gives

$$2ik_p\left(\frac{\tilde{P}}{q} + \frac{d\tilde{P}}{dz}\right) + \frac{k_p^2\rho^2\tilde{P}}{q^2}\left(\frac{dq}{dz} - 1\right) = 0. \tag{F.7}$$

Since we must satisfy Eq. (F.7) for all ρ we have

$$\frac{dq}{dz} = 1 \tag{F.8a}$$

and

$$\frac{d\tilde{P}}{dz} + \frac{\tilde{P}}{q} = 0. \tag{F.8b}$$

The solution of Eq. (F.8a) is just the propagation law

$$q(z) = z + q_0 \tag{F.9}$$

where q_0 is a complex constant (that can also depend on the frequency, ω). Placing this solution into Eq. (F.8b) then also gives

$$\tilde{P}(z) = \frac{P_0}{z + q_0} = \frac{P_0}{q(z)}. \tag{F.10}$$

Thus, we see that a propagating Gaussian beam is given by

$$
\begin{aligned}
p &= \frac{P_0}{q(z)} \exp\left[ik_p z\right] \exp\left[\frac{ik_p \rho^2}{2q(z)}\right] \\
&= \frac{P_0}{z + q_0} \exp\left[ik_p z\right] \exp\left[\frac{ik_p \rho^2}{2(z + q_0)}\right]
\end{aligned}
\tag{F.11}
$$

(where we have omitted writing explicitly the $\exp(-i\omega t)$ term, a convention we will follow throughout the remainder of this Appendix). To put this beam expression in a more understandable form, let the constant q_0 be represented in the form

$$q_0 = -\left(z_0 + \frac{i\pi w_0^2}{\lambda_p}\right), \tag{F.12}$$

where (z_0, w_0) have the dimensions of a length and $\lambda_p = 2\pi/k_p$ is the wavelength. The distance $z_c = \pi w_0^2/\lambda_p$ is called the *confocal distance* (or *confocal parameter*). In terms of these parameters, therefore, the Gaussian beam is given by

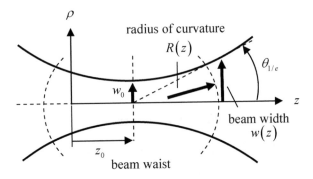

Fig. F.1. A Gaussian beam of circular cross-section propagating in the z-direction, showing the wave front curvature and the beam width. The beam waist is located at $z = z_0$ where the beam width is w_0. The half angle divergence of the beam at a large distance from the beam waist is defined by the angle $\theta_{1/e}$.

$$p = \frac{P_0}{(z - z_0) - iz_c} \exp\left[ik_p z\right]$$

$$\cdot \exp\left[\frac{ik_p \rho^2 (z - z_0)/2}{(z - z_0)^2 + (z_c)^2} - \frac{\rho^2 (z_c / w_0)^2}{(z - z_0)^2 + (z_c)^2}\right]. \tag{F.13}$$

Now, define a *beam width* parameter, $w(z)$, and a *beam wave front curvature* parameter, $R(z)$, as

$$\frac{1}{\left[w(z)\right]^2} = \frac{(z_c / w_0)^2}{(z - z_0)^2 + (z_c)^2}$$

$$\frac{1}{R(z)} = \frac{(z - z_0)}{(z - z_0)^2 + (z_c)^2}. \tag{F.14}$$

Then the Gaussian beam becomes

$$p = \frac{P_0}{(z - z_0) - iz_c} \exp\left[ik_p z\right] \exp\left[\frac{ik_p \rho^2}{2R(z)} - \frac{\rho^2}{(w(z))^2}\right]. \tag{F.15}$$

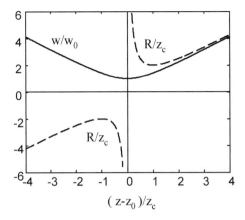

Fig. F.2. A plot of the normalized beam width, w/w_0, and the normalized radius of curvature, R/z_c, versus the normalized distance, $(z-z_0)/z_c$.

Figure F.1 shows a side view of the Gaussian beam represented by Eq. (F.15) and Fig. F.2 shows a plot of both the normalized beam width and normalized curvature parameters. We see that the beam wave front curvature, $R(z)$, is infinite at the location $z = z_0$. Therefore at that location the wave front of the Gaussian beam is planar. For $z > z_0$ the curvature is positive and the Gaussian beam is a diverging beam while for $z < z_0$ the beam is a converging beam. From Eq. (F.15) we see that the amplitude of the Gaussian beam in the ρ-direction is a Gaussian function whose width is $w(z)$, where the width is defined as the radial distance to which the beam amplitude drops by a factor e^{-1} from its on-axis value. Figures F.1 and F.2 show that the minimum beam width also occurs at $z = z_0$ which is called the location of the *beam waist*. At the beam waist from Eq. (F.14) it follows that $w(z_0) = w_0$.

Physically, the confocal parameter $z_c = \pi w_0^2 / \lambda_p$ is the axial distance from the beam waist to where the Gaussian beam remains reasonably well collimated (i.e. where the beam width is approximately a constant) [F.2]. This can be seen from Eq. (F.14) where at $z - z_0 = z_c$ we find $w = 1.414 w_0$ so that the beam width is only 40% larger than at the waist. However, at larger distances from the beam waist the beam width

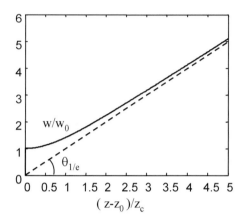

Fig. F.3. The normalized beam width versus normalized distance, $(z - z_0)/z_c$ and the corresponding asymptotic beam growth angle.

grows considerably wider as the beam diverges. If we compute the half angle to the e^{-1} width point in the beam, $\theta_{1/e}$, (see Figs. F.1 and F.3) where

$$\theta_{1/e} = \lim_{z \to \infty} \left\{ \tan^{-1} \left[\frac{w(z)}{(z - z_0)} \right] \right\} = \tan^{-1} \left(\frac{\lambda_p}{\pi w_0} \right) = \tan^{-1} \left(\sqrt{\frac{\lambda_p}{\pi z_c}} \right) \qquad (F.16)$$

we see that if the wavelength, λ, is much smaller than the beam width, w_0, then the *asymptotic beam growth angle*, $\theta_{1/e}$, is very small. For example, if we consider a Gaussian beam with a waist size $w_0 = 3$ mm radiating into water at 5 MHz, then the wavelength $\lambda_p = 0.3$ mm and $\theta_{1/e} = 1.8$ degrees. This shows that a Gaussian beam of roughly the same size as an ultrasonic NDE transducer that propagates at MHz frequencies will be highly collimated. A typical NDE transducer beam at these frequencies also is highly collimated (see, for example, Fig. 8.3). It is this fact that makes it possible to take a relatively few Gaussian beams of different widths and waist locations and accurately synthesize the wave field of an NDE transducer, as shown in Chapter 9. In contrast, it takes a superposition of many more spherical waves or plane waves to model a transducer wave field since neither of those wave types are collimated beams like the Gaussian beam.

It is interesting to compare Eq. (F.15) with the paraxial approximation for the propagation of a spherical wave in the neighborhood of the z-axis. This case is examined in Chapter 9 as part of the discussion of the paraxial approximation. For such a spherical wave we have (see Eq. (9.6))

$$p = \frac{A}{z} \exp\left[ik_p z\right] \exp\left(\frac{ik_p \rho^2}{2z}\right).$$ (F.17)

Both Eq. (F.15) and (F.17) have a varying "amplitude" term, a plane wave propagation term, $\exp\left(ik_p z\right)$, and a phase term that is quadratic in the radial distance, ρ, from the propagation axis. In fact, we can view the Gaussian beam as representing a spherical wave propagating from a complex source point in the paraxial approximation [F.3]. Such complex point sources can be used as a means of forming wave solutions that do not rely on the paraxial approximation, but we will not discuss those solutions here. Note that if we let $q_0 = 0$ in the Gaussian beam it reduces exactly to the spherical wave of Eq. (F.17).

Whereas the amplitude of a spherical wave becomes infinite at the source location $z = 0$ (see Eq. (F.17)), the Gaussian beam (see Eq. (F.15)) remains well behaved everywhere. In fact, as shown in Chapter 9, a Gaussian beam is never singular, even after propagation and reflection/refraction in multiple media. The same is not true for spherical or plane waves which at high frequencies can become singular at focal points or caustics after either of those wave types interact with curved interfaces. This non-singular behavior of Gaussian beams is also a feature of this wave type that makes it a better building block than spherical or plane waves to generate more complex wave fields.

F.2 Quasi-Plane Wave Conditions and the Paraxial Approximation

If the pressure is given by Eq. (F.11) then the velocity, v_z, in the direction of the propagating Gaussian beam is given by

$$v_z = \frac{1}{i\omega\rho}\frac{\partial p}{\partial z}$$

$$= \frac{1}{i\omega\rho}\left[ik_p - \frac{q'}{q} - \frac{ik_p\rho^2}{2}\left(\frac{q'}{q^2}\right)\right]p,$$

(F.18)

where $q' = dq/dz = 1$. We will now consider the conditions under which the magnitude of the first term on the right side of Eq. (F.18) is much larger than the other two terms. First consider the condition

$$\left|\frac{q'}{q}\right| << k_p.$$

(F.19)

This condition is equivalent to requiring that

$$k_p|q| = k_p\left|(z - z_0) - iz_c\right| >> 1.$$

(F.20)

Because $|q| \geq z_c$ Eq. (F.20) will certainly be satisfied if we require the stronger condition $k_p|z_c| >> 1$ which gives

$$2\pi^2\left(\frac{w_0}{\lambda_p}\right)^2 >> 1.$$

(F.21)

Now consider the second condition

$$\frac{k_p\rho^2}{2}\left|\frac{q'}{q^2}\right| << k_p.$$

(F.22)

Since most of the energy in a Gaussian beam is contained within a beam width, w, Eq. (F.22) will be satisfied for all ρ within that distance if we set $\rho = w$ in Eq. (F.22) and require

$$\frac{w^2}{2}\left|\frac{1}{q^2}\right| << 1.$$

(F.23)

But since $w \geq w_0$ Eq. (F.23) implies

$$\frac{w_0^2}{2}\left|\frac{1}{q^2}\right| << 1 \qquad \text{or, equivalently,} \qquad \frac{2|q^2|}{w_0^2} >> 1.$$

Again, since $|q| \geq z_c$ the above inequality will certainly be satisfied if we require the stronger condition

$$2\left(\frac{z_c}{w_0}\right)^2 = 2\pi^2\left(\frac{w_0}{\lambda_p}\right)^2 \gg 1 \tag{F.24}$$

This is just the same result as obtained in Eq. (F.21). Thus, we see that as long as the beam waist size is much larger than a wavelength, the velocity, v_z, in the Gaussian beam given by Eq. (F.18) reduces to

$$v_z = \frac{p}{\rho c_p}, \tag{F.25}$$

which is a relationship also true for plane waves (see Appendix D). Thus, in terms of the pressure-velocity relationship we can view a Gaussian beam as behaving like a quasi-plane wave.

These results are also useful for examining the requirement given by Eq. (F.5) for the paraxial approximation to be valid for a Gaussian beam. Since we have

$$P = \frac{P_0}{q} \exp\left[ik_p \rho^2 / 2q\right] \tag{F.26}$$

it follows that

$$\frac{dP}{dz} = -\left(\frac{q'}{q} + \frac{ik\rho^2}{2}\frac{q'}{q^2}\right)P. \tag{F.27}$$

But the terms appearing in the brackets in Eq. (F.27) are the same terms we have just analyzed, so under the same condition given by either Eq. (F.21) or Eq. (F.24) we have

$$\left|\frac{dP}{dz}\right| \ll \left|2k_p P\right|. \tag{F.28}$$

Using Eq. (F.28) we see that Eq. (F.5) can also be written as

$$\left|\frac{\partial^2 P}{\partial z^2}\right| \ll \left|4k_p^2 P\right|. \tag{F.29}$$

If we differentiate Eq. (F.27) once more and assume that again Eq. (F.19) and (F.22) are satisfied for the terms of those forms that appear in the expression for d^2P/dz^2, then Eq. (F.29) will be satisfied if for the one remaining term we have

$$\left| \frac{k_p^2 \rho^4}{4} \left(\frac{q'}{q^2} \right)^2 P \right| << 4k_p^2 |P|. \tag{F.30}$$

[The details are not given here as they are very similar to those just presented for proving Eq. (F.25)]. Thus, the paraxial condition of Eq. (F.29) becomes

$$\left| \left(\frac{\rho^2}{2} \frac{q'}{q^2} \right)^2 \right| << 4 \tag{F.31}$$

which is certainly satisfied if we require

$$\left(\frac{\rho^2}{2} \left| \frac{q'}{q^2} \right| \right)^2 << 1. \tag{F.32}$$

But if we take the square root of both sides of Eq. (F.32) we obtain Eq. (F.22) again so that the condition for the paraxial approximation to be valid for a Gaussian beam is once more either Eq . (F.21) or Eq. (F.24).

We can also view the conditions of Eqs. (F.21), (F.24) in terms of the asymptotic beam growth angle. Placing this paraxial approximation condition into Eq. (F.16) gives

$$\theta_{1/e} = \tan^{-1}\left(\frac{\lambda_p}{\pi w_0} \right) << \tan^{-1}\left(\sqrt{2} \right)$$

which is satisfied if $\theta_{1/e} << 54.7°$. If we keep the beam growth angle to about half this angle ($\theta_{1/e} \cong 30°$) we might expect the propagating Gaussian beam will not violate significantly the paraxial condition. Angular values of this size are consistent with the angular limits on the paraxial approximation discussed in Chapter 9 using plane waves and spherical waves.

F.3 Transmission/Reflection of a Gaussian Beam

As shown previously a circularly symmetrical Gaussian beam is completely determined by the amplitude, $\tilde{P}(z) = P_0/q(z)$, and the phase parameter,

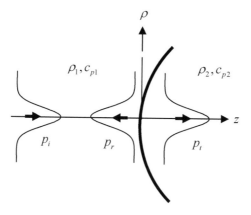

Fig. F.4. Transmission and reflection of a circular cross-section compressional wave Gaussian beam at a spherically curved interface between two media.

$q(z)$. The *propagation law* of Eq. (F.9) describes how both these amplitude and phase terms change as this Gaussian beam propagates. Here we want to define the corresponding *transmission and reflection laws* when the symmetric Gaussian beam strikes a spherically curved interface of radius, R_0, at normal incidence, as shown in Fig. F.4. In this case, both axially symmetric transmitted and reflected Gaussian beams are generated. The incident, transmitted, and reflected Gaussian beams can all be written in general as

$$p_i = \tilde{P}_i(z)\exp\left[ik_{p1}z + i\omega t_0\right]\exp\left[\frac{ik_{p1}\rho^2}{2q_i(z)}\right]$$

$$p_r = \tilde{P}_r(z_2)\exp\left[ik_{p1}z_2 + i\omega t_0\right]\exp\left[\frac{ik_{p1}\rho^2}{2q_r(z_2)}\right] \qquad \text{(F.33)}$$

$$p_t = \tilde{P}_t(z)\exp\left[ik_{p2}z + i\omega t_0\right]\exp\left[\frac{ik_{p2}\rho^2}{2q_t(z)}\right],$$

where (p_i, p_r, p_t) are the incident, reflected and transmitted wave pressures and $k_{pm} = \omega/c_{pm}$ ($m = 1,2$) are the wave numbers for the first and second media, respectively, as shown in Fig. F.4. The z-coordinate here is taken with its origin at the interface (see Fig. F.4). Both the incident and transmitted waves are propagating in the $+z$ direction, but the reflected wave is propagating in the $z_2 = -z$ direction. Typically, the incident

Gaussian beam will have started out from some fixed position at time $t = 0$ located at a distance, D, from the interface in medium one so that the term, $t_0 = D/c_{p1}$, which appears in all the beams of Eq. (F.33) simply represents the common time delay for all these waves.

At the curved interface, Σ, the boundary conditions require that pressure, p, and the normal velocity, v_z, must be continuous so that we have

$$p_i(\Sigma) + p_r(\Sigma) = p_t(\Sigma)$$
$$v_{iz}(\Sigma) + v_{rz}(\Sigma) = v_{tz}(\Sigma). \tag{F.34}$$

Because we showed in the paraxial approximation the pressure and velocity in the Gaussian beams must satisfy Eq. (F.25), the boundary conditions of Eq. (F.34) can be re-written as

$$p_i(\Sigma) + p_r(\Sigma) = p_t(\Sigma)$$
$$\frac{p_i(\Sigma)}{\rho_1 c_{p1}} - \frac{p_r(\Sigma)}{\rho_1 c_{p1}} = \frac{p_t(\Sigma)}{\rho_2 c_{p2}}, \tag{F.35}$$

where the minus sign arises in Eq. (F.35) since $p_r = \rho_1 c_{p1} v_{z_2} = -\rho_1 c_{p1} v_z$. We will not attempt to satisfy the boundary conditions of Eq. (F.35) exactly, but consistent with the paraxial approximation where all these Gaussian beams are considered as quasi-plane waves confined to a region near the z-axis, we will match the amplitude ($\tilde{P}$) terms in Eq. (F.34) only at the point $z = 0$ and the phase terms to second order in the distance, ρ, from the z-axis. Thus, the boundary conditions of Eq. (F.35) for the amplitude terms become

$$\tilde{P}_i(0) + \tilde{P}_r(0) = \tilde{P}_t(0)$$
$$\frac{\tilde{P}_i(0)}{\rho_1 c_{p1}} - \frac{\tilde{P}_r(0)}{\rho_1 c_{p1}} = \frac{\tilde{P}_t(0)}{\rho_2 c_{p2}} \tag{F.36}$$

(where we can cancel all the phase terms in Eq. (F.36) since they will all be made common in the following discussion). Solving these equations we find

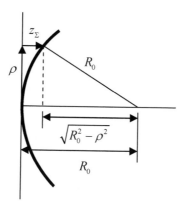

Fig. F.5. The geometry of the spherical interface.

$$\tilde{P}_r(0) = \frac{\rho_2 c_{p2} - \rho_1 c_{p1}}{\rho_1 c_{p1} + \rho_2 c_{p2}} \tilde{P}_i(0) = R_p \tilde{P}_i(0)$$

$$\tilde{P}_t(0) = \frac{2\rho_2 c_{p2}}{\rho_1 c_{p1} + \rho_2 c_{p2}} \tilde{P}_i(0) = T_p \tilde{P}_i(0),$$

(F.37)

where (R_p, T_p) are the plane wave reflection and transmission coefficients (based on pressure ratios – see Appendix D).

Now consider the matching of the phase terms of Eq. (F.35). On the interface, Σ, from Fig. F.5 we see that

$$z_\Sigma = R_0 - \sqrt{R_0^2 - \rho^2}$$

$$\cong R_0 - R_0 \left[1 - \frac{\rho^2}{2R_0^2} + ... \right]$$

$$= \frac{\rho^2}{2R_0}$$

(F.38)

so matching the incident and transmitted Gaussian beam phase terms to second order we have from Eq. (F.33)

$$\frac{ik_{p1}\rho^2}{2R_0} + \frac{ik_{p1}\rho^2}{2q_i(0)} + i\omega t_0 = \frac{ik_{p2}\rho^2}{2R_0} + \frac{ik_{p2}\rho^2}{2q_t(0)} + i\omega t_0,$$

(F.39)

which gives the *transmission law*

$$\frac{1}{q_t(0)} = \left(\frac{c_{p2}}{c_{p1}} - 1\right)\frac{1}{R_0} + \frac{c_{p2}}{c_{p1}}\frac{1}{q_i(0)}. \tag{F.40}$$

Similarly, matching the incident and reflected Gaussian beam phase terms in Eq. (F.33) we find

$$\frac{ik_{p1}\rho^2}{2R_0} + \frac{ik_{p1}\rho^2}{2q_i(0)} + i\omega t_0 = -\frac{ik_{p1}\rho^2}{2R_0} + \frac{ik_{p1}\rho^2}{2q_r(0)} + i\omega t_0 \tag{F.41}$$

to obtain the *reflection law*

$$\frac{1}{q_r(0)} = \frac{2}{R_0} + \frac{1}{q_i(0)}. \tag{F.42}$$

If we let $\left(R_i(0), w_i(0)\right)$ be the wave front curvature and beam width of the incident Gaussian beam at the interface, respectively and similarly define $\left(R_t(0), w_t(0)\right)$ and $\left(R_r(0), w_r(0)\right)$ for the transmitted and reflected beams, since

$$\frac{1}{q_m(0)} = \frac{1}{R_m(0)} + \frac{i\lambda_m}{\pi\left[w_m(0)\right]^2} \qquad m = (i,t,r) \tag{F.43}$$

taking the real and imaginary parts of the transmission and reflection laws show that

$$w_t(0) = w_r(0) = w_i(0) \tag{F.44}$$

i.e. the widths of all the beams at the interface are the same and also

$$\frac{1}{R_t(0)} = \left(\frac{c_{p2}}{c_{p1}} - 1\right)\frac{1}{R_0} + \frac{c_{p2}}{c_{p1}}\frac{1}{R_i(0)}$$

$$\frac{1}{R_r(0)} = \frac{2}{R_0} + \frac{1}{R_i(0)} \tag{F.45}$$

showing how the wave front curvatures of the incident and transmitted/reflected Gaussian beams are related at the interface. For a planar interface, these relations simply reduce to

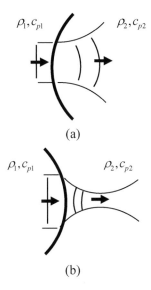

(a)

(b)

Fig. F.6. A Gaussian beam incident on a curved interface where the waist of the incident beam is located at the interface and the corresponding transmitted Gaussian beam and its wave front curvature is shown for **(a)** $c_{p2} > c_{p1}$ and $R_0 > 0$, **(b)** $c_{p2} > c_{p1}$ and $R_0 < 0$.

$$\frac{1}{R_t(0)} = \frac{c_{p2}}{c_{p1}} \frac{1}{R_i(0)}$$
$$\frac{1}{R_r(0)} = \frac{1}{R_i(0)}.$$

(F.46)

From Eq. (F.45) we can gain some understanding of the effects of the curvature of the interface (and the wave speeds) on the transmitted wave if we consider the case where the waist of the incident beam occurs at the interface so that $1/R_i(0) = 0$. Then if we have $c_{p2} > c_{p1}$ and $R_0 > 0$ we see from Eq. (F.45) that $R_t(0) > 0$. In this case the transmitted Gaussian beam is a diverging beam as shown in Fig. F.6 (a). This type of interface is therefore a *defocusing interface* for the transmitted wave. If instead we have $c_{p2} > c_{p1}$ and $R_0 < 0$ we find $R_t(0) < 0$ and the transmitted Gaussian beam is a converging beam as shown in Fig. F.6 (b). In this case the interface acts as a *focusing interface* for the transmitted wave. But for

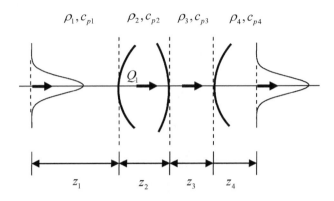

Fig. F.7. The transmission of a circularly symmetric Gaussian beam across multiple spherically curved interfaces.

$c_{p2} < c_{p1}$ the interfaces shown in Fig. F.6 (a), (b) are instead focusing and defocusing interfaces, respectively, for the transmitted wave. These same focusing or defocusing characteristics of curved interfaces were also discussed in Chapter 8, section 8.12. If we again let $1/R_i(0) = 0$ and examine the reflected wave, Eq. (F.45) shows that regardless of the wave speeds we have $R_r(0) > 0$ if $R_0 > 0$ and $R_r(0) < 0$ if $R_0 < 0$, which results in a diverging (defocused) and converging (focused) reflected Gaussian beam, respectively.

F.4 Gaussian Beams at Multiple Interfaces and ABCD Matrices

In the last section we developed the transmission/reflection laws for a symmetrical Gaussian beam at normal incidence to a spherically curved interface. We also have previously obtained the propagation law for a Gaussian beam (Eq. (F.9)). If an axially symmetrical Gaussian beam interacts with multiple spherically curved interfaces at normal incidence we can use those laws and the plane wave transmission/reflection coefficients to obtain the final form of the Gaussian beam (see Fig. F.7, where the beam is shown in undergoing multiple transmissions only, but we will also consider here multiple reflections as well). For example, if a

Gaussian beam starts at $z = 0$ with a pressure amplitude $\tilde{P}(0)$ and phase parameter $q_1(0)$, then after propagation through a distance z_1 we have:

$$p = \tilde{P}(0)\frac{q_1(0)}{q_1(z_1)}\exp\left(ik_{p1}z_1\right)\exp\left[\frac{ik_{p1}}{2}\frac{\rho^2}{q_1(z_1)}\right]. \tag{F.47}$$

If this beam then is transmitted across an interface at a point Q_1 (see Fig. F.7) where $z = z_1$ and propagates from Q_1 a distance z_2 in a second medium we have

$$
\begin{aligned}
p(z_2,\omega) &= \tilde{P}(0)\frac{q_2(Q_1)}{q_2(z_2)}T_{12}\frac{q_1(0)}{q_1(Q_1)} \\
&\cdot\exp\left(ik_{p1}z_1 + ik_{p2}z_2\right)\exp\left[\frac{ik_{p2}}{2}\frac{\rho^2}{q_2(z_2)}\right],
\end{aligned}
\tag{F.48}
$$

where $q_m(z)$ is the q-parameter for the mth media and we will take the z-coordinate for each medium to have as its origin the starting point for the Gaussian beam in that medium. Since point Q_1 is both the ending point for the beam in medium one and the starting point for the beam in medium two we have $q_1(Q_1)=q(z_1)$, $q_2(Q_1)=q_2(0)$ so we can also write Eq. (F.48) as

$$
\begin{aligned}
p(z_2,\omega) &= \tilde{P}(0)\frac{q_2(0)}{q_2(z_2)}T_{12}\frac{q_1(0)}{q_1(z_1)} \\
&\cdot\exp\left(ik_{p1}z_1 + ik_{p2}z_2\right)\exp\left[\frac{ik_{p2}}{2}\frac{\rho^2}{q_2(z_2)}\right].
\end{aligned}
\tag{F.49}
$$

Obviously this same process can be continued for additional transmissions (or reflections). After the interaction with M interfaces, for example, we could write the beam in medium $M+1$ as

$$
\begin{aligned}
p(z_{M+1},\omega) &= \tilde{P}(0)\frac{q_{M+1}(0)}{q_{M+1}(z_{M+1})}\prod_{m=1}^{M}\tilde{T}_{m\,m+1}\frac{q_m(0)}{q_m(z_m)} \\
&\cdot\exp\left[i\sum_{m=1}^{M+1}k_{pm}z_m\right]\exp\left[\frac{ik_{pM+1}}{2}\frac{\rho^2}{q_{M+1}(z_{M+1})}\right],
\end{aligned}
\tag{F.50}
$$

where $\tilde{T}_{m\,m+1}$ is either a transmission or reflection coefficient depending on whether we are considering a transmitted or reflected wave at the mth interface between medium m and $m+1$. The propagation and transmission/reflection laws developed previously then can be written for all $M+1$ media as

Propagation laws: (for $M+1$ media)

$$q_m(z_m) = q_m(0) + z_m \qquad (m=1, M+1) \quad \text{(F.51)}$$

Transmission laws: (for M interfaces)

$$q_{m+1}(Q_m) = \frac{q_m(Q_m)}{\left(\dfrac{c_{p\,m+1}}{c_{p\,m}} - 1\right)\dfrac{q_m(Q_m)}{(R_0)_m} + \dfrac{c_{p\,m+1}}{c_{p\,m}}} \qquad (m=1, M) \quad \text{(F.52)}$$

Reflection laws: (for M interfaces)

$$q_{m+1}(Q_m) = \frac{q_m(Q_m)}{\dfrac{2}{(R_o)_m} q_m(Q_m) + 1} \qquad (m=1, M) \quad \text{(F.53)}$$

If we let a final value of q after propagation/transmission/reflection be q_f and an initial value before propagation/transmission/reflection be q_i, then all these laws can be written in the form [F.1], [F.2]

$$q_f = \frac{Aq_i + B}{Cq_i + D} \qquad \text{(F.54)}$$

or, equivalently,

$$\frac{1}{q_f} = \frac{D(1/q_i) + C}{B(1/q_i) + A}, \qquad \text{(F.55)}$$

and the ABCD parameters can be placed in an ABCD matrix that defines each law:

Propagation laws:

$$\begin{bmatrix} A^d & B^d \\ C^d & D^d \end{bmatrix} = \begin{bmatrix} 1 & z_m \\ 0 & 1 \end{bmatrix} \qquad \text{(F.56)}$$

Transmission laws:

$$
\begin{bmatrix} A^t & B^t \\ C^t & D^t \end{bmatrix} = \begin{bmatrix} 1 & 0 \\ \dfrac{\left(c_{p\,m+1}/c_{p\,m}-1\right)}{\left(R_0\right)_m} & \dfrac{c_{p\,m+1}}{c_{p\,m}} \end{bmatrix}
\tag{F.57}
$$

Reflection laws:

$$
\begin{bmatrix} A^r & B^r \\ C^r & D^r \end{bmatrix} = \begin{bmatrix} 1 & 0 \\ \dfrac{2}{\left(R_0\right)_m} & 1 \end{bmatrix}
\tag{F.58}
$$

A remarkable feature of writing the laws in this fashion is that even after multiple propagations and transmissions/reflections the final and starting q-values can still be related in the forms of Eq. (F.54) and Eq. (F.55) as

$$
q_f = \frac{A^G q_i + B^G}{C^G q_i + D^G}
\tag{F.59}
$$

and

$$
\frac{1}{q_f} = \frac{D^G\left(1/q_i\right)+C^G}{B^G\left(1/q_i\right)+A^G},
\tag{F.60}
$$

where the "global" ABCD matrix components appearing in Eq. (F.59) and Eq. (F.60) can be obtained from a matrix multiplication of all the individual propagation, transmission, and reflection ABCD matrices that define a particular set of beam propagations or interface interactions. One can easily prove this fact by merely placing Eq. (F.54) for one ABCD matrix into Eq. (F.54) involving a second ABCD matrix and showing that the resulting equation again is in the same form of Eq. (F.54) but with ABCD elements corresponding to the matrix multiplication of the original two matrices. For example, after propagation of a beam in medium one followed by a transmission across an interface we would have

$$
\begin{bmatrix} A^G & B^G \\ C^G & D^G \end{bmatrix} = \begin{bmatrix} A^t & B^t \\ C^t & D^t \end{bmatrix} \begin{bmatrix} A^d & B^d \\ C^d & D^d \end{bmatrix}
\tag{F.61}
$$

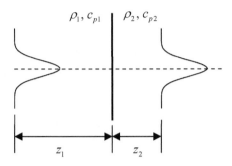

Fig. F.8. Propagation of a Gaussian beam across a plane interface.

and this same process can be continued for any number of interactions [Note: the order of the matrix multiplications is important. In the above example propagation occurs first, followed by transmission, but in multiplying the ABCD matrices this order is reversed]. Using ABCD matrices in this fashion makes it very easy to follow a Gaussian beam through multiple   interfaces and similar ABCD matrices are commonly used in the laser science field to describe the interaction a Gaussian laser beam with multiple optical elements such as lenses, mirrors, etc. As a simple example, consider the propagation of a Gaussian beam through a distance z_1 in medium one followed by transmission across a plane interface, and then propagation through a distance z_2 in medium two (see Fig. F.8). In this case we have

$$\begin{bmatrix} A^G & B^G \\ C^G & D^G \end{bmatrix} = \begin{bmatrix} 1 & z_2 \\ 0 & 1 \end{bmatrix}\begin{bmatrix} 1 & 0 \\ 0 & c_{p2}/c_{p1} \end{bmatrix}\begin{bmatrix} 1 & z_1 \\ 0 & 1 \end{bmatrix}$$

$$= \begin{bmatrix} 1 & z_1 + \left(c_{p2}/c_{p1}\right)z_2 \\ 0 & \left(c_{p2}/c_{p1}\right) \end{bmatrix}$$

(F.62)

so that Eq. (F.59) yields

$$q_2\left(z_2\right) = \frac{c_{p1}}{c_{p2}}\left[q_1\left(0\right) + z_1 + \frac{c_{p2}}{c_{p1}}z_2\right].$$

(F.63)

This shows that the phase terms of the incident beam propagating, which is given by:

$$\frac{ik_{p1}}{2}\frac{\rho^2}{q_1(0)+z_1}$$

(F.64)

becomes for the transmitted beam

$$\frac{ik_{p1}}{2}\frac{\rho^2}{q_1(0)+\left(z_1+\left(c_{p2}/c_{p1}\right)z_2\right)},$$

(F.65)

which looks exactly like the incident beam term with the replacement $z_1 \rightarrow z_1 + \left(c_{p2}/c_{p1}\right)z_2$. This same behavior is discussed in Chapter 8, section 8.5 when examining the on-axis pressure for a circular piston transducer in the paraxial approximation.

In a single medium problem the phase term in the Gaussian beam can be written as

$$\frac{ik_{p1}}{2}\frac{\rho^2}{q_1(z_1)}=\frac{ik_{p1}}{2}\frac{\rho^2}{\left[\dfrac{A_1^d q_1(0)+B_1^d}{C_1^d q_1(0)+D_1^d}\right]},$$

(F.66)

where the propagation ABCD matrix components for medium one are $A_1^d = D_1^d = 1, B_1^d = z_1, C_1^d = 0$. By using the global ABCD matrix formed from the individual ABCD matrices for a multiple medium problem, the phase term in Eq. (F.50) can also be written in the same form where

$$\frac{ik_{pM+1}}{2}\frac{\rho^2}{q_{M+1}(z_{M+1})}=\frac{ik_{pM+1}}{2}\frac{\rho^2}{\left[\dfrac{A^G q_1(0)+B^G}{C^G q_1(0)+D^G}\right]}.$$

(F.67)

In a single medium case the amplitude coefficient of the Gaussian beam contains the term

$$\frac{q_1(0)}{q_1(z_1)}=\frac{q_1(0)}{\left[A_1^d q_1(0)+B_1^d\right]}.$$

(F.68)

The similar part of the amplitude coefficient in Eq. (F.50) for a multiple medium problem contains a series of products of the same form given on the left side of Eq. (F.68). Consider, for example, the first two products given by:

$$\frac{q_2(0)\, q_1(0)}{q_2(z_2)\, q_1(z_1)} \tag{F.69}$$

and the global ABCD matrix corresponding to propagation in medium one, transmission across the first interface, and propagation in medium two:

$$\begin{bmatrix} A^G & B^G \\ C^G & D^G \end{bmatrix} = \begin{bmatrix} A_2^d & B_2^d \\ C_2^d & D_2^d \end{bmatrix}\begin{bmatrix} A_1^t & B_1^t \\ C_1^t & D_1^t \end{bmatrix}\begin{bmatrix} A_1^d & B_1^d \\ C_1^d & D_1^d \end{bmatrix}, \tag{F.70}$$

where $\left(A_m^d, B_m^d, C_m^d, D_m^d\right)$ are the ABCD matrix components for propagation in medium m, and $\left(A_m^t, B_m^t, C_m^t, D_m^t\right)$ are the ABCD components for transmission across the mth interface. We have

$$\frac{1}{q_2(z_2)} = \frac{1}{A_2^d q_2(0) + B_2^d} \tag{F.71}$$

$$q_2(0) = \frac{q_1(z_1)}{C_1^t q_1(z_1) + D_1^t}$$

so combining these two relations we find

$$\frac{1}{q_2(z_2)} = \frac{C_1^t q_1(z_1) + D_1^t}{A' q_1(z_1) + B'} = \frac{\left[q_1(z_1)/q_2(0)\right]}{A' q_1(z_1) + B'}, \tag{F.72}$$

where

$$\begin{bmatrix} A' & B' \\ C' & D' \end{bmatrix} = \begin{bmatrix} A_2^d & B_2^d \\ C_2^d & D_2^d \end{bmatrix}\begin{bmatrix} A_1^t & B_1^t \\ C_1^t & D_1^t \end{bmatrix}$$

$$= \begin{bmatrix} A_2^d & B_2^d \\ C_2^d & D_2^d \end{bmatrix}\begin{bmatrix} 1 & 0 \\ C_1^t & D_1^t \end{bmatrix}. \tag{F.73}$$

We then can substitute $q_1(z_1) = A_1^d q_1(0) + B_1^d$ into $A' q_1(z_1) + B'$ to obtain

$$A' q_1(z_1) + B' = A^G q_1(0) + B^G \tag{F.74}$$

in terms of the global matrix elements $A^G = A' A_1^d$, $B^G = A' A_1^d + B'$. This also follows by writing Eq. (F.70) as

$$\begin{bmatrix} A^G & B^G \\ C^G & D^G \end{bmatrix} = \begin{bmatrix} A' & B' \\ C' & D' \end{bmatrix} \begin{bmatrix} A_1^d & B_1^d \\ C_1^d & D_1^d \end{bmatrix}$$

$$= \begin{bmatrix} A' & B' \\ C' & D' \end{bmatrix} \begin{bmatrix} A_1^d & B_1^d \\ 0 & 1 \end{bmatrix}. \tag{F.75}$$

Thus, from Eqs. (F.72) and (F.74) we can write

$$\frac{q_2(0)}{q_2(z_2)} \frac{q_1(0)}{q_1(z_1)} = \frac{q_1(0)}{A^G q_1(0) + B^G}, \tag{F.76}$$

which has exactly the same form as for the single medium case (Eq. (F.68)). We can continue this process and consider all the other pairs of amplitude terms in Eq. (F.50) in exactly the same manner and so obtain

$$\frac{q_{M+1}(0)}{q_{M+1}(z_{M+1})} \prod_{m=1}^{M} \frac{q_m(0)}{q_m(z_m)} = \frac{q_1(0)}{A^G q_1(0) + B^G}, \tag{F.77}$$

where now $\left(A^G, B^G\right)$ are elements of the global ABCD matrix for all the media and interfaces involved in going from medium one to medium $M+1$.

We can also define a global transmission/reflection coefficient, $\mathcal{T} = \prod_{m=1}^{M} \tilde{T}_{m\,m+1}$ and a propagation delay term $t_0 = \sum_{m=1}^{M+1} z_m / c_{pm}$ and write Eq. (F.50) as

$$p(z_{M+1}, \omega) = \mathcal{T} \tilde{P}(0) \frac{q_1(0)}{A^G q_1(0) + B^G}$$

$$\cdot \exp[i\omega t_0] \exp\left[\frac{ik_{p\,M+1}}{2} \frac{\rho^2}{A^G q_1(0) + B^G} \frac{}{C^G q_1(0) + D^G} \right], \tag{F.78}$$

which is in exactly the same form as for the propagation of a Gaussian beam in a single medium where using the same notation we have:

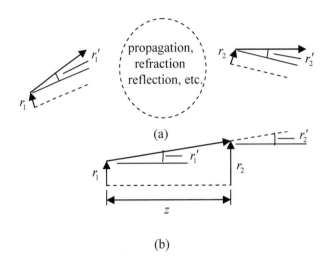

Fig. F.9. (a) A paraxial geometrical ray before and after a general ray interaction, and (b) the special case of propagation of the ray over a distance, z. The central ray is the dashed line and the paraxial ray is assumed to have a small distance from the central ray and a small slope relative to the central ray.

$$p\left(z_1, \omega\right) = \tilde{P}(0) \frac{q_1(0)}{A_1^d q_1(0) + B_1^d}$$

$$\cdot \exp\left[ik_{p1}z_1\right] \exp\left[\frac{ik_{p1}}{2} \frac{\rho^2}{\dfrac{A_1^d q_1(0) + B_1^d}{C_1^d q_1(0) + D_1^d}}\right]. \tag{F.79}$$

In Chapter 9 it is shown that even in more general Gaussian beam problems one can use ABCD matrices, but for those cases the scalar $\left(A, B, C, D\right)$ components are replaced by 2x2 matrices $\left(\mathbf{A}, \mathbf{B}, \mathbf{C}, \mathbf{D}\right)$.

The ABCD matrices used here for our Gaussian beam problems are closely related to the same ABCD matrices used in geometrical optics to facilitate the tracing of paraxial rays through optical elements [F.1], [F.2]. Consider, for example a central ray before and after a given interaction (such as propagation through a lens or reflection from a mirror, etc.) as shown in Fig. F.9 (a) and a nearby (paraxial) ray. Let $\left(r_1, r_1'\right)$ be the displacement and slope of the paraxial ray from the central ray before an interaction and let (r_2, r_2') be the same quantities for the paraxial ray after an interaction (see Fig. F.9 (a)). For a paraxial ray that is close to the central ray and at a small slope to that ray it is reasonable to assume that these quantities are linearly related to one another, i.e.

$$\begin{pmatrix} r_2 \\ r_2' \end{pmatrix} = \begin{bmatrix} A & B \\ C & D \end{bmatrix} \begin{pmatrix} r_1 \\ r_1' \end{pmatrix}. \tag{F.80}$$

However, if we define curvatures $R_1 = r_1 / r_1'$, $R_2 = r_2 / r_2'$ we find

$$R_2 = \frac{AR_1 + B}{CR_1 + D}, \tag{F.81}$$

which is of the same form as Eq. (F.59) for the q- parameter in a Gaussian beam. This is perhaps not surprising since the wave front curvature of the Gaussian beam, $R(z)$, is related to $q(z)$ through

$$\mathrm{Re}\left(\frac{1}{q(z)} \right) = \frac{1}{R(z)} \tag{F.82}$$

(where Re denotes "real part of") so that we can view the use of the ABCD matrices for Gaussian beam problems as the extension of the geometrical optics relations for real ray curvatures to corresponding complex q-values that define the Gaussian beam. To demonstrate in a simple case that the geometrical optics ABCD matrices are indeed the same as our Gaussian beam matrices, consider the ABCD matrix for propagation of a ray through a distance, z, as shown in Fig. F.9 (b). Then since $r_2' = r_1'$ and $r_2 = r_1 + r_1' z$ (for small slopes) we have

$$\begin{pmatrix} r_2 \\ r_2' \end{pmatrix} = \begin{bmatrix} 1 & z \\ 0 & 1 \end{bmatrix} \begin{pmatrix} r_1 \\ r_1' \end{pmatrix}, \tag{F.83}$$

which is identical to the propagation ABCD matrix of Eq. (F.56). In Chapter 9, this same example is discussed in a more general context (see section 9.4).

F.5 Multi-Gaussian Beam Modeling

In 1988 Wen and Breazeale [F.4] showed that by the superposition of only 10 Gaussian beams, one could generate an accurate model of the radiated wave field of a circular planar piston transducer. Since commercial ultrasonic NDE transducers can often be modeled as piston transducers, this multi-Gaussian beam model is a very effective tool for simulating the sound beams generated in NDE tests. Here we will briefly outline Wen and Breazeale's multi-Gaussian beam model and relate it to our previous Gaussian beam discussions. Chapter 9 also gives many more details of multi-Gaussian beam models.

At $z = 0$ for a single Gaussian beam the pressure is given by (see Eq. (F.11))

$$p = \frac{P_0}{q_0} \exp\left[\frac{ik_p \rho^2}{2q_0} \right].$$
(F.84)

Wen and Breazeale wrote Eq. (F.81) instead as

$$\frac{p}{\rho c_p v_0} = A \exp\left[-B\rho^2 / a^2 \right]$$
(F.85)

and used a non-linear least squares optimization procedure to determine a set of 10 complex A and B coefficients that produced a constant velocity, v_0, on the face of a circular piston transducer of radius a located at $z = 0$, as discussed in more detail in Chapter 9. The wave field generated by the superposition of 10 Gaussian beams with the starting forms of Eq. (F.81) is shown in Chapter 9 to match well the exact wave field of the piston transducer except close to the transducer face. Note that these A, B coefficients represent Gaussian beams of different waist locations, widths and amplitudes since

$$A = \frac{-P_0 / \rho c_p v_0}{z_0 + iz_c}$$

$$B = \frac{ik_p a^2}{2(z_0 + iz_c)}$$
(F.86)

from which it follows that

$$-q_0 = z_0 + i z_c = \frac{i k_p a^2 / 2}{B}$$

$$\frac{P_0}{\rho c_p v_0} = -A \frac{i k_p a^2 / 2}{B}.$$

(F.87)

A multi-Gaussian beam model of a transducer uses the A, B coefficients directly to synthesize the transducer wave field so there is no advantage in expressing the wave field in terms of Gaussian waist locations and width parameters, as is commonly done in the laser science literature. Instead, using Eq. (F.87) we can write the propagating Gaussian beam in a single medium (see Eq. (F.11)) in terms of A and B directly:

$$p = \frac{\rho c_p v_0 A}{1 + i B z / D_R} \exp\left[i k_p z \right] \exp\left[\frac{i k_p \rho^2}{2} \frac{(i B / D_R)}{(1 + i B z / D_R)} \right],$$

(F.88)

where $D_R = k_p a^2 / 2$ is called the *Rayleigh distance* for the piston transducer, a quantity that is analogous to the confocal parameter for a Gaussian beam. Using the ten A, B coefficients of Wen and Breazeale then yields a multi-Gaussian transducer beam model for a single medium given by

$$p = \sum_{n=1}^{10} \frac{\rho c_p v_0 A_n}{1 + i B_n z / D_R} \exp\left[i k_p z \right] \exp\left[\frac{i k_p \rho^2}{2} \frac{(i B_n / D_R)}{(1 + i B_n z / D_R)} \right],$$

(F.89)

which is the form used in Chapter 9 (see Eq. (9.134)). If we define starting values for each of the Gaussian beams in Eq. (F.89) as

$$\left[q_1(0) \right]_n = \frac{-i k_p a^2 / 2}{B_n}$$

$$\tilde{P}_n(0) = \rho c_p v_0 A_n$$

(F.90)

then using Eq. (F.78) we have a very simple model for the field of a piston transducer after multiple transmissions or reflections:

$$p(z_{M+1},\omega) = \sum_{n=1}^{10} \mathcal{T} \tilde{P}_n(0) \frac{\left[q_1(0)\right]_n}{A^G \left[q_1(0)\right]_n + B^G}$$

$$\cdot \exp\left[i\omega t_0\right] \exp\left[\frac{ik_{p\,M+1}}{2} \frac{\rho^2}{\dfrac{A^G \left[q_1(0)\right]_n + B^G}{C^G \left[q_1(0)\right]_n + D^G}}\right]. \tag{F.91}$$

In Chapter 9, use is made of the Wen and Breazeale coefficients and the corresponding $(\mathbf{A}, \mathbf{B}, \mathbf{C}, \mathbf{D})$ matrices in this same manner to obtain transducer wave fields much more complex multiple media problems.

F.6 References

F.1 Goldsmith PF (1998) Quasi-optical systems. IEEE Press, Piscataway, NJ

F.2 Siegman AE (1986) Lasers. University Science Books, Mill Valley, CA

F.3 Heyman E, Felsen LB (2001) Gaussian beam and pulsed-beam dynamics: complex-source and complex-spectrum formulations within and beyond paraxial asymptotics. J. Opt. Soc. Am. 18: 1588-1611

F.4 Wen JJ, Breazeale MA (1988) A diffraction beam field expressed as the superposition of Gaussian beams. J. Acoust. Soc. Am. 83: 1752-1756

F.5 Huang D, Breazeale MA (1999) A Gaussian finite-element method for description of sound diffraction. J. Acoust. Soc. Am. 106: 1771-1781

F.7 Exercises

1. Consider the propagation of a central ray and a paraxial ray at oblique incidence across a plane interface where the paraxial ray lies in the x-z plane (see Fig. F.10). Relate the distances (x_1, x_2) to each other to first order in terms of the angles (θ_1, θ_2) directly from the geometry. Also, using Snell's law, which must be satisfied for both the central ray and the paraxial ray, relate the slopes (x_1', x_2') to each other to first order in terms of (θ_1, θ_2) and (c_{p1}, c_{p2}). Combining these results, obtain the ABCD matrix for this case, where

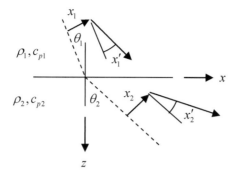

Fig. F.10. A central ray (dashed line) and a nearby paraxial ray (arrows) being transmitted at oblique incidence across a planar interface.

$$\begin{Bmatrix} x_2 \\ x_2' \end{Bmatrix} = \begin{bmatrix} A_x & B_x \\ C_x & D_x \end{bmatrix} \begin{Bmatrix} x_1 \\ x_1' \end{Bmatrix}$$

Since there is no change in direction for the central ray in the y-direction, the values (y_2, y_2') and (y_1, y_1') for the paraxial ray are related by a corresponding ABCD matrix valid near normal incidence. From your previous results let $\theta_1, \theta_2 \to 0$ to show that in this case

$$\begin{Bmatrix} y_2 \\ y_2' \end{Bmatrix} = \begin{bmatrix} A_y & B_y \\ C_y & D_y \end{bmatrix} \begin{Bmatrix} y_1 \\ y_1' \end{Bmatrix} = \begin{bmatrix} 1 & 0 \\ 0 & c_{p2}/c_{p1} \end{bmatrix} \begin{Bmatrix} y_1 \\ y_1' \end{Bmatrix}$$

Because there are different ABCD matrices in the x- and y-directions, an incident Gaussian beam of circular cross-section, where the phase term is given at the interface by

$$\exp\left[\frac{ik_{p1}}{2} \frac{\rho^2}{q_i} \right] = \exp\left[\frac{ik_{p1}}{2} \frac{\left(x^2 + y^2\right)}{q_i} \right]$$

will be changed, upon transmission through the interface into a Gaussian beam of elliptical cross section, where the phase term is:

$$\exp\left[\frac{ik_{p2}}{2}\left(\frac{x^2}{\dfrac{A_x q_i + B_x}{C_x q_i + D_x}} + \frac{y^2}{c_{p1}q_i}\right)\right]$$

Thus, for oblique incidence problems we can no longer consider only circular cross-section Gaussian beam solutions of the paraxial wave equation but must treat more general solutions for elliptical cross-section Gaussian beams. For oblique incidence on curved interfaces the transmitted Gaussian beam can also be rotated, resulting in Gaussian beams with phase terms containing both quadratic and mixed products of the coordinates, i.e. (x^2, xy, y^2). Chapter 9 treats these more general cases by seeking Gaussian beam solutions to the paraxial wave equation given by

$$p = P(z)\exp\left(ik_{p1}z\right)\exp\left(\frac{i\omega}{2}\mathbf{X}^T\mathbf{M}_p(z)\mathbf{X}\right)$$

where $\mathbf{X} = [x, y]^T$ and $\mathbf{M}_p$ is a 2x2 symmetrical matrix. For a circular cross-section Gaussian beam then we have

$$\mathbf{M}_p(z) = \begin{bmatrix} \dfrac{1}{c_{p1}q(z)} & 0 \\ 0 & \dfrac{1}{c_{p1}q(z)} \end{bmatrix}$$

2. Wen and Breazeale also defined 15 Gaussian beam coefficients (A_n, B_n) that improve on the modeling of a circular planar piston transducer in comparison to their original 10 coefficients [F.5]. Use the MATLAB function gauss_c15 that returns those 15 coefficients and write a MATLAB script that obtains the normalized pressure field, $p/\rho_1 c_{p1} v_0$, for a 6.35 mm radius piston transducer radiating through spherically curved water-steel interface ($R_0 = 76$ mm) at a frequency of 5 MHz (see Fig. F.11) and plots the magnitude of the on-axis normalized pressure versus the distance z_2 in the steel from $z_2 = 0$ to $z_2 = 50$ mm. Modify the script and consider the same case but where $R_0 = -76$ mm.

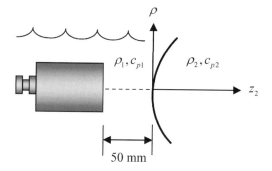

Fig. F.11. A circular planar piston transducer radiating a sound beam through a spherically curved fluid-solid interface.

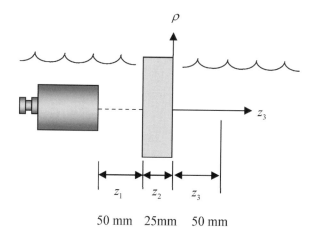

Fig. F.12. Radiation of an immersion transducer through an aluminum plate.

3. Rewrite the scripts of problem 2 so that they display a 2-D image of the magnitude of the normalized pressure (i.e. normalized pressure versus (ρ, z_2)) in the steel for both the defocusing and focusing interfaces considered there.

4. Use the ABCD matrices and the 15 coefficients of Wen and Breazeale contained in the MATLAB function gauss_c15 to write a MATLAB script that obtains the normalized pressure, $p/\rho_1 c_{p1} v_0$, in a sound beam that is directly transmitted (with no reflections) from a 10 MHz, 6.35 mm radius planar piston transducer through the aluminum plate shown in Fig. F.12. The script should plot the magnitude of the normalized pressure versus ρ at $z_3 = 50$ mm.

5. Modify the script of problem 4 so that the normalized pressure transmitted through the plate is evaluated at many frequencies for $\rho = 0, z_3 = 50$ mm and is multiplied at each frequency by the MATLAB function spectrum1 written for exercise 1 in Appendix A, where the center frequency fc = 10 MHz and the bandwidth bw = 4 MHz. Evaluate this product at 1024 positive frequencies ranging from zero to 100 MHz and use the Fourier transform IFourierT defined in Appendix A to obtain a time-domain pulse. Plot that pulse versus time. In evaluating the normalized pressure, ignore the $\exp\left[ik_{p1}z_1 + ik_{p2}z_2 + ik_{p3}z_3\right]$ propagation term which simply produces a time delay.

G MATLAB Functions and Scripts

A number of MATLAB functions and scripts are described in the text. The MATLAB code listings for all these functions/scripts are available on the web at www.springer.com/978-0-387-49061-8. The web site also has the MAT-files that contain the experimental data used in various model/ experimental comparisons. In this Appendix we will summarize the MATLAB functions and scripts discussed in the text and give code listings for those functions which are not explicitly defined elsewhere. Note that a number of the MATLAB functions used in the exercises are not given here, but they can be found on the web site. In some cases those MATLAB functions are given in p-code form instead of ordinary open text m-files so that they can be used by students as unknown "black boxes" in some of the exercise problems given at the end of the Chapters. Those p-code functions were generated in MATLAB release 7.0 so that they will not work with earlier versions of MATLAB. If this poses a problem, there are alternate p-code versions of the same functions on the web that were generated in release 6.5 and are identified by having a "65" in their function name.

G.1 Fourier Analysis Functions

```
Vf = FourierT(vt , dt);
vt = IFourierT(Vf , dt);
y = s_space(a, b, M);
y = c_shift(vt, N);
y = t_shift(t, N);
y = Wiener_filter(O, I, e);
Vf =lp_filter(f, fstart, fend);
y = system_f (f, amp, fc, bw);
```

The functions Fourier_T(vt, dt) and IFourierT(Vf, dt) perform the Fast Fourier transform and its inverse on a set of sampled values in the time and frequency domains, respectively. Besides those sampled values the sampling interval in the time domain, dt, is the only other input parameter to these functions. These discrete Fourier transforms implement the Fourier transform and its inverse as defined in Appendix A. Code listings of both these functions can be found in Appendix A.

The function s_space(a, b, M) is a utility function that produces a set of M evenly spaced sampled values from a to $(b - dx)$, where $dx = (b - a)/M$ is the sample spacing. These are precisely the sampled values that are used in Fourier analysis so that this function is used primarily to generate the time and frequency axes to use in conjunction with FourierT and IFourierT. This function is discussed in Chapter 12 and the code listing for the function can be found in section G.8.

The function c_shift(vt, N) moves the last N components of the vector vt into the first N component places and shifts the remaining components of vt to follow those N components. This type of shift is called a circular shift. This shift is sometimes needed since IFourierT always generates a set of sampled time domain values over the time interval [0, T), where $T = 1/df$ is the length of the total time window and df is the sample interval in the frequency domain. However, if the sampled time domain function values are non-zero before time $t = 0$, these sampled values at negative times will appear in the upper half of the window and the function will appear to be "split". This splitting can be removed by applying c_shift to the sampled values with a large enough value for N. For example, consider the following eight function values:

```
>> f= [ 1 1 0 0 0 0 2 2];
```

If we apply c_shift to this function with N = 2 we obtain

```
>> fs =c_shift(f,2)

fs =

    2   2   1   1   0   0   0   0
```

The use of c_shift on a sampled time-domain signal will also mean that the corresponding sampling times will be incorrect. In some cases this is not significant, but if one wants to also change the time axis appropriately to preserve the original time values then the function t_shift(t, N) can also be used in conjunction with c_shift. As a simple example of the action of t_shift, consider the following eight time domain values:

```
>> t = [ 0  0.1  0.2  0.3  0.4  0.5  0.6  0.7];
```

Now, apply t_shift to this sampled time axis with N = 2:

>> t_shift(t, 2)

ans =

  -0.2000   -0.1000   0   0.1000   0.2000   0.3000   0.4000
0.5000

      In most cases t_shift is used in conjunction with c_shift in plotting a sampled function. For example, the MATLAB command plot(t_shift(t, N), c_shift(V, N)) will do a circular shift of the last N sampled values contained in the vector V and also modify the sampled values of the time axis contained in the vector t appropriately so that the original time origin is changed appropriately in the resulting plot. The code listings for both c_shift and t_shift are given in section G.8.

      The function Wiener_filter(O, I, e) is described in Appendix C where its code listing is also given. This function takes the sampled frequency domain values contained in vectors O and I and performs a deconvolution. A direct deconvolution would simply be an element by element division, i.e. in MATLAB we would compute

>> G = O./I ;

      The Wiener filter function modifies this division process and desensitizes it to noise, as discussed in Appendix C. The constant, e, which is the other input to this function, is used in the Wiener filter to represent the noise level present. Generally, small values such as e = 0.01 to 0.05 work well in many ultrasonic NDE problems.

      The function lp_filter(f, fstart, fend) generates a low-pass filter that is unity below a frequency value, fstart, and smoothly goes to zero at the value, fend. Above fend the function is zero. Multiplying a model-based function (that is defined in the frequency domain) by this low-pass filter will remove the high frequency content and allow one to perform an inverse FFT on the product as long as fend is chosen below the Nyquist frequency. The code listing for this function is given in section G.8.

      The function system_f (f, amp, fc, bw) models the behavior of a system function in the frequency domain with a Gaussian that is defined by its amplitude, center frequency, and bandwidth. The code listing for this function is given in section G.8.

G.2 Setup Functions

setup = setup_maker;
display_setup;

The function setup_maker provides a way of storing all the input parameters needed to generate the ultrasonic measurement models described in Chapter 12. The function places a default set of values for all these parameters in a MATLAB structure called setup which then can be modified by the user to produce any set of parameters needed to describe a particular ultrasonic system configuration. When modifying the setup structure, it is convenient to be able to examine its contents to check that the proper parameters are present. This can be easily done with the function display_setup which lists all the current setup parameters. There are no input arguments for either of these functions. The code listings for both setup_maker and display_setup are given in Chapter 12.

G.3 Ultrasonic Beam Modeling Functions

[A, B] = gauss_c15;
T = fluid_solid(setup);
T =smooth_solid(setup);
V = init_z(setup);
[Vf, setup] = MGbeam(setup);
[Vi, setup] = I_MGbeam(setup);

The multi-Gaussian beam model described in Chapter 9 and implemented in software in Chapter 12 uses a set 15 complex-valued amplitude and phase coefficients (A,B) to model the sound beam generated by an ultrasonic transducer. Those 15 coefficients are returned by the function gauss_c15. There are no input arguments for gauss_c15. The code listing for gauss_c15 is given in Chapter 12.

When a sound beam passes through an interface, changes in the amplitude of the beam are controlled by the plane wave transmission coefficient as discussed in Chapter 9. The expressions for the plane wave transmission coefficients for a fluid-solid interface are obtained in Appendix D and those expressions are coded in the MATLAB function fluid_solid, whose code listing is given in Chapter 12. The only input argument of the function fluid_solid is the setup structure. This function extracts the necessary material and geometry parameters needed from that structure and returns the appropriate plane wave transmission coefficient for the incident and transmitted wave types specified in setup.

In pulse-echo angle beam testing a P-wave transducer is placed on a solid wedge instead of being in a fluid. This wedge is then placed in "smooth" contact with the surface of the component being tested, as described in Appendix D. The expressions for the transmission coefficients for such a setup are given in Appendix D. The function smooth_solid, whose code listing is given in section G.8, performs identically to the fluid_solid function but returns instead the appropriate transmission coefficient for the angle beam testing setup.

In performing ultrasonic beam modeling studies, one may want to perform beam calculations at a single frequency for multiple locations in the beam field or synthesize a pulse by performing beam field calculations at many frequencies for a single location or multiple locations. Thus, the setup parameters setup.f, setup.geom.z1, setup.geom.z2, setup.geom.x2, and setup.geom.y2 may be scalars, vectors, or matrices depending on the type of study one wants to perform. The function init_z(setup) decides what the largest size of matrix is present for these parameters and simply outputs an empty array of values of that size. That empty array is then filled with beam field (velocity) values when the actual beam model calculations are performed by a beam model function. This pre-allocation of an empty array is done for efficiency. The code listing for the init_z function is given in Chapter 12.

The function MGbeam(setup) uses the multi-Gaussian beam theory described in Chapter 9 and the specific implementation described in Chapter12 to return the complex-valued velocity amplitude of the transducer sound field generated in a pulse-echo immersion test with input parameters as specified in the setup structure. In performing these calculations MG_beam also uses the functions gauss_c15, fluid_solid, and init_z described previously. The only input argument to the MGbeam function is the setup structure. The outputs of MGBeam are the beam velocity amplitude and a new setup structure that contains updated values for setup.wave.c1, setup.wave.c2, and setup.wave.T12 parameters (see Chapter 12). The code listing for MGbeam is given in Chapter 12.

The function I_MGbeam(setup) uses all the same inputs and functions as MGbeam but instead of the velocity amplitude returned by MGbeam, this function returns a spatial integral of the square of the velocity amplitude, as required by the measurement model for long cylindrical reflectors such as a side-drilled hole (see Chapter 12). Like MGbeam, I_MGbeam also returns a new setup structure that contains updated values for setup.wave.c1, setup.wave.c2, and setup.wave.T12 parameters. The code listing for I_MGbeam is given in Chapter12.

G.4 Flaw Scattering Functions

A = A_void(setup);
A = A_crack(setup);
A = A_SDH(setup);
A = A_void_Psep(setup);
A = A_void_Ssep(setup);
A = A_SDH_Psep(setup);
A = A_SDH_Ssep(setup);
A = A_unity(setup);

As discussed in Chapters 10, 11 and 12 a component of the vector far-field scattering amplitude of a flaw is a quantity that can be used to characterize the flaw response. This quantity appears explicitly as part of an ultrasonic measurement model when the beam variations over the flaw surface are negligible. The functions A_void(setup), A_crack(setup), and A_SDH(setup) return the pulse-echo far-field scattering amplitude component for a void, crack, and a side-drilled hole, respectively, using the Kirchhoff approximation. The incident waves can either be P-waves or S-waves. Code listings for all three of these functions are given in Chapter 12.

Spherical and cylindrical shaped flaws are the only two geometries where one can obtain exact separation of variables solutions for the far- field scattering amplitude of a flaw in a solid. The functions A_void_Psep(setup) and A_void_Ssep(setup) return the pulse-echo scattering amplitudes for a spherical void for incident P-waves or SV-waves, respectively, using the method of separation of variables. The functions A_SDH_Psep(setup) and A_SDH_Ssep(setup) return the pulse-echo scattering amplitudes for a cylindrical void (side-drilled hole) for P-waves and SV-waves, respectively, when the incident wave direction is perpendicular to the axis of the hole. These two functions use a 2-D separation of variables solution for the hole and convert it to a 3-D scattering amplitude normalized by the length of the hole using the relationship described in Chapter 10. The code listings for all four functions that implement these separation of variables solutions are given in section G.8.

When determining the far-field scattering amplitude of a flaw experimentally, one needs to deconvolve a measured flaw response with all those terms in the measurement model except the far-field scattering amplitude term, as discussed in Chapter 13. Those terms can be generated by measurement model function with the far-field scattering amplitude response set equal to one at all frequencies. The function A_unity(setup) simply returns these needed values of unity. The code listing for this function is given in Chapter 13.

G.5 Ultrasonic Measurement Modeling Functions

```
y = attenuate(setup);
s = systf(setup);
s = exp_systf(setup);
[Vf, setup] = TG_PE_MM(setup);
[Vi, setup] = SDH_PE_MM(setup);
```

An ultrasonic measurement model requires an ultrasonic beam model and flaw scattering model to account for the beam propagation and scattering effects present in an ultrasonic measurement. Since the beam model functions MGbeam and I_MGbeam predict the beam amplitudes in ideal (lossless) media, material attenuation effects must be included separately. The function attenuate(setup) returns a frequency dependent attenuation factor that allows us to include these losses based on measured attenuation coefficients placed in the setup structure. The code listing for this function is given in Chapter 12.

An ultrasonic measurement model also requires a specification of the system function that characterizes all the electrical and electromechanical components present in the measurement system. For simulation studies, one can use a model-based system function that mimics the behavior of a real (measured) system function. The function systf(setup) is such a function that returns a purely model-based system function determined by specified amplitude, center frequency, and bandwidth parameters in the setup structure. In contrast, the function exp_systf (setup) uses the measured voltage in a reference experiment to determine the system function experimentally. This sampled voltage and the corresponding sampled time axis must be contained in a MAT-file whose name is contained in the setup structure. The function exp_systf then uses these measured values in combination with other parameters of the reference setup contained in the setup structure to return the system function. Code listings for systf and exp_systf are given in Chapter 12.

The function TG_PE_MM(setup) generates the pulse-echo response of a flaw in an immersion setup, as described in Chapter 12, using the Thompson-Gray measurement model. This measurement model is suitable for modeling the response of a flaw when the beam variations over the flaw surface are negligible, as discussed in Chapter 11. This function returns the output voltage (in the frequency domain) and a new setup structure that contains updated values for setup.wave.c1, setup.wave.c2, and setup.wave.T12 parameters. The code listing for this function is given in Chapter 12.

The function SDH_PE_MM(setup) similarly returns the pulse-echo output voltage (in the frequency domain) for a side-drilled hole in a an immersion setup, as discussed in Chapter 12. This measurement model assumes the beam variations are negligible over the cross-sectional area of the side-drilled hole but accounts for the beam variations over the entire length of the hole. This function also returns a new setup structure that contains updated values for setup.wave.c1, setup.wave.c2, and setup.wave. T12 parameters. The code listing for this function is given in Chapter 12.

G.6 Miscellaneous Functions

```
y= pulserVT(V0, t0, a1, a2, t) ;
y =fresnel_int(x) ;
```

The function pulserVT implements Eq. (2.3) of Chapter 2 which uses the four parameters (V0, t0, a1, a2) to model the open-circuit output voltage of a spike pulser or square wave pulser versus time. The code listing for this function is given in section G.8.The function fresnel_int computes the Fresnel integral, where the argument, x, is the upper limit of that integral. As shown in Chapter 8, this integral appears in modeling rectangular transducers. The code listing for this function is given in section G.8.

G.7 MATLAB Script Examples

```
TG_sphere_example1
TG_sphere_example2
TG_sphere_example3
FBH_example1
SDH_example1
SDH_deconvolve1
```

In Chapters 12 and 13 scripts that implement a number of measurement model examples are given. The script TG_sphere_example1, for example, uses the Thompson-Gray measurement model to calculate the time domain pulse-echo P-wave response of an on-axis spherical pore interrogated by a 12.7 mm diameter, 5 MHz planar probe through a fluid-solid interface at normal incidence. The code for this script is given in Code Listing 12.11. In this case, a model-based system function is used in the calculations. The script TG_sphere_example2 models the same spherical pore considered in TG_sphere_example1 but uses an experimentally determined system function instead to synthesize the time-domain signal. The code for this script is given in Code Listing 12.13. The script TG_sphere_example3 also calculates the time domain response of the same pore contained in the previous two scripts but replaces the planar probe with a 12.46 mm diameter, 172.9 mm focal length focused probe (both of which values are measured effective parameters) and uses an experimentally determined system function for this probe. The modeled response is then compared to a measured signal. The code for this script is given in Code Listing 12.14.

FBH_example1 is a script that illustrates an example of a measurement model calculation where the beam variations over the face of the flaw must be accounted for. In this case the script calculates the time domain pulse-echo P-wave response of an on-axis #8 flat-bottom hole interrogated by a 12.7 mm diameter, 5 MHz planar probe through a fluid-solid interface at normal incidence and compares the modeled response to an experimentally measured signal. The code for this script is given in Code Listing 12.15.

The script SDH_example1 calculates the pulse-echo P-wave time domain response of an on-axis 1 mm diameter side-drilled hole interrogated by a 12.7 mm diameter, 5 MHz planar probe through a fluid-solid interface at normal incidence. The script uses an experimentally determined system function and compares the modeled response to an experimentally measured signal. The code for the script is given in Code Listing 12.19.

In Chapter 13, the script SDH_deconvolve1 demonstrates how a model-based approach can be used to extract the scattering amplitude of a side-drilled hole from a measured signal. This script uses the side-drilled hole measurement model and the measured pulse-echo P-wave time-domain response of an on-axis 1 mm diameter side-drilled hole interrogated by a 12.7 mm diameter, 5 MHz planar probe through a fluid-solid interface at normal incidence to obtain an experimental far field scattering amplitude for the hole by deconvolution.

This experimental result is plotted versus frequency and compared to the theoretical scattering amplitude calculated by the method of separation of variables. The code for this script is given in Code Listing 13.2.

G.8 Code Listings of Some Supporting Functions

Many of the MATLAB functions that implement the examples discussed in this book are given in the Chapters and Appendices. The previous sections describe where those functions can be found. This section gives MATLAB code listings for functions that are not given elsewhere in the text. All of the MATLAB functions, scripts, and experimental data files are on the web at www.springer.com/978-0-387-49061-8.

Code Listing G.1. The MATLAB function s_space for generating sampled values for use in Fourier analysis.

```
function  y = s_space(xstart, xend, num)
% S_SPACE(XSTART,XEND, NUM) generates num evenly spaced sampled
% values from xstart to (xend - dx), where dx is the sample
% spacing. This is useful in FFT analysis where we generate
% sampled periodic functions. Example: generate 1000
% sampled frequencies from 0 to 100MHz via f =s_space(0,100,1000);
% In this case the last value of f will be 99.9 MHz and the
% sampling interval will be 100/1000 =0.1 MHz.
%
ye =linspace(xstart, xend, num+1);
y=ye(1:num);
```

Code Listing G.2. A circular shift function to use with FFT operations

```
function y = c_shift(x, n)
% C_SHIFT moves the last n components of the vector x
% into the first n component places and shifts the
% remaining components of x to follow those n components,
% i.e. this is a circular shift.  Note: x must be row or column vector
%
[nr,nc]= size(x);
if nr == 1
   len = nc;
   y = [x(len-n+1 : end), x(1:len -n)];
elseif nc == 1
   len = nr;
   y = [x(len-n+1 : end); x(1:len -n)];
```

```
else
    error(' c_shift only works with vectors')
end
```

Code Listing G.3. A time shift function to be used with c_shift to preserve the appropriate time axis values.

```
function  y = t_shift(x, n)
% T-SHIFT is used with the C_SHIFT function to change the time axis
% values appropriately so that the time axis is shifted along with the
% function.
% Example use:  plot(t_shift(t, 100), c_shift(fun, 100))
[nr,nc]= size(x);
dx = x(2) -x(1);
if nr == 1
    len = nc;
    y = [x(len-n+1 : end)-x(end)-dx+x(1), x(1:len -n)];
elseif nc == 1
    len = nr;
    y = [x(len-n+1 : end)-x(end)-dx+x(1); x(1:len -n)];
else
    error(' t_shift only works with vectors')
end
```

Code Listing G.4. A low-pass filter for use in Fourier analysis where we have to remove the frequencies above a certain value.

```
function  Vf =lp_filter(f, fstart, fend)
% LP_FILTER(f, fstart, fend) generates a low-pass filter
% which has a value of 1.0 below the frequency value
% fstart and tapers to zero at frequencies above the
% value fend with a cosine function.
% The calling sequence is:
% Vf = lp_filter(f, fstart, fend)

if fend > f(end)
    error( 'fend exceeds max frequency')
end
if fend < fstart
```

```
    error(' fend must be greater than fstart')
end
const = ones(size(f)).*(f < fstart);
taper = cos(pi.*(f-fstart)./(2*(fend-fstart))).*(f >= fstart & f <= fend);
Vf = const + taper;
```

Code Listing G.5. A function that simulates the band-limited behavior of a system function, i.e. where the frequency response is maximum at a particular frequency and has an extent in the frequency domain defined by a bandwidth parameter.

```
function y = system_f (f, amp, fc, bw)
% SYSTEM_F(f, amp, fc, bw) returns the system function as modeled by a
% Gaussian window function of amplitude amp
% centered at frequency fc and with a bandwidth bw defined to
% be the spread in frequency at the half amplitude point in the Gaussian.
% The Gaussian is tapered to zero at frequencies below fc with a
% sine function to guarantee the dc value is always zero.
% For small fc and large bw, this tapering will distort the Gaussian
% The calling sequence for this function is: y =system_f(f, amp, fc, bw);

% compute the 'a' parameter and define system function above and below the
% center frequency
a = sqrt(log(2))/(pi*bw);
s1 = exp(-(2*a*pi*(f - fc)).^2).*(f > fc);
s2 = exp(-(2*a*pi*(f - fc)).^2).*sin(pi*f/(2*fc)).*(f <= fc);
% combine terms to obtain total system function
y = amp*(s1 + s2);
```

Code Listing G.6. A function for calculating the transmission coefficient for refracted P-waves or S-waves at the interface between two solids in smooth (shear stress free) contact. The incident wave must be a P-wave.

```
function T12 = smooth_solid(setup)
% SMOOTH_SOLID(SETUP) computes the P-P (tpp)
% and P-S (tps) transmission coefficients based on velocity ratios
% for two solids in smooth contact. It obtains the necessary input
% parameters from the setup structure and then returns the
% appropriate transmission coefficient
```

```
% Note: If setup.matl.cs1 = 0 the values returned are for a fluid-solid interface.

% get setup parameters
type1 =setup.type1;
type2 =setup.type2;
inc= setup.geom.i_ang;
d1 = setup.matl.d1;
d2 =setup.matl.d2;
cp1 = setup.matl.cp1;
cs1 =setup.matl.cs1;
cp2 =setup.matl.cp2;
cs2 =setup.matl.cs2;


% consistency check (if incident wave in medium 1 is an S-wave
% then can't use this fluid-solid trans. coefficient)

if strcmp(type1, 's')
    error('wrong wave type for medium 1')
end
iang = (inc.*pi)./180;  %change degrees to radians

%calculate sines and cosines of all incident and refracted angles
sinp1 = sin(iang);
cosp1 = sqrt(1-sinp1.^2);
sins1 = (cs1/cp1)*sin(iang);
coss1= sqrt(1-sins1.^2);
sinp2 = (cp2/cp1)*sin(iang);
sins2 =(cs2/cp1)*sin(iang);
    % take into account cosines of refracted angles may be imaginary beyond
    % critical angles
cosp2= (i*sqrt(sinp2.^2 - 1)).*(sinp2 >= 1) + ...
    (sqrt(1 - sinp2.^2)).*(sinp2 < 1);
coss2 = (i*sqrt(sins2.^2 - 1)).*(sins2 >= 1) + ...
    (sqrt(1 - sins2.^2)).*(sins2 < 1);

%calculate transmission coefficients
denom1 = (cp1/cp2).*(cosp2./cosp1).*...
    (4.*((cs1/cp1)^2).*(sins1.*coss1.*sinp1.*cosp1) + ...
    1 - 4.*(sins1.^2).*(coss1.^2));
denom2 = (d2/d1).*(4.*((cs2/cp2)^2).*(sins2.*coss2.*sinp2.*cosp2) ...
                + 1 - 4.*(sins2.^2).*(coss2.^2));
    denom = denom1 + denom2;

tpp = ((2*cp1/cp2).*(1-2*sins1.^2).*(1-2*sins2.^2))./denom;
tps = -((4*cp1*cs2/(cp2^2)).*sinp2.*cosp2.*(1-2*sins1.^2))./denom;
%select appropriate coefficient
```

```
if strcmp(type2, 'p')
   T12 = tpp;
elseif strcmp(type2, 's')
   T12 = tps;
else
   error('wrong wave type specification')
end
```

Code Listing G.7. A function that uses the separation of variables method to calculate the far field pulse-echo P-wave scattering of a spherical void.

```
function Aout = A_void_Psep(setup)
% A_VOID_PSEP  computes the far field P-wave scattering amplitude, Aout ,
% for a spherical void of radius b in an elastic solid (pulse echo)
% using the method of separation of variables. the only input parameter is
% the setup structure. The complex scattering amplitude, Aout,
% is returned (in mm).
% The calling sequence is:
% Aout = A_void_Psep(setup)

% get input parameters
f=setup.f;
b =setup.flaw.b;
cp = setup.matl.cp2;
cs =setup.matl.cs2;

cr = cp/cs;            % ratio of P- and S- wave speeds
kp = 2000*pi*b*f./cp;    % non-dimensional wave number, P-waves
kp = kp + .0001*(kp == 0);

% break P-wave wave number into two regions: kp < 2 and kp >= 2
indc = find(kp < 2.);
kpd =kp(indc);
ind2 =find(kp >= 2.);
kpu = kp(ind2);
% S-wave wave numbers over same ranges
ksd =cr*kpd;
ksu =cr*kpu;

% use relatively small, fixed number of terms for kp <2
num = 10;
% compute scattering amplitude over kp <2 for sphere of radius b
```

```
A1 =sca(kpd, ksd, num, b);

% use much larger number of terms for kp >= 2
num2= 10 + round(kpu(end));
% compute scattering amplitude over kp >= 2 for sphere of radius b
A2 = sca(kpu,ksu, num2, b);

% combine two ranges
Aout= [A1  A2];
% force zero frequency scattering amplitude to zero exactly
Aout(1) =0;

% subfunction for calculating scattering amplitude with a given number
% of terms in the series. Generally, ten terms should be adequate
% for kp < 2 and a number of terms that is proportional to the max
% kp-value should be adequate for large kp values. However, if the
% max kp-value is very large, the number of terms used here based on this
% value may be too large for the values just above kp =2, resulting in the
% round-offs that cause the function to return NaNs at those lower
% frequencies. This function has been tested up to kp = 90 without problems
% of this sort.

function A = sca(xp,xs, numb, b)
An = zeros(size(xp));      % initialize array of zeros

% First compute the normalized scattering amplitude A/b.

% xp = P- wave number, xs = S- wave number, k is an integer.
% Uses spherical Bessel functions and spherical Hankel functions
% of order k defined by sphJ(k,x), sphH(k, x)

for k = 0:numb
e3 = (2.*k+1).*((k.^2 - k - xs.^2./2).*sphJ(k, xp) +2.*xp.*sphJ(k+1,xp));
e4 = (2.*k+1).*((k-1).*sphJ(k, xp) - xp.*sphJ(k+1, xp));
e32 = -k.*(k+1).*((k-1).*sphH(k, xs) - xs.*sphH(k+1, xs));
e31 = (k.^2 - k - xs.^2./2).*sphH(k, xp) + 2.*xp.*sphH(k+1, xp);
e41 = (k - 1).*sphH(k, xp) -xp.*sphH(k+1, xp);
e42 = -(k.^2 -1 - xs.^2./2).*sphH(k, xs) - xs.*sphH(k+1, xs);
if k == 0
  c = e3./e31;
else
c = (e3.*e42 - e4.*e32)./(e31.*e42 - e41.*e32);
end
An = An + ((-1.)^k)*c;
end
% Now, put the b factor back, insert i/kp term which multiplies entire result
```

```
A = i*b*An./xp;
```

Code Listing G.8. A function that uses the separation of variables method to calculate the far field pulse-echo S-wave scattering of a spherical void.

```
function Aout = A_void_Ssep(setup)
% A_VOID_SSEP computes the far field SV-wave scattering amplitude, Aout ,
% for a spherical void of radius b in an elastic solid (pulse echo)
% using the method of separation of variables. the only input parameter is
% the setup structure. The complex scattering amplitude, Aout,
% is returned (in mm).
% The calling sequence is:
% Aout = A_void_Ssep(setup)

% get input parameters
f=setup.f;
b =setup.flaw.b;
cp = setup.matl.cp2;
cs =setup.matl.cs2;

cr = cp/cs;              % ratio of P- and S- wave speeds
ks = 2000*pi*b*f./cs;   % non-dimensional wave number, S-waves
ks = ks + .001*(ks == 0);

% break S-wave wave number into two regions: ks < 5 and ks >= 5
indc = find(ks < 5);
ksd =ks(indc);
ind2 =find(ks >= 5);
ksu = ks(ind2);
% P-wave wave numbers over same ranges
kpd =ksd./cr;
kpu =ksu./cr;

% use relatively small, fixed number of terms for ks <5
num = 10;
% compute scattering amplitude over ks < 5 for sphere of radius b
A1 =sca(kpd, ksd, num, b);

% use much larger number of terms for ks >= 5
num2= 10 + round(ksu(end));
% compute scattering amplitude over ks >= 5 for sphere of radius b
A2 = sca(kpu,ksu, num2, b);
```

% combine two ranges
Aout= [A1 A2];
% force zero frequency scattering amplitude to zero exactly
Aout(1) =0;

% subfunction for calculating scattering amplitude with a given number
% of terms in the series. Generally, ten terms should be adequate
% for ks < 5 and a number of terms that is proportional to the max
% ks-value should be adequate for large ks values. However, if the
% max ks-value is very large, the number of terms used here based on this
% value may be too large for the values just above ks=5, resulting in the
% round-offs that cause the function to return NaNs at those lower
% frequencies. this function has been tested up to ks = 50 without problems
% of this sort.

```
function A = sca(xp,xs, numb, b)
An = zeros(size(xp));       % initialize array of zeros
```

% First compute the normalized scattering amplitude A/b.

% xp = P- wave number, xs = S- wave number, k is an integer.
% Uses spherical Bessel functions and spherical Hankel functions
% of order k defined by sphJ(k,x), sphH(k, x)

```
for k = 1:numb
j12 = k.*(k+1).*((k-1).*sphJ(k, xs)-xs.*sphJ(k+1,xs));
h12 = k.*(k+1).*((k-1).*sphH(k, xs) -xs.*sphH(k+1,xs));
h13 =((k.^2-k-xs.^2./2).*sphH(k,xp) +2.*xp.*sphH(k+1, xp));
j41 =((k-1).*sphJ(k, xs) -xs.*sphJ(k+1, xs))./2;
h41 = ((k-1).*sphH(k, xs) -xs.*sphH(k+1, xs))./2;
j42 = (k.^2 -1 -xs.^2./2).*sphJ(k, xs) + xs.*sphJ(k+1, xs);
h42 =((k.^2 -1 -xs.^2./2).*sphH(k, xs) + xs.*sphH(k+1, xs));
h43 = ((k-1).*sphH(k, xp) - xp.*sphH(k+1, xp));

c = (h13.*j42 -j12.*h43 )./(h13.*h42 -h12.*h43) -j41./h41;

An = An + (-1)^k.*((2.*k+1)./2).*c./(-i.*xs);
end
```

% Now, put the b factor back
A = b*An;

Code Listing G.9. A function that uses the separation of variables method to calculate the normalized 3-D far field pulse-echo P-wave scattering amplitude of a cylindrical void, $A_{3D}\left(\mathbf{e}_i^p;-\mathbf{e}_i^p\right)/L$.

```matlab
function Ascatt = A_SDH_Psep(setup)
% A_SDH_PSEP computes the separation of variables solution
% for the 3-D non-dimensional pulse-echo P-wave
% scattering amplitude, Ascatt, for a side-drilled hole
% of radius b (in mm).
% The function returns the scattering amplitude, A, divided
% by the length, L, i.e. Ascatt = A/L so that a value for L
% does not need to be specified. The only input to the function
% is the setup structure. The calling sequence is:
% Ascatt =A_SDH_Psep(setup);

% get setup parameters
f =setup.f;
b = setup.flaw.b;
cp =setup.matl.cp2;
cs = setup.matl.cs2;
%
cr = cp/cs;              % ratio of P- and S- wave speeds
kp = 2000*pi*b*f./cp;     % non-dimensional wave number, P-waves
kp = kp + .0001*(kp == 0);

% break P-wave wave number into two regions: kp < 2 and kp >= 2
indc = find(kp < 2.);
kpd =kp(indc);
ind2 =find(kp >= 2.);
kpu = kp(ind2);
% S-wave wave numbers over same ranges
ksd =cr*kpd;
ksu =cr*kpu;

% use relatively small, fixed number of terms for kp <2
num = 10;
% compute normalized scattering amplitude over kp <2 for sphere of radius b
A1 =sca(kpd, ksd, num);

% use much larger number of terms for kp >= 2
num2= 10 + round(kpu(end));
% compute normalized scattering amplitude over kp >= 2 for sphere of radius b
A2 = sca(kpu,ksu, num2);
% combine two ranges
```

```
Ascatt= [A1  A2];
% force zero frequency normalized scattering amplitude to zero exactly
Ascatt(1) =0;

% subfunction for calculating normalized scattering amplitude for a
% side- drilled hole with a given number of terms in the series.
% Generally, ten terms should be adequate
% for kp < 2 and a number of terms that is proportional to the max
% kp-value should be adequate for large kp values. However, if the
% max kp-value is very large, the number of terms used here based on this
% value may be too large for the values just above kp =2, resulting in the
% round-offs that cause the function to return NaNs at those lower
% frequencies. This function has been tested up to kp = 90 without problems
% of this sort.
% This function uses Hankel functions of type m, order n given by the MATLAB
% function besselh(n, m, x)

function A = sca(kp,ks, numb)
% initialize arrays
An = zeros(size(kp));
Ckp1 =zeros(size(kp));
Ckp2 =zeros(size(kp));
Cks1 =zeros(size(kp));
Dkp1 =zeros(size(kp));
Dkp2 =zeros(size(kp));
Dks1 =zeros(size(kp));
c =zeros(size(kp));

%calculate the series
for n = 0:numb

Ckp1 =(n^2 +n -(ks.^2/2)).*besselh(n, 1,kp) -((2*n).*besselh(n,1,kp)...
 -kp.*besselh(n+1,1,kp));
Ckp2 =(n^2 +n -(ks.^2/2)).*besselh(n, 2,kp) -((2*n).*besselh(n,2,kp)...
 -kp.*besselh(n+1,2,kp));
Cks1 =(n^2 +n -(ks.^2/2)).*besselh(n, 1,ks) -((2*n).*besselh(n,1,ks)...
 -ks.*besselh(n+1,1,ks));
Dkp1 = (n^2 +n).*besselh(n,1,kp) -n*((2*n).*besselh(n,1,kp)...
 -kp.*besselh(n+1,1,kp));
Dkp2 = (n^2 +n).*besselh(n,2,kp) -n*((2*n).*besselh(n,2,kp)...
 -kp.*besselh(n+1,2,kp));
Dks1 = (n^2 +n).*besselh(n,1,ks) -n*((2*n).*besselh(n,1,ks)....
 -ks.*besselh(n+1,1,ks));

if n == 0
    c = 1+ Ckp2./Ckp1;
```

```
else
c = 2*(1+(Ckp2.*Cks1 -Dkp2.*Dks1)./(Ckp1.*Cks1 - Dkp1.*Dks1));
end
An = An + ((-1.)^n)*c;
end
% Now, put the external factor in
A = (i/(2*pi))*An;
```

Code Listing G.10. A function that uses the separation of variables method to calculate the normalized 3-D far field pulse-echo SV-wave scattering amplitude of a cylindrical void, $A_{3D}\left(\mathbf{e}_i^s;-\mathbf{e}_i^s\right)/L$.

```
function Ascatt = A_SDH_Ssep(setup)
% A_SDH_SSEP computes the separation of variables solution
% for the 3-D non-dimensional pulse-echo SV-wave
% scattering amplitude, Ascatt, for a side-drilled hole
% of radius b (in mm).
% The function returns the scattering amplitude, A, divided
% by the length, L, i.e. Ascatt = A/L so that a value for L
% does not need to be specified. The only input to the function
% is the setup structure. The calling sequence is:
% Ascatt =A_SDH_Ssep(setup);
%

% get input parameters
f=setup.f;
b =setup.flaw.b;
cp = setup.matl.cp2;
cs =setup.matl.cs2;

cr = cp/cs;              % ratio of P- and S- wave speeds
ks = 2000*pi*b*f./cs;    % non-dimensional wave number, S-waves
ks = ks + .001*(ks == 0);

% break S-wave wave number into two regions: ks < 5 and ks >= 5
indc = find(ks < 5);
ksd =ks(indc);
ind2 =find(ks >= 5);
ksu = ks(ind2);
% P-wave wave numbers over same ranges
kpd =ksd./cr;
kpu =ksu./cr;
```

% use relatively small, fixed number of terms for ks <5
num = 10;
% compute scattering amplitude over ks < 5 for sphere of radius b
A1 =sca(kpd, ksd, num);

% use much larger number of terms for ks >= 5
num2= 10 + round(ksu(end));
% compute scattering amplitude over ks >= 5 for sphere of radius b
A2 = sca(kpu,ksu, num2);

% combine two ranges
Ascatt= [A1 A2];
% force zero frequency scattering amplitude to zero exactly
Ascatt(1) =0;

% subfunction for calculating normalized scattering amplitude for a
% side-drilled hole with a given number of terms in the series.
% Generally, ten terms should be adequate
% for ks < 5 and a number of terms that is proportional to the max
% ks-value should be adequate for large ks values. However, if the
% max ks-value is very large, the number of terms used here based on this
% value may be too large for the values just above ks = 5, resulting in the
% round-offs that cause the function to return NaNs at those lower
% frequencies. This function has been tested up to ks = 50 without
% problems of this sort.
% This function uses Hankel functions of type m, order n defined by the
% MATLAB function besselh(n, m, x)

function A = sca(kp, ks, numb)

% initialize arrays
An = zeros(size(kp));
Cnp1 =zeros(size(kp));
Cns1 =zeros(size(kp));
Cns2 =zeros(size(kp));
Dnp1 =zeros(size(kp));
Dnp2 =zeros(size(kp));
Dns1 =zeros(size(kp));
c =zeros(size(kp));

% Calculate series
for n = 0:numb

Cnp1 =(n^2 +n -(ks.^2/2)).*besselh(n, 1,kp) -((2*n).*besselh(n,1,kp)...
 -kp.*besselh(n+1,1,kp));

```
Cns2 =(n^2 +n -(ks.^2/2)).*besselh(n, 2,ks) -((2*n).*besselh(n,2,ks)...
-ks.*besselh(n+1,2,ks));
Cns1 =(n^2 +n -(ks.^2/2)).*besselh(n, 1,ks) -((2*n).*besselh(n,1,ks)...
-ks.*besselh(n+1,1,ks));
Dnp1 = (n^2 +n).*besselh(n,1,kp) -n*((2*n).*besselh(n,1,kp)...
-kp.*besselh(n+1,1,kp));
Dns2 = (n^2 +n).*besselh(n,2,ks) -n*((2*n).*besselh(n,2,ks)...
-ks.*besselh(n+1,2,ks));
Dns1 = (n^2 +n).*besselh(n,1,ks) -n*((2*n).*besselh(n,1,ks)...
-ks.*besselh(n+1,1,ks));

if n == 0
   c = 1+ Cns2./Cns1;
else
c = 2*(1+(Cns2.*Cnp1 -Dns2.*Dnp1)./(Cnp1.*Cns1 - Dnp1.*Dns1));
end
An = An + ((-1.)^n)*c;
end
% Now, put the external factor in
A = (i/(2*pi))*An;
```

Code Listing G.11. A function that models the open-circuit voltage output versus time of a pulser.

```
function V = pulserVT(V0, t0, a1, a2, t)
% PULSERVT(V0, t0, a1, a2, t) models the open-circuit voltage of
% a spike or square wave pulser using the four parameters V0, t0,
% a1, and a2. The parameter V0 controls the amplitude and the other
% parameters control the rise and fall characteristics of the pulse.
% The input parameter t is a set of sampled times.
t = t + eps*( t ==0);
Vinf = V0/(1-exp(-a1*t0));
V = -Vinf*(1- exp(-a1*t)).*(t <= t0) -V0*exp(-a2*(t -t0)).*(t > t0);
```

Code Listing G.12. A function that computes the Fresnel integral.

```
function y=fresnel_int(x)
%FRESNEL_INT(X) computes the Fresnel integral defined as the integral
%from t = 0 to t = x of the function exp(i*pi*t^2/2). Uses the approximate
```

```
%expressions given by Abramowitz and Stegun, Handbook of Mathematical
%Functions, Dover Publications, 1965, pp. 301-302.
%The calling sequence is: y = fresnel_int(x)

%separate arguments into positive and negative values, change sign
%of the negative values
xn =-x(x<0);
xp=x(x >=0);

%compute cosine and sine integrals of the negative values, using the
%oddness property of the cosine and sign integrals
[cn,sn] =cs_int(xn);
cn= -cn;
sn = -sn;

%compute cosine and sine integrals of the positive values

[cp, sp]=cs_int(xp);

%combine cosine and sine integrals for positive and negative
%values and return the complex Fresnel integral
ct =[cn cp];
st =[sn sp];
y=ct+i*st;

%CS_INT(XI) calculates approximations of the cosine and sine integrals
%for positive values of xi only(see Abramowitz and Segun reference above)
function [c, s]=cs_int(xi)
f =(1+0.926.*xi)./(2+1.792.*xi +3.104.*xi.^2);        % f function (see ref.)
g=1./(2+4.142.*xi+3.492.*xi.^2+6.67.*xi.^3);         % g function (see ref.)
c=0.5 +f.*sin(pi.*xi.^2./2) -g.*cos(pi.*xi.^2./2);  % cosine integral approx.
s = 0.5 -f.*cos(pi.*xi.^2./2)-g.*sin(pi.*xi.^2./2); % sine integral approx.
```

Index

Printed in the United States of America

ISBN-13: 978-1492346135

ISBN-10: 1492346136

http://pythonforbiologists.com

Set in PT Serif and `Source Code Pro`

SARVESH

KNOXVILLE, 08/15/2016

sarvesh.2786@gmail.com.

About the author

Martin started his programming career by learning Perl during the course of his PhD in evolutionary biology, and started teaching other people to program soon after. Since then he has taught introductory programming to hundreds of biologists, from undergraduates to PIs, and has maintained a philosophy that programming courses must be friendly, approachable, and practical.

In his academic career, Martin mixed research and teaching at the University of Edinburgh, culminating in a two year stint as Lecturer in Bioinformatics. He now runs programming courses for biological researchers as a full time freelancer.

You can get in touch with Martin at

martin@pythonforbiologists.com

Martin's other works include *Advanced Python for Biologists* and *Python for complete beginners*.

Table of Contents

1:   Introduction and environment

Why have a programming book for biologists?

If you're reading this book, then you probably don't need to be convinced that programming is becoming an increasingly essential part of the tool kit for biologists of all types. You might, however, need to be convinced that a book like this one, developed especially for biologists, can do a better job of teaching you to program than a general-purpose introductory programming book. Here are a few of the reasons why I think that is the case.

A biology-specific programming book allows us to use examples and exercises that use biological problems. This serves two important purposes: firstly, it provides motivation and demonstrates the types of problems that programming can help to solve. Experience has shown that beginners make much better progress when they are motivated by the thought of how the programs they write will make their life easier! Secondly, by using biological examples, the code and exercises throughout the book can form a library of useful code snippets, which we can refer back to when we want to solve real-life problems. In biology, as in all fields of programming, the same problems tend to recur time and time again, so it's very useful to have this collection of examples to act as a reference – something that's not possible with a general-purpose programming book.

A biology-specific programming book can also concentrate on the parts of the language that are most useful to biologists. A language like Python has many features and in the course of learning it we inevitably have to concentrate on some and miss others out. The set of features which are important to us in biology are slightly different to those which are most useful for general-purpose programming – for example, we are much more interested in manipulating text (including things like DNA and protein sequences) than the average programmer. Also, there are several features of Python that would not normally be discussed in an introductory programming book, but which are very useful to biologists (for example, regular expressions and subprocesses). Having a biology-specific textbook allows us to include these features, along with explanations of why they are particularly useful to us.

A related point is that a textbook written just for biologists allows us to introduce features in a way that allows us to start writing useful programs right away. We can do this by taking into account the sorts of problems that repeatedly crop up in biology, and prioritising the features that are best at solving them. This book has been designed so that you should be able to start writing small but useful programs using only the tools in the first couple of chapters.

Why Python?

The importance of programming languages is often overstated. What I mean by that is that people who are new to programming tend to worry far too much about what language to learn. The choice of programming language does matter, of course, but it matters far less than most people think it does. To put it another way, choosing the "wrong" programming language is very unlikely to mean the difference between failure and success when learning. Other factors (motivation, having time to devote to learning, helpful colleagues) are far more important, yet receive less attention.

The reason that people place so much weight on the *"what language should I learn?"* question is that it's a big, obvious question, and it's not difficult to find people who will give you strong opinions on the subject. It's also the first big question that beginners have to answer once they've decided to learn programming, so it assumes a great deal of importance in their minds.

There are three main reasons why choice of programming language is not as important as most people think it is. Firstly, nearly everybody who spends any significant amount of time programming as part of their job will eventually end up using multiple languages. Partly this is just down to the simple constraints of various languages – if you want to write a web application you'll probably do it in Javascript, if you want to write a graphical user interface you'll probably use something like Java, and if you want to write low-level algorithms you'll probably use C.

Secondly, learning a first programming language gets you 90% of the way towards learning a second, third, and fourth one. Learning to think like a programmer in the way that you break down complex tasks into simple ones is a skill that cuts across all languages – so if you spend a few months learning Python and then discover that you really need to write in C, your time won't have been wasted as you'll be able to pick it up much quicker.

Thirdly, the kinds of problems that we want to solve in biology are generally amenable to being solved in any language, even though different programming languages are good at different things. In other words, as a beginner, your choice of language is vanishingly unlikely to prevent you from solving the problems that you need to solve.

Having said all of the above, when learning to program we *do* need to pick a language to work in, so we might as well pick one that's going to make the job easier. Python is such a language for a number of reasons:

- It has a consistent syntax, so you can generally learn one way of doing things and then apply it in multiple places

- It has a sensible set of built in libraries for doing lots of common tasks

- It is designed in such a way that there's an obvious way of doing most things

- It's one of the most widely used languages in the world, and there's a lot of advice, documentation and tutorials available on the web

- It's designed in a way that lets you start to write useful programs as soon as possible

- Its use of indentation, while annoying to people who aren't used to it, is great for beginners as it enforces a certain amount of readability

Python also has a couple of points to recommend it to biologists and scientists specifically:

- It's widely used in the scientific community

- It has a couple of very well designed libraries for doing complex scientific computing (although we won't encounter them in this book)

- It lend itself well to being integrated with other, existing tools

- It has features which make it easy to manipulate strings of characters (for example, strings of DNA bases and protein amino acid residues, which we as biologists are particularly fond of)

Python vs. Perl

For biologists, the question *"what language should I learn"* often really comes down to the question *"should I learn Perl or Python?"*, so let's answer it head on. Perl and Python are both perfectly good languages for solving a wide variety of biological problems. However, after extensive experience teaching both Perl and Python to biologists, I've come the conclusion that Python is an easier language to learn by virtue of being more **consistent** and more **readable**.

An important thing to understand about Perl and Python is that they are *incredibly* similar (despite the fact that they look very different), so the point above about learning a second language applies doubly. Many Python and Perl features have a one-to-one correspondence, and so if you find that you have to work in Perl after learning Python you'll find it quite familiar.

How to use this book

Programming books generally fall into two categories; reference-type books, which are designed for looking up specific bits of information, and tutorial-type books, which are designed to be read cover-to-cover. This book is an example of the latter – code samples in later chapters often use material from previous ones, so you need to make sure you read the chapters in order. Exercises or examples from one chapter are sometimes used to illustrate the need for features that are introduced in the next.

There are a number of fundamental programming concepts that are relevant to material in multiple different chapters. In this book, rather than introduce these concepts all in one go, I've tried to explain them as they become necessary. This results in a tendency for earlier chapters to be longer than later ones, as they involve the introduction of more new concepts.

A certain amount of jargon is necessary if we want to talk about programs and programming concepts. I've tried to define each new technical term at the point where it's introduced, and then use it thereafter with occasional reminders of the meaning.

Chapters tend to follow a predictable structure. They generally start with a few paragraphs outlining the motivation behind the features that it will cover – why do they exist, what problems do they allow us to solve, and why are they useful in biology specifically? These are followed by the main body of the chapter in which we discuss the relevant features and how to use them. The length of the chapters varies quite a lot – sometimes we want to cover a topic briefly, other times we need more depth. This section ends with a brief recap outlining what we have learned, followed by exercises and solutions (more on that topic below).

Formatting

A book like this has lots of special types of text – we'll need to look at examples of Python code and output, the contents of files, and technical terms. Take a minute to note the typographic conventions we'll be using.

In the main text of this book, **bold type** is used to emphasize important points and *italics* for technical terms and filenames. Where code is mixed in with normal text it's written in a `monospaced font like this` with a grey background. Occasionally there are footnotes[1] to provide additional information that is interesting to know but not crucial to understanding, or to give links to web pages.

---

1   Like this.

Example Python code is highlighted with a solid border and the name of the matching example file is written just underneath the example to the right:

```
Some example code goes here
```

example.py

Not every bit of code has a matching example file – much of the time we'll be building up a Python program bit by bit, in which case there will be a single example file containing the finished version of the program. The example files are in separate folders, one for each chapter, to make them easy to find.

Sometimes it's useful to refer to a specific line of code inside an example. For this, we'll use numbered circles like this❶:

```
a line of example code
another line of example code
this is the important line❶
here is another line
```

Example output (i.e. what we see on the screen when we run the code) is highlighted with a dotted border:

```
Some output goes here
```

Often we want to look at the code and the output it produces together. In these situations, you'll see a solid-bordered code block followed immediately by a dotted-bordered output block.

Other blocks of text (usually file contents or typed command lines) don't have any kind of border and look like this:

```
contents of a file
```

Often when looking at larger examples, or when looking at large amounts of output, we don't need to see the whole thing. In these cases, I'll use ellipses (...) to indicate that some text has been missed out.

I have used UK English spelling throughout, which I hope will not prove distracting to US readers.

In programming, we use different types of brackets for different purposes, so it's important to have different names for them. Throughout this book, I will use the word *parentheses* to refer to (), *square brackets* to refer to [], and *curly brackets* to refer to {}.

Exercises and solutions

The final part of each chapter is a set of exercises and solutions. The number and complexity of exercises differ greatly between chapters depending on the nature of the material. As a rule, early chapters have a large number of simple exercises, while later chapters have a small number of more complex ones. Many of the exercise problems are written in a deliberately vague manner and the exact details of how the solutions work is up to you (very much like real-life programming!) You can always look at the solutions to see one possible way of tackling the problem, but there are often multiple valid approaches.

The exercises are probably the most important part of the book – when learning programming, it's vital that you practice writing programs from scratch rather than simply reading examples. I strongly recommend that you try tackling the exercises yourself before reading the solutions; there really is no substitute for practical experience when learning to program.

I also encourage you to adopt an attitude of curious experimentation when working on the exercises – if you find yourself wondering if a particular variation on a problem is solvable, or if you recognize a closely-related problem from your own work, try solving it! Continuous experimentation is a key part of developing as a programmer, and the quickest way to find out what a particular function or feature will do is to try it.

The example solutions to exercises are written in a different way to most programming textbooks: rather than simply present the finished solution, I have outlined the thought processes involved in solving the exercises and shown how the solution is built up step-by-step. Hopefully this approach will give you an insight into the problem-solving mindset that programming requires. It's probably a good idea to read through the solutions even if you successfully solve the exercise problems yourself, as they sometimes suggest an approach that is not immediately obvious.

As with the code example files, the input files (for those exercises that use them) and the solutions are separated into different folders, one per chapter.

You can download the exercise files from the following URL:

```
http://pythonforbiologists.com/index.php/exercise-files/
```

Some of the exercises require you to start with a given bit of code. For these, you'll find plain text files with the starting code in the relevant folder.

When learning to code, dealing with textbook exercises can be frustrating because it's often difficult to see the connection between the exercise problems and the type of programming you want to do. To help with this feeling, I've included a section at the end of each set of solutions entitled *What have we learned?* where I attempt to explain the relevance of the exercises to the wider world of programming. While you're working on the exercises, don't worry if they don't seem particularly relevant to your own research– they have been designed to help you practice the skills you will need regardless of the types of programs you want to write.

Getting in touch

Learning to program is a difficult task, and my one goal in writing this book is to make it as easy and accessible as possible to get started. So, if you find anything that is hard to understand, or you think may contain an error, please get in touch – just drop me an email at

```
martin@pythonforbiologists.com
```

and I promise to get back to you. If you find the book useful, then please also consider leaving an Amazon review to help other people find it.

Setting up your environment

All that you need in order to follow the examples and exercises in this book is a standard Python installation and a text editor. All the code in this book will run on either Linux, Mac or Windows machines. The slight differences between operating systems are explained in the text (mostly in chapter 9).

Python 2 vs. Python 3

As will quickly become clear if you spend any amount of time on the official Python website, there are two versions of Python currently available. The Python world is, at the time of writing, in the middle of a transition from version 2 to version 3. A discussion of the pros and cons of each version is well beyond the scope of this book[1], but here's what you need to know: install Python 3 if possible, but if you end up with Python 2, don't worry – all the code examples in the book will work with both versions.

If you're going to use Python 2, there is just one thing that you have to do in order to make some of the code examples work: include this line at the start of all your programs:

```
from __future__ import division
```

We won't go into the explanation behind this line, except to say that it's necessary in order to correct a small quirk with the way that Python 2 handles division of numbers.

---

1   You might encounter writing online that makes the 2 to 3 changeover seem like a big deal, and it is – but only for existing, large projects. When writing code from scratch, as you'll be doing when learning, you're unlikely to run into any problems.

Depending on what version you use, you might see slight differences between the output in this book and the output you get when you run the code on your computer. I've tried to note these differences in the text where possible.

Installing Python

The process of installing Python depends on the type of computer you're running on.

If you're using **Windows**, start by going to this page:

```
https://www.python.org/downloads/windows/
```

then follow the link at the top of the page to the latest release. From here you can download and run the Windows installer.

If you're using **Mac OS X**, head to this page:

```
https://www.python.org/downloads/mac-osx/
```

then follow the link at the top of the page to the latest release. From here you can download and run the OS X installer.

If you're running a mainstream **Linux** distribution like Ubuntu, Python is probably already installed. If your Linux installation doesn't already have Python installed, try installing it with your package manager – the command will probably be either

```
sudo apt-get install python idle
```

or

```
sudo yum install python idle
```

Editing and running Python programs

In order to learn Python, we need two things: the ability to **edit** Python programs, and the ability to **run** them and view the output. There are two different ways to do this – using a text editor from the command line, or using Python's graphical editor program.

Using the command line

If you're already comfortable using the command line, then this will probably be the easiest way to get started. Firstly, you'll need to be able to open a new terminal. If you're using Windows, you can do this by running the *command prompt* program. If you're using OS X, run the *terminal* program from inside the *Utilities* folder. If you're using Linux, you probably already know how to open a new terminal – the program is probably called something like *Terminal Emulator*.

Since a Python program is just a text file, you can create and edit it with any text editor of your choice. Note that by a text editor I **don't** mean a word processor – do **not** try to edit Python programs with Microsoft Word, LibreOffice Writer, or similar tools, as they tend to insert special formatting marks that Python cannot read.

When choosing a text editor, there is one feature that is essential[1] to have, and one which is nice to have. The essential feature is something that's usually called *tab emulation*. The effect of this feature at first seems quite odd; when enabled, it replaces any tab characters that you type with an equivalent number of space characters (usually set to four). The reason why this is useful is discussed at length in chapter 4, but here's a brief explanation: Python is very fussy about your use of tabs and spaces, and unless you are very disciplined when typing, it's easy to end up with a mixture of tabs and spaces in your programs. This causes very infuriating problems, because they look the same to you, but not to Python! Tab emulation fixes the problem by making it effectively impossible for you to type a tab character.

---

1   OK, so it's not strictly essential, but you will find life much easer if you have it.

The feature that is nice to have is *syntax highlighting*. This will apply different colours to different parts of your Python code, and can help you spot errors more easily.

Recommended text editors are **Notepad++** for Windows[1], **TextWrangler** for Mac OSX[2], and **gedit** for Linux[3], all of which are freely available.

To run a Python program from the command line, just type the name of the Python executable (*python.exe* on Windows, *python* on OS X and Linux) followed by the name of the Python file you've created.

If any of the above doesn't work or seems complicated, just use the graphical editor as described in the next section.

Using a graphical editor

Python comes with a program called IDLE which provides a friendly graphical interface for writing and running Python code. IDLE is an example of an **Integrated Development Environment** (sometimes shortened to IDE).

IDLE works identically on Windows, OS X and Linux. To create a new Python file, just start the IDLE program and select *New File* from the *File* menu. This will open a new window in which you can type and edit Python code. When you want to run your Python program, use the *File* menu to save it (remember that the filename should end with .py) then select *Run Module* from the *Run* menu. The output will appear in the *Python Shell* window.

You can also use IDLE as a text editor – for example, to view input and output files. Just select *Open* from the *File* menu and pick the file that you want to view. To open a non-Python file, you'll have to select *All files* from the *Files of type* drop-down menu.

---

1   http://notepad-plus-plus.org/
2   http://www.barebones.com/products/TextWrangler/
3   https://projects.gnome.org/gedit/

Reading the documentation

Part of the teaching philosophy that I've used in writing this book is that it's better to introduce a few useful features and functions rather than overwhelm you with a comprehensive list. The best place to go when you do want a complete list of the options available in Python is the official documentation

http://www.python.org/doc/

which, compared to many languages, is very readable.

2:   Printing and manipulating text

Why are we so interested in working with text?

Open the first page of a book about learning Python[1], and the chances are that the first examples of code you'll see will involve **numbers**. There's a good reason for that: numbers are generally simpler to work with than text – there are not too many things you can do with them (once you've got basic arithmetic out of the way) and so they lend themselves well to examples that are easy to understand. It's also a pretty safe bet that the average person reading a programming book is doing so because they need to do some number-crunching.

So what makes this book different – why is this first chapter about text rather than numbers? The answer is that, as biologists, we have a particular interest in dealing with text rather than numbers (though of course, we'll need to learn how to manipulate numbers too). Specifically, we're interested in particular types of text that we call *sequences* – the DNA and protein sequences that constitute much of the data that we deal with in biology.

There are other reasons that we have a greater interest in working with text than the average novice programmer. As scientists, the programs that we write often need to work as part of a pipeline, alongside other programs that have been written by other people. To do this, we'll often need to write code that can **understand** the output from some other program (we call this *parsing*) or **produce** output in a format that another program can operate on. Both of these tasks require manipulating text.

I've hinted above that computers consider numbers and text to be different in some way. That's an important idea, and one that we'll return to in more detail later. For now, I want to introduce an important piece of jargon – the word *string*. String is the word we use to refer to a piece of text in a computer program (it just means a string of characters). From this point on we'll use the word *string* when

---

1   Or indeed, any other programming language.

we're talking about computer code, and we'll reserve the word *sequence* for when we're discussing biological sequences like DNA and protein.

Printing a message to the screen

The first thing we're going to learn is how to print[1] a message to the screen. Here's a line of Python code that will cause a friendly message to be printed. Quick reminder: solid lines indicate Python code, dotted lines indicate output.

```
print("Hello world")
```

hello_world.py

Let's take a look at the various bits of this line of code, and give some of them names:

The whole line is called a *statement*.

`print()` is the name of a *function*. The function tells Python, in vague terms, what we want to do – in this case, we want to print some text. The function name is always[2] followed by parentheses[3].

The bits of text inside the parentheses are called the *arguments* to the function. In this case, we just have one argument (later on we'll see examples of functions that take more than one argument, in which case the arguments are separated by commas).

The arguments tell Python what we want to do more specifically – in this case, the argument tells Python exactly what it is we want to print: the message "Hello World".

---

1   When we talk about printing text inside a computer program, we are not talking about producing a document on a printer. The word "print" is used for any occasion when our program outputs some text – in this case, the output is displayed in your terminal.

2   This is not strictly true, but it's easier to just follow this rule than worry about the exceptions.

3   There are several different types of brackets in Python, so for clarity we will always refer to *parentheses* when we mean these: (), *square brackets* when we mean these: [] and *curly brackets* when we mean these: {}.

Assuming you've followed the instructions in chapter 1 and set up your Python environment, type the line of code above into your favourite text editor, save it, and run it. You should see a single line of output like this:

```
Hello world
```

Quotes are important

In normal writing, we only surround a bit of text in quotes when we want to show that they are being spoken. In Python, however, strings are **always** surrounded by quotes. That is how Python is able to tell the difference between the instructions (like the function name) and the data (the thing we want to print). We can use either single or double quotes for strings – Python will happily accept either. The following two statements behave exactly the same:

```
print("Hello world")
print('Hello world')
```

different_quotes.py

Let's take a look at the output to prove it[1]:

```
Hello world
Hello world
```

You'll notice that the output above doesn't contain quotes – they are part of the code, not part of the string itself. If we **do** want to include quotes in the output, the easiest thing to do[2] is use the other type of quotes for surrounding the string:

---

1   From this point on, I won't tell you to create a new file, enter the text, and run the program for each example – I will simply show you the output – but I encourage you to try the examples yourself.

2   The alternative is to place a backslash character (\) before the quote – this is called *escaping* the quote and will prevent Python from trying to interpret it.

```
print("She said, 'Hello world'")
print('He said, "Hello world"')
```

printing_quotes.py

The above code will give the following output:

```
She said, 'Hello world'
He said, "Hello world"
```

Be careful when writing and reading code that involves quotes – you have to make sure that the quotes at the beginning and end of the string match up.

Use comments to annotate your code

Occasionally, we want to write some text in a program that is for humans to read, rather than for the computer to execute. We call this type of line a *comment*. To include a comment in your source code, start the line with a hash symbol[1]:

```
# this is a comment, it will be ignored by the computer
print("Comments are very useful!")
```

comment.py

You're going to see a lot of comments in the source code examples in this book, and also in the solutions to the exercises. Comments are a very useful way to document your code, for a number of reasons:

- You can put the explanation of what a particular bit of code does right next to the code itself. This makes it much easier to find the documentation for a line of code that is in the middle of a large program, without having to search through a separate document.

- Because the comments are part of the source code, they can never get mixed up or separated. In other words, if you are looking at the source

---

1   This symbol has many names – you might know it as number sign, pound sign, octothorpe, sharp (from musical notation), cross, or pig-pen.

code for a particular program, then you automatically have the documentation as well. In contrast, if you keep the documentation in a separate file, it can easily become separated from the code.

- Having the comments right next to the code acts as a reminder to update the documentation whenever you change the code. The only thing worse than undocumented code is code with old documentation that is no longer accurate!

Don't make the mistake, by the way, of thinking that comments are only useful if you are planning on showing your code to somebody else. When you start writing your own code, you will be amazed at how quickly you forget the purpose of a particular section or statement. If you are working on a solution to one of the exercises in this book on Friday afternoon, then come back to it on Monday morning, it will probably take you quite a while to pick up where you left off.

Comments can help with this problem by giving you hints about the purpose of code, meaning that you spend less time trying to understand your old code, thus speeding up your progress. A side benefit is that writing a comment for a bit of code reinforces your understanding at the time you are doing it. A good habit to get into is writing a quick one-line comment above any line of code that does something interesting:

```
# print a friendly greeting
print("Hello world")
```

You'll see this technique used a lot in the code examples in this book, and I encourage you to use it for your own code as well.

Error messages and debugging

It may seem depressing soon in the book to be talking about errors! However, it's worth pointing out at this early stage that **computer programs almost never work correctly the first time**. Programming languages are not like natural languages – they have a very strict set of rules, and if you break any of them, the computer will not attempt to guess what you intended, but instead will stop running and present you with an error message. You're going to be seeing a lot of these error messages in your programming career, so let's get used to them as soon as possible.

Forgetting quotes

Here's one possible error we can make when printing a line of output – we can forget to include the quotes:

```
print(Hello world)
```

missing_quotes.py

This is easily done, so let's take a look at the output we'll get if we try to run the above code[1]:

```
  File "error.py", line 1❶
    print(Hello world)
                     ^❷
SyntaxError❸: invalid syntax
```

Looking at the output, we see that the error occurs on the first line of the file ❶. Python's best guess at the location of the error is just before the close parentheses ❷. Depending on the type of error, this can be wrong by quite a bit, so don't rely on it too much!

The type of error is a `SyntaxError` ❸, which mean that Python can't understand the code – it breaks the rules in some way (in this case, the rule that strings must be surrounded by quotation marks). We'll see different types of errors later in this book.

---

1 The output that you see might be very slightly different from this, depending on a bunch of factors like your operating system and the exact version of Python you are using.

Spelling mistakes

What happens if we miss-spell the name of the function?:

```
prin("Hello world")
```

spelling.py

We get a different type of error – a `NameError` – and the error message is a bit more helpful:

```
Traceback (most recent call last):
  File "error.py", line 1, in <module>
    prin("Hello world")❶
NameError: name 'prin'❷ is not defined
```

This time, Python doesn't try to show us where on the line the error occurred, it just shows us the whole line❶ . The error message tells us which word Python doesn't understand❷, so in this case, it's quite easy to fix.

Splitting a statement over two lines

What if we want to print some output that spans multiple lines? For example, we want to print the word "Hello" on one line and then the word "World" on the next line – like this:

```
Hello
World
```

We might try putting a new line in the middle of our string like this:

```
print("Hello
World")
```

but that won't work and we'll get the following error message:

```
File "error.py", line 1
    print("Hello ❶
              ^
SyntaxError: EOL while scanning string literal❷
```

Python finds the error when it gets to the end of the first line of code❶. The error message❷ is a bit more cryptic than the others. *EOL* stands for End Of Line, and *string literal* means a string in quotes. So to put this error message in plain English: "I started reading a string in quotes, and I got to the end of the line before I came to the closing quotation mark".

If splitting the line up doesn't work, then how do we get the output we want.....?

Printing special characters

The reason that the code above didn't work is that Python got confused about whether the new line was part of the string (which is what we wanted) or part of the source code (which is how it was actually interpreted). What we need is a way to include a new line as part of a string, and luckily for us, Python has just such a tool built in. To include a new line, we write a backslash followed by the letter n – Python knows that this is a special character and will interpret it accordingly. Here's the code which prints "Hello world" across two lines:

```
# how to include a newline in the middle of a string
print("Hello\nworld")
```

print_newline.py

Notice that there's no need for a space before or after the newline.

There are a few other useful special characters as well, all of which consist of a backslash followed by a letter. The only ones which you are likely to need for the exercises in this book are the tab character (\t) and the carriage return character (\r). The tab character can sometimes be useful when writing a program that will produce a lot of output. The carriage return character works a bit like a newline in that it puts the cursor back to the start of the line, but doesn't actually start a new line, so you can use it to overwrite output – this is sometimes useful for long-running programs.

Storing strings in variables

OK, we've been playing around with the `print()` function for a while; let's introduce something new. We can take a string and assign a name to it using an equals sign – we call this a *variable*:

```
# store a short DNA sequence in the variable my_dna
my_dna = "ATGCGTA"
```

The variable `my_dna` now points to the string `"ATGCGTA"`. We call this assigning a variable, and once we've done it, we can use the variable name instead of the string itself – for example, we can use it in a `print()` statement[1]:

```
# store a short DNA sequence in the variable my_dna
my_dna = "ATGCGTA"

# now print the DNA sequence
print(my_dna)
```

print_variable.py

Notice that when we use the variable in a `print()` statement, we don't need any quotation marks – the quotes are part of the string, so they are already "built in" to the variable `my_dna`. Also notice that this example includes a blank line to separate the different bits and make it easier to read. We are allowed to put as many blank lines as we like in our programs when writing Python – the computer will ignore them.

A common error is to include quotes around a variable name:

---

1   If it's not clear why this is useful, don't worry – it will become much more apparent when we look at some longer examples.

```
my_dna = "ATGCGTA"
print("my_dna")
```

but if we do this, then Python prints the name of the variable rather than its contents:

```
my_dna
```

We can change the value of a variable as many times as we like once we've created it:

```
my_dna = "ATGCGTA"
print(my_dna)

# change the value of my_dna
my_dna = "TGGTCCA"
```

Here's a very important point that trips many beginners up: variable names are **arbitrary** – that means that we can pick whatever we like to be the name of a variable. So our code above would work in exactly the same way if we picked a different variable name:

```
# store a short DNA sequence in the variable banana
banana = "ATGCGTA"

# now print the DNA sequence
print(banana)
```

What makes a good variable name? Generally, it's a good idea to use a variable name that gives us a clue as to what the variable refers to. In this example, my_dna is a good variable name, because it tells us that the content of the variable is a DNA sequence. Conversely, banana is a bad variable name, because it doesn't really tell us anything about the value that's stored. As you read through the code examples in this book, you'll get a better idea of what constitutes good and bad variable names.

This idea – that names for things are arbitrary, and can be anything we like – is a theme that will occur many times in this book, so it's important to keep it in mind. Occasionally you will see a variable name that **looks like** it has some sort of relationship with the value it points to:

```
my_file = "my_file.txt"
```

but don't be fooled! Variable names and strings are separate things.

I said above that variable names can be anything we want, but it's actually not quite that simple – there are some rules we have to follow. We are only allowed to use letters, numbers, and underscores, so we can't have variable names that contain odd characters like £, ^ or %. We are not allowed to start a name with a number (though we can use numbers in the middle or at the end of a name). Finally, we can't use a word that's already built in to the Python language like "print"[1].

It's also important to remember that variable names are case-sensitive, so `my_dna`, `MY_DNA`, `My_DNA` and `My_Dna` are all different variables. Technically this means that you could use all four of those names in a Python program to store different values, but please don't do this – it is very easy to become confused when you use very similar variable names.

Tools for manipulating strings

Now we know how to store and print strings, we can take a look at a few of the facilities that Python has for manipulating them. Python has many built in tools for carrying out common operations, and in this next section we'll take a look at them one-by-one. In the exercises at the end of this chapter, we'll look at how we can use multiple different tools together in order to carry out more complex operations.

---

1   Strictly speaking, that's not true – we **can** overwrite a built in name, but it will cause a lot of problems, so don't do it.

Concatenation

We can concatenate (stick together) two strings using the + symbol[2]. This symbol will join together the string on the left with the string on the right:

```python
my_dna = "AATT" + "GGCC"
print(my_dna)
```

print_concatenated.py

Let's take a look at the output:

```
AATTGGCC
```

In the above example, the things being concatenated were strings, but we can also use variables that point to strings:

```python
upstream = "AAA"
my_dna = upstream + "ATGC"
# my_dna is now "AAAATGC"
```

We can even join multiple strings together in one go:

```python
upstream = "AAA"
downstream = "GGG"
my_dna = upstream + "ATGC" + downstream
# my_dna is now "AAAATGCGGG"
```

It's important to realize that the result of concatenating two strings together is itself a string. So it's perfectly OK to use a concatenation inside a print statement:

```python
print("Hello" + " " + "world")
```

As we'll see in the rest of the book, using one tool inside another is quite a common thing to do in Python.

---

2   We call this the *concatenation operator*.

Finding the length of a string

Another useful built in tool in Python is the `len()` function (`len` is short for length). Just like the `print()` function, the `len()` function takes a single argument (take a quick look back at when we were discussing the `print()` function for a reminder about what arguments are) which is a string. However, the behaviour of `len()` is quite different to that of `print()`. Instead of outputting text to the screen, `len()` outputs a value that can be stored – we call this the *return value*. In other words, if we write a program that uses `len()` to calculate the length of a string, the program will run but we won't see any output:

```
# this line doesn't produce any output
len("ATGC")
```

If we want to actually use the return value, we need to store it in a variable, and then do something useful with it (like printing it):

```
dna_length = len("AGTC")
print(dna_length)
```

print_length.py

There's another interesting thing about the `len()` function: the result (or *return value*) is not a string, it's a number. This is a very important idea so I'm going to write it out in bold: **Python treats strings and numbers differently.**

We can see that this is the case if we try to concatenate together a number and a string. Consider this short program which calculates the length of a DNA sequence and then prints a message telling us the length:

```
# store the DNA sequence in a variable
my_dna = "ATGCGAGT"

# calculate the length of the sequence and store it in a variable
dna_length = len(my_dna)

# print a message telling us the DNA sequence lenth
print("The length of the DNA sequence is " + dna_length)
```

When we try to run this program, we get the following error:

```
    print("The length of the DNA sequence is " + dna_length)
TypeError: cannot concatenate 'str' and 'int' objects❶
```

The error message❶ is short but informative: "cannot concatenate 'str' and 'int' objects". Python is complaining that it doesn't know how to concatenate a string (which it calls str for short) and a number (which it calls int – short for integer). Strings and numbers are examples of *types* – different kinds of information that can exist inside a program[1].

Happily, Python has a built in solution – a function called str() which turns a number[2] into a string so that we can print it. Here's how we can modify our program to use it – I've removed the comments from this version to make it a bit more compact:

```
my_dna = "ATGCGAGT"
dna_length = len(my_dna)

print("The length of the DNA sequence is " + str(dna_length))
```

print_dna_length.py

The only thing we have changed is that we've replace dna_length with str(dna_length) inside the print() statement[3]. Notice that because we're

---

1   If you want to read more, there's a full explanation of how types work in the chapter on object-oriented programming in *Advanced Python for Biologists*.
2   Or a value of any non-string type, but we'll come to that later.

using one function (`str()`) inside another function (`print()`), our statement now ends with two closing parentheses.

Let's take a moment to refresh our memory of all the new terms we've learned by writing out what we need to know about the `str()` function:

> `str()` *is a function which takes one argument* (whose type is *number*),
> *and returns a value (whose type is string) representing that number.*

If you're unsure about the meanings of any of the words in italics, skip back to the earlier parts of this chapter where we discussed them. Understanding how types work is key to avoiding many of the frustrations which new programmers typically encounter, so make sure the idea is clear in your mind before moving on with the rest of this book.

Sometimes we need to go the other way – we have a string that we need to turn into a number. The function for doing this is called `int()`, which is short for integer. It takes a string as its argument and returns a number:

```
number = 3 + int('4')
# number is now 7
```

We won't need to use `int()` for a while, but once we start reading information from files later on in the book it will become very useful.

Changing case

We can convert a string to lower case by using a new type of syntax – a *method* that belongs to strings. A *method* is like a *function*, but instead of being built in to the Python language, it belongs to a particular *type*. The method we are talking about here is called `lower()`, and we say that it belongs to the *string* type. Here's how we use it:

---

3   If you experiment with some of the code here, you might discover that you can also print a
     number directly without using `str()` – but only if you don't try to concatenate it.

```
my_dna = "ATGC"
# print my_dna in lower case
print(my_dna.lower())
```

print_lower.py

Notice how using a method looks different to using a function. When we use a function like `print()` or `len()`, we write the **function** name first and the **arguments** go in parentheses:

```
print("ATGC")
len(my_dna)
```

When we use a method, we write the name of the **variable** first, followed by a period, then the name of the method, then the method arguments in parentheses. For the example we're looking at here, `lower()`, there is no argument, but we still need to put the opening and closing parentheses.

It's important to notice that the `lower()` method does not actually change the variable; instead it returns a copy of the variable in lower case. We can prove that it works this way by printing the variable before and after running `lower()`. Here's the code to do so:

```
my_dna = "ATGC"

# print the variable
print("before: " + my_dna)

# run the lower method and store the result
lowercase_dna = my_dna.lower()

# print the variable again
print("after: " + my_dna)
```

print_before_and_after.py

and here's the output we get:

```
before: ATGC
after: ATGC
```

Just like the `len()` function, in order to actually do anything useful with the `lower()` method, we need to store the result (or print it right away).

Because the `lower()` method belongs to the string type, we can only use it on variables that are strings. If we try to use it on a number:

```
my_number = len("AGTC")
# my_number is 4
print(my_number.lower())
```

we will get an error that looks like this:

```
AttributeError: 'int' object has no attribute 'lower'
```

The error message is a bit cryptic, but hopefully you can grasp the meaning: something that is a number (an `int`, or integer) does not have a `lower()` method. This is a good example of the importance of types in Python code: **we can only use methods on the type that they belong to.**[1]

Before we move on, let's just mention that there is another method that belongs to the string type called `upper()` – you can probably guess what it does!

Replacement

Here's another example of a useful method that belongs to the string type: `replace()`. `replace()` is slightly different from anything we've seen before – it takes two arguments (both strings) and returns a copy of the variable where all occurrences of the first string are replaced by the second string. That's quite a long-winded description, so here are a few examples to make things clearer:

---

1   For details of how this is actually implemented, see the chapter on object-oriented programming in *Advanced Python for Biologists*.

```
protein = "vlspadktnv"

# replace valine with tyrosine
print(protein.replace("v", "y"))

# we can replace more than one character
print(protein.replace("vls", "ymt"))

# the original variable is not affected
print(protein)
```

`replace.py`

And this is the output we get:

```
ylspadktny
ymtpadktnv
vlspadktnv
```

Notice that in the first line out output, both "v" characters have been replaced with "y". We'll take a look at more tools for carrying out string replacement in chapter 7.

Extracting part of a string

What do we do if we have a long string, but we only want a short portion of it? This is known as taking a *substring*, and it has its own notation in Python. To get a substring, we follow the variable name with a pair of square brackets which enclose a start and stop position, separated by a colon. Again, this is probably easier to visualize with a couple of examples – let's reuse our protein sequence from before:

```
protein = "vlspadktnv"

# print positions three to five
print(protein[3:5])

# positions start at zero, not one
print(protein[0:6])

# if we miss out the last number, it goes to the end of the string
print(protein[2:])
```

print_substrings.py

and here's the output:

```
pa
vlspad
spadktnv
```

There are two important things to notice here. Firstly, we actually start counting from position zero, rather than one – in other words, position 3 is actually the fourth character[1]. This explains why the first character of the first line of output is p and not s as you might think. Secondly, the positions are **inclusive** at the start, but **exclusive** at the stop. In other words, the expression protein[3:5] gives us everything starting at the fourth character, and stopping just before the sixth character (i.e. characters four and five).

If we just give a single number in the square brackets, we'll just get a single character:

```
protein = "vlspadktnv"
first_residue = protein[0]
```

We'll learn a lot more about this type of notation, and what we can do with it, in chapter 4.

---

1   This seems very annoying when you first encounter it, but we'll see later why it's necessary.

Counting and finding substrings

A very common job in biology is to count the number of times some pattern occurs in a DNA or protein sequence. In computer programming terms, what that translates to is counting the number of times a *substring* occurs in a *string*. The method that does the job is called `count()`. It takes a single argument whose type is string, and returns the number of times that the argument is found in the variable. The return type is a number, so be careful about how you use it!

Let's use our protein sequence one last time as an example. Remember that we have to use our old friend `str()` to turn the counts into strings so that we can print them.

```python
protein = "vlspadktnv"
# count amino acid residues
valine_count = protein.count('v')
lsp_count = protein.count('lsp')
tryptophan_count = protein.count('w')

# now print the counts
print("valines: " + str(valine_count))
print("lsp: " + str(lsp_count))
print("tryptophans: " + str(tryptophan_count))
```

count_amino_acids.py

The output shows how the `count()` method behaves:

```
valines: 2
leucines: 1
tryptophans: 0
```

A closely related problem to counting substrings is finding their location. What if instead of counting the number of proline residues in our protein sequence we want to know where they are? The `find()` method will give us the answer, at least for simple cases. `find()` takes a single string argument, just like `count()`, and returns a number which is the position at which that substring first appears in the string (in computing, we call that the *index* of the substring).

Remember that in Python we start counting from zero rather than one, so position 0 is the first character, position 4 is the fifth character, etc. A couple of examples:

```
protein = "vlspadktnv"
print(str(protein.find('p')))
print(str(protein.find('kt')))
print(str(protein.find('w')))
```

find_amino_acids.py

And the output:

```
3
6
-1
```

Notice the behaviour of `find()` when we ask it to locate a substring that doesn't exist – we get -1 as the answer.

Both `count()` and `find()` have a pretty serious limitation: you can only search for **exact** substrings. If you need to count the number of occurrences of a variable protein motif, or find the position of a variable transcription factor binding site, they will not help you. The whole of chapter 7 is devoted to tools that can do those kinds of jobs.

Of the tools we've discussed in this section, three – `replace()`, `count()` and `find()` – require at least two strings to work, so be careful that you don't get confused about the order – remember that:

```
my_dna.count(my_motif)
```

is **not** the same as:

```
my_motif.count(my_dna)
```

Splitting up a string into multiple bits

An obvious question which biologists often ask when learning to program is "how do we split a string (e.g. a DNA sequence) into multiple pieces?" That's a common job in biology, but unfortunately we can't do it yet using the tools from this chapter. We'll talk about various different ways of splitting strings in chapter 4. I mention it here just to reassure you that we will learn how to do it eventually!

Recap

We started this chapter talking about strings and how to work with them, but along the way we had to take a lot of diversions, all of which were necessary to understand how the different string tools work. Thankfully, that means that we've covered most of the nuts and bolts of the Python language, which will make future chapters go much more smoothly.

We've learned about some general features of the Python programming language like

- the difference between *functions*, *statements* and *arguments*
- the importance of *comments* and how to use them
- how to use Python's error messages to fix bugs in our programs
- how to store *values* in *variables*
- the way that *types* work, and the importance of understanding them
- the difference between *functions* and *methods*, and how to use them both

And we've encountered some tools that are specifically for working with strings:

- concatenation
- different types of quotes
- special characters
- changing the case of a string

- finding and counting substrings

- replacing bits of a string with something new

- extracting bits of a string to make a new string

Many of the above topics will crop up again in future chapters, and will be discussed in more detail, but you can always return to this chapter if you want to brush up on the basics. The exercises for this chapter will allow you to practice using the string manipulation tools and to become familiar with them. They'll also give you the chance to practice building bigger programs by using the individual tools as building blocks.

Exercises

Reminder: the descriptions of the exercises are brief and may be kind of ambiguous – just like requirements for programs you will write in real life! Try the exercises yourself before you look at the solutions, but make sure to read the solutions even if you find the exercises easy, as they contain extra details that may be useful.

All of the exercises in this chapter involve starting with a DNA sequence string. To save you typing out the whole sequence each time, start each exercise by copying and pasting the relevant line in the *exercise_inputs.txt* file in the chapter 2 exercises folder.

Calculating AT content

Here's a short DNA sequence:

```
ACTGATCGATTACGTATAGTATTTGCTATCATACATATATATCGATGCGTTCAT
```

Write a program that will print out the AT content of this DNA sequence (i.e. the proportion of bases that are either A or T). Hint: you can use normal mathematical symbols like add (+), subtract (-), multiply (*), divide (/) and parentheses to carry out calculations on numbers in Python.

Reminder: if you're using Python 2 rather than Python 3, include this line at the top of your program:

```
from __future__ import division
```

Complementing DNA

Here's a short DNA sequence:

```
ACTGATCGATTACGTATAGTATTTGCTATCATACATATATATCGATGCGTTCAT
```

Write a program that will print the complement of this sequence. To find the complement we replace each base with its pair: A with T, T with A, C with G and G with C.

Restriction fragment lengths

Here's a short DNA sequence:

```
ACTGATCGATTACGTATAGTAGAATTCTATCATACATATATATCGATGCGTTCAT
```

The sequence contains a recognition site for the EcoRI restriction enzyme, which cuts at the motif G*AATTC (the position of the cut is indicated by an asterisk). Write a program which will calculate the size of the two fragments that will be produced when the DNA sequence is digested with EcoRI.

Splicing out introns, part one

Here's a short section of genomic DNA:

```
ATCGATCGATCGATCGACTGACTAGTCATAGCTATGCATGTAGCTACTCGATCGATCGATCGATC
GATCGATCGATCGATCGATCATGCTATCATCGATCGATATCGATGCATCGACTACTAT
```

It comprises two exons and an intron. The first exon runs from the start of the sequence to the sixty-third character, and the second exon runs from the ninety-first character to the end of the sequence. Write a program that will print just the coding regions of the DNA sequence.

Splicing out introns, part two

Using the data from part one, write a program that will calculate what percentage of the DNA sequence is coding.

Reminder: if you're using Python 2 rather than Python 3, include this line at the top of your program:

```
from __future__ import division
```

Splicing out introns, part three

Using the data from part one, write a program that will print out the original genomic DNA sequence with coding bases in uppercase and non-coding bases in lowercase.

Solutions

Calculating AT content

This exercise is going to involve a mixture of strings and numbers. Let's remind ourselves of the easiest way to calculate AT content:

$$AT\ content = \frac{A+T}{length}$$

There are three numbers we need to figure out: the number of A characters, the number of T characters, and the length of the sequence. We know that we can get the length of the sequence using the `len()` function, and we can count the number of A and T using the `count()` method. Here are a few lines of code that we think will calculate the numbers we need:

```
my_dna = "ACTGATCGATTACGTATAGTATTTGCTATCATACATATATATCGATGCGTTCAT"
length = len(my_dna)
a_count = my_dna.count('A')
t_count = my_dna.count('T')
```

At this point, it seems sensible to check that these lines work before we go any further. So rather than diving straight in and doing some calculations, let's print out these numbers so that we can eyeball them and see if they look approximately right. We'll have to remember to turn the numbers into strings using `str()` so that we can print them:

```
my_dna = "ACTGATCGATTACGTATAGTATTTGCTATCATACATATATATCGATGCGTTCAT"
length = len(my_dna)
a_count = my_dna.count('A')
t_count = my_dna.count('T')

print("length: " + str(length))
print("A count: " + str(a_count))
print("T count: " + str(t_count))
```

Let's take a look at the output from this program:

```
length: 54
A count: 16
T count: 21
```

That looks about right, but how do we know if it's exactly right? We could go through the sequence manually base by base, and verify that there are sixteen As and eighteen Ts, but that doesn't seem like a great use of our time: also, what would we do if the sequence were 51 kilobases rather than 51 bases? A better idea is to run the exact same code with a much shorter test sequence, to verify that it works before going ahead and running it on the larger sequence.

Here's a version that uses a very short test sequence with one of each of the four bases:

```
test_dna = "ATGC"
length = len(test_dna)
a_count = test_dna.count('A')
t_count = test_dna.count('T')

print("length: " + str(length))
print("A count: " + str(a_count))
print("T count: " + str(t_count))
```

and here's the output:

```
length: 4
A count: 1
T count: 1
```

Everything looks OK – we can probably go ahead and run the code on the long sequence. But wait; we know that the next step is going to involve doing some calculations using the numbers. If we switch back to the long sequence now, then we'll be in the same position as we were before – we'll end up with an answer for the AT content, but we won't know if it's the right one.

A better plan is to stick with the short test sequence until we've written the whole program, and check that we get the right answer for the AT content (we can easily see by glancing at the test sequence that the AT content is 0.5). Here goes – we'll use the add and divide symbols from the exercise hint:

```
test_dna = "ATGC"
length = len(test_dna)
a_count = test_dna.count('A')
t_count = test_dna.count('T')

at_content = a_count + t_count / length
print("AT content is " + str(at_content))
```

The output from this program looks like this:

```
AT content is 1.25
```

That doesn't look right. Looking back at the code we can see what has gone wrong – in the calculation, the division has taken precedence over the addition, so what we have actually calculated is:

$$A + \frac{T}{length}$$

To fix it, all we need to do is add some parentheses around the addition, so that the line becomes:

```
at_content = (a_count + t_count) / length
```

Now we get the correct output for the test sequence:

```
AT content is 0.5
```

and we can go ahead and run the program using the longer sequence, confident that the code is working and that the calculations are correct. Here's the final version:

```
my_dna = "ACTGATCGATTACGTATAGTATTTGCTATCATACATATATATCGATGCGTTCAT"
length = len(my_dna)
a_count = my_dna.count('A')
t_count = my_dna.count('T')

at_content = (a_count + t_count) / length
print("AT content is " + str(at_content))
```

at_content.py

and the final output:

```
AT content is 0.6851851851851852
```

Complementing DNA

This one seems pretty straightforward – we need to take our sequence and replace A with T, T with A, C with G, and G with C. We'll have to make four separate calls to `replace()`, and use the return value for each on as the input for the next tone. Let's try it:

```
my_dna = "ACTGATCGATTACGTATAGTATTTGCTATCATACATATATATCGATGCGTTCAT"

# replace A with T
replacement1 = my_dna.replace('A', 'T')

# replace T with A
replacement2 = replacement1.replace('T', 'A')

# replace C with G
replacement3 = replacement2.replace('C', 'G')

# replace G with C
replacement4 = replacement3.replace('G', 'C')

# print the result of the final replacement
print(replacement4)
```

When we take a look at the output, however, something seems wrong:

```
ACACAACCAAAACCAAAACAAAAACCAAACAAACAAAAAAAACCAACCCAACAA
```

We can see just by looking at the original sequence that the first letter is A, so the first letter of the printed sequence should be its complement, T. But instead the first letter is A. In fact, all of the bases in the printed sequence are either A or C. This is definitely not what we want!

Let's try and track the problem down by printing out all the intermediate steps as well:

```
my_dna = "ACTGATCGATTACGTATAGTATTTGCTATCATACATATATATCGATGCGTTCAT"

replacement1 = my_dna.replace('A', 'T')
print(replacement1)

replacement2 = replacement1.replace('T', 'A')
print(replacement2)

replacement3 = replacement2.replace('C', 'G')
print(replacement3)

replacement4 = replacement3.replace('G', 'C')
print(replacement4)
```

The output from this program makes it clear what the problem is:

```
TCTGTTCGTTTTCGTTTTGTTTTTGCTTTCTTTCTTTTTTTTCGTTGCGTTCTT
ACAGAACGAAAACGAAAAGAAAAAGCAAACAAACAAAAAAAACGAAGCGAACAA
AGAGAAGGAAAAGGAAAAGAAAAAGGAAAGAAAGAAAAAAAAAGGAAGGGAAGAA
ACACAACCAAAACCAAAACAAAAACCAAACAAACAAAAAAAACCAACCCAACAA
```

The first replacement (the result of which is shown in the first line of the output) works fine – all the As have been replaced with Ts (for example, look at the first character – it's A in the original sequence and T in the first line of the output).

The second replacement is where it starts to go wrong: all the Ts are replaced by As, **including those that were there as a result of the first replacement**. So during the first two replacements, the first character is changed from A to T and then straight back to A again.

How are we going to get round this problem? One option is to pick a temporary alphabet of four letters and do each replacement twice:

```
my_dna = "ACTGATCGATTACGTATAGTATTTGCTATCATACATATATATCGATGCGTTCAT"

replacement1 = my_dna.replace('A', 'H')
replacement2 = replacement1.replace('T', 'J')
replacement3 = replacement2.replace('C', 'K')
replacement4 = replacement3.replace('G', 'L')
replacement5 = replacement4.replace('H', 'T')
replacement6 = replacement5.replace('J', 'A')
replacement7 = replacement6.replace('K', 'G')
replacement8 = replacement7.replace('L', 'C')
print(replacement8)
```

This gets us the result we are looking for. It avoids the problem with the previous program by using another letter to stand in for each base while the replacements are being done. For example, A is first converted to H and then later on H is converted to T.

Here's a slightly more elegant way of doing it. We can take advantage of the fact that the `replace()` method is case sensitive, and make all the replaced bases lower case. Then, once all the replacements have been carried out, we can simply call `upper()` and change the whole sequence back to upper case. Let's take a look at how this works:

```
my_dna = "ACTGATCGATTACGTATAGTATTTGCTATCATACATATATATCGATGCGTTCAT"
replacement1 = my_dna.replace('A', 't')
print(replacement1)
replacement2 = replacement1.replace('T', 'a')
print(replacement2)
replacement3 = replacement2.replace('C', 'g')
print(replacement3)
replacement4 = replacement3.replace('G', 'c')
print(replacement4)
print(replacement4.upper())
```

complement_dna.py

The output lets us see exactly what's happening – notice that in this version of
the program we print the final string twice, once as it is and then once converted
to upper case:

```
tCTGtTCGtTTtCGTtTtGTtTTTGCTtTCtTtCtTtTtTtTCGtTGCGTTCtT
tCaGtaCGtaatCGatatGataaaGCataCtatCtatatataCGtaGCGaaCta
tgaGtagGtaatgGatatGataaaGgatagtatgtatatatagGtaGgGaagta
tgactagctaatgcatatcataaacgatagtatgtatatatagctacgcaagta
TGACTAGCTAATGCATATCATAAACGATAGTATGTATATATAGCTACGCAAGTA
```

We can see that as the program runs, each base in turn is replaced by its
complement in lower case. Since the next replacement is only looking for upper
case characters, bases don't get changed back as they did in the first version of
our program.

Restriction fragment lengths

Let's start this exercise by solving the problem manually. If we look through the
DNA sequence we can spot the EcoRI site at position 21. Here's the sequence with
the base positions labelled above and the EcoRI motif in bold:

```
         1         2         3         4         5
0123456789012345678901234567890123456789012345678901234
ACTGATCGATTACGTATAGTAGAATTCTATCATACATATATATCGATGCGTTCAT
```

Since the EcoRI enzyme cuts the DNA between the G and first A, we can figure out that the first fragment will run from position 0 to position 21, and the second fragment from position 22 to the last position, 54. Therefore the lengths of the two fragments are 22 and 33.

Writing a program to figure out the lengths is just a question of applying the same logic. We'll use the `find()` method to figure out the position of the start of the EcoRI motif, then add one to account for the fact that the positions start counting from zero – this will give us the length of the first fragment. From there we can get the length of the second fragment by finding the length of the input sequence and subtracting the length of the first fragment:

```
my_dna = "ACTGATCGATTACGTATAGTAGAATTCTATCATACATATATATCGATGCGTTCAT"
frag1_length = my_dna.find("GAATTC") + 1
frag2_length = len(my_dna) - frag1_length
print("length of fragment one is " + str(frag1_length))
print("length of fragment two is " + str(frag2_length))
```

fragment_lengths.py

The output from this program confirms that it agrees with the answer we got manually:

```
length of fragment one is 22
length of fragment two is 33
```

It's worth noting that this program assumes that the DNA sequence definitely does contain the restriction site we're looking for. If we try the same program using a DNA sequence which **doesn't** contain the site, it will report a fragment of length 0 and a fragment whose length is equal to the total length of the DNA sequence. While this is not strictly wrong, it's a little misleading – if we were going to use this program for real life work, we'd probably prefer to have slightly different behaviour depending on whether or not the DNA sequence contained the motif we're looking for. We'll talk about how to implement that type of behaviour in chapter 6.

Splicing out introns, part one

In this exercise, we're being asked to produce a program that does the job of a spliceosome – splits a DNA sequence at two specified locations to make three pieces, then join the outer two pieces together[1]. We can start by defining a variable to hold the DNA sequence:

```
my_dna =
"ATCGATCGATCGATCGACTGACTAGTCATAGCTATGCATGTAGCTACTCGATCGATCGATCGATCGATCGATC
GATCGATCGATCATGCTATCATCGATCGATATCGATGCATCGACTACTAT"
```

Because the sequence is quite long, this single statement actually runs over three lines – although, of course, if you open this code in a text editor it might look different depending on your set up.

The next step in solving this exercise is to extract the two exons from our DNA sequence. We'll have to use the substring notation from earlier in the chapter, and we'll need to take care with the numbers.

The first bit of the sequence goes from the first character to the sixty-third character, so we might be tempted to write a line like this:

```
exon1 = my_dna[1:63]
```

However, remember that when we take a substring like this the numbers are inclusive at the start, but exclusive at the end, so our stop position needs to be one higher:

```
exon1 = my_dna[1:64]
```

The second exon starts at the ninety-first base and goes to the end of the DNA sequence. There are a number of different ways we could express this. One is to figure out the position of the last character by using the `len()` function to get the length of the DNA sequence:

---

1   We know that that's not really how a splicosome works, but it's fine as a conceptual model.

```
exon2 = my_dna[91:len(my_dna)]
```

But we already know that leaving out the stop position causes the substring to go all the way to the end of the string, so we can just write it like this:

```
exon2 = my_dna[91:]
```

which is easier to read.

Putting all these ideas together gives us this program:

```
my_dna =
"ATCGATCGATCGATCGACTGACTAGTCATAGCTATGCATGTAGCTACTCGATCGATCGATCGATCGATCGATC
GATCGATCGATCATGCTATCATCGATCGATATCGATGCATCGACTACTAT"
exon1 = my_dna[1:64]
exon2 = my_dna[91:]
print(exon1 + exon2)
```

The output from this code looks vaguely right:

```
TCGATCGATCGATCGACTGACTAGTCATAGCTATGCATGTAGCTACTCGATCGATCGATCGATCATCGATCGAT
ATCGATGCATCGACTACTAT
```

but when we look more closely we can see that something is not right. The printed coding sequence is supposed to start at the very first character of the input sequence, but it's starting at the second. We have forgotten to take into account the fact that Python starts counting from zero. Let's try again:

```
my_dna =
"ATCGATCGATCGATCGACTGACTAGTCATAGCTATGCATGTAGCTACTCGATCGATCGATCGATCGATCGATC
GATCGATCGATCATGCTATCATCGATCGATATCGATGCATCGACTACTAT"
exon1 = my_dna[0:63]
exon2 = my_dna[90:]
print(exon1 + exon2)
```

introns1.py

Now the output looks correct – the coding sequence starts at the very beginning of the input sequence:

```
ATCGATCGATCGATCGACTGACTAGTCATAGCTATGCATGTAGCTACTCGATCGATCGATCGATCATCGATCGA
TATCGATGCATCGACTACTAT
```

Splicing out introns, part two

This is a straightforward piece of number crunching. There are a couple of ways to go about it. We could use the exon start/stop coordinates to calculate the length of the coding portion of the sequence. However, since we've already written the code to generate the coding sequence, we can simply calculate the length of it, and then divide by the length of the input sequence:

```
my_dna =
"ATCGATCGATCGATCGACTGACTAGTCATAGCTATGCATGTAGCTACTCGATCGATCGATCGATCGATCGATC
GATCGATCGATCATGCTATCATCGATCGATATCGATGCATCGACTACTAT"
exon1 = my_dna[0:63]
exon2 = my_dna[90:]
coding_length = len(exon1 + exon2)
total_length = len(my_dna)
print(coding_length / total_length)
```

The output shows that we're nearly right:

```
0.780487804878
```

We have calculated the coding proportion as a fraction, but the exercise called for a percentage. We can easily fix this by multiplying by 100. Notice that the symbol for multiplication is not x, as you might think, but *. The final code:

```
my_dna =
"ATCGATCGATCGATCGACTGACTAGTCATAGCTATGCATGTAGCTACTCGATCGATCGATCGATCGATCGATC
GATCGATCGATCATGCTATCATCGATCGATATCGATGCATCGACTACTAT"
exon1 = my_dna[0:63]
exon2 = my_dna[90:]
coding_length = len(exon1 + exon2)
total_length = len(my_dna)
print(100 * coding_length / total_length)
```

introns2.py

gives the correct output:

```
78.0487804878
```

although we probably don't really require that number of significant figures. In chapter 5 we will learn how to format the output nicely.

Splicing out introns, part three

This sounds quite tricky, but we have already done the hard bit in part one. All we need to do is extract the intron sequence as well as the exons, convert it to lower case, then concatenate the three sequences to recreate the original genomic sequence:

```
my_dna =
"ATCGATCGATCGATCGACTGACTAGTCATAGCTATGCATGTAGCTACTCGATCGATCGATCGATCGATCGATC
GATCGATCGATCATGCTATCATCGATCGATATCGATGCATCGACTACTAT"

exon1 = my_dna[0:63]
intron = my_dna[63:90]
exon2 = my_dna[90:]

print(exon1 + intron.lower() + exon2)
```

introns3.py

This program nicely illustrates the benefit of having the substring positions inclusive and the start and exclusive at the end: notice how the last position for `exon1` (90) is also the first position for `intron`.

Looking at the output, we see an upper case DNA sequence with a lower case section in the middle, as expected:

```
ATCGATCGATCGATCGACTGACTAGTCATAGCTATGCATGTAGCTACTCGATCGATCGATCGAtcgatcgatcg
atcgatcgatcatgctATCATCGATCGATATCGATGCATCGACTACTAT
```

When we are applying several transformations to text, as in this exercise, there are usually a number of different ways we can write the program. For example, we could store the lower case version of the intron, rather than converting it to lower case when printing:

```
intron = my_dna[63:90].lower()
```

Or we could avoid using variables for the introns and exons all together, and do everything in one big `print()` statement:

```
print(my_dna[0:63] + my_dna[63:90].lower() + my_dna[90:])
```

This last option is very concise, but a bit harder to read than the more verbose way.

As the exercises in this book get longer, you'll notice that there are more and more different ways to write the code – you may end up with solutions that look very different to the example solutions. When trying to choose between different ways to write a program, always favour the solution that is clearest in intent and easiest to read.

What have we learned?

On the surface, these exercises are about manipulating DNA sequences. It's unlikely that you'll have to solve these exact same problems in your own programs.

On a deeper level, however, the exercises are about learning to break down problems into individual steps which can be solved using the tools available to us. Even the simplest of problems requires using several different tools in the right order. The remainder of the exercises in this book – and nearly all the programs you will write in the future – will require you to break down problems in this way.

We've also learned a few specific lessons. In the first exercise, we saw how it's important to test code using simple inputs in order to check that it's giving the right answer. In the complementing DNA exercise, we saw that it can be challenging to keep track of the changes that are made to a variable as a program runs. In the last few exercises concerning introns and exons, we saw how existing code can often be modified and reused to solve a slightly different problem.

We will return to these themes in future exercises.

3:   Reading and writing files

Why are we so interested in working with files?

As we start this chapter, we find ourselves once again doing things in a slightly different order to most programming books. The majority of introductory programming books won't consider working with external files until much further along, so why are we introducing it now?

The answer, as was the case in the last chapter, lies in the particular jobs that we want to use Python for. The data that we as biologists work with is stored in files, so if we're going to write useful programs we need a way to get the data out of files and into our programs (and *vice versa*). As you were going through the exercises in the previous chapter, it may have occurred to you that copying and pasting a DNA sequence directly into a program each time we want to use it is not a very good approach to take, and you'd be right. The sequences we were working with in the exercises were very short; obviously real life data will be much longer. Also, it seems inelegant to have the data we want to work on mixed up with the code that manipulates it. In this chapter we'll see a better way to do it.

We're lucky in biology in that many of the types of data that we work with are stored in text[1] files which are relatively simple to process using Python. Chief among these, of course, are DNA and protein sequence data, which can be stored in a variety of formats.[2] But there are many other types of data – sequencing reads, quality scores, SNPs, phylogenetic trees, read maps, geographical sample data, genetic distance matrices – which we can access from within our Python programs.

Another reason for our interest in file input/output is the need for our Python programs to work as part of a pipeline or work flow involving other, existing tools.

---

1   i.e. files which you can open in a text editor and read, as opposed to binary files which cannot be read directly.

2   In this book we'll mostly be talking about FASTA format as it's the simplest and most common format, but there are many more.

When it comes to using Python in the real world, we often want Python to either accept data from, or provide data to, another program. Often the easiest way to do this is to have Python read, or write, files in a format that the other program already understands.

Reading text from a file

Firstly, a quick note about what we mean by text. In programming, when we talk about *text files*, we are not necessarily talking about something that is human readable. Rather, we are talking about a file that contains characters and lines – something that you could open and view in a text editor, regardless of whether you could actually make sense of the file or not. Examples of text files which you might have encountered include:

- FASTA files of DNA or protein sequences
- files containing output from command-line programs (e.g. BLAST)
- FASTQ files containing DNA sequencing reads
- HTML files
- and of course, Python code itself

In contrast, most files that you encounter day-to-day will be *binary files* – ones which are not made up of characters and lines, but of bytes. Examples include:

- image files (JPEGs and PNGs)
- audio files
- video files
- compressed files (e.g. ZIP files)

If you're not sure whether a particular file is text or binary, there's a very simple way to tell – just open it up in a text editor. If the file displays without any problem, then it's text (regardless of whether you can make sense of it or not). If you get an error or a warning from your text editor, or the file displays as a collection of indecipherable characters, then it's binary.

The examples and exercises in this chapter are a little different from those in the previous one, because they rely on the existence of the files that we are going to manipulate. If you want to try running the examples in this chapter, you'll need to make sure that there is a file in your working directory called *dna.txt* which has a single line containing a DNA sequence. The easiest way to do this is to run the examples while in the *chapter_3* folder inside the exercises download[1].

Using open to read a file

In Python, as in the physical world, we have to open a file before we can read what's inside it. The Python function that carries out the job of opening a file is very sensibly called `open()`. It takes one argument – a string which contains the name of the file – and returns a *file object:*

```
my_file = open("dna.txt")
```

A *file object* is a new type which we haven't encountered before, and it's a little more complicated than the string and number types that we saw in the previous chapter. With strings and numbers it was easy to understand what they represented – a single bit of text, or a single number. A file object, in contrast, represents something a bit less tangible – it represents a file on your computer's hard drive.

The way that we use file objects is a bit different to strings and numbers as well. If you glance back at the examples from the previous chapter you'll see that most of the time when we want to use a variable containing a string or number we just use the variable name:

---

1   If you haven't downloaded the example files yet, you'll find the link in the Introduction.

```
my_string = 'abcdefg'
print(my_string)
my_number = 42
print(my_number + 1)
```

In contrast, when we're working with file objects most of our interaction will be through *methods*. This style of programming will seem unusual at first, but as we'll see in this chapter, the file type has a well thought-out set of methods which let us do lots of useful things.

The first thing we need to be able to do is to read the contents of the file. The file type has a `read()` method which does this. It doesn't take any arguments, and the return value is a string, which we can store in a variable. Once we've read the file contents into a variable, we can treat them just like any other string – for example, we can print them:

```
my_file = open("dna.txt")
file_contents = my_file.read()
print(file_contents)
```

`print_file_contents.py`

Files, contents and filenames

When learning to work with files it's very easy to get confused between a *file object*, a *filename*, and the *contents* of a file. Take a look at the following bit of code:

```
my_file_name = "dna.txt"❶
my_file = open(my_file_name)❷
my_file_contents = my_file.read()❸
```

What's going on here? First, we store the string *dna.txt* in the variable `my_file_name`❶. Next, we use the variable `my_file_name` as the argument to the `open()` function, and store the resulting file object in the variable my_file❷. Finally, we call the read() method on the variable my_file, and store the resulting string in the variable my_file_contents❸.

The important thing to understand about this code is that there are three separate variables which have different types and which are storing three very different things:

- my_file_name is a string, and it stores the name of a file on disk.

- my_file is a file object, and it represents the file itself.

- my_file_contents is a string, and it stores the text that is in the file.

Remember that variable names are arbitrary – the computer doesn't care what you call your variables. So this piece of code is exactly the same as the previous example:

```
apple = "dna.txt"
banana = open(apple)
grape = banana.read()
```

except it is harder to understand! In contrast, the filename (*dna.txt*) is **not** arbitrary – it must correspond to the name of a file on the hard drive of your computer.

A common error is to try to use the read() method on the wrong thing. Recall that read() is a method that only works on file objects. If we try to use the read() method on the filename:

```
my_file_name = "dna.txt"
my_contents = my_file_name.read()
```

read_error.py

we'll get an AttributeError – Python will complain that strings don't have a read() method[1]:

---

1   From now on, I'll just show the relevant bits of output when discussing error message.

```
AttributeError: 'str' object has no attribute 'read'
```

Another common error is to use the *file object* when we meant to use the *file contents*. If we try to print the file object:

```
my_file_name = "dna.txt"
my_file = open(my_file_name)
print(my_file)
```

print_file_object.py

we won't get an error, but we'll get an odd looking line of output:

```
<open file 'dna.txt', mode 'r' at 0x7fc5ff7784b0>
```

We won't discuss the meaning of this line now: just remember that if you try to print the contents of a file but instead you get some output that looks like the above, you have almost definitely printed the file **object** rather than the file **contents**.

Dealing with newlines

Let's take a look at the output we get when we try to print some information from a file. We'll use the *dna.txt* file from the *chapter_3* exercises folder. This file contains a single line with a short DNA sequence. Open the file in a text editor and take a look at it.

We're going to write a simple program to read the DNA sequence from the file and print it out along with its length. Putting together the file functions and methods from this chapter, and the material we saw in the previous chapter, we get the following code:

```
# open the file
my_file = open("dna.txt")

# read the contents
my_dna = my_file.read()

# calculate the length
dna_length = len(my_dna)

# print the output
print("sequence is " + my_dna +  " and length is " + str(dna_length))
```

print_seq_and_length.py

When we look at the output, we can see that there are two things wrong.

```
sequence is ACTGTACGTGCACTGATC
 and length is 19
```

Firstly, the output has been split over two lines, even though we didn't ask for it. And secondly, the length is wrong – there are only 18 characters in the DNA string.

Both of these problems have the same explanation: Python has included the newline character at the end of the *dna.txt* file as part of the contents. In other words, the variable my_dna has a newline character at the end of it. If we could view the my_dna variable directly[1], we would see that it looks like this:

```
'ACTGTACGTGCACTGATC\n'
```

This explains why the output from our program is split over two lines – the newline character is part of the string we are trying to print. It also explains why the length is wrong – Python is including the newline character when it counts the number of characters in the string.

---

1   In fact, we can do this – there's a function called repr() that returns a representation of a variable.

r strip (handwritten annotation at top)

The solution is also simple. Because this is such a common problem, strings have a method for removing newlines from the end of them. The method is called rstrip(), and it takes one string argument which is the character that you want to remove. In this case, we want to remove the newline character (\n). Here's a modified version of the code – note that the argument to rstrip() is itself a string so needs to be enclosed in quotes:❶

```
my_file = open("dna.txt")
my_file_contents = my_file.read()

# remove the newline from the end of the file contents
my_dna = my_file_contents.rstrip("\n")❶

dna_length = len(my_dna)
print("sequence is " + my_dna +  " and length is " + str(dna_length))
```

print_length_and_seq2.py

and now the output looks just as we expected:

```
sequence is ACTGTACGTGCACTGATC and length is 18
```

In the code above, we first read the file contents and then removed the newline, in two separate steps:

```
my_file_contents = my_file.read()
my_dna = my_file_contents.rstrip("\n")
```

but it's more common to read the contents and remove the newline all in one go, like this:

```
my_dna = my_file.read().rstrip("\n")
```

This is a bit tricky to read at first as we are using two different methods (read() and rstrip()) in the same statement. The key is to read it from left to right – we take the my_file variable and use the read() method on it, then we take the

61

output of that method (which we know is a string) and use the `rstrip()` method on it. The result of the `rstrip()` method is then stored in the `my_dna` variable.

If you find it difficult to write the whole thing as one statement like this, just break it up and do the two things separately – your programs will run just as well.

Missing files

What happens if we try to read a file that doesn't exist?

```
my_file = open("nonexistent.txt")
```

We get a new type of error that we've not seen before:

```
IOError: [Errno 2] No such file or directory: 'nonexistent.txt'
```

If you encounter this error, you've probably got the filename wrong, or are working in the wrong folder. Ideally, we'd like to be able to check if a file exists **before** we try to open it – we'll see how to do that in chapter 9.

Writing text to files

All the example programs that we've seen so far in this book have produced output straight to the screen. That's great for exploring new features and when working on programs, because it allows you to see the effect of changes to the code right away. It has a few drawbacks, however, when writing code that we might want to use in real life.

Printing output to the screen only works well when there isn't very much of it[1]. It's great for short programs and status messages, but quickly becomes cumbersome for large amounts of output. Some terminals struggle with large amounts of text, or worse, have a limited scrollback capability which can cause

---

1   Linux users may be aware that we can redirect terminal output to a file using shell redirection, which can get around some of these problems.

the first bit of your output to disappear. It's not easy to search in output that's being displayed at the terminal, and long lines tend to get wrapped. Also, for many programs we want to send different bits of output to different files, rather than having it all dumped in the same place.

Most importantly, terminal output vanishes when you close your terminal program. For small programs like the examples in this book, that's not a problem – if you want to see the output again you can just rerun the program. If you have a program that requires a few hours to run, that's not such a great option.

Opening files for writing

In the previous section, we saw how to open a file and read its contents. We can also open a file and *write* some data to it, but we have to use the `open()` function in a slightly different way. To open a file for writing, we use a two-argument version of the `open()` function, where the second argument is a short string describing what we want to do to the file[1]. This second argument can be "r" for reading, "w" for writing, or "a" for appending. If we leave out the second argument (like we did for all the examples above), Python uses the default, which is "r" for reading.

The difference between "w" and "a" is subtle, but important. If we open a file that already exists using the mode "w", then we will overwrite the current contents with whatever data we write to it. If we open an existing file with the mode "a", it will add new data onto the end of the file, but will **not** remove any existing content. If there doesn't already exist a file with the specified name, then "w" and "a" behave identically – they will both create a new file to hold the output.

Quite a lot of Python functions and methods have these optional arguments. For the purposes of this book, we will only mention them when they are directly relevant to what we're doing. If you want to see all the optional arguments for a particular method or function, the best place to look is the official Python documentation – see chapter 1 for details.

---

1   We call this the *mode* of the file, and there are a few more options not mentioned here – mostly for working with binary files.

Once we've opened a file for writing, we can use the file `write()` method to write some text to it. `write()` works a lot like `print()` – it takes a single string argument - but instead of printing the string to the screen it writes it to the file.

Here's how we use `open()` with a second argument to open a file and write a single line of text to it:

```
my_file = open("out.txt", "w")
my_file.write("Hello world")
```

write.py

Because the output is being written to the file in this example, you won't see any output on the screen if you run it. To check that the code has worked, you'll have to run it, then open the file *out.txt* in your text editor and check that its contents are what you expect[1].

Remember that with `write()`, just like with `print()`, we can use **any** string as the argument. This also means that we can use any method or function that **returns** a string. The following are all perfectly OK:

```
# write "abcdef"
my_file.write("abc" + "def")

# write "8"
my_file.write(str(len('AGTGCTAG')))

# write "TTGC"
my_file.write("ATGC".replace('A', 'T'))

# write "atgc"
my_file.write("ATGC".lower())

# write contents of my_variable
my_file.write(my_variable)
```

---

1 .txt is the standard filename extension for a plain text file. Later in this book, when we generate output files with a particular format, we'll use different filename extensions.

Closing files

There's one more important file method to look at before we finish this chapter – close(). Unsurprisingly, this is the opposite of open() (but note that it's a *method*, whereas open() is a *function*). We should call close() after we're done reading or writing to a file – we won't go into the details here, but it's a good habit to get into as it avoids some types of bugs that can be tricky to track down[1]. close() is an unusual method as it takes no arguments (so it's called with an empty pair of parentheses) and doesn't return any useful value:

```python
my_file = open("out.txt", "w")
my_file.write("Hello world")
# remember to close the file
my_file.close()
```

close_file.py

Paths and folders

So far, we have only dealt with opening files in the same folder that we are running our program. What if we want to open a file from a different part of the file system?

The open() function is quite happy to deal with files from anywhere on your computer, as long as you give the full path (i.e. the sequence of folder names that tells you the location of the file). Just give the **path to the file** as the argument rather than the **name of the file**. The format of the file path looks different depending on your operating system. If you're using Windows, the path will look like this:

```python
my_file = open("c:/windows/Desktop/myfolder/myfile.txt")
```

---

1   Specifically, it helps to ensure that output to a file is flushed, which is necessary when we want to make a file available to another program as part of our work flow.

Notice that the folder names are separated by forward slashes rather than the back slashes that Windows normally uses. This is to avoid problems with special characters like the ones we saw in chapter 2.

If you're using a Mac or Linux machine, then the path will look slightly different:

```
my_file = open("/home/martin/myfolder/myfile.txt")
```

Recap

We've taken a whole chapter to introduce the various ways of reading and writing to files, because it's such an important part of building programs that are useful in real life. We've seen how working with file contents is always a two step process – we must open a file before reading or writing – and looked at several common pitfalls.

We've also introduced a couple of new concepts that are more widely applicable. We've encountered our first example of an optional argument in the `open()` function (we'll see more of these in future chapters). We've also encountered the first example of a complex data type – the file object – and seen how we can do useful things with it by calling its various methods, in contrast to the simple strings and numbers that we've been working with in the previous chapter. In future chapters, we'll learn about more of these complex data types and how to use them.

Exercises

Splitting genomic DNA

Look in the *chapter_3* folder for a file called *genomic_dna.txt* – it contains the same piece of genomic DNA that we were using in the final exercise from chapter 2. Write a program that will split the genomic DNA into coding and non-coding parts, and write these sequences to two separate files.

Hint: use your solution to the last exercise from chapter 2 as a starting point.

Writing a FASTA file

FASTA file format is a commonly used DNA and protein sequence file format. A single sequence in FASTA format looks like this:

```
>sequence_name
ATCGACTGATCGATCGTACGAT
```

Where `sequence_name` is a header that describes the sequence (the greater-than symbol indicates the start of the header line). Often, the header contains an accession number that relates to the record for the sequence in a public sequence database. A single FASTA file can contain multiple sequences, like this[1]:

```
>sequence_one
ATCGATCGATCGATCGAT
>sequence_two
ACTAGCTAGCTAGCATCG
>sequence_three
ACTGCATCGATCGTACCT
```

---

1   Real world FASTA files sometimes spread the sequence across multiple lines, but to keep things simple we'll just put the sequence on one line for this exercise.

Write a program that will create a FASTA file for the following three sequences – make sure that all sequences are in uppercase and only contain the bases A, T, G and C.

Sequence header	DNA sequence
ABC123	ATCGTACGATCGATCGATCGCTAGACGTATCG
DEF456	actgatcgacgatcgatcgatcacgact
HIJ789	ACTGAC-ACTGT--ACTGTA----CATGTG

Writing multiple FASTA files

Use the data from the previous exercise, but instead of creating a single FASTA file, create three new FASTA files – one per sequence. The names of the FASTA files should be the same as the sequence header names, with the extension *.fasta*.

Solutions

Splitting genomic DNA

We have a head-start on this problem, because we have already tackled a similar problem in the previous chapter. Let's remind ourselves of the solution we ended up with for that exercise:

```
my_dna =
"ATCGATCGATCGATCGACTGACTAGTCATAGCTATGCATGTAGCTACTCGATCGATCGATCGATCGATCGATC
GATCGATCGATCATGCTATCATCGATCGATATCGATGCATCGACTACTAT"
exon1 = my_dna[0:63]
intron = my_dna[63:90]
exon2 = my_dna[90:]
print(exon1 + intron.lower() + exon2)
```

What changes do we need to make? Firstly, we need to read the DNA sequence from a file instead of writing it in the code:

```
dna_file = open("genomic_dna.txt")
my_dna = dna_file.read()
```

Secondly, we need to create two new file objects to hold the output:

```
coding_file = open("coding_dna.txt", "w")
noncoding_file = open("noncoding_dna.txt", "w")
```

Finally, we need to concatenate the two exon sequences and write them to the coding DNA file, and write the intron sequence to the non-coding DNA file:

```
coding_file.write(exon1 + exon2)
noncoding_file.write(intron)
```

Let's put it all together, with some blank lines to separate out the different parts of the program:

```
# open the file and read its contents
dna_file = open("genomic_dna.txt")
my_dna = dna_file.read()

# extract the different bits of DNA sequence
exon1 = my_dna[0:62]
intron = my_dna[62:90]
exon2 = my_dna[90:]

# open the two output files
coding_file = open("coding_dna.txt", "w")
noncoding_file = open("noncoding_dna.txt", "w")

# write the sequences to the output files
coding_file.write(exon1 + exon2)
noncoding_file.write(intron)
```

genomic_dna.py

Writing a FASTA file

Let's start this problem by thinking about the variables we're going to need. We have three DNA sequences in total, so we'll need three variables to hold the sequence headers, and three more to hold the sequences themselves:

```
header_1 = "ABC123"
header_2 = "DEF456"
header_3 = "HIJ789"
seq_1 = "ATCGTACGATCGATCGATCGCTAGACGTATCG"
seq_2 = "actgatcgacgatcgatcgatcacgact"
seq_3 = "ACTGAC-ACTGT--ACTGTA----CATGTG"
```

FASTA format has alternating lines of header and sequence, so before we try any sequence manipulation, let's try to write a program that produces the lines in the right order. Rather than writing to a file, we'll print the output to the screen for now – that will make it easier to see the output right away. Once we've got it working, we'll switch over to file output. Here's a few lines which will print data to the screen:

```
print(header_1)
print(seq_1)
print(header_2)
print(seq_2)
print(header_3)
print(seq_3)
```

and here's what the output looks like:

```
ABC123
ATCGTACGATCGATCGATCGCTAGACGTATCG
DEF456
actgatcgacgatcgatcgatcacgact
HIJ789
ACTGAC-ACTGT--ACTGTA----CATGTG
```

Not far off – the lines are in the right order, but we forgot to include the greater-than symbol at the start of the header. Also, we don't really need to print the header and the sequence separately for each sequence – we can include a newline character in the print string in order to get them on separate lines. Here's an improved version of the code:

```
print('>' + header_1 + '\n' + seq_1)
print('>' + header_2 + '\n' + seq_2)
print('>' + header_3 + '\n' + seq_3)
```

and the output looks better too:

```
>ABC123
ATCGTACGATCGATCGATCGCTAGACGTATCG
>DEF456
actgatcgacgatcgatcgatcacgact
>HIJ789
ACTGAC-ACTGT--ACTGTA----CATGTG
```

Next, let's tackle the problems with the sequences. The second sequence is in lower case, and it needs to be in upper case – we can fix that using the `upper()` string method. The third sequence has a bunch of gaps that we need to remove.

We haven't come across a `remove()` method.... but we do know how to replace one character with another. If we replace all the gap characters with an empty string, it will be the same as removing them[1]. Here's a version that fixes both sequences:

```
print('>' + header_1 + '\n' + seq_1)
print('>' + header_2 + '\n' + seq_2.upper())
print('>' + header_3 + '\n' + seq_3.replace('-', ''))
```

Now the printed output is perfect:

```
>ABC123
ATCGTACGATCGATCGATCGCTAGACGTATCG
>DEF456
ACTGATCGACGATCGATCGATCACGACT
>HIJ789
ACTGACACTGTACTGTACATGTG
```

The final step is to switch from printed output to writing to a file. We'll `open()` a new file, and change the three `print()` lines to `write()`:

```
output = open("sequences.fasta", "w")
output.write('>' + header_1 + '\n' + seq_1)
output.write('>' + header_2 + '\n' + seq_2.upper())
output.write('>' + header_3 + '\n' + seq_3.replace('-', ''))
```

After making these changes the code doesn't produce any output on the screen, so to see what's happened we'll need to take a look at the *sequences.fasta* file:

```
>ABC123
ATCGTACGATCGATCGATCGCTAGACGTATCG>DEF456
ACTGATCGACGATCGATCGATCACGACT>HIJ789
ACTGACACTGTACTGTACATGTG
```

---

1   An empty string is just a pair of quotation marks with nothing in between them.

This doesn't look right – the second and third lines have been joined together, as have the fourth and fifth. What has happened?

It looks like we've uncovered a difference between the `print()` function and the `write()` method. `print()` automatically puts a newline at the end of the string, whereas `write()` doesn't. This means we've got to be careful when switching between them! The fix is quite simple: we'll add a newline onto the end of each string that gets written to the file:

```
output = open("sequences.fasta", "w")
output.write('>' + header_1 + '\n' + seq_1 + '\n')
output.write('>' + header_2 + '\n' + seq_2.upper() + '\n')
output.write('>' + header_3 + '\n' + seq_3.replace('-', '') + '\n')
```

The arguments for the `write()` statements are getting quite complicated, but they are all made up of simple building blocks. For example the last one, if we translated it into English, would read

> "a greater-than symbol, followed by the variable header_3, followed by a newline, followed by the variable seq_3 with all hyphens replaced with nothing, followed by another newline".

Here's the final code, including the variable definition at the beginning, with blank lines and comments:

```
# set the values of all the header variables
header_1 = "ABC123"
header_2 = "DEF456"
header_3 = "HIJ789"

# set the values of all the sequence variables
seq_1 = "ATCGTACGATCGATCGATCGCTAGACGTATCG"
seq_2 = "actgatcgacgatcgatcgatcacgact"
seq_3 = "ACTGAC-ACTGT-ACTGTA----CATGTG"

# make a new file to hold the output
output = open("sequences.fasta", "w")

# write the header and sequence for seq1
output.write('>' + header_1 + '\n' + seq_1 + '\n')

# write the header and uppercase sequences for seq2
output.write('>' + header_2 + '\n' + seq_2.upper() + '\n')

# write the header and sequence for seq3 with hyphens removed
output.write('>' + header_3 + '\n' + seq_3.replace('-', '') + '\n')
```

writing_a_fasta_file.py

There's an exercise that uses different techniques to solve a very similar problem at the end of the chapter on functional programming in Advanced Python for Biologists – if you find yourself carrying out this type of process in real life code, then it's probably worth a look.

Writing multiple FASTA files

We can solve this problem with a slight modification of our solution to the previous exercise. We'll need to create three new files to hold the output, and we'll construct the name of each file by using string concatenation:

```
output_1 = open(header_1 + ".fasta", "w")
output_2 = open(header_2 + ".fasta", "w")
output_3 = open(header_3 + ".fasta", "w")
```

Remember, the first argument to `open()` is a string, so it's fine to use a concatenation because we know that the result of concatenating two strings is also a string.

We'll also change the `write()` statements so that we have one for each of the output files. We need to be careful with the number here in order to make sure that we get the right sequence in each file. Here's the final code, with comments.

```python
# set the values of all the header variables
header_1 = "ABC123"
header_2 = "DEF456"
header_3 = "HIJ789"

# set the values of all the sequence variables
seq_1 = "ATCGTACGATCGATCGATCGCTAGACGTATCG"
seq_2 = "actgatcgacgatcgatcgatcacgact"
seq_3 = "ACTGAC-ACTGT-ACTGTA----CATGTG"

# make three files to hold the output
output_1 = open(header_1 + ".fasta", "w")
output_2 = open(header_2 + ".fasta", "w")
output_3 = open(header_3 + ".fasta", "w")

# write one sequence to each output file
output_1.write('>' + header_1 + '\n' + seq_1 + '\n')
output_2.write('>' + header_2 + '\n' + seq_2.upper() + '\n')
output_3.write('>' + header_3 + '\n' + seq_3.replace('-', '') + '\n')
```

writing_multiple_fasta_files.py

Looking at the code above, it seems like there's a lot of redundancy there. Each of the four sections of code – setting the header values, setting the sequence values, creating the output files, and writing data to the output files – consists of three nearly identical statements. Although the solution works, it seems to involve a lot of unnecessary typing!

Also, having so much nearly identical code seems likely to cause errors if we need to change something. In the next chapter, we'll examine some tools which will allow us to start removing some of that redundancy.

What have we learned?

The exercises for this chapter have been about the simple mechanics of reading and writing files. The majority of programs that you'll want to write will involve files in some capacity, so exercises like these are good practice.

We've also encountered a few specific problems that are commonly encountered when working with files. We've seen how to create files using a variable as part of the filename. We've also seen the difference between `print()` and `write()` in the way that they handle line endings.

4: Lists and loops

Why do we need lists and loops?

Think back over the exercises that we've seen in the previous two chapters – they've all involved dealing with one bit of information at a time. In chapter 2, we used string manipulation tools to process single sequences, and in chapter 3, we practised reading and writing files one at a time. The closest we got to using multiple pieces of data was during the final exercise in chapter 3, where we were dealing with three DNA sequences.

If that's all that Python allowed us to do, it wouldn't be a very helpful tool for biology. In fact, there's a good chance that you're reading this book because you want to be able to write programs to help you deal with large datasets. A very common situation in biological research is to have a large collection of data (DNA sequences, SNP positions, gene expression measurements) that all need to be processed in the same way. In this chapter, we'll learn about the fundamental programming tools that will allow our programs to do this.

So far we have learned about several different data types (strings, numbers, and file objects), all of which store a single bit of information[1]. When we've needed to store multiple bits of information (for example, the three DNA sequences in the chapter 3 exercises) we have simply created more variables to hold them:

```
# set the values of all the sequence variables
seq_1 = "ATCGTACGATCGATCGATCGCTAGACGTATCG"
seq_2 = "actgatcgacgatcgatcgatcacgact"
seq_3 = "ACTGAC-ACTGT-ACTGTA----CATGTG"
```

The limitations of this approach became clear quite quickly as we looked at the solution code – it only worked because the number of sequences were small, and we knew the number in advance. If we were to repeat the exercise with three

---

1   We know that files are slightly different to strings and numbers because they can store a lot of information, but each file object still only refers to a single file.

hundred or three thousand sequences, the vast majority of the code would be given over to storing variables and it would become completely unmanageable. And if we were to try and write a program that could process an unknown number of input sequences (for instance, by reading them from a file), we wouldn't be able to do it. To make our programs able to process multiple pieces of data, we need an entirely new type of structure which can hold many pieces of information at the same time – a *list*.

We've also dealt exclusively with programs whose statements are executed from top to bottom in a very straightforward way. This has great advantages when first starting to think about programming – it makes it very easy to follow the flow of a program. The downside of this sequential style of programming, however, is that it leads to very redundant code like we saw at the end of the previous chapter:

```
# make three files to hold the output
output_1 = open(header_1 + ".fasta", "w")
output_2 = open(header_2 + ".fasta", "w")
output_3 = open(header_3 + ".fasta", "w")
```

Again; it was only possible to solve the exercise in this manner because we knew in advance the number of output files we were going to need. Looking at the code, it's clear that these three lines consist of essentially the same statement being executed multiple times, with some slight variations. This idea of repetition-with-variation is incredibly common in programming problems, and Python has built in tools for expressing it – *loops*.

Creating lists and retrieving elements

To make a new list, we put several strings or numbers[1] inside square brackets, separated by commas:

```
apes = ["Homo sapiens", "Pan troglodytes", "Gorilla gorilla"]
conserved_sites = [24, 56, 132]
```

---

1   Or in fact, any other type of value or variable.

Each individual item in a list is called an *element*. To get a single element from the list, write the variable name followed by the *index* of the element you want in square brackets:

```python
apes = ["Homo sapiens", "Pan troglodytes", "Gorilla gorilla"]
conserved_sites = [24, 56, 132]
print(apes[0])
first_site = conserved_sites[2]
```

create_list.py

If we want to go in the other direction – i.e. we know which element we want but we don't know the index – we can use the `index()` method:

```python
apes = ["Homo sapiens", "Pan troglodytes", "Gorilla gorilla"]
chimp_index = apes.index("Pan troglodytes")
# chimp_index is now 1
```

Remember that in Python we start counting from zero rather than one, so the first element of a list is always at index zero. If we give a negative number, Python starts counting from the **end** of the list rather than the beginning – so it's easy to get the last element from a list:

```python
last_ape = apes[-1]
```

What if we want to get more than one element from a list? We can give a start and stop position, separated by a colon, to specify a range of elements:

```python
ranks = ["kingdom","phylum", "class", "order", "family"]
lower_ranks = ranks[2:5]
# lower ranks are class, order and family
```

sublist.py

Does this look familiar? It's the exact same notation that we used to get substrings back in chapter 2, and it works in exactly the same way – numbers are **inclusive** at the start and **exclusive** at the end. The fact that we use the same notation for strings and lists hints at a deeper relationship between the two types. In fact, what we were doing when extracting substrings in chapter 2 was **treating a string as though it were a list of characters**. This idea – that we can treat a variable as though it were a list when it's not – is a powerful one in Python and we'll come back to it later in this chapter.

Working with list elements

To add another element onto the end of an existing list, we can use the `append()` method:

```
apes = ["Homo sapiens", "Pan troglodytes", "Gorilla gorilla"]
print(apes)
apes.append("Pan paniscus")
print(apes)
```

`append()` is an interesting method because it actually changes the variable on which it's used – in the above example, the `apes` list goes from having three elements to having four:

```
['Homo sapiens', 'Pan troglodytes', 'Gorilla gorilla']
['Homo sapiens', 'Pan troglodytes', 'Gorilla gorilla', 'Pan paniscus']
```

We can get the length of a list by using the `len()` function, just like we did for strings:

```
apes = ["Homo sapiens", "Pan troglodytes", "Gorilla gorilla"]
print("There are " + str(len(apes)) + " apes")
apes.append("Pan paniscus")
print("Now there are " + str(len(apes)) + " apes")
```

`list_length.py`

The output shows that the number of elements in `apes` really has changed:

```
There are 3 apes
Now there are 4 apes
```

We can concatenate two lists just as we did with strings, by using the plus symbol:

```
apes = ["Homo sapiens", "Pan troglodytes", "Gorilla gorilla"]
monkeys = ["Papio ursinus", "Macaca mulatta"]
primates = apes + monkeys

print(str(len(apes)) + " apes")
print(str(len(monkeys)) + " monkeys")
print(str(len(primates)) + " primates")
```

concatenate_lists.py

As we can see from the output, this doesn't change either of the two original lists – it makes a brand new list which contains elements from both:

```
3 apes
2 monkeys
5 primates
```

If we want to add elements from a list onto the end of an existing list, changing it in the process, we can use the `extend()` method. `extend()` behaves like `append()` but takes a *list* as its argument rather than a single *element*.

Here are two more list methods that change the variable they're used on: `reverse()` and `sort()`. Both `reverse()` and `sort()` work by changing the order of the elements in the list. If we want to print out a list to see how this works, we need to used `str()` (just as we did when printing out numbers):

```
ranks = ["kingdom","phylum", "class", "order", "family"]
print("at the start : " + str(ranks))

ranks.reverse()
print("after reversing : " + str(ranks))

ranks.sort()
print("after sorting : " + str(ranks))
```

reverse_and_sort.py

If we take a look at the output, we can see how the order of the elements in the list is changed by these two methods:

```
at the start : ['kingdom', 'phylum', 'class', 'order', 'family']
after reversing : ['family', 'order', 'class', 'phylum', 'kingdom']
after sorting : ['class', 'family', 'kingdom', 'order', 'phylum']
```

By default, Python sorts strings in alphabetical order and numbers in ascending numerical order[1].

Writing a loop

Imagine we wanted to take our list of apes:

```
apes = ["Homo sapiens", "Pan troglodytes", "Gorilla gorilla"]
```

and print out each element on a separate line, like this:

```
Homo sapiens is an ape
Pan troglodytes is an ape
Gorilla gorilla is an ape
```

One way to do it would be to just print each element separately:

---

1   We can sort in other ways too – take a look at the functional programming chapter in *Advanced Python for Biologists*.

```
print(apes[0] + " is an ape")
print(apes[1] + " is an ape")
print(apes[2] + " is an ape")
```

but this is very repetitive and relies on us knowing the number of elements in the list. What we need is a way to say something along the lines of

> For each element in the list of apes, print out the element, followed by the words "is an ape".

Python's loop syntax allows us to express those instructions like this:

```
for ape in apes:
    print(ape + " is an ape")
```

loop.py

Let's take a moment to look at the different parts of this loop. We start by writing `for x in y`, where `y` is the name of the list we want to process and `x` is the name we want to use for the current element each time round the loop.

`x` is just a variable name (so it follows all the rules that we've already learned about variable names), but it behaves slightly differently to all the other variables we've seen so far. In all previous examples, we create a variable and store something in it, and then the value of that variable doesn't change unless we change it ourselves. In contrast, when we create a variable to be used in a loop, we don't set its value – the value of the variable will be automatically set to each element of the list in turn, and it will be different each time round the loop.

This first line of the loop ends with a colon, and all the subsequent lines (just one, in this case) are indented. Indented lines can start with any number of tab or space characters, but they must all be indented in the same way. This pattern – a line which ends with a colon, followed by some indented lines – is very common in Python, and we'll see it in several more places throughout this book. A group of indented lines is often called a *block* of code[1].

In this case, we refer to the indented block as the *body* of the loop, and the lines inside it will be executed once for each element in the list. To refer to the current element, we use the variable name that we wrote in the first line. The body of the loop can contain as many lines as we like, and can include all the functions and methods that we've learned about, with one important exception: we're not allowed to change the list while inside the body of the loop[2].

Here's an example of a loop with a more complicated body:

```python
apes = ["Homo sapiens", "Pan troglodytes", "Gorilla gorilla"]

for ape in apes:
    name_length = len(ape)
    first_letter = ape[0]
    print(ape + " is an ape. Its name starts with " + first_letter)
    print("Its name has " + str(name_length) + " letters")
```

complex_loop.py

The body of the loop in the code above has four statements, two of which are `print()` statements, so each time round the loop we'll get two lines of output. If we look at the output we can see all six lines:

```
Homo sapiens is an ape. Its name starts with H
Its name has 12 letters
Pan troglodytes is an ape. Its name starts with P
Its name has 15 letters
Gorilla gorilla is an ape. Its name starts with G
Its name has 15 letters
```

Why is the above approach better than printing out these six lines in six separate statements? Well, for one thing, there's much less redundancy – here we only

---

1   If you're familiar with any other programming languages, you might know code blocks as things that are surrounded with curly brackets – the indentation does the same job in Python.

2   Changing the list while looping can cause Python to become confused about which elements have already been processed and which are yet to come.

needed to write two `print()` statements. This also means that if we need to make a change to the code, we only have to make it once rather than three separate times. Another benefit of using a loop here is that if we want to add some elements to the list, we don't have to touch the loop code at all. Consequently, it doesn't matter how many elements are in the list, and it's not a problem if we don't know how many are going to be in it at the time when we write the code.

Many problems that can be solved with loops can also be solved using a tool called list comprehensions – see the chapter on comprehensions in *Advanced Python for Biologists* for details.

Indentation errors

Unfortunately, introducing tools like loops that require an indented block of code also introduces the possibility of a new type of error – an `IndentationError`. Notice what happens when the indentation of one of the lines in the block does not match the others❶:

```
apes = ["Homo sapiens", "Pan troglodytes", "Gorilla gorilla"]

for ape in apes:
    name_length = len(ape)
  first_letter = ape[0]❶
    print(ape + " is an ape. Its name starts with " + first_letter)
    print("Its name has " + str(name_length) + " letters")
```

indentation_error.py

When we run this code, we get an error message before the program even starts to run:

```
IndentationError: unindent does not match any outer indentation level
```

When you encounter an `IndentationError`, go back to your code and check that all the lines in the block match up. Also check that you are using either tabs or spaces for indentation, not both. The easiest way to do this, as mentioned in chapter 1, is to enable *tab emulation* in your text editor.

Using a string as a list

We've already seen how a string can pretend to be a list – we can use list index notation to get individual characters or substrings from inside a string. Can we also use loop notation to **process** a string as though it were a list? Yes – if we write a loop statement with a string in the position where we'd normally find a list, Python treats **each character** in the string as a separate element. This allows us to very easily process a string one character at a time:

```python
name = "python"
for character in name:
    print("one character is " + character)
```

string_as_list.py

In this case, we're just printing each individual character:

```
one character is p
one character is y
one character is t
one character is h
one character is o
one character is n
```

The process of repeating a set of instructions for each element of a list (or character in a string) is called *iteration*, and we often talk about *iterating over* a list or string.

Splitting a string to make a list

So far in this chapter, all our lists have been written manually. However, there are plenty of functions and methods in Python that produce lists as their output. One such method that is particularly interesting to biologists is the `split()` method which works on strings. `split()` takes a single argument, called the *delimiter*, and splits the original string wherever it sees the delimiter, producing a list. Here's an example:

```
names = "melanogaster,simulans,yakuba,ananassae"
species = names.split(",")
print(str(species))
```

split.py

We can see from the output that the string has been split wherever there was a comma leaving us with a list of strings:

```
['melanogaster', 'simulans', 'yakuba', 'ananassae']
```

Of course, once we've created a list in this way we can iterate over it using a loop, just like any other list.

Iterating over lines in a file

Another very useful thing that we can iterate over is a file. Just as a string can pretend to be a list for the purposes of looping, a file object can do the same trick[1]. When we treat a string as a list, each character becomes an individual element, but when we treat a file as a list, each **line** becomes an individual element. This makes processing a file line by line very easy:

```
file = open("some_input.txt")
for line in file:
    # do something with the line
```

Notice that in this example we are iterating over the file object, **not** over the file contents. If we iterate over the file contents like this:

---

1   If you're interested in how this "pretending" actually works, look up the Python documentation for *iterators* – but be prepared to do quite a bit of reading!

```
file = open("some_input.txt")
contents = file.read()
for line in contents:
    # warning: line contains just a single character!
```

then each time round the loop we will be dealing with a single character, which is probably not what we want. A good way to avoid this mistake is to ask yourself, whenever you open a file, whether you want to get the contents as one big string (in which case you should use `read()`) or line-by-line (in which case you should iterate over the file object).

Another common pitfall is to iterate over the same file object twice:

```
file = open("some_input.txt")

# print the length of each line
for line in file:
    print("The length is " + str(len(line)))

# print the first character of each line
for line in file:
    print("The first character is " + line[0])
```

If we run this code, we'll find that the second `for` loop never gets executed. The reason for this is that file objects are *exhaustible*. Once we have iterated over a file object, Python "remembers" that it is already at the end of the file, so when we try to iterate over it again, there are no lines remaining to be read. One way round this problem is to close and reopen the file each time we want to iterate over it:

```
# print the length of each line
file = open("some_input.txt")
for line in file:
    print("The length is " + str(len(line)))
file.close()

# print the first character of each line
file = open("some_input.txt")
for line in file:
    print("The first character is " + line[0])
file.close()
```

A better approach is to read the lines of the file into a list, then iterate over the list (which we can safely do multiple times). The file object `readlines()` method returns a list of all the lines in a file, and we can use it like this:

```
# first store a list of lines in the file
file = open("some_input.txt")
all_lines = file.readlines()

# print the lengths
for line in all_lines:
    print("The length is " + str(len(line)))

# print the first characters
for line in all_lines:
    print("The first character is " + line[0])
```

Take this approach so that you don't have to close everything.

readlines.py

Looping with ranges

Sometimes we want to iterate over a list of numbers. Imagine we have a protein sequence:

```
protein = "vlspadktnv"
```

and we want to print out the first three residues, then the first four residues, etc. etc.:

```
vls
vlsp
vlspa
vlspad
...etc...
```

We can use the substring notation that we learned in chapter 2 to extract the bit of the name we want to print. If we try writing it without a loop, then we get very repetitive code:

```
print(protein[0:1])
print(protein[0:2])
print(protein[0:3])
...
```

Looking at this code, the structure of the problem becomes clear: each time we print out a line, the end position of the substring needs to increase by one. Obviously we need a loop to do this, but what are we going to iterate over? We can't just iterate over the protein string, because that will give us individual residues, which is not what we want. We can manually assemble a list of stop positions, and iterate over that:

```
stop_positions = [3,4,5,6,7,8,9,10]
for stop in stop_positions:
    substring = protein[0:stop]
    print(substring)
```

but this seems cumbersome, and only works if we know the length of the protein sequence in advance.

A better solution is to use the range() function. range() is a built in Python function that generates lists of numbers[1]. The behaviour of the range function depends on how many arguments we give it. Below are a few examples, with the output following directly after the code.

---

1   In Python 3, range() returns a range object rather than a list, but you can still iterate over it using exactly the same code.

With a **single** argument, `range()` will count up from zero to that number, excluding the number itself:

```
for number in range(6):
    print(number)
```

range1.py

```
0
1
2
3
4
5
```

With **two** numbers, `range()` will count up from the first number (inclusive[1]) to the second (exclusive):

```
for number in range(3, 8):
    print(number)
```

range2.py

```
3
4
5
6
7
```

With **three** numbers, `range()` will count up from the first to the second with the step size given by the third:

---

1 The rules for ranges are the same as for array notation – inclusive on the low end, exclusive on the high end – so you only have to memorize them once!

```
for number in range(2, 14, 4):
    print(number)
```

range3.py

```
2
6
10
```

Recap

In this chapter we've seen several tools that work together to allow our programs to deal elegantly with multiple pieces of data. **Lists** let us store many elements in a single variable, and **loops** let us process those elements one by one. In learning about loops, we've also been introduced to the block syntax and the importance of indentation in Python.

We've also seen several useful ways in which we can use the notation we've learned for working with lists with other types of data. Depending on the circumstances, we can treat *strings*, *files*, and *ranges* as if they were lists. This is a very helpful feature of Python, because once we've become familiar with the syntax for working with lists, we can use it in many different place. Learning about these tools has also helped us make sense of some interesting behaviour that we observed in earlier chapters.

Lists are the first example we've encountered of structures that can hold multiple pieces of data. We'll encounter another such structure – the *dict* – in chapter 8. In fact, Python has several more such data types – you'll find a full survey of them in the chapter on complex data structures in *Advanced Python for Biologists*.

Exercises

Note: all the files mentioned in these exercises can be found in the *chapter_4* folder of the exercises download.

Processing DNA in a file

The file *input.txt* contains a number of DNA sequences, one per line. Each sequence starts with the same 14 base pair fragment – a sequencing adapter that should have been removed. Write a program that will (a) trim this adapter and write the cleaned sequences to a new file and (b) print the length of each sequence to the screen.

Multiple exons from genomic DNA

The file *genomic_dna.txt* contains a section of genomic DNA, and the file *exons.txt* contains a list of start/stop positions of exons. Each exon is on a separate line and the start and stop positions are separated by a comma. Write a program that will extract the exon segments, concatenate them, and write them to a new file.

This is a tricky exercise with several parts: your program will have to:

- read the exon file line by line
- split each exon line into two numbers
- turn those numbers into integers
- extract the matching part of the genomic DNA sequence
- concatenate all the exon sequences together

Solutions

Processing DNA in a file

This seems a bit more complicated than previous exercises – we are being asked to write a program that does two things at once! – so lets tackle it one step at a time. First, we'll write a program that simply reads each sequence from the file and prints it to the screen:

```
file = open("input.txt")
for dna in file:
    print(dna)
```

We can see from the output that we've forgotten to remove the newlines from the ends of the DNA sequences – there is a blank line between each line of output:

```
ATTCGATTATAAGCTCGATCGATCGATCGATCGATCGATCGATCGATCGATC

ATTCGATTATAAGCACTGATCGATCGATCGATCGATCGATGCTATCGTCGT

ATTCGATTATAAGCATCGATCACGATCTATCGTACGTATGCATATCGATATCGATCGTAGTC

ATTCGATTATAAGCACTATCGATGATCTAGCTACGATCGTAGCTGTA

ATTCGATTATAAGCACTAGCTAGTCTCGATGCATGATCAGCTTAGCTGATGATGCTATGCA
```

but we'll ignore that for now. The next step is to remove the first 14 bases of each sequence. We know that we want to take a substring from each sequence, starting at the fifteenth character, and continuing to the end. We can use the trick we learned in chapter 2 and leave off the stop position when getting the substring, since the sequences are all slightly different lengths.

Here's what the code looks like with the substring part added. Remember that the 15th character is at position 14❶, since we start counting from zero:

```
file = open("input.txt")
for dna in file:
    trimmed_dna = dna[14:] ❶
    print(trimmed_dna)
```

As before, we are simply printing the trimmed DNA sequence to the screen, and from the output we can confirm that the first 14 bases have been removed from each sequence:

TCGATCGATCGATCGATCGATCGATCGATCGATCGATC

ACTGATCGATCGATCGATCGATCGATGCTATCGTCGT

ATCGATCACGATCTATCGTACGTATGCATATCGATATCGATCGTAGTC

ACTATCGATGATCTAGCTACGATCGTAGCTGTA

ACTAGCTAGTCTCGATGCATGATCAGCTTAGCTGATGATGCTATGCA

Now that we know our code is working, we'll switch from printing to the screen to writing to a file. We'll have to open the file **before** the loop, then write the trimmed sequences to the file **inside** the loop, then close the file **after** the loop:

```
file = open("input.txt")
output = open("trimmed.txt", "w")
for dna in file:
    last_character_position = len(dna)
    trimmed_dna = dna[14:last_character_position]
    output.write(trimmed_dna)
output.close()
```

Opening the *trimmed.txt* file, we can see that the result looks good. It didn't matter that we never removed the newlines, because they appear in the correct place in the output file anyway:

```
TCGATCGATCGATCGATCGATCGATCGATCGATCGATC
ACTGATCGATCGATCGATCGATCGATGCTATCGTCGT
ATCGATCACGATCTATCGTACGTATGCATATCGATATCGATCGTAGTC
ACTATCGATGATCTAGCTACGATCGTAGCTGTA
ACTAGCTAGTCTCGATGCATGATCAGCTTAGCTGATGATGCTATGCA
```

Now the final step – printing the lengths to the screen – requires just one more line of code. Here's the final program in full, with comments:

```python
# open the input file
file = open("input.txt")

# open the output file
output = open("trimmed.txt", "w")

# go through the input file one line at a time
for dna in file:

    # get the substring from the 15th character to the end
    trimmed_dna = dna[14:]

    # print out the trimmed sequence
    output.write(trimmed_dna)

    # print out the length to the screen
    print("processed sequence with length " + str(len(trimmed_dna)))

# close the output file
output.close()
```

remove_adapter.py

There are a couple of easy ways to get this type of program wrong. If we open and close the file **inside** the loop instead of **before❶**:

```
file = open("input.txt")

for dna in file:
    output = open("trimmed.txt", "w")
    trimmed_dna = dna[14:]
    output.write(trimmed_dna)❶
    print("processed sequence with length " + str(len(trimmed_dna)))
    output.close()
```

Then we'll overwrite the output file each time and end up with an output file that only contains the last sequence. If we open the file **before** the loop, but close it **inside** the loop❶:

```
file = open("input.txt")
output = open("trimmed.txt", "w")

for dna in file:
    trimmed_dna = dna[14:]
    output.write(trimmed_dna)
    print("processed sequence with length " + str(len(trimmed_dna)))
    output.close()❶
```

Then we'll get an error on the second loop iteration when we try to write to a closed file. And if we write to the file **after** the loop rather than **inside**:

```
file = open("input.txt")
output = open("trimmed.txt", "w")

for dna in file:
    trimmed_dna = dna[14:]
    print("processed sequence with length " + str(len(trimmed_dna)))

output.write(trimmed_dna)❶
output.close()
```

then we'll only see the last processed sequence.

Multiple exons from genomic DNA

This is very similar to the exercises from the previous two chapters, and so our solution to it is going to look very similar. Let's concentrate on the new bit of the problem first – reading the file of exon locations. As before, we can start by opening the file and printing each line to the screen:

```
exon_locations = open("exons.txt")
for line in exon_locations:
    print(line)
```

This gives us a loop in which we are dealing with a different exon each time round. If we look at the output, we can see that we still have a newline at the end of each line, but we'll not worry about that for now:

```
5,58

72,133

190,276

340,398
```

Now we have to split each line into a start and stop position. The `split()` method is probably a good choice for this job – let's see what happens when we split each line using a comma as the delimiter:

```
exon_locations = open("exons.txt")
for line in exon_locations:
    positions = line.split(',')
    print(positions)
```

The output shows that each line, when split, turns into a list of two elements:

```
['5', '58\n']
['72', '133\n']
['190', '276\n']
['340', '398\n']
```

The second element of each list has a newline on the end, because we haven't removed them. Let's try assigning the start and stop position to sensible variable names, and printing them out individually:

```
exon_locations = open("exons.txt")
for line in exon_locations:
    positions = line.split(',')
    start = positions[0]
    stop = positions[1]
    print("start is " + start + ", stop is " + stop)
```

The output shows that this approach works – the start and stop variables take different values each time round the loop:

```
start is 5, stop is 58

start is 72, stop is 133

start is 190, stop is 276

start is 340, stop is 398
```

Now let's try putting these variables to use. We'll read the genomic sequence from the file all in one go using `read()` – there's no need to process each line separately, as we just want the entire contents. Then we'll use the exon coordinates to extract one exon each time round the loop, and print it to the screen:

```
genomic_dna = open("genomic_dna.txt").read()
exon_locations = open("exons.txt")
for line in exon_locations:
    positions = line.split(',')
    start = positions[0]
    stop = positions[1]
    exon = genomic_dna[start:stop]
    print("exon is: " + exon)
```

Unfortunately, when we run this code we get an error:

```
    exon = genomic_dna[start:stop]
TypeError: slice indices must be integers or None or have an __index__
method
```

What has gone wrong? Recall that the result of using `split()` on a string is a list of strings – this means that the `start` and `stop` variables in our program are also strings (because they're just individual elements of the `positions` list). The problem comes when we try to use them as numbers when getting a substring. Fortunately, it's easily fixed – we just have to use the `int()` function to turn our strings into numbers:

```
    start = int(positions[0])
    stop = int(positions[1])
```

and the program works as intended.

Next step: doing something useful with the exons, rather than just printing them to the screen. The exercise description says that we have to concatenate the exon sequences to make a long coding sequence. If we had all the exons in separate variables, then this would be easy:

```
coding_seq = exon1 + exon2 + exon3 + exon4
```

but instead we have a single `exon` variable that stores one exon at a time. Here's one way to get the complete coding sequence: before the loop starts we'll create a new variable called `coding_sequence` and assign it to an empty string. Then, each time round the loop, we'll add the current exon on to the end, and store the result back in the same variable. When the loop has finished, the `coding_sequence` variable will contain all the exons. This is what the code looks like:

```
genomic_dna = open("genomic_dna.txt").read()
exon_locations = open("exons.txt")
coding_sequence = ""❶

for line in exon_locations:
    positions = line.split(',')
    start = int(positions[0])
    stop = int(positions[1])
    exon = genomic_dna[start:stop]
    coding_sequence = coding_sequence + exon❷
    print("coding sequence is : " + coding_sequence)❸
```

Before we start processing the exons file, we create the `coding_sequence` variable❶, and each time round the loop we add the current `exon` on to the end. This looks odd because the `coding_sequence` variable is on both the left and right side of the equals sign, but is actually quite a common technique. The trick to understanding this line is to read the right side of the statement first i.e.

> *Concatenate the current* `coding_sequence` *and the current* `exon`, *then store the result of that concatenation in* `coding_sequence`.

At the end of the loop, instead of printing the exon we're now printing the coding sequence❸, and we can see from the output how the coding sequence is gradually built up as we go round the loop:

```
coding sequence is : CGTACCGTCGACGATGCTACGATCGTCGATCGTAGTCGATCATCGATCGATCG
coding sequence is :
CGTACCGTCGACGATGCTACGATCGTCGATCGTAGTCGATCATCGATCGATCGCGATCGATCGATATCGATCGA
TATCATCGATGCATCGATCATCGATCGATCGATCGATCGA
coding sequence is :
CGTACCGTCGACGATGCTACGATCGTCGATCGTAGTCGATCATCGATCGATCGCGATCGATCGATATCGATCGA
TATCATCGATGCATCGATCATCGATCGATCGATCGATCGACGATCGATCGATCGTAGCTAGCTAGCTAGATCGA
TCATCATCGTAGCTAGCTCGACTAGCTACGTACGATCGATGCATCGATCGTA
coding sequence is :
CGTACCGTCGACGATGCTACGATCGTCGATCGTAGTCGATCATCGATCGATCGCGATCGATCGATATCGATCGA
TATCATCGATGCATCGATCATCGATCGATCGATCGATCGACGATCGATCGATCGTAGCTAGCTAGCTAGATCGA
TCATCATCGTAGCTAGCTCGACTAGCTACGTACGATCGATGCATCGATCGTACGATCGATCGATCGATCGATCG
ATCGATCGATCGATCGTAGCTAGCTACGATCG
```

The final step is to save the coding sequence to a file. We can do this at the end of the program with three lines of code. Here's the final code with comments:

[handwritten note: understand this completely]

```
# open the genomic dna file and read the contents
genomic_dna = open("genomic_dna.txt").read()

# open the exons locations file
exon_locations = open("exons.txt")

# create a variable to hold the coding sequence
coding_sequence = ""

# go through each line in the exon locations file
for line in exon_locations:

    # split the line using a comma
    positions = line.split(',')

    # get the start and stop positions
    start = int(positions[0])
    stop = int(positions[1])

    # extract the exon from the genomic dna
    exon = genomic_dna[start:stop]

    # append the exon to the end of the current coding sequence
    coding_sequence = coding_sequence + exon

# write the coding sequence to an output file
output = open("coding_sequence.txt", "w")
output.write(coding_sequence)
output.close()
```

write_coding_sequence.py

[handwritten note: Problem:- can we extract sequences for a fasta file that are smaller than 500 amino acids??]

This is the most complicated code we've written so far, so take a moment to read through and make sure that you understand what's going on. Remember that there are three files involved in this program – the genomic sequence file, the exon locations, and the output file.

What have we learned?

In the real world, we're unlikely to run into either of these exact problems: adapter trimming and coding sequence extraction are easily carried out by existing tools. However, if we forget for a moment about the specifics and think

103

about the general theme of the problems – bringing together data from two different sources, processing them, and saving the result – it should be clear that many different programming problems fall into this category.

We'll see many more examples of lists and loops in the remainder of this book, and have plenty more opportunities to practice using them in the exercises. However, these exercises have illustrated a few important points. We've seen how switching between printing and writing output can be a useful strategy when working on code and how, when using loops and files together, we have to be particularly careful of the order in which we do things. The second exercise gave us an insight into the problems associated with splitting up files into chunks of data – particularly when we want to treat some of the data as numbers.

5: Writing our own functions

Why do we want to write our own functions?

Take a look back at the very first exercise in this book – the one in chapter 2 where we had to write a program to calculate the AT content of a DNA sequence. Let's remind ourselves of the code:

```
my_dna = "ACTGATCGATTACGTATAGTATTTGCTATCATACATATATATCGATGCGTTCAT"

length = len(my_dna)
a_count = my_dna.count('A')
t_count = my_dna.count('T')
at_content = (a_count + t_count) / length

print("AT content is " + str(at_content))
```

If we discount the first line (whose job is to store the input sequence) and the last line (whose job is to print the result), we can see that it takes four lines of code to calculate the AT content[1]. This means that every place in our code where we want to calculate the AT content of a sequence, we need these same four lines – and we have to make sure we copy them exactly, without any mistakes.

It would be much simpler if Python had a built in function (let's call it `get_at_content()`) for calculating AT content. If that were the case, then we could just run `get_at_content()` in the same way we run `print()`, or `len()`, or `open()`. Although, sadly, Python does not have such a built in function, it does have the next best thing – a way for us to create our own functions.

Creating our own function to carry out a particular job has many benefits. It allows us to reuse the same code many times within a program without having to copy it out each time. Additionally, if we find that we have to make a change to

---

1 We could, of course, compress this down to a single line:
   `at_content = (my_dna.count('A') + my_dna.count('T')) / len(my_dna)`
   but it would be much less readable.

the code, we only have to do it in one place. Splitting our code into functions also allows us to tackle larger problems, as we can work on different bits of the code independently. We can also reuse code across multiple programs.

Defining a function

Let's go ahead and create our `get_at_content()` function. Before we start typing, we need to figure out what the inputs (the number and types of the *function arguments*) and outputs (the type of the *return value*) are going to be. For this function, that seems pretty obvious – the input is going to be a single DNA sequence, and the output is going to be a decimal number. To translate these into Python terms: the function will take a single argument of type *string*, and will return a value of type *number*[1]. Here's the code:

```
def get_at_content(dna):
    length = len(dna)
    a_count = dna.count('A')
    t_count = dna.count('T')
    at_content = (a_count + t_count) / length
    return at_content
```

define_function.py

Reminder: if you're using Python 2 rather than Python 3, include this line at the top of your program:

```
from __future__ import division
```

The first line of the function definition contains several different elements. We start with the word `def`, which is short for *define* (writing a function is called *defining* it). Following that we write the name of the function, followed by the names of the argument variables in parentheses. Just like we saw before with

---

1   In fact, we can be a little bit more specific: we can say that the return value will be of type `float` – a floating point number (i.e. one with a decimal point).

normal variables, the function name and the argument names are arbitrary – we can call them whatever we like.

The first line ends with a colon, just like the first line of the loops that we were looking at in the previous chapter. And just like loops, this line is followed by a *block* of indented lines – the *function body*. The function body can have as many lines of code as we like, as long as they all have the same indentation. Within the function body, we can refer to the arguments by using the variable names from the first line. In this case, the variable `dna` refers to the sequence that was passed in as the argument to the function.

The last line of the function causes it to return the AT content that was calculated in the function body. To `return` from a function, we simply write return followed by the value that the function should output.

There are a couple of important things to be aware of when writing functions. Firstly, we need to make a clear distinction between *defining* a function, and *running* it (we refer to running a function as *calling* it). The code we've written above will not cause anything to happen when we run it, because we've not actually asked Python to execute the `get_at_content()` function – we have simply defined what it is. The code in the function will not be executed until we call the function like this:

```
get_at_content("ATGACTGGACCA")
```

If we simply call the function like that, however, then the AT content will vanish once it's been calculated. In order to use the function to do something useful, we must either store the result in a variable:

```
at_content = get_at_content("ATGACTGGACCA")
```

Or use it directly:

```
print("AT content is " + str(get_at_content("ATGACTGGACCA")))
```

Secondly, it's important to understand that the argument variable `dna` does not hold any particular value when the function is defined[1]. Instead, its job is to hold whatever value is given as the argument when the function is called. In this way it's analogous to the loop variables we saw in the previous chapter: loop variables hold a different value each time round the loop, and function argument variables hold a different value each time the function is called.

Finally, be aware that any variables that we create as part of the function only exist inside the function, and cannot be accessed outside. If we try to use a variable that's created inside❶ the function from outside❷:

```
def get_at_content(dna):
    length = len(dna)
    a_count = dna.count('A')❶
    t_count = dna.count('T')
    at_content = (a_count + t_count) / length
    return at_content

print(a_count)❷
```

We'll get an error:

```
NameError: name 'a_count' is not defined
```

Calling and improving our function

Let's write a small program that uses our new function, to see how it works. We'll try both storing the result in a variable before printing it❶ and printing it directly❷:

---

1   Indeed, it doesn't actually exist when it's defined, only when it runs.

```
def get_at_content(dna):
    ...

my_at_content = get_at_content("ATGCGCGATCGATCGAATCG")
print(str(my_at_content))❶

print(get_at_content("ATGCATGCAACTGTAGC"))❷
print(get_at_content("aactgtagctagctagcagcgta"))
```

calling_function.py

Looking at the output, we can see that the first function call works fine – the AT content is calculated to be 0.45, is stored in the variable `my_at_content`, then printed. However, the output for the next two calls is not so great. The second function call produces a number with way too many figures after the decimal point, and the third function call, with the input sequence in lower case, gives a result of 0.0, which is definitely not correct:

```
0.45
0.5294117647058824
0.0
```

We'll fix these problems by making a couple of changes to the `get_at_content()` function. We can add a rounding step in order to limit the number of significant figures in the result. Python has a built in `round()` function that takes two arguments – the number we want to round, and the number of significant figures. We'll call the `round()` function on the result before we return it❶. And we can fix the lower case problem by converting the input sequence to uppercase❷ before starting the calculation. Here's the new version of the function, with the same three function calls:

```
def get_at_content(dna):
    length = len(dna)
    a_count = dna.upper().count('A')❷
    t_count = dna.upper().count('T')
    at_content = (a_count + t_count) / length
    return round(at_content, 2)❶

my_at_content = get_at_content("ATGCGCGATCGATCGAATCG")
print(str(my_at_content))
print(get_at_content("ATGCATGCAACTGTAGC"))
print(get_at_content("aactgtagctagctagcagcgta"))
```

improved_function.py

and now the output is just as we want:

```
0.45
0.53
0.52
```

We can make the function even better though: why not allow it to be called with the number of significant figures as an argument[1]? In the above code, we've picked two significant figures, but there might be situations where we want to see more. Adding the second argument is easy; we just add it to the argument variable list❶ on the first line of the function definition, and then use the new argument variable in the call to `round()`❷. We'll throw in a few calls to the new version of the function with different arguments to check that it works:

---

1   An even better solution would be to specify the number of significant figures in the string representation of the number when it's printed.

```
def get_at_content(dna, sig_figs):❶
    length = len(dna)
    a_count = dna.upper().count('A')
    t_count = dna.upper().count('T')
    at_content = (a_count + t_count) / length
    return round(at_content, sig_figs)❷

test_dna = "ATGCATGCAACTGTAGC"
print(get_at_content(test_dna, 1))
print(get_at_content(test_dna, 2))
print(get_at_content(test_dna, 3))
```

two_arguments.py

The output confirms that the rounding works as intended:

```
0.5
0.53
0.529
```

Encapsulation with functions

Let's pause for a moment and consider what happened in the previous section. We wrote a function, and then wrote some code that used that function. In the process of writing the code that used the function, we discovered a couple of problems with our original function definition. **We were then able to go back and change the function definition, without having to make any changes to the code that used the function.**

I've written that last sentence in bold, because it's incredibly important. It's no exaggeration to say that understanding the implications of that sentence is the key to being able to write larger, useful programs. The reason it's so important is that it describes a programming phenomenon that we call *encapsulation*.

Encapsulation just means dividing up a complex program into little bits which we can work on independently. In the example above, the code is divided into two parts – the part where we define the function, and the part where we use it – and we can make changes to one part without worrying about the effects on the other part.

This is a very powerful idea, because without it, the size of programs we can write is limited to the number of lines of code we can hold in our brain at one time. Some of the example code in the solutions to exercises in the previous chapter were starting to push at this limit already, even for relatively simple problems. By contrast, using functions allows us to build up a complex program from small building blocks, each of which individually is small enough to understand in its entirety.

Because using functions is so important, future solutions to exercises will use them when appropriate, even when it's not explicitly mentioned in the problem text. I encourage you to get into the habit of using functions in your solutions too.

Functions don't always have to take an argument

There's nothing in the rules of Python to say that your function **must** take an argument. It's perfectly possible to define a function with no arguments:

```python
def get_a_number():
    return 42
```

but such functions tend not to be very useful. For example, we can write a version of get_at_content() that doesn't require any arguments by setting the value of the dna variable inside the function:

```python
def get_at_content():
    dna = "ACTGATGCTAGCTA"
    length = len(dna)
    a_count = dna.upper().count('A')
    t_count = dna.upper().count('T')
    at_content = (a_count + t_count) / length
    return round(at_content, 2)
```

but that's obviously not very useful, as it calculates the AT content for the exact same sequence every time it's called!

Occasionally you may be tempted to write a no-argument function that works like this:

```
def get_at_content():
    length = len(dna)
    a_count = dna.upper().count('A')
    t_count = dna.upper().count('T')
    at_content = (a_count + t_count) / length
    return round(at_content, 2)

dna = "ACTGATCGATCG"❶
print(get_at_content())
```

At first this seems like a good idea – it works because the function gets the value of the dna variable that is set[1] before the function call❶. However, this breaks the encapsulation that we worked so hard to achieve. The function now only works if there is a variable called dna set in the bit of the code where the function is called, so the two pieces of code are no longer independent.

If you find yourself writing code like this, it's usually a good idea to identify which variables from outside the function are being used inside it, and turn them into arguments.

---

1   It doesn't matter that the variable is set **after** the function is defined – all that matters it that it's set **before** the function is called.

Functions don't always have to return a value

Consider this variation of our function – instead of **returning** the AT content, this function **prints** it to the screen:

```
def print_at_content(dna):
    length = len(dna)
    a_count = dna.upper().count('A')
    t_count = dna.upper().count('T')
    at_content = (a_count + t_count) / length
    print(str(round(at_content, 2)))
```

When you first start writing functions, it's very tempting to do this kind of thing. You think:

> OK, I need to calculate and print the AT content – I'll write a function that does both.

The trouble with this approach is that it results in a function that is less flexible. Right now you want to print the AT content to the screen, but what if you later discover that you want to write it to a file, or use it as part of some other calculation? You'll have to write more functions to carry out these tasks.

The key to designing flexible functions is to recognize that the job *calculate and print the AT content* is actually two separate jobs – **calculating** the AT content, and **printing** it. Try to write your functions in such a way that they just do one job. You can then easily write code to carry out more complicated jobs by using your simple functions as building blocks.

Functions can be called with named arguments

What do we need to know about a function in order to be able to use it? We need to know what the return value and type is, and we need to know the number and type of the arguments. For the examples we've seen so far in this book, we also need to know the **order** of the arguments. For instance, to use the open() function we need to know that the name of the file comes first, followed by the mode of the file. And to use our two-argument version of `get_at_content()` as described above, we need to know that the DNA sequence comes first, followed by the number of significant figures.

There's a feature in Python called *keyword arguments* which allows us to call functions in a slightly different way. Instead of giving a list of arguments in parentheses:

```
get_at_content("ATCGTGACTCG", 2)
```

we can supply a list of argument variable names and values like this:

```
get_at_content(dna="ATCGTGACTCG", sig_figs=2)
```

This style of calling functions[1] has several advantages. It doesn't rely on the order of arguments, so we can use whichever order we prefer. These two statements behave identically:

```
get_at_content(dna="ATCGTGACTCG", sig_figs=2)
get_at_content(sig_figs=2, dna="ATCGTGACTCG")
```

It's also clearer to read what's happening when the argument names are given explicitly.

We can even mix and match the two styles of calling – the following are all identical:

```
get_at_content("ATCGTGACTCG", 2)
get_at_content(dna="ATCGTGACTCG", sig_figs=2)
get_at_content("ATCGTGACTCG", sig_figs=2)
```

keyword_arguments.py

Although we're not allowed to start off with keyword arguments then switch back to normal – this will cause an error:

---

1   It works with methods too, including all the ones we've seen so far.

```
get_at_content(dna="ATCGTGACTCG", 2)
```

Keyword arguments can be particularly useful for functions and methods that have a lot of optional arguments, and we'll use them where appropriate in the examples and exercise solutions in the rest of this book.

Function arguments can have defaults

We've encountered function arguments with defaults before, when we were discussing opening files in chapter 3. Recall that the open() function takes two arguments – a filename and a mode string – but that if we call it with **just** a filename it uses a default value for the mode string. We can easily take advantage of this feature in our own functions – we simply specify the default value in the first line of the function definition❶. Here's a version of our get_at_content() function where the default number of significant figures is two:

```
def get_at_content(dna, sig_figs=2): ❶
    length = len(dna)
    a_count = dna.upper().count('A')
    t_count = dna.upper().count('T')
    at_content = (a_count + t_count) / length
    return round(at_content, sig_figs)
```

Now we have the best of both worlds. If the function is called with two arguments, it will use the number of significant figures specified; if it's called with one argument, it will use the default value of two significant figures. Let's see some examples:

```
get_at_content("ATCGTGACTCG")
get_at_content("ATCGTGACTCG", 3)
get_at_content("ATCGTGACTCG", sig_figs=4)
```

default_argument_values.py

The function takes care of filling in the default value for sig_figs for the first function call where none is supplied:

```
0.45
0.455
0.4545
```

Function argument defaults allow us to write very flexible functions which can have varying numbers of arguments. It only makes sense to use them for arguments where a sensible default can be chosen – there's no point specifying a default for the `dna` argument in our example. They are particularly useful for functions where some of the optional arguments are only going to be used infrequently.

Testing functions

When writing code of any type, it's important to periodically check that your code does what you intend it to do. If you look back over the solutions to exercises from the first few chapters, you can see that we generally test our code at each step by printing some output to the screen and checking that it looks OK. For example in chapter 2, when we were first calculating AT content, we used a very short test sequence to verify that our code worked before running it on the real input.

The reason we used a test sequence was that, because it was so short, we could easily work out the answer manually and compare it to the answer given by our code. This idea – running code on a test input and comparing the result to an answer **that we know to be correct**[1] – is such a useful one that Python has a built in tool for expressing it: `assert`. An assertion consists of the word `assert`, followed by a call to our function, then **two** equals signs, then the result that we expect[2].

For example, we know that if we run our `get_at_content()` function on the DNA sequence "ATGC" we should get an answer of 0.5. This assertion will test whether that's the case:

---

1   Think of it as similar to running a positive control in a wet-lab experiment.
2   In fact, assertions can include any conditional statement; we'll learn about those in the next chapter.

```
assert get_at_content("ATGC") == 0.5
```

Notice the two equals signs – we'll learn the reason behind that in the next chapter. The way that assertion statements work is very simple; if an assertion turns out to be false (i.e. if Python executes our function on the input "ATGC" and the answer **isn't** 0.5) then the program will stop and we will get an `AssertionError`.

Assertions are useful in a number of ways. They provide a means for us to check whether our functions are working as intended and therefore help us track down errors in our programs. If we get some unexpected output from a program that uses a particular function, and the assertion tests for that function all pass, then we can be confident that the error doesn't lie in the function but in the code that calls it.

They also let us modify a function and check that we haven't introduced any errors. If we have a function that passes a series of assertion tests, and we make some changes to it, we can rerun the assertion tests and, assuming they all pass, be confident that we haven't broken the function[1].

Assertions are also useful as a form of documentation. By including a collection of assertion tests alongside a function, we can show exactly what output is expected from a given input.

Finally, we can use assertions to test the behaviour of our function for unusual inputs. For example, what is the expected behaviour of `get_at_content()` when given a DNA sequence that includes unknown bases (usually represented as `N`)? A sensible way to handle unknown bases would be to exclude them from the AT content calculation – in other words, the AT content for a given sequence shouldn't be affected by adding a bunch of unknown bases. We can write an assertion that expresses this:

```
assert get_at_content("ATGCNNNNNNNNNNN") == 0.5
```

---

1 This idea is very similar to a process in software development called *regression testing*.

This assertions fails for the current version of `get_at_content()`. However, we can easily modify the function to remove all **N** characters before carrying out the calculation❶:

```
def get_at_content(dna, sig_figs=2):
    dna = dna.replace('N', '')❶
    length = len(dna)
    a_count = dna.upper().count('A')
    t_count = dna.upper().count('T')
    at_content = (a_count + t_count) / length
    return round(at_content, sig_figs)
```

and now the assertion passes.

It's common to group a collection of assertions for a particular function together to test for the correct behaviour on different types of input. Here's an example for `get_at_content()` which shows a range of different types of behaviour:

```
assert get_at_content("A") == 1
assert get_at_content("G") == 0
assert get_at_content("ATGC") == 0.5
assert get_at_content("AGG") == 0.33
assert get_at_content("AGG", 1) == 0.3
assert get_at_content("AGG", 5) == 0.33333
```

test_function.py

When we have a collection of tests like this, we often refer to it as a *test suite*.

Recap

In this chapter, we've seen how packaging code into functions helps us to manage the complexity of large programs and promote code reuse. We learned how to define and call our own functions along with various new ways to supply arguments to functions. We also looked at a couple of things that are possible in Python, but rarely advisable – writing functions without arguments or return values. Finally, we explored the use of assertions to test our functions, and discussed how we can use them to catch errors before they become a problem.

The remaining chapters in this book will make use of functions in both the examples and the exercise solutions, so make sure you are comfortable with the new ideas from this chapter before moving on.

This chapter has covered the basics of writing and using functions, but there's much more we can do with them – in fact, there's a whole style of programming (*functional programming*) which revolves around the manipulation of functions. You'll find a discussion of this in the chapter in *Advanced Python for Biologists* called, unsurprisingly, *functional programming*.

Exercises

Both parts of the exercise for this chapter require you to test your answers with a collection of `assert` statements. Rather than typing them all out, you'll find the assert lines in a file called *assert_statements.txt* inside the chapter 5 exercises folder.

Percentage of amino acid residues, part one

Write a function that takes two arguments – a protein sequence and an amino acid residue code – and returns the percentage of the protein that the amino acid makes up. Use the following assertions to test your function:

```
assert my_function("MSRSLLLRFLLFLLLLPPLP", "M") == 5
assert my_function("MSRSLLLRFLLFLLLLPPLP", "r") == 10
assert my_function("msrslllrfllflllllpplp", "L") == 50
assert my_function("MSRSLLLRFLLFLLLLPPLP", "Y") == 0
```

You'll have to change the function name `my_function` in the `assert` statements to whatever you decide to call your function.

Reminder: if you're using Python 2 rather than Python 3, include this line at the top of your program:

```
from __future__ import division
```

Percentage of amino acid residues, part two

Modify the function from part one so that it accepts a list of amino acid residues rather than a single one. If no list is given, the function should return the percentage of hydrophobic amino acid residues (A, I, L, M, F, W, Y and V). Your function should pass the following assertions:

```
assert my_function("MSRSLLLRFLLFLLLLPPLP", ["M"]) == 5
assert my_function("MSRSLLLRFLLFLLLLPPLP", ['M', 'L']) == 55
assert my_function("MSRSLLLRFLLFLLLLPPLP", ['F', 'S', 'L']) == 70
assert my_function("MSRSLLLRFLLFLLLLPPLP") == 65
```

Solutions

Percentage of amino acid residues, part one

This is a similar problem to ones that we've tackled before, but we'll have to pay attention to the details. Let's start with a piece of code that does the calculation for a specific protein sequence and amino acid code, and then turn it into a function. Calculating the percentage is very similar to calculating the AT content, but we will need to multiply the result by 100 to get a percentage rather than a fraction:

```
protein = "MSRSLLLRFLLFLLLLPPLP"
aa = "R"

aa_count = protein.count(aa)
protein_length = len(protein)
percentage = aa_count * 100 / protein_length
print(percentage)
```

Now we'll make this code into a function by turning the two variables `protein` and `aa` into arguments, and returning the percentage rather than printing it. We'll add in the assertions at the end of the program to test if the function is doing its job:

```
def get_aa_percentage(protein, aa):
    aa_count = protein.count(aa)
    protein_length = len(protein)
    percentage = aa_count * 100 / protein_length
    return percentage

# test the function with assertions
assert get_aa_percentage("MSRSLLLRFLLFLLLLPPLP", "M") == 5
assert get_aa_percentage("MSRSLLLRFLLFLLLLPPLP", "r") == 10
assert get_aa_percentage("msrslllrfllfllllpplp", "L") == 50
assert get_aa_percentage("MSRSLLLRFLLFLLLLPPLP", "Y") == 0
```

Running the code shows that one of the assertions is failing – the error message tells us which assertion is the failed one:

```
    assert get_aa_percentage("MSRSLLLRFLLFLLLLPPLP", "r") == 10
AssertionError
```

Our function fails to work when the protein sequence is in upper case, but the amino acid residue code is in lower case. Looking at the assertions, we can make an educated guess that the next one (with the protein in lower case and the amino acid in upper case) is probably going to fail as well. Let's try to fix both of these problems by converting both the protein and the amino acid string to upper case at the start of the function. We'll use the same trick as we did before of converting a string to upper case and then storing the result back in the same variable❶:

```
def get_aa_percentage(protein, aa):

    # convert both inputs to upper case
    protein = protein.upper()❶
    aa = aa.upper()

    aa_count = protein.count(aa)
    protein_length = len(protein)
    percentage = aa_count * 100 / protein_length
    return percentage
```

amino_acids1.py

Now all the assertions pass without error.

Percentage of amino acid residues, part two

This exercise involves something that we've not seen before: a function that takes a list as one of its arguments. As in the previous exercise, we'll pick one of the assertion cases and write the code to solve it first, then turn the code into a function.

124

There are actually two ways to approach this problem. We can use a loop to go through each of the given amino acid residues in turn, counting up the number of times they occur in the protein sequence, to get a total count. Or, we can treat the protein sequence string as a list (as described in the previous chapter) and ask, for each position, whether the character at that position is a member of the list of amino acid residues that we're looking for. We'll use the first method here; in the next chapter we'll learn about the tools necessary to implement the second approach.

We'll need some way to keep a running total of matching amino acids as we go round the loop, so we'll create a new variable outside the loop and update it each time round. The code inside the loop will be quite similar to that from the previous exercise. Here's the code with some `print()` statements so we can see exactly what is happening:

```python
protein = "MSRSLLLRFLLFLLLLPPLP"
aa_list = ['M', 'L', 'F']

# the total variable will hold the total number of matching residues
total = 0
for aa in aa_list:
    print("counting number of " + aa)
    aa = aa.upper()
    aa_count = protein.count(aa)

    # add the number for this residue to the total count
    total = total + aa_count
    print("running total is " + str(total))

percentage = total * 100 / len(protein)
print("final percentage is " + str(percentage))
```

When we run the code, we can see how the running total increases each time round the loop:

```
counting number of M
running total is 1
counting number of L
running total is 11
counting number of F
running total is 13
final percentage is 65.0
```

Now let's take the code and, just like before, turn the protein string and the amino acid list into arguments to create a function:

```
def get_aa_percentage(protein, aa_list):
    protein = protein.upper()
    protein_length = len(protein)
    total = 0
    for aa in aa_list:
        aa = aa.upper()
        aa_count = protein.count(aa)
        total = total + aa_count
    percentage = total * 100 / protein_length
    return percentage
```

This function passes all the assertion tests except the last one, which tests the behaviour when run with only one argument. In fact, Python never even gets as far as testing the result from running the function, as we get an error indicating that the function didn't complete:

```
TypeError: get_aa_percentage() takes exactly 2 arguments (1 given)
```

Fixing the error takes only one change: we add a default value for `aa_list` in the first line of the function definition:

126

```
def get_aa_percentage(protein, aa_list=['A','I','L','M','F','W','Y','V']):

    protein = protein.upper()
    protein_length = len(protein)
    total = 0
    for aa in aa_list:
        aa = aa.upper()
        aa_count = protein.count(aa)
        total = total + aa_count
    percentage = total * 100 / protein_length
    return percentage

assert get_aa_percentage("MSRSLLLRFLLFLLLLPPLP", ["M"]) == 5
assert get_aa_percentage("MSRSLLLRFLLFLLLLPPLP", ['M', 'L']) == 55
assert get_aa_percentage("MSRSLLLRFLLFLLLLPPLP", ['F', 'S', 'L']) == 70
assert get_aa_percentage("MSRSLLLRFLLFLLLLPPLP") == 65
```

amino_acids2.py

and now all the assertions pass.

If you wrote the exercise a slightly different way, you might have had a bit more trouble with one of the assertions. In the above code, we calculate the percentage by first multiplying the total by 100, then dividing by the length:

```
percentage = total * 100 / protein_length
```

However, if we do it the other way around:

```
percentage = (total / protein_length) * 100
```

we get an incorrect answer for the second assertion. A full explanation of why this happens is beyond the scope of this book: the short explanation is that working with floating point numbers in Python can sometimes introduce rounding errors. For the full story, take a look at this page in the Python documentation:

```
https://docs.python.org/2/tutorial/floatingpoint.html
```

What have we learned?

Although this exercise is about analysing protein sequences, the principles of separating code in functions applies to all types of programming problems. The process of modifying a function – as we did in the second part of the exercise – is one that's very commonly used in real life programming.

These exercises were also our first encounter with what we might think of as a *specification* – the assertions that described how our function was supposed to work. Having a clear idea of the intended behaviour of a function is a great help when working on real life programming problems.

6: Conditional tests

Programs need to make decisions

If we look back at the examples and exercises in previous chapters, something that stands out is the lack of decision making. We've gone from doing simple calculations on individual bits of data to carrying out more complicated procedures on collections of data, but each individual piece of data (a sequence, a base, a species name, an exon) has been treated identically.

Real life problems, however, often require our programs to act as decision makers: to examine a property of some bit of data and decide what to do with it. In this chapter, we'll see how to do that using *conditional statements*. Conditional statements are features of Python that allow us to build decision points in our code. They allow our programs to decide which out of a number of possible courses of action to take – instructions like "*print the name of the sequence if it's longer than 300 bases*" or "*group two samples together if they were collected less than 10 metres apart*".

Before we can start using conditional statements, however, we need to understand *conditions*.

Conditions, True and False

A *condition* is simply a bit of code that can produce a true or false answer. The easiest way to understand how conditions work in Python is to try out a few examples. The following example prints out the result of testing (or *evaluating*) a bunch of different conditions – some mathematical examples, some using string methods, and one for testing if a value is included in a list:

```
print(3 == 5)
print(3 > 5)
print(3 <=5)
print(len("ATGC") > 5)
print("GAATTC".count("T") > 1)
print("ATGCTT".startswith("ATG"))
print("ATGCTT".endswith("TTT"))
print("ATGCTT".isupper())
print("ATGCTT".islower())
print("V" in ["V", "W", "L"])
```

print_conditions.py

If we look at the output, we can see that each of the conditions gives a true/false answer:

```
False
False
True
False
True
True
False
True
False
True
```

But what's actually being printed here? At first glance, it looks like we're printing the strings "True" and "False", but those strings don't appear anywhere in our code. What is actually being printed is the special built in values that Python uses to represent true and false – they are capitalized so that we know they're these special values.

Incidentally, we can show that these values are special by trying to print them. The following code runs without errors (note the absence of quotation marks):

```
print(True)
print(False)
```

whereas trying to print arbitrary unquoted words:

```
print(Hello)
```

always causes a `NameError`.

There's a wide range of things that we can include in conditions, and it would be impossible to give an exhaustive list here. The basic building blocks are:

- equals (represented by `==`)
- greater and less than (represented by `>` and `<`)
- greater and less than or equal to (represented by `>=` and `<=`)
- not equal (represented by `!=`)
- is a value in a list (represented by `in`)
- are two objects the same[1] (represented by `is`)

Many data types also provide methods that return `True` or `False` values, which are often a lot more convenient to use than the building blocks above. We've already seen a few in the code sample above: for example, strings have a `startswith()` method that returns `True` if the string on which the method is called starts with the string given as an argument. We'll mention these true/false methods when they come up.

Notice that the test for equality is **two** equals signs, not one. Forgetting the second equals sign will cause an error.

Now that we know how to express tests as conditions, let's see what we can do with them.

if statements

The simplest kind of conditional statement is an `if` statement. Hopefully the syntax is fairly simple to understand:

---

1 A discussion of what this actually means in Python is beyond the scope of this book, so we'll avoid using this comparison for the chapter.

```
expression_level = 125
if expression_level > 100:
    print("gene is highly expressed")
```

We write the word **if**, followed by a condition, and end the first line with a colon. There follows a block of indented lines of code (the *body* of the if statement), which will only be executed if the condition is true. This colon-plus-block pattern should be familiar to you from the chapters on loops and functions.

Most of the time, we want to use an **if** statement to test a property of some variable whose value we don't know at the time when we are writing the program. The example above is obviously useless, as the value of the `expression_level` variable is not going to change!

Here's a slightly more interesting example – we'll define a list of gene accession names[1] and print out just the ones that start with "a":

```
accs = ['ab56', 'bh84', 'hv76', 'ay93', 'ap97', 'bd72']
for accession in accs:
    if accession.startswith('a'):
        print(accession)
```

print_accessions.py

Looking at the output allows us to check that this works as intended:

```
ab56
ay93
ap97
```

If you take a close look at the code above, you'll see something interesting – the lines of code inside the loop are indented (just as we've seen before), but the line of code inside the **if** statement is indented **twice** – once for the loop, and once for the **if**. This is the first time we've seen multiple levels of indentation, but it's

---

1   *Accession names* (sometimes shortened to *accessions*) are just unique alphanumeric identifiers that we use to refer to sequences in databases.

very common once we start working with larger programs. Whenever we have one loop or `if` statement nested inside another, we'll have this type of indentation.

Python is quite happy to have as many levels of indentation as needed, but you'll need to keep careful track of which lines of code belong at which level. If you find yourself writing a piece of code that requires more than three levels of indentation, it's generally an indication that that piece of code should be turned into a function.

else statements

Closely related to the `if` statement is the `else` clause[1]. The examples above use a *yes/no* type of decision-making: should we print the gene accession number or not? Often we need an *either/or* type of decision, where we have two possible actions to take. To do this, we can add an `else` clause after the end of the body of an `if` statement:

```
expression_level = 125
if expression_level > 100:
    print("gene is highly expressed")
else:
    print("gene is lowly expressed")
```

The `else` clause doesn't have any condition of its own – rather, the else statement body is executed when the condition of the `if` statement is false.

Here's an example which uses `if` and `else` to split up a list of accession names into two different files – accessions that start with "a" go into the first file, and all other accessions go into the second file:

---

1   We call `else` a *clause* rather than a *statement* because technically it's a part of the `if` statement – you never see an `else` on its own.

```
file1 = open("one.txt", "w")
file2 = open("two.txt", "w")

accs = ['ab56', 'bh84', 'hv76', 'ay93', 'ap97', 'bd72']

for accession in accs:
    if accession.startswith('a'):
        file1.write(accession + "\n")
    else:
        file2.write(accession + "\n")
```

write_accessions.py

Notice how there are multiple indentation levels as before, but that the `if` and `else` statements are at the **same** level.

elif statements

What if we have more than two possible branches? For example, say we want three files of accession names: ones that start with "a", ones that start with "b", and all others. We could have a second `if` statement nested inside the `else` clause of the first `if` statement:

```
file1 = open("one.txt", "w")
file2 = open("two.txt", "w")
file3 = open("three.txt", "w")

accs = ['ab56', 'bh84', 'hv76', 'ay93', 'ap97', 'bd72']

for accession in accs:
    if accession.startswith('a'):
        file1.write(accession + "\n")
    else:
        if accession.startswith('b'):
            file2.write(accession + "\n")
        else:
            file3.write(accession + "\n")
```

write_accessions_nested.py

This works, but is difficult to read – we can quickly see that we need an extra level of indentation for every additional choice we want to include. To get round this, Python has an `elif` statement, which merges together `else` and `if` and allows us to rewrite the above example in a much more elegant way:

```python
accs = ['ab56', 'bh84', 'hv76', 'ay93', 'ap97', 'bd72']
for accession in accs:
    if accession.startswith('a'):
        file1.write(accession + "\n")
    elif accession.startswith('b'):
        file2.write(accession + "\n")
    else:
        file3.write(accession + "\n")
```

write_accessions_elif.py

Notice how this version of the code only needs two levels of indention. In fact, using `elif` we can have any number of branches without adding any extra indentation:

```python
for accession in accs:
    if accession.startswith('a'):
        file1.write(accession + "\n")
    elif accession.startswith('b'):
        file2.write(accession + "\n")
    elif accession.startswith('c'):
        file3.write(accession + "\n")
    elif accession.startswith('d'):
        file4.write(accession + "\n")
    elif accession.startswith('e'):
        file5.write(accession + "\n")
    else:
        file6.write(accession + "\n")
```

Note the order of the statements in the example above; we always start with an `if` and end with an `else`, and all the `elif` statements go in the middle. This kind of `if/elif/else` structure is very useful when we have several mutually-exclusive options. In the example above, only one branch can be true for each accession number – a string can't start with both "a" and "b". If we have a situation where the branches are not mutually exclusive – i.e. where more than one branch can be taken – then we simply need a series of if statements:

```python
for accession in accs:
    if accession.startswith('a'):
        file1.write(accession + "\n")
    if accession.endswith('z'):
        file2.write(accession + "\n")
    if len(accession) == 4:
        file3.write(accession + "\n")
    if accession.count('j') > 5:
        file4.write(accession + "\n")
```

In the example above, a single accession can satisfy more than one condition – a string can start with "a" **and** end with "z" – so it makes sense to use multiple `if` statements.

while loops

Here's one final thing we can do with conditions: use them to determine when to exit a loop. In chapter 4 we learned about loops that *iterate over* a collection of elements (like a list, a string or a file). Python has another type of loop called a `while` loop. Rather than running a set number of times, a `while` loop runs until some condition is met. For example, here's a bit of code that increments a `count` variable by one each time round the loop, stopping when the `count` variable reaches ten:

```python
count = 0
while count<10:
    print(count)
    count = count + 1
```

Because normal loops in Python are so powerful[1], `while` loops are used much less frequently than in other languages, so we won't discuss them further.

---

1   The example code here could be better accomplished with a `for` loop and a `range()`.

Building up complex conditions

What if we wanted to express a condition that was made up of several parts? Imagine we want to go through our list of accessions and print out only the ones that start with "a" and end with "3". We could use two nested `if` statements:

```
accs = ['ab56', 'bh84', 'hv76', 'ay93', 'ap97', 'bd72']
for accession in accs:
    if accession.startswith('a'):
        if accession.endswith('3'):
            print(accession)
```

but this brings in an extra, unneeded level of indention. A better way is to join the two conditions with `and` to make a complex expression:

```
accs = ['ab56', 'bh84', 'hv76', 'ay93', 'ap97', 'bd72']
for accession in accs:
    if accession.startswith('a') and accession.endswith('3'):
        print(accession)
```

accessions_and.py

This version is nicer in two ways: it doesn't require the extra level of indentation, and the condition reads in a very natural way. We can also use `or` to join up two conditions, to produce a complex condition that will be true if either of the two simple conditions are true:

```
accs = ['ab56', 'bh84', 'hv76', 'ay93', 'ap97', 'bd72']
for accession in accs:
    if accession.startswith('a') or accession.startswith('b'):
        print(accession)
```

accessions_or.py

We can even join up complex conditions to make more complex conditions – here's an example which prints accessions if they start with either "a" or "b" and end with "4":

```
accs = ['ab56', 'bh84', 'hv76', 'ay93', 'ap97', 'bd72']
for acc in accs:
    if (acc.startswith('a') or acc.startswith('b')) and acc.endswith('4'):
        print(acc)
```

accessions_complex.py

Notice how we have to include parentheses in the above example to avoid ambiguity. If we have three simple conditions represented by X, Y and Z, then the complex condition

```
(X or Y) and Z
```

is not the same as the complex condition

```
X and (Y or Z)
```

Finally, we can negate any type of condition by prefixing it with the word not. This example will print out accessions that start with "a" and **don't** end with 6:

```
accs = ['ab56', 'bh84', 'hv76', 'ay93', 'ap97', 'bd72']
for acc in accs:
    if acc.startswith('a') and not acc.endswith('6'):
        print(acc)
```

accessions_not.py

By using a combination of and, or and not (along with parentheses where necessary) we can build up arbitrarily complex conditions. This kind of use for conditions – identifying elements in a list – can often be done better using either the filter() function, or a *list comprehension*. You'll find examples of each in the chapters on functional programming and comprehensions respectively in *Advanced Python for Biologists*.

These three words are collectively known as *boolean operators* and crop up in a lot of places. For example, imagine you want to search a protein sequence database for full length cytochrome oxidase subunit one proteins. You could simply search using the query

```
COX1
```

but you would encounter two big problems: any sequences that were labelled as COI rather than COX1 would be missing from the results list, and the results list would contain partial sequences. To get around these problems, you might construct a query like this:

```
COX1 or COI and not partial
```

which uses the same tools and logic as we've just seen in Python.

Writing true/false functions

Sometimes we want to write a function that can be used in a condition. This is very easy to do – we just make sure that our function always returns either `True` or `False`. Remember that `True` and `False` are built in values in Python, so they can be passed around, stored in variables, and returned, just like numbers or strings.

Here's a function that determines whether or not a DNA sequence is AT-rich (we'll say that a sequence is AT-rich if it has an AT content of more than 0.65):

```python
def is_at_rich(dna):
    length = len(dna)
    a_count = dna.upper().count('A')
    t_count = dna.upper().count('T')
    at_content = (a_count + t_count) / length
    if at_content > 0.65:
        return True
    else:
        return False
```

boolean_function.py

We'll test this function on a few sequences to see if it works:

```
print(is_at_rich("ATTATCTACTA"))
print(is_at_rich("CGGCAGCGCT"))
```

The output shows that the function returns `True` or `False` just like the other conditions we've been looking at:

```
True
False
```

Therefore we can use our function in an `if` statement:

```
if is_at_rich(my_dna):
    # do something with the sequence
```

Because the last four lines of our function are devoted to evaluating a condition and returning True or False, we can write a slightly more compact version. In this example we evaluate the condition, and then return the result right away❶:

```
def is_at_rich(dna):
    length = len(dna)
    a_count = dna.upper().count('A')
    t_count = dna.upper().count('T')
    at_content = (a_count + t_count) / length
    return at_content > 0.65❶
```

This is a little more concise, and also easier to read once you're familiar with the idiom.

Recap

In this short chapter, we've dealt with two things: conditions, and the statements that use them.

We've seen how simple conditions can be joined together to make more complex ones, and how the concepts of truth and falsehood are built in to Python on a fundamental level. We've also seen how we can incorporate `True` and `False` in our own functions in a way that allows them to be used as part of conditions.

We've been introduced to four different tools that use conditions – `if`, `else`, `elif`, and `while` – in approximate order of usefulness. You'll probably find, in the programs that you write and in your solutions to the exercises in this book, that you use `if` and `else` very frequently, `elif` occasionally, and `while` almost never.

Exercises

In the *chapter_6* folder in the exercises download, you'll find a text file called *data.csv*, containing some made-up data for a number of genes. Each line contains the following fields for a single gene in this order: species name, sequence, gene name, expression level. The fields are separated by commas (hence the name of the file – **csv** stands for **Comma Separated Values**). Think of it as a representation of a table in a spreadsheet – each line is a row, and each field in a line is a column. All the exercises for this chapter use the data in this file.

This is a multi part exercise which involves extracting and printing data from the file. The nature of this type of problem means that it's quite easy to get a program that runs without errors, but doesn't quite produce the correct output, so be sure to check your solutions manually.

Reminder: if you're using Python 2 rather than Python 3, include this line at the top of your programs:

```
from __future__ import division
```

Several species

Print out the gene names for all genes belonging to *Drosophila melanogaster* or *Drosophila simulans*.

Length range

Print out the gene names for all genes between 90 and 110 bases long.

AT content

Print out the gene names for all genes whose AT content is less than 0.5 and whose expression level is greater than 200.

Complex condition

Print out the gene names for all genes whose name begins with "k" or "h" except those belonging to *Drosophila melanogaster*.

High low medium

For each gene, print out a message giving the gene name and saying whether its AT content is high (greater than 0.65), low (less than 0.45) or medium (between 0.45 and 0.65).

Solutions

Several species

These exercises are somewhat more complicated than previous ones, and they're going to require material from multiple different chapters to solve. The first problem is to deal with the format of the data file. Open it in a text editor and take a look before continuing.

We know that we're going to have to open the file (chapter 3) and process the contents line by line (chapter 4). To deal with each line, we'll have to split it to make a list of columns (chapter 4), then apply the condition (this chapter) in order to figure out whether or not we should print it. Here's a program that will read each line from the file, split it using commas as the delimiter, then assign each of the four columns to a variable and print the gene name:

```python
data = open("data.csv")

for line in data:

    columns = line.rstrip("\n").split(",")
    species = columns[0]
    sequence = columns[1]
    name = columns[2]
    expression = columns[3]

    print(name)
```

Notice that we use `rstrip()` to remove the newline from the end of the current line before splitting it. We know the order of the fields in the line because they were mentioned in the exercise description, so we can easily assign them to the four variables. This program doesn't do anything useful, but we can check the output to confirm that it gets the names right:

```
kdy647
jdg766
kdy533
hdt739
hdu045
teg436
```

Now we can add in the condition. We want to print the name if the species is either *Drosophila melanogaster* or *Drosophila simulans*. If the species name is neither of those two, then we don't want to do anything. This is a yes/no decision, so we need an `if` statement:

```python
data = open("data.csv")

for line in data:

    columns = line.rstrip("\n").split(",")
    species = columns[0]
    sequence = columns[1]
    name = columns[2]
    expression = columns[3]

    if species=="Drosophila melanogaster" or species=="Drosophila simulans":
        print(name)
```

several_species.py

We can check the output we get:

```
kdy647
jdg766
kdy533
```

against the contents of the file, and confirm that the program is working.

Length range

We can reuse a large part of the code from the previous exercise to help solve this one. We have another complex condition: we only want to print names for genes whose length is between 90 and 110 bases – in other words, genes whose length is greater than 90 and less than 110. We'll have to calculate the length using the `len()` function. Once we've done that the rest of the program is quite straightforward:

```python
data = open("data.csv")

for line in data:

    columns = line.rstrip("\n").split(",")
    species = columns[0]
    sequence = columns[1]
    name = columns[2]
    expression = columns[3]

    if len(sequence) > 90 and len(sequence) < 110:
        print(name)
```

length_range.py

A useful shorthand for checking if a variable is between to values is this:

```python
if 90 < len(sequence) < 110:
    # do something
```

which we could also use.

AT content

This exercise has a complex condition like the others, but it also requires us to do a bit more calculation – we need to be able to calculate the AT content of each sequence. Rather than starting from scratch, we'll simply use the function that we wrote in the previous chapter and include it at the start of the program. Once we've done that, it's a case of using the output from `get_at_content()` as part of the condition. We must be careful to convert the fourth column – the expression level – into an integer❶ so that it can be compared:

```
# our function to get AT content
def get_at_content(dna):
    length = len(dna)
    a_count = dna.upper().count('A')
    t_count = dna.upper().count('T')
    at_content = (a_count + t_count) / length
    return at_content

data = open("data.csv")
for line in data:
    columns = line.rstrip("\n").split(",")
    species = columns[0]
    sequence = columns[1]
    name = columns[2]
    expression = int(columns[3])❶
    if get_at_content(sequence) < 0.5 and expression > 200:
        print(name)
```

at_content.py

Complex condition

There are no calculations to carry out for this exercise – the complexity comes from the fact that there are three components to the condition, and they have to be joined together in the right way:

```
data = open("data.csv")

for line in data:
    columns = line.rstrip("\n").split(",")
    species = columns[0]
    sequence = columns[1]
    name = columns[2]
    expression = columns[3]

    if (name.startswith('k') or name.startswith('h')) and species !=
"Drosophila melanogaster":
        print(name)
```

complex_condition.py

The line containing the `if` statement is quite long, so it wraps around onto the next line on this page, but it's still just a single line in the program file. There are two different ways to express the requirement that the name is not *Drosophila melanogaster*. In the above example we've used the not-equals sign (`!=`) but we could also have used the `not` boolean operator like this:

```
if (name.startswith('k') or name.startswith('h')) and not species ==
"Drosophila melanogaster":
    print(name)
```

For these long, complex conditions it can sometimes be useful to split them over multiple lines to make them easier to read. We're allowed to do this in Python, as long as we put an extra set of parentheses around the whole thing:

```
if ((name.startswith('k') or name.startswith('h'))
    and not species == "Drosophila melanogaster"):
    print(name)
```

High low medium

Now we come to an exercise that requires the use of multiple branches. We have three different printing options for each gene – high, low and medium – so we'll need an `if..elif..else` section to handle the conditions. We'll use the `get_at_content()` function as before:

```
# our function to get AT content
def get_at_content(dna):
    length = len(dna)
    a_count = dna.upper().count('A')
    t_count = dna.upper().count('T')
    at_content = (a_count + t_count) / length
    return at_content

data = open("data.csv")
for line in data:
    columns = line.rstrip("\n").split(",")
    species = columns[0]
    sequence = columns[1]
    name = columns[2]
    expression = columns[3]

    if get_at_content(sequence) > 0.65:
        print(name + " has high AT content")
    elif get_at_content(sequence) < 0.45:
        print(name + " has low AT content")
    else:
        print(name + " has medium AT content")
```

high_low_medium.py

Checking the output confirms that the conditions are working:

```
kdy647 has high AT content
jdg766 has medium AT content
kdy533 has medium AT content
hdt739 has low AT content
hdu045 has medium AT content
teg436 has medium AT content
```

What have we learned?

As with many of the exercises in this book, the output from these probably isn't very exciting – this is all made up data, after all. However, take a moment to think about the general type of problem we have been solving – taking a collection of structured data and filtering it based on various criteria. Many real life problems fall into this category – imagine, for example, taking a list of SNPs and finding just the ones that are different between two populations, or taking a list of contigs and finding just the ones that contain your gene of interest.

It's also interesting to note that the solutions to these exercises will work for input files of any size with absolutely no changes in the code. To keep the examples simple, we have been working with a data file with just six rows, but we could take a similar file containing details of millions of genes and, as long as the order of the fields remained the same, our solutions would work just as well.

Finally, it's worth noting that most spreadsheet programs (e.g. Google Sheets, Microsoft Excel, GNU calc, etc.) are capable of saving files in CSV format. If you have some information stored in a spreadsheet, try saving it as CSV, then come up with a couple of simple questions and try writing programs to answer them.

7: Regular expressions

The importance of patterns in biology

A lot of what we do when writing programs for biology can be described as searching for *patterns* in *strings*. The obvious examples come from the analysis of biological sequence data – remember that DNA, RNA and protein sequences are just strings. Many of the things we want to look for in biological sequences can be described in terms of patterns:

- protein domains

- DNA transcription factor binding motifs

- restriction enzyme cut sites

- degenerate PCR primer sites

- runs of mononucleotides

However, it's not just sequence data that can have interesting patterns. As we discussed in chapter 3, most of the other types of data we have to deal with in biology comes in the form of strings[1] inside text files – things like:

- read mapping locations

- geographical sample coordinates

- taxonomic names

- gene names

- gene accession numbers

- BLAST search results

In previous chapters, we've looked at some programming tasks that involve pattern recognition in strings. We've seen how to count individual amino acid

---

1   Note that although many of the things in this list are numerical data, they're still read in to Python programs as strings and need to be manipulated as such.

residues (and even groups of amino acid residues) in protein sequences (chapter 5), and how to identify restriction enzyme cut sites in DNA sequences (chapter 2). We've also seen how to examine parts of gene names and match them against individual characters (chapter 6).

The common theme among all these problems is that they involve searching for a **fixed** pattern. But there are many problems that we want to solve that require more flexible patterns. For example:

- given a DNA sequence, what's the length of the poly-A tail?

- given a gene accession name, extract the part between the third character and the underscore

- given a protein sequence, determine if it contains this highly redundant protein domain motif

Because these types of problems crop up in so many different fields, there's a standard set of tools in Python[1] for dealing with them: *regular expressions*. Regular expressions[2] are a topic that might not be covered in a general purpose programming book, but because they're so useful in biology, we're going to devote the whole of this chapter to looking at them.

Although the tools for dealing with regular expressions are built in to Python, they are not made automatically available when you write a program. In order to use them we must first talk about modules.

Modules in Python

The functions and data types that we've discussed so far in this book have been the basic ones that are likely to be needed in pretty much every program – tools for dealing with strings and numbers, for reading and writing files, and for manipulating lists of data. As such, they are automatically made available when we start to create a Python program. If we want to open a file, we simply write a statement that uses the `open()` function.

---

1   And in many other languages and utilities.
2   The name is often abbreviated to *regex*.

However, there's another category of tools in Python which are more specialized. Regular expressions are one example, but there is a large list of specialized tools which are very useful when you need them[1], but are not likely to be needed for the majority of programs. Examples include tools for doing advanced mathematical calculations, for downloading data from the web, for running external programs, and for manipulating dates. Each collection of specialized tools – really just a collection of specialized *functions* and *data types* – is called a *module*.

For reasons of efficiency, Python doesn't automatically make these modules available in each new program, as it does with the more basic tools. Instead, we have to explicitly load each module of specialized tools that we want to use inside our program. To load a module we use the `import` statement[2]. For example, the module that deals with regular expressions is called `re`, so if we want to write a program that uses the regular expression tools we must include the line:

```
import re
```

at the top of our program. When we then want to use one of the tools from a module, we have to prefix it with the module name[3]. For example, to use the regular expression `search()` function (which we'll discuss later in this chapter) we have to write:

```
re.search(pattern, string)
```

rather than simply:

```
search(pattern, string)
```

If we forget to import the module which we want to use, or forget to include the module name as part of the function call, we will get a `NameError`.

---

1 Indeed, this is one of the great strengths of the Python language.
2 This is the reason for the `from __future__ import division` statement that we have to include if we're using Python 2.
3 There are ways round this, but we won't consider them in this book.

We'll encounter various other modules in the rest of this book. For the rest of this chapter specifically, all code examples will require the `import re` statement in order to work. For clarity, we won't include it, so if you want try running any of the code in this chapter, you'll need to add it at the start of your code[1].

Raw strings

Writing regular expression patterns, as we'll see in the next section of this chapter, requires us to type a lot of special characters. Recall from chapter 2 that certain combinations of characters are interpreted by Python to have a special meaning. For example, `\n` means *start a new line*, and `\t` means *insert a tab character*.

Unfortunately, there are a limited number of special characters to go round, so some of the characters that have a special meaning in regular expressions clash with the characters that **already** have a special meaning. Python's way round this problem is to have a special rule for strings: if we put the letter `r` immediately before the opening quotation mark, then any special characters inside the string are ignored:

```
print(r"\t\n")
```

The r stands for *raw*, which is Python's description for a string where special characters are ignored. Notice that the `r` goes **outside** the quotation marks – it is not part of the string itself. We can see from the output that the above code prints out the string just as we've written it:

```
\t\n
```

without any tabs or newlines. You'll see this special *raw* notation used in all the regular expression code examples in this chapter – even when it's not strictly necessary – because it's a good habit to get in to.

---

1   The example files already have this line.

Searching for a pattern in a string

We'll start off with the simplest regular expression tool. `re.search()` is a true/false function that determines whether or not a pattern appears somewhere in a string. It takes two arguments, both strings. The first argument is the pattern that you want to search for, and the second argument is the string that you want to search in. For example, here's how we test if a DNA sequence contains an EcoRI restriction site:

```
dna = "ATCGCGAATTCAC"
if re.search(r"GAATTC", dna):
    print("restriction site found!")
```

ecor1.py

Notice that we've used the raw notation for the pattern string, even though it's not strictly necessary since it doesn't contain any special characters.

Alternation

Now that we've seen a simple example of how to use `re.search()`, let's look at something a bit more interesting. This time, we'll check for the presence of an AvaII recognition site, which can have two different sequences: GGACC and GGTCC. One way to do this would be to use the techniques we learned in the previous chapter to make a complex condition using `or`:

```
dna = "ATCGCGAATTCAC"
if re.search(r"GGACC", dna) or re.search(r"GGTCC", dna):
    print("restriction site found!")
```

But a better way is to capture the variation in the AvaII site using a regular expression. One useful feature of regular expressions is called alternation. To represent a number of different alternatives, we write the alternatives inside parentheses separated by a pipe character. In the case of AvaII, there are two alternatives for the third base – it can be either A or T – so the pattern looks like this:

`GG(A|T)CC`

Writing the pattern as a raw string and putting it inside a call to `re.search()` gives us the code:

```
dna = "ATCGCGAATTCAC"
if re.search(r"GG(A|T)CC", dna):
    print("restriction site found!")
```

ava2.py

Notice the power of what we've done here; we've written a single pattern which captures all the variation in the sequence in one string.

Character groups

The BisI restriction enzyme cuts at an even wider range of motifs – the pattern is GCNGC, where N represents any base. We can use the same alternation technique to represent this pattern:

`GC(A|T|G|C)GC`

However, there's another regular expression feature that lets us write the pattern more concisely. A pair of square brackets with a list of characters inside them can represent any one of these characters. So the pattern `GC[ATGC]GC` will match `GCAGC`, `GCTGC`, `GCGGC` and `GCCGC`. Here's a program that checks for the presence of a BisI restriction site using character groups:

```
dna = "ATCGCGAATTCAC"
if re.search(r"GC[ATGC]GC", dna):
    print("restriction site found!")
```

bis1.py

Taken together, alternation and character groups do a pretty good job of capturing the kind of variation that we're interested in for biological programming. Before we move on, here are two short cuts that deal with specific, common scenarios.

If we want a character in a pattern to match **any** character in the input, we can use a period or dot. For example, the pattern `GC.GC` would match all four possibilities in the BisI example above. However, the period would also match any character which is not a DNA base, or even a letter. Therefore, the whole pattern would also match `GCFGC`, `GC&GC` and `GC9GC`, which may not be what we want, so be careful when using this feature.

Sometimes it's easier, rather than listing all the acceptable characters, to specify the characters that we **don't** want to match. Putting a caret ^ at the start of a character group like this

`[^XYZ]`

will negate it, and match any character that **isn't** in the group. The example above will match any character other than X, Y or Z.

Quantifiers

The regular expression features discussed above let us describe variation in the individual characters of patterns. Another group of features, *quantifiers*, let us describe variation in the number of times a section of a pattern is repeated.

A question mark immediately following a character means that that character is optional – it can match either **zero or one times**. So in the pattern `GAT?C` the `T` is optional, and the pattern will match either `GATC` or `GAC`. If we want to apply a question mark to more than one character, we can group the characters in parentheses. For example, in the pattern GGG(AAA)?TTT the group of three `A`s is optional, so the pattern will match either `GGGAAATTT` or `GGGTTT`.

A plus sign immediately following a character or group means that the character or group **must** be present but can be repeated any number of times – in other words, it will match **one or more times**. For example, the pattern `GGGA+TTT` will match three Gs, followed by one or more As, followed by three Ts. So it will match GGGATTT, GGGAATT, GGGAAATT, etc. but **not** GGGTTT.

An asterisk immediately following a character or group means that the character or group is optional, but can also be repeated. In other words, it will match **zero or more times**. For example, the pattern GGGA*TTT will match three Gs, followed by zero or more As, followed by three Ts. So it will match GGGTTT, GGGATTT, GGGAATTT, etc. It's the most flexible quantifier

If we want to match a specific number of repeats, we can use curly brackets. Following a character or group with a **single** number inside curly brackets will match **exactly** that number of repeats. For example, the pattern GA{5}T will match GAAAAAT but not GAAAAT or GAAAAAAT. Following a character or group with a **pair of numbers** inside curly brackets separated with a comma allows us to specify an acceptable range of repeats.

For example, the pattern GA{2,4}T means G, followed by between 2 and 4 As, followed by T. So it will match GAAT, GAAAT and GAAAAT but not GAT or GAAAAAT.

Just like with substrings, we can leave out the lower or upper limits. A{3,} will match three or more As, and G{,7} will match up to seven Gs.

Positions

The final set of regular expression tools we're going to look at don't represent characters at all, but rather positions in the input string. The caret symbol ^ matches the **start** of a string[1], and the dollar symbol $ matches the **end** of a string. The pattern ^AAA will match AAATTT but not GGGAAATTT. The pattern GGG$ will match AAAGGG but not AAAGGGCCC.

---

1 Remember that we've already seen this symbol inside character groups – it has a different meaning depending on where it is.

Combining

The real power of regular expressions comes from combining these tools. We can use quantifiers together with alternations and character groups to specify very flexible patterns. For example, here's a complex pattern to identify full-length eukaryotic messenger RNA sequences:

```
^AUG[AUGC]{30,1000}A{5,10}$
```

Reading the pattern from left to right, it will match:

- an ATG start codon at the beginning of the sequence

- followed by between 30 and 1000 bases which can be A, T, G or C

- followed by a poly-A tail of between 5 and 10 bases at the end of the sequence

As you can see, regular expressions can be quite tricky to read until you're familiar with them! However, it's well worth investing a bit of time learning to use them, as the same notation is used across multiple different tools. The regular expression skills that you learn in Python are transferable to other programming languages, command line tools, and text editors.

The features we've discussed above are the ones most useful in biology, and are sufficient to tackle all the exercises at the end of the chapter. However, there are many more regular expression features available in Python. If you want to become a regular expression master, it's worth reading up on *greedy vs. minimal quantifiers*, *back-references*, *lookahead* and *lookbehind assertions*, and *built in character classes*.

Before we move on to look at some more sophisticated uses of regular expressions, it's worth noting that there's a method similar to `re.search()` called `re.match()`. The difference is that `re.search()` will identify a pattern occurring **anywhere** in the string, whereas `re.match()` will only identify a pattern if it matches the **entire** string. Most of the time we want the former behaviour.

More ways to use patterns

In all the examples we've seen so far, we used `re.search()` as the condition in an `if` statement to decide whether or not a string contained a pattern. However, there are lots more interesting things we can do with regular expression patterns.

Extracting the part that matched

Often in our programs we want to find out not only **if** a pattern matched, but **what part** of the string was matched. To do this, we need to store the result of using `re.search()`, then use the `group()` method on the resulting object.

When introducing the `re.search()` function above I said that it was a true/false function. That's not *exactly* correct though – if it finds a match, it doesn't return `True`, but rather an object that is evaluated as true in a conditional context[1] (if the distinction doesn't seem important to you, then you can safely ignore it). The value that's actually returned is a *match object* – a new data type that we've not encountered before. Like a file object (see chapter 3), a match object doesn't represent a simple thing, like a number or string. Instead, it represents the results of a regular expression search. And just like a file object, a match object has a number of useful methods for getting data out of it.

One such method is the `group()` method. If we call this method on the result of a regular expression search, we get the portion of the input string that matched the pattern. Here's an example: imagine we want to take a DNA sequence and determine whether or not it contains any ambiguous bases – i.e. any bases that are not A, T, G or C. We can use a negated character group to write a regular expression that will match any non-ATGC base:

`[^ATGC]`

and test the sequence like this:

---

1   If a match isn't found, then the same thing applies; the function doesn't return `False`, but a different built in value – `None` – that evaluates as false. If this doesn't make sense, don't worry about it.

```
dna = "ATCGCGYAATTCAC"

if re.search(r"[^ATGC]", dna):
    print("ambiguous base found!")
```

The code above tells us that the DNA sequence contained a non-ATGC base, but it doesn't tell us exactly **what** the base was. To do that, we need to call the `group()` method on the match object like this:

```
dna = "CGATNCGGAACGATC"
m = re.search(r"[^ATGC]", dna)

# m is now a match object
if m:
    print("ambiguous base found!")
    ambig = m.group()
    print("the base is " + ambig)
```

extract_match.py

The output from this program:

```
ambiguous base found!
the base is N
```

tells us not only that the sequence contained an ambiguous base, but that the ambiguous base was N.

Extracting multiple groups

What if we want to extract more than one bit of the pattern? Say we want to match a scientific name like *Homo sapiens* or *Drosophila melanogaster*. The pattern is relatively simple: multiple characters, followed by a space, followed by multiple characters:

`.+ .+`

To match multiple characters we're using a period (meaning *any character*) followed by an asterisk (meaning *repeated at least once but possibly multiple times*).

Now let's say that we want to extract the genus name and species name into separate variables. We add parentheses around the parts of the pattern that we want to store:

```
(.+) (.+)
```

This is called *capturing* part of the pattern. We can now refer to the captured bits of the pattern by supplying an argument to the `group()` method. `group(1)` will return the bit of the string matched by the section of the pattern in the first set of parentheses, `group(2)` will return the bit matched by the second, etc.:

```
scientific_name = "Homo sapiens"

m = re.search("(.+) (.+)", scientific_name)

if m:
    genus = m.group(1)
    species = m.group(2)
    print("genus is " + genus + ", species is " + species)
```

extract_groups.py

The output shows how the two bits of the same pattern were stored in different variables. Note that the space, which was part of the pattern but **not** part of the captured groups, isn't in either of the two variables:

```
genus is Homo, species is sapiens
```

If you're keeping count, you'll realize that we now have **three** different roles for parentheses in regular expressions:

- surrounding the alternatives in an alternation

- grouping parts of a pattern for use with a quantifier

- defining parts of a pattern to be extracted after the match

Getting match positions

As well as containing information about the **contents** of a match, the match object also holds information about the **position** of the match. The `start()` and `end()` methods get the positions of the start and end of the pattern on the string. Let's go back to our ambiguous base example and find the position of the ambiguous base:

```
dna = "CGATNCGGAACGATC"
m = re.search(r"[^ATGC]", dna)

if m:
    print("ambiguous base found!")
    print("at position " + str(m.start()))
```

positions.py

Remember that we start counting from zero, so in this case, the match starting at the fifth base has a start position of four:

```
ambiguous base found!
at position 4
```

Multiple matches

An obvious limitation of the above example is that it can only find a single ambiguous base, because `re.search()` can only find a single match. To process multiple matches, we need to switch to `re.finditer()`, which returns a list of match objects[1] which we can process in a loop:

---

1 Strictly speaking, it doesn't return a *list* but an *iterator object*. The distinction doesn't matter to us, since we can still use it in a `for` loop.

```
dna = "CGCTCNTAGATGCGCRATGACTGCAYTGC"

matches = re.finditer(r"[^ATGC]", dna)
for m in matches:
    base = m.group()
    pos  = m.start()
    print(base + " found at position " + str(pos))
```

finditer.py

We can see from the output that we now find all three parts of the string that match the pattern:

```
N found at position 5
R found at position 15
Y found at position 25
```

Getting multiple matches as strings

A common scenario is where we want to get a list of all the parts of a string that match a given pattern. Here's a regular expression pattern that matches runs of A and T longer than five bases:

`[AT]{6,}`

Here's a DNA sequence with the bits that we want to extract in bold:

ACTGC**ATTATAT**CGTACG**AAATTATA**CGCGCG

We could extract the bits of the string that match the pattern using `re.finditer()` and `group()`:

```
dna = "CTGCATTATATCGTACGAAATTATACGCGCG!"

matches = re.finditer(r"[AT]{6,}", dna)

result = []
for m in matches:
    result.append(m.group())

print(result)
```

but because this is a common problem, there's a special method for dealing with it called `re.findall()`. Just like the other methods we've seen, `re.findall()` takes the pattern and the string as arguments, but rather than returning a list of match objects it returns a list of strings. We can rewrite our code like this:

```
dna = "CTGCATTATATCGTACGAAATTATACGCGCG!"
result = re.findall(r"[AT]{6,}", dna)
print(result)
```

findall.py

Splitting a string using a regular expression

Occasionally it can be useful to split a string using a regular expression pattern as the delimiter. The normal string `split()` method doesn't allow this, but the `re` module has a `split()` function of its own that takes a regular expression pattern as an argument. The first argument is the pattern, the second argument is the string to be split.

Imagine we have a consensus DNA sequence that contains ambiguity codes, and we want to extract all runs of contiguous unambiguous bases. We need to split the DNA string wherever we see a base that isn't A, T, G or C:

```
dna = "ACTNGCATRGCTACGTYACGATSCGAWTCG"
runs = re.split(r"[^ATGC]", dna)
print(runs)
```

split.py

The output shows how the function works – the return value is a list of strings:

```
['ACT', 'GCAT', 'GCTACGT', 'ACGAT', 'CGA', 'TCG']
```

Notice that the bits of the string that matched the pattern are excluded from the output (just like the delimiters are excluded from the output when we use the normal `split()` method).

Recap

In this chapter we learned about **regular expressions**, and the **functions** and **methods** that use them.

We started with a brief introduction to two concepts that, while not part of the regular expression tools, are necessary in order to use them – modules and raw strings. We got a brief overview of features that can be used in regular expression patterns, and a quick look at the range of different things we can do with them. Just as regular expressions themselves can range from simple to complex, so can their uses. We can use regular expressions for simple tasks – like determining whether or not a sequence contains a particular motif – or for complicated ones, like identifying messenger RNA sequences by using complex patterns.

Before we move on to the exercises, it's important to recognize that for any given pattern, there are probably multiple ways to describe it using a regular expression. Near the start of the chapter, we came up with the pattern

`GG(A|T)CC`

to describe the AvaII restriction enzyme recognition site, but the same pattern could also be written as:

- `GG[AT]CC`
- `(GGACC|GGTCC)`
- `(GGA|GGT)CC`

- `G{2}[AT]C{2}`

As with other situations where there are multiple different ways to write the same thing, it's best to be guided by what is clearest to read.

Exercises

Accession names

Here's a list of made up gene accession names:

```
accessions = ['xkn59438', 'yhdck2', 'eihd39d9', 'chdsye847', 'hedle3455',
'xjhd53e', '45da', 'de37dp']
```

Copy and paste this line from the *accessions.txt* file in the *Chapter_7* exercises folder.

Write a program that will print only the accession names that satisfy the following criteria – treat each criterion separately:

- contain the number 5
- contain the letter d or e
- contain the letters d and e in that order
- contain the letters d and e in that order with a single letter between them
- contain both the letters d and e in any order
- start with x or y
- start with x or y and end with e
- contain three or more digits in a row
- end with d followed by either a, r or p

Double digest

In the *Chapter_7* exercises folder, there's a file called *dna.txt* which contains a made up DNA sequence. Predict the fragment lengths that we will get if we digest the sequence with two made up restriction enzymes – AbcI, whose recognition site is **ANT/AAT**, and AbcII, whose recognition site is **GCRW/TG**. The forward slashes (/) in the recognition sites represent the place where the enzyme cuts the DNA.

Solutions

Accession names

Obviously, the bulk of the work here is going to be coming up with the regular expression patterns to select each subset of the accession names. Here's the easy bit – storing the accession names in a list and then processing them in a loop (the first line wraps round because it's too long to fit on the page):

```
accs = ["xkn59438", "yhdck2", "eihd39d9", "chdsye847", "hedle3455",
"xjhd53e", "45da", "de37dp"]
for acc in accs:
    # print if it passes the test
```

Now we can tackle the different criteria one by one. For each example, the code (bordered by solid lines) is followed immediately by the output (bordered by dotted lines). Because the programs are so short, the solutions are all in a single file called *accession_names.py*.

The first criterion is straightforward – accession names that contain the number **5**. We don't even have to use any fancy regular expression features:

```
for acc in accs:
    if re.search(r"5", acc):
        print(acc)
```

```
xkn59438
hedle3455
xjhd53e
45da
```

Now for accession names that contain the letters **d** or **e**. We can use either alternation or a character group. Here's a solution using alternation:

170

```
for acc in accs:
    if re.search(r"(d|e)", acc):
        print(acc)
```

```
yhdck2
eihd39d9
chdsye847
hedle3455
xjhd53e
45da
de37dp
```

The next one – accession names that contain both the letters **d** and **e**, in that order – is a bit more tricky. We can't just use a simple alternation or a character group, because they match **any** of their constituent parts, and we need both **d** and **e**. One way to think of the pattern is **d**, followed by some other letters and numbers, followed by **e**. We have to be careful with our quantifiers, however – at first glance the pattern `d.+e` looks good, but it will fail to match the accession where **e** follows **d** directly. To allow for the fact that the two letters might be right next to each other, we need to use the asterisk:

```
for acc in accs:
    if re.search(r"d.*e", acc):
        print(acc)
```

```
chdsye847
hedle3455
xjhd53e
de37dp
```

The next requirement – **d**, followed by a single letter, followed by **e** – is actually easier to write a pattern for, even though it sounds more complicated. We simply remove the asterisk, and the period will now match any single character:

```
for acc in accs:
    if re.search(r"d.e", acc):
        print(acc)
```

```
hedle3455
```

The next requirement – **d** and **e** in any order – is more difficult. We could do it with an alternation using the pattern `(d.*e|e.*d)`, which translates as *d then e, or e then d*. In this case, I think it's clearer to carry out two separate regular expression searches and combine them into a complex condition:

```
for acc in accs:
    if re.search(r"d.*e", acc) or re.search(r"e.*d", acc):
        print(acc)
```

```
hedle3455
de37dp
```

To find accessions that start with either **x** or **y**, we need to combine an alternation with a start-of-string anchor:

```
for acc in accs:
    if re.search(r"^(x|y)", acc):
        print(acc)
```

```
xkn59438
yhdck2
xjhd53e
```

We can modify this quite easily to add the requirement that the accession ends with **e**. As before, we need to use `.*` in the middle to match any number of any character, resulting in quite a complex pattern:

```
for acc in accs:
  if re.search(r"^(x|y).*e$", acc):
      print(acc)
```

```
xjhd53e
```

To match three or more numbers in a row, we need a more specific quantifier – the curly brackets – and a character group which contains all the numbers:

```
for acc in accs:
    if re.search(r"[0123456789]{3,100}", acc):
        print(acc)
```

```
    xkn59438
    chdsye847
    hedle3455
```

We can actually make this a bit more concise. The character group of all digits is such a common one that there's a built in shorthand for it: `\d`. We can also take advantage of a shorthand in the curly bracket quantifier – if we leave off the upper bound, then it matches with no upper limit. The more concise version:

```
for acc in accs:
    if re.search(r"\d{3,}", acc):
        print(acc)
```

```
    xkn59438
    chdsye847
    hedle3455
```

The final requirement is quite simple and only requires a character group and an end-of-string anchor to solve:

```
for acc in accs:
    if re.search(r"d[arp]$", acc):
        print(acc)
```

```
45da
de37dp
```

Double digest

This is a hard problem, and there are several ways to approach it. Let's simplify it by first figuring out what the fragment lengths would be if we digested the sequence with just a single restriction enzyme[1]. We'll open and read the file all in one go (there's no need to process it line by line as it's just a single sequence), then we'll use `re.finditer()` to figure out the positions of all the cut sites.

The patterns themselves are relatively simple: N means any base, so the pattern for the *AbcI* site is `A[ATGC]TAAT`. The ambiguity code R means A or G and the code W means A or T, so the pattern for *AbcII* is `GC[AG][AT]TG`. Here's the code to calculate the start positions of the matches for *AbcI*:

```
dna = open("dna.txt").read().rstrip("\n")

print("AbcI cuts at:")
for match in re.finditer(r"A[ATGC]TAAT", dna):
    print(match.start())
```

The output from this looks good:

---

1   For the purposes of this exercise, we are of course ignoring all the interesting chemical kinetics of restriction enzymes and assuming that all enzymes cut with complete specificity and efficiency.

```
AbcI cuts at:
1140
1625
```

but it's not quite right – it's telling us the positions of the start of each match, but the enzyme actually cuts 3 bases downstream of the start. To get the true position of the cut site, we need to add three to the start of each match:

```
dna = open("dna.txt").read().rstrip("\n")

print("AbcI cuts at:")
for match in re.finditer(r"A[ATGC]TAAT", dna):
    print(match.start() + 3)
```

```
AbcI cuts at:
1143
1628
```

Now we've got the cut positions, how are we going to work out the fragment sizes? One way is to go through each cut site in order and measure the distance between it and the previous one – that will give us the length of a single fragment. To make this work we'll have to add "imaginary" cut sites at the very start and end of the sequence:

```
dna = open("dna.txt").read().rstrip("\n")

all_cuts = [0]❶
for match in re.finditer(r"A[ATGC]TAAT", dna):❷
    all_cuts.append(match.start() + 3)❸
all_cuts.append(len(dna))❹

print(all_cuts)
```

Let's take a moment to examine what's going on in this program. We start by creating a new list variable called `all_cuts` to hold the cut positions❶. At this point, the `all_cuts` variable only has one element: zero, the position of the start of the sequence. Next, for each match to the pattern❷, we take the start position, add three to it to get the cut position, and append that number to the `all_cuts` list❸. Finally, we append the position of the last character in the DNA string to the `all_cuts` list❹. When we print the `all_cuts` list, we can see that it contains the position of the start and end of the string, and the internal positions of the cut sites:

```
[0, 1143, 1628, 2012]
```

Now we can write a second loop to go through the `all_cuts` list and, for each cut position, work out the size of the fragment that will be created by figuring out the distance to the previous cut site (i.e. the previous element in the list). This is a bit tricky: normally if we wanted to iterate over the cut positions we'd do something like this:

```
for pos in all_cuts:
    # do something
```

but in this scenario that won't work – to calculate each fragment size we need **both** the current **and** previous cut positions, which we have no way to get using a normal for loop.

To make this work we have to start at the second element of the list (because the first element has no previous element) and we have to work with the **index** of each element, rather than the element itself. We'll use the `range()` function to generate the list of indexes that we want to process – we need to go from index 1 (i.e. the second element of the list) to the last index (which is the length of the list):

```
for i in range(1,len(all_cuts)):❶
    this_cut_position = all_cuts[i]❷
    previous_cut_position = all_cuts[i-1]❸
    fragment_size = this_cut_position - previous_cut_position❹
    print("one fragment size is " + str(fragment_size))
```

The loop variable i is used to store each value that is generated by the range function❶. For each value of i we get the cut position at that index❷ and the cut position at the previous index❸ and then figure out the distance between them❹. The output shows how, for two cuts, we get three fragments:

```
one fragment size is 1143
one fragment size is 485
one fragment size is 384
```

Now for the final part of the solution: how do we do the same thing for two different enzymes? We can add in the second enzyme pattern with the appropriate cut site offset and append the cut positions to the `all_cuts` variable:

```
dna = open("dna.txt").read().rstrip("\n")
all_cuts = [0]

# add cut positions for AbcI
for match in re.finditer(r"A[ATGC]TAAT", dna):
    all_cuts.append(match.start() + 3)

# add cut positions for AbcII
for match in re.finditer(r"GC[AG][AT]TG", dna):
    all_cuts.append(match.start() + 4)

# add the final position
all_cuts.append(len(dna))
print(all_cuts)
```

but look what happens when we print the elements of `all_cuts`:

```
[0, 1143, 1628, 488, 1577, 2012]
```

We get zero, then the two cut positions for the first enzyme in ascending order, then the two cut positions for the second enzyme in ascending order, then the position of the end of the sequence. The method for turning a list of cut positions into fragment sizes that we developed above isn't going to work with this list, because it relies on the list of positions being in ascending order. If we try it with the list of cut positions produced by the above code, we'll end up with obviously incorrect fragment sizes:

```
one fragment size is 1143
one fragment size is 485
one fragment size is -1140
one fragment size is 1089
one fragment size is 434
```

Happily, Python's built in `sort()` function can come to the rescue. All we need to do is sort the list of cut positions before processing it, and we get the right answers. Here's the complete, final code:

```python
dna = open("dna.txt").read().rstrip("\n")
all_cuts = [0]

# add cut positions for AbcI
for match in re.finditer(r"A[ATGC]TAAT", dna):
    all_cuts.append(match.start() + 3)

# add cut positions for AbcII
for match in re.finditer(r"GC[AG][AT]TG", dna):
    all_cuts.append(match.start() + 4)

# add the final position
all_cuts.append(len(dna))
sorted_cuts = sorted(all_cuts)
print(sorted_cuts)


for i in range(1,len(sorted_cuts)):
    this_cut_position = sorted_cuts[i]
    previous_cut_position = sorted_cuts[i-1]
    fragment_size = this_cut_position - previous_cut_position
    print("one fragment size is  " + str(fragment_size))
```

double_digest.py

8:   Dictionaries

Storing paired data

Suppose we want to count the number of As in a DNA sequence. Carrying out the calculation is quite straightforward – in fact it's one of the first things we did in chapter 2:

```
dna = "ATCGATCGATCGTACGCTGA"
a_count = dna.count("A")
```

How will our code change if we want to generate a complete list of base counts for the sequence? We'll add a new variable for each base:

```
dna = "ATCGATCGATCGTACGCTGA"
a_count = dna.count("A")
t_count = dna.count("T")
g_count = dna.count("G")
c_count = dna.count("C")
```

and now our code is starting to look rather repetitive. It's not too bad for the four individual bases, but what if we want to generate counts for the 16 dinucleotides:

```
dna = "ATCGATCGATCGTACGCTGA"
aa_count = dna.count("AA")
at_count = dna.count("AT")
ag_count = dna.count("AG")
...etc. etc.
```

or the 64 trinucleotides:

```
dna = "ATCGATCGATCGTACGCTGA"
aaa_count = dna.count("AAA")
aat_count = dna.count("AAT")
aag_count = dna.count("AAG")
...etc. etc.
```

For trinucleotides and longer, the situation is particularly bad. The DNA sequence is 20 bases long, so it only contains 18 overlapping trinucleotides in total:

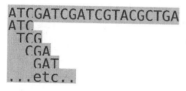

So there can be, at most, 18 unique trinucleotides in the sequence (and for a repetitive sequence, many fewer unique trinucleotides). This means that at least 46 out of our 64 variables will hold the value zero.

One possible way round this is to store the values in a list. Let's look at an example involving dinucleotides. If we create a list of the 16 possible dinucleotides we can iterate over it, calculate the count for each one, and store all the counts in a list[1]. Take a look at the code – the list of dinucleotides is quite long so it's been split over four lines to make it easier to read:

---

1   For this example, we are just going to write out the dinucleotides as a list in the code in order to keep things simple. For a discussion of how to generate lists of DNA sequences of any length – not just dinucleotides! – see the start of the chapter on recursion in *Advanced Python for Biologists*.

```
dna = "ATGATCGATCGAGTGA"
dinucleotides = ['AA','AT','AG','AC',
                 'TA','TT','TG','TC',
                 'GA','GT','GG','GC',
                 'CA','CT','CG','CT']
all_counts = []
for dinucleotide in dinucleotides:
    count = dna.count(dinucleotide)
    print("count is " + str(count) + " for " + dinucleotide)
    all_counts.append(count)
print(all_counts)
```

count_dinucleotides.py

Although the code is above is quite compact, and doesn't require huge numbers of variables, the output shows two problems with this approach:

```
count is 0 for AAA
count is 1 for AAT
count is 0 for AAG
count is 0 for AAC
count is 0 for ATA
count is 0 for ATT
count is 1 for ATG
count is 2 for ATC
...
[0, 1, 0, 0, 0, 0, 1, 2, 0, 0, 0, 0, 0, 0, 1, 0, 0, 0, 0, 1, 0, 0, 0, 0,
2, 0, 0, 0, 0, 0, 2, 0, 0, 2, 0, 0, 1, 0, 0, 0, 0, 0, 0, 0, 0, 1, 0, 0, 0,
0, 0, 0, 0, 0, 1, 0, 1, 1, 0, 1, 0, 0, 0, 0]
```

Firstly, the data are still incredibly sparse – the vast majority of the counts are zero. Secondly, the counts themselves are now disconnected from the dinucleotides. If we want to look up the count for a single dinucleotide – for example, TG – we first have to figure out that TG was the 7th dinucleotide in the list. Only then can we get the element at the correct index:

```
print("count for TG is " + str(all_counts[7]))
```

We can try various tricks to get round this problem. What if we used the `index()` method to figure out the position of the dinucleotide we are looking for in the list?

```
i = dinucleotides.index('TG')
print(all_counts[i])
```

This works because we have two lists of the same length, with a one-to-one correspondence between the elements:

```
print(dinucleotides)
print(all_counts)
```

```
['AA', 'AT', 'AG', 'AC', 'TA', 'TT', 'TG', 'TC', 'GA', 'GT', 'GG', 'GC',
'CA', 'CT', 'CG', 'CT']
[2, 2, 0, 2, 0, 0, 2, 0, 3, 0, 0, 0, 0, 0, 1, 0]
```

This is a little bit nicer, but still has major drawbacks. We're still storing all those zeros, and now we have two lists to keep track of. We need to be incredibly careful when manipulating either of the two lists to make sure that they stay perfectly synchronized – if we make any change to one list but not the other, then there will no longer be a one-to-one correspondence between elements and we'll get the wrong answer when we try to look up a count.

This approach is also slow[1]. To find the index of a given dinucleotide in the `dinucleotides` list, Python has to look at each element one at a time until it finds the one we're looking for. This means that as the size of the list grows[2], the time taken to look up the count for a given element will grow alongside it.

If we take a step back and think about the problem in more general terms, what we need is a way of storing pairs of data (in this case, dinucleotides and their

---

1   As a rule, we've avoided talking about performance in this book, but we'll break the rule in this case.

2   For instance, imagine carrying out the same exercise with the approximately one million unique 10-mers.

counts) in a way that allows us to efficiently look up the count for any given dinucleotide. This problem of storing paired data is incredibly common in programming. We might want to store:

- protein sequence names and their sequences

- DNA restriction enzyme names and their motifs

- codons and their associated amino acid residues

- colleagues' names and their email addresses

- sample names and their co-ordinates

- words and their definitions

All these are examples of what we call *key-value pairs*. In each case we have pairs of *keys* and *values*:

Key	Value
dinucleotide	count
name	protein sequence
name	restriction enzyme motif
codon	amino acid residue
sample	coordinates
word	definition

The last example in this table – words and their definitions – is an interesting one because we have a tool in the physical world for storing this type of data: a dictionary. Python's tool for solving this type of problem is also called a dictionary (usually abbreviated to *dict*) and in this chapter we'll see how to create and use them.

Creating a dictionary

The syntax for creating a dictionary is similar to that for creating a list, but we use curly brackets rather than square ones. Each pair of data, consisting of a key and a value, is called an *item*. When storing items in a dictionary, we separate them with commas. Within an individual item, we separate the key and the value with a colon. Here's a bit of code that creates a dictionary of restriction enzymes (using data from the previous chapter) with three items:

```
enzymes = { 'EcoRI':r'GAATTC','AvaII':r'GG(A|T)CC', 'BisI':r'GC[ATGC]GC' }
```

In this case, the keys and values are both strings[1]. Splitting the dictionary definition over several lines makes it easier to read:

```
enzymes = {
    'EcoRI' : r'GAATTC',
    'AvaII' : r'GG(A|T)CC',
    'BisI'  : r'GC[ATGC]GC'
}
```

and doesn't affect how the code works. To retrieve a bit of data from the dictionary – i.e. to look up the motif for a particular enzyme – we write the name of the dictionary, followed by the key in square brackets:

```
print(enzymes['BisI'])
```

The code looks very similar to using a list, but instead of giving the index of the element we want, we're giving the *key* for the *value* that we want to retrieve.

Dictionaries are a very useful way to store data, but they come with some restrictions. The only types of data we are allowed to use as keys are strings and numbers[2], so we can't, for example, create a dictionary where the keys are file

---

1 The values are actually raw strings, but that's not important.
2 Not strictly true; we can use any immutable type, but that is beyond the scope of this book.

objects. Values can be whatever type of data we like. Also, keys **must** be unique – we can't store multiple values for the same key.

Building dicts

In real life programs, it's relatively rare that we'll want to create a dictionary all in one go like in the example above. More often, we'll want to create an empty dictionary, then add key/value pairs to it (just as we often create an empty list and then add elements to it).

To create an empty dictionary we simply write a pair of curly brackets on their own, and to add elements, we use the square brackets notation on the left hand side of an assignment. Here's a bit of code that stores the restriction enzyme data one item at a time:

```
enzymes = {}
enzymes['EcoRI'] = r'GAATTC'
enzymes['AvaII'] =  r'GG(A|T)CC'
enzymes['BisI'] =  r'GC[ATGC]GC'
```

We can delete a key from a dictionary using the pop() method. pop() actually returns the value and deletes the key at the same time:

```
# remove the EcoRI enzyme from the dict
enzymes.pop('EcoRI')
```

Let's take another look at the dinucleotide count example from the start of the module. Here's how we store the dinucleotides and their counts in a dict:

```
dna = "AATGATGAACGAC"
dinucleotides = ['AA','AT','AG','AC',
                 'TA','TT','TG','TC',
                 'GA','GT','GG','GC',
                 'CA','CT','CG','CT']
all_counts = {}
for dinucleotide in dinucleotides:
    count = dna.count(dinucleotide)
    print("count is " + str(count) + " for " + dinucleotide)
    all_counts[dinucleotide] = count
print(all_counts)
```

dinucleotide_dict.py

We can see from the output that the dinucleotides and their counts are stored together in the `all_counts` variable:

```
{'AA': 2, 'AC': 2, 'GT': 0, 'AG': 0, 'TT': 0, 'CG': 1, 'GG': 0, 'GC': 0,
'AT': 2, 'GA': 3, 'TG': 2, 'CT': 0, 'CA': 0, 'TC': 0, 'TA': 0}
```

We still have a lot of repetitive counts of zero, but looking up the count for a particular dinucleotide is now very straightforward:

```
print(all_counts['TA'])
```

We no longer have to worry about either "memorizing" the order of the counts or maintaining two separate lists.

Let's now see if we can find a way of avoiding storing all those zero counts. We can add an `if` statement that ensures that we only store a count if it's greater than zero:

```
dna = "AATGATGAACGAC"
dinucleotides = ['AA','AT','AG','AC',
                 'TA','TT','TG','TC',
                 'GA','GT','GG','GC',
                 'CA','CT','CG','CT']
all_counts = {}
for dinucleotide in dinucleotides:
    count = dna.count(dinucleotide)
    if count > 0:
        all_counts[dinucleotide] = count
print(all_counts)
```

nonzero_dinucleotides.py

When we look at the output from the above code, we can see that the amount of
data we're storing is much smaller – just the counts for the dinucleotides that
actually occur in the sequence:

```
{'AA': 2, 'AC': 2, 'CG': 1, 'AT': 2, 'GA': 3, 'TG': 2}
```

Now we have a new problem to deal with. Looking up the count for a given
dinucleotide works fine when the count is positive:

```
print(all_counts['TA'])
```

But when the count is zero, the dinucleotide doesn't appear as a key in the dict:

```
print(all_counts['TC'])
```

so we will get a `KeyError` when we try to look it up:

```
KeyError: 'TC'
```

There are two possible ways to fix this. We can check for the existence of a key in
a dict (just like we can check for the existence of an element in a list), and only try
to retrieve it once we know it exists:

```
if 'TC' in all_counts:
    print(all_counts('TC'))
else
    print(0)
```

Alternatively, we can use the dict's get() method. get() usually works just like using square brackets: the following two lines do exactly the same thing:

```
print(all_counts['TC'])
print(all_counts.get('TC'))
```

The thing that makes get() really useful, however, is that it can take an optional second argument, which is the default value to be returned if the key isn't present in the dict. In this case, we know that if a given dinucleotide doesn't appear in the dict then its count is zero, so we can give zero as the default value and use get() to print out the count for any dinucleotide:

```
print("count for TG is " + str(all_counts.get('TG', 0)))
print("count for TT is " + str(all_counts.get('TT', 0)))
print("count for GC is " + str(all_counts.get('GC', 0)))
print("count for CG is " + str(all_counts.get('CG', 0)))
```

As we can see from the output, we now don't have to worry about whether or not any given dinucleotide appears in the dict – get() takes care of everything and returns zero when appropriate:

```
count for TG is 2
count for TT is 0
count for GC is 0
count for CG is 1
```

More generally, assuming we have a dinucleotide string store in the variable dn, we can run a line of code like this:

```
print("count for " + dn + " is " + str(all_counts.get(dn, 0)))
```

and be sure of getting the right answer.

Iterating over a dictionary

What if, instead of looking up a single item from a dictionary, we want to do something for all items? For example, imagine that we wanted to take our `all_counts` dict variable from the code above and print out all dinucleotides where the count was 2. One way to do it would be to iterate over the list of dinucleotides, looking up the count for each one and deciding whether or not to print it[1]:

```
for dinucleotide in dinucleotides:
    if all_counts.get(dinucleotide, 0) == 2:
        print(dinucleotide)
```

As we can see from the output, this works perfectly well:

```
AA
AT
AC
TG
```

For this example, this approach works because we have a list of the dinucleotides already written as part of the program. Most of the time when we create a dict, however, we'll do it using some other method which doesn't require an explicit list of the keys. For example, here's a different way to generate a dict of dinucleotide counts which uses two nested `for` loops to enumerate all the possible dinucleotides:

---

1   Strictly speaking, in this example there's no need to build a dict at all – we could just check the count and print a line if it's equal to two – but most programs that use dicts will be a bit more complex.

```
dna = "AATGATGAACGAC"
bases = ['A','T','G','C']
all_counts = {}
for base1 in bases:
    for base2 in bases:
        dinucleotide = base1 + base2
        count = dna.count(dinucleotide)
        if count > 0:
            all_counts[dinucleotide] = count
```

loops_dinucleotides.py

The resulting dict is just the same as in our previous examples, but because we haven't got a list of dinucleotides handy, we have to take a different approach to find all the dinucleotides where the count is two. Fortunately, the information we need – the list of dinucleotides that occur at least once – is stored in the dict as the keys.

Iterating over keys

When used on a dict, the `keys()` method returns a list of all the keys in the dict:

```
print(all_counts.keys())
```

Looking at the output[1] confirms that this is the list of dinucleotides we want to consider (remember that we're looking for dinucleotides with a count of two, so we don't need to consider ones that aren't in the dict as we already know that they have a count of zero):

```
['AA', 'AC', 'CG', 'AT', 'GA', 'TG']
```

To find all the dinucleotides that occur exactly twice in the DNA sequence we can take the output of `keys()` and iterate over it, keeping the body of the loop the same as before:

---

1   If you're using Python 3 you might see slightly different output here, but all the code examples will work just the same.

```
for dinucleotide in all_counts.keys():
    if all_counts.get(dinucleotide) == 2:
        print(dinucleotide)
```

This version prints exactly the same set of dinucleotides as the approach that used our list:

```
AA
AC
AT
TG
```

Before we move on, take a moment to compare the output immediately above this paragraph with the output from the version that used the list from earlier in this section. You'll notice that while the **set of dinucleotides** is the same, the **order in which they appear** is different. This illustrates an important point about dicts – they are *inherently unordered*. That means that when we use the `keys()` method to iterate over a dict, we can't rely on processing the items in the same order that we added them. This is in contrast to lists, which always maintain the same order when looping. If we want to control the order in which keys are printed we can use the `sorted()` function to sort the list before processing it:

```
for dinucleotide in sorted(all_counts.keys()):
    if all_counts.get(dinucleotide) == 2:
        print(dinucleotide)
```

Iterating over items

In the example code above, the first thing we need to do inside the loop is to look up the value for the current key. This is a very common pattern when iterating over dicts – so common, in fact, that Python has a special shorthand for it. Instead of doing this:

```
for key in my_dict.keys():
    value = my_dict.get(key)
    # do something with key and value
```

We can use the `items()` method to iterate over pairs of data, rather than just keys:

```
for key, value in my_dict.items():
    # do something with key and value
```

The `items()` method does something slightly different from all the other methods we've seen so far in this book; rather than returning a **single value**, or a **list of values**, it returns a **list of pairs of values**[1]. That's why we have to give two variable names at the start of the loop. Here's how we can use the `items()` method to process our dict of dinucleotide counts just like before:

```
for dinucleotide, count in all_counts.items():
    if count == 2:
        print(dinucleotide)
```

This method is generally preferred for iterating over items in a dict, as it is very readable.

Lookup vs. iteration

Before we finish this chapter; a word of warning: don't make the mistake of iterating over all the items in a dict in order to look up a single value. Imagine we want to look up the number of times the dinculeotide AT occurs in our example above. It's tempting to use the `items()` method to write a loop that looks at each item in the dict until we find the one we're looking for:

---

1   Each pair is actually a *tuple* – see the chapter on complex data structures in *Advanced Python for Biologists* for a full explanation.

```
for dinucleotide, count in all_counts.items():
    if dinucleotide == 'AT':
        print(count)
```

and this will work, but it's completely unnecessary (and slow). Instead, simply use the `get()` method to ask for the value associated with the key you want:

```
print(all_counts.get('AT'))
```

Recap

We started this chapter by examining the problem of storing paired data in Python. After looking at a couple of unsatisfactory ways to do it using tools that we've already learned about, we introduced a new type of data structure – the dict – which offers a much nicer solution to the problem of storing paired data.

Later in the chapter, we saw that the real benefit of using dicts is the efficient lookup they provide. We saw how to create dicts and manipulate the items in them, and several different ways to look up values for known keys. We also saw how to iterate over all the items in dictionary.

In the process, we uncovered a few restrictions on what dicts are capable of – we're only allowed to use a couple of different data types for keys, they must be unique, and we can't rely on their order. Just as a physical dictionary allows us to rapidly look up the definition for a word but not the other way round, Python dictionaries allow us to rapidly look up the value associated with a key, but not the reverse.

Exercises

DNA translation

Here's a dict which represents the genetic code – the keys are codons and the values are amino acid residues:

```
gencode = {
    'ATA':'I', 'ATC':'I', 'ATT':'I', 'ATG':'M',
    ...
    'TGC':'C', 'TGT':'C', 'TGA':'_', 'TGG':'W'}
```

The full version of this dict is in a file called *genetic_code.txt* in the Chapter_8 exercise folder.

Use this dict to write a program which will translate a DNA sequence into protein. You'll have to figure out how to:

- split the DNA sequence into codons

- look up the amino acid residue for each codon

- join all the amino acids to give a protein

Test your program on a couple of different inputs to see what happens. How does your program cope with a sequence whose length is not a multiple of 3? How does it cope with a sequence that contains unknown bases?

Solutions

DNA translation

The description of this exercise is very short, but it hides quite a bit of complexity! To translate a DNA sequence we need to carry out a number of different steps. First, we have to split up the sequence into codons. Then, we need to go through each codon and translate it into the corresponding amino acid residue. Finally, we need to create a protein sequence by adding all the amino acid residues together.

We'll start off by figuring out how to split a DNA sequence into codons. Because this exercise is quite tricky, we'll pick a very short test DNA sequence to work on – just three codons:

```
dna = "ATGTTCGGT"
```

How are we going to split up the DNA sequence into groups of three bases? It's tempting to try to use the `split()` method, but remember that the `split()` method only works if the things you want to split are separated by a delimiter. In our case, there's nothing separating the codons, so `split()` will not help us.

Something that might be able to help us is substring notation. We know that this allows us to extract part of a string, so we can do something like this:

```
dna = "ATGTTCGGT"
codon1 = dna[0:3]
codon2 = dna[3:6]
codon3 = dna[6:9]
print(codon1, codon2, codon3)
```

As we can see from the output, this works:

```
('ATG', 'TTC', 'GGT')
```

but it's not a great solution, as we have to fill in the numbers manually. Since the numbers follow a very predictable pattern, it should be possible to generate them automatically. The start position for each substring is initially zero, then goes up by three for each successive codon. The stop position is just the start position plus three.

Recall that the job of the `range()` function is to generate sequences of numbers. In order to generate the sequence of substring start positions, we need to use the three-argument version of `range()`, where the first argument is the number to start at, the second argument is the number to finish at, and the third argument is the step size. For our DNA sequence above, the number to start at is zero, and the step size is three. The number to finish at it not six but seven, because ranges are exclusive at the finish. This bit of code shows how we can use the `range()` function to generate the list of start positions:

```
for start in range(0,7,3):
    print(start)
```

```
0
3
6
```

To find the stop position for a given start position we just add three, so we can easily split our DNA into codons using a loop:

```
dna = "ATGTTCGGT"
for start in range(0,7,3):
    codon = dna[start:start+3]
    print("one codon is" + codon)
```

```
one codon is ATG
one codon is TTC
one codon is GGT
```

Now that we know how to split a DNA sequence up into codons, let's turn our attention to the problem of translating those codons. Given the dict from the exercise file, we can look up the amino acid for a given codon using either of the two methods that we learned about:

```
print(gencode['CAT'])
print(gencode.get('GTC'))
```

```
H
V
```

If we look up the amino acid for each codon inside the loop of our original code, we can print both the codon and the amino acid translation[1]:

```
dna = "ATGTTCGGT"
for start in range(0, len(dna), 3):
    codon = dna[start:start+3]
    aa = gencode.get(codon)
    print("one codon is " + codon)
    print("the amino acid is " + aa)
```

```
one codon is ATG
the amino acid is M
one codon is TTC
the amino acid is F
one codon is GGT
the amino acid is G
```

This is starting to look promising. The final step is to actually do something with the amino acid residues rather than just printing them. A nice idea is to take our cue from the way that a ribosome behaves and add each new amino acid residue onto the end of a protein to create a gradually growing string:

---

1   From now on, we won't include the statement which creates the dictionary in our code samples as it takes up too much room, so if you want to try running these yourself you'll need to add it back at the top.

```
dna = "ATGTTCGGT"

protein = ""❶
for start in range(0,len(dna),3):
    codon = dna[start:start+3]
    aa = gencode.get(codon)
    protein = protein + aa❷

print("protein sequence is " + protein)
```

In the above code, we create a new variable to hold the protein sequence immediately before we start the loop❶, then add a single character onto the end of that variable each time round the loop❷. By the time we finish the loop, we have built up the complete protein sequence and we can print it out:

```
protein sequence is MFG
```

This looks like a very useful bit of code, so let's turn it into a function. Our function will take one argument – the DNA sequence as a string – and will return a string containing the protein sequence[1]:

```
def translate_dna(dna):

    protein = ""
    for start in range(0,len(dna),3):
        codon = dna[start:start+3]
        aa = gencode.get(codon)
        protein = protein + aa
    return protein
```

We can now call our function by passing in a DNA sequence, and getting back a protein sequence:

---

1   You'll notice that this function relies on the gencode variable which is defined outside the function – something that I told you not to do in chapter 5. This is an exception to the rule: defining the gencode variable inside the function means that it would have to be created anew each time we wanted to translate a DNA sequence.

198

```
print(translate_dna("ATGTTCGGT"))
```

```
MFG
```

Now that we have a working function, we can address the last few bits of the problem. What happens when we call our function using a sequence whose length isn't a multiple of three?

```
print(translate_dna("ATGTTCGGTA"))
```

We get an error message:

```
TypeError: cannot concatenate 'str' and 'NoneType' objects
```

In order to see why this happens, we need to look at the individual codons, which we can do by adding a `print()` statement to our function:

```
def translate_dna(dna):

    protein = ""
    for start in range(0,len(dna),3):
        codon = dna[start:start+3]
        print("codon is " + codon)
        aa = gencode.get(codon)
        protein = protein + aa
    return protein

print(translate_dna("ATGTTCGGTA"))
```

Now the output shows the problem:

```
codon is ATG
codon is TTC
codon is GGT
codon is A
```

We have an incomplete codon at the end of our sequence. Clearly we need to modify the second argument to `range()` – the position to finish the sequence of numbers – in order to take into account the length of the DNA sequence. At this point, we have to confront the problem of what to do if we're given a DNA sequence whose length is not an exact multiple of three. Clearly, we cannot translate an incomplete codon, so we want the start position of the final codon to equal to the length of the DNA sequence minus two. This guarantees that there will always be two more characters following the position of the final codon start – i.e. enough for a complete codon.

Here's the modified version of our function, in which we calculate the start position of the last valid codon and store it in a variable❶ which we can later use as the argument to `range()`❷:

```
def translate_dna(dna):
    last_codon_start = len(dna) - 2❶
    protein = ""
    for start in range(0,last_codon_start,3):❷
        codon = dna[start:start+3]
        aa = gencode.get(codon)
        protein = protein + aa
    return protein
```

Now we can translate sequences of any length:

```
print(translate_dna("ATGTTCGGT"))
print(translate_dna("ATGTTCGGTA"))
print(translate_dna("ATGTTCGGTAA"))
print(translate_dna("ATGTTCGGTAAA"))
```

and our function will ignore the extra bases on the end until there's enough for one full extra codon:

```
MFG
MFG
MFG
MFGK
```

Finally, let's see what happens when we throw in an undetermined base:

```
print(translate_dna("ATGTTNCGGT"))
```

Again, we get an error:

```
TypeError: cannot concatenate 'str' and 'NoneType' objects
```

because Python can't find a value in the dict for the second codon **TTN**. How should we fix this? We could add an `if` statement to the function which only translates the DNA sequence if it doesn't contain any unambiguous bases, but that seems a little too conservative – there are plenty of situations in which we might want to generate a protein sequence for a DNA sequence that has unknown bases. We could add an `if` statement inside the loop which only translates a given codon if it doesn't contain any ambiguous bases, but that would lead to protein translations of an incorrect length – we know that the codon **TTN** will translate to an amino acid, we just don't know which one it will be.

The most sensible solution seems to be to translate any codon with an unknown base into the symbol for an unknown amino acid residue, which is **X**. The optional second argument to the `get()` function makes it very easy:

```
def translate_dna(dna):
    last_codon_start = len(dna) - 2
    protein = ""
    for start in range(0,last_codon_start,3):
        codon = dna[start:start+3]
        aa = gencode.get(codon, 'X')
        protein = protein + aa
    return protein
```

dna_translation.py

Let's test it on three different sequences:

```
# input sequence is easy
print(translate_dna("ATGTTCGGT"))

# input sequence has incomplete codons at the end
print(translate_dna("ATCGATCGAT"))

# input sequence contains N
print(translate_dna("ACGANCGAT"))
```

```
MFG
IDR
TXD
```

Now that we know our function works for any input DNA sequence, we can turn these three test sequences into assert statements to make a small test suite:

```
# input sequence is easy
assert(translate_dna("ATGTTCGGT")) == "MFG"

# input sequence has incomplete codons at the end
assert(translate_dna("ATCGATCGAT")) == "IDR"

# input sequence contains N
assert(translate_dna("ACGANCGAT")) == "TXD"
```

dna_translation.py

What have we learned?

In this exercise we've implemented one of the core algorithms of bioinformatics. Many of the tools that we rely on for sequence analysis involve DNA to protein translation somewhere in the pipeline. The genetic code is a particularly good example of data that can be represented in a dict. Codons are unique (though amino acids may not be) and we want to look up amino acids based on codon, not the other way around, so the dict is a good fit.

The techniques that we've practised in this exercise will be useful whenever we need to look up one piece of information based on another. The complications that we encountered – in particular, how to deal with keys that are not in the dict – are likely to occur in many different programming situations.

9: Files, programs, and user input

File contents and manipulation

Reading from and writing to files was one of the first things we looked at in this book, back in chapter 3. For some programs, however, we're not just concerned with the contents of files, but with files and folders themselves. This is especially likely to be the case for programs that have to operate as part of a work flow involving other tools and software. For example, we may need to copy, move, rename and delete files, or we may need to process all files in a certain folder.

Although it seems like a simple task (after all, the file manager tools that come with your operating system can carry most of them out), file manipulation in a language like Python is actually quite tricky. That's because the code that we write has to function identically on different operating systems – including Windows, Linux and Mac machines – which may handle files quite differently.

Thankfully, Python includes a couple of modules[1] that take care of these differences for us and provide us with a set of useful functions for manipulating files. The modules' names are `os` (short for Operating System) and `shutil` (short for SHell UTILities). In the next section we'll see how they can be used to carry out various common (but important) tasks.

A note on the code examples

Since the code examples in this chapter unavoidably involve interaction with the operating system, some of the details will be operating system specific. In other words, the examples in this chapter might not run on your computer without some tweaking. In addition to the paths being different (refer back to chapter 3), the success of the code examples for many functions relies on the files and folders actually being present on the computer on which the examples are run. The code examples in this chapter will use Linux style paths, and will refer to folders and

---

1   Take a look back at chapter 7 for a reminder of how modules work.

files on my computer, so if you want to try running them, you'll probably need to change the paths to refer to files on your own computer.

Basic file manipulation

To rename an existing file, we import the `os` module, then use the `os.rename()` function. The `os.rename()` function takes two arguments, both strings. The first is the current name of the file, the second is the new name:

```
import os
os.rename("old.txt", "new.txt")
```

The above code assumes that the file *old.txt* is in the folder where we are running our Python program (also known as the *working directory*). If it's elsewhere in the filesystem, then we have to give the complete path:

```
os.rename("/home/martin/biology/old.txt", "/home/martin/biology/new.txt")
```

If we specify a different folder, but the same filename in the second argument, then the function will move the file from one folder to another:

```
os.rename("/home/martin/biology/old.txt", "/home/martin/python/old.txt")
```

Of course, we can move and rename a file in one step if you like:

```
os.rename("/home/martin/biology/old.txt", "/home/martin/python/new.txt")
```

`os.rename()` works on folders as well as files:

```
os.rename("/home/martin/old_folder", "/home/martin/new_folder")
```

If we try to move a file to a folder that doesn't exist we'll get an error. We need to create the new folder first with the `os.mkdir()` function:

```
os.mkdir("/home/martin/python")
```

If we need to create a bunch of directories all in one go, we can use the `os.mkdirs()` function (note the s on the end of the name):

```
os.mkdirs("/a/long/path/with/lots/of/folders")
```

To copy a file or folder we use the `shutil` module. We can copy a single file with `shutil.copy()`:

```
shutil.copy("/home/martin/original.txt", "/home/martin/copy.txt")
```

or a folder with `shutil.copytree()`:

```
shutil.copytree("/home/martin/original_folder",
"/home/martin/copy_folder")
```

To test whether a file or folder exists, use `os.path.exists()`. This function returns `True` or `False`, so we can use it as a condition:

```
if os.path.exists("/home/martin/email.txt"):
    print("You have mail!")
```

Deleting files and folders

There are different functions for deleting files, empty folders, and non-empty folders. To delete a single file, use `os.remove()`:

```
os.remove("/home/martin/unwanted_file.txt")
```

To delete an empty folder, use `os.rmdir()`:

```
os.rmdir("/home/martin/empty")
```

To delete a folder and all the files in it, use `shutil.rmtree()`:

```
shutil.rmtree("home/martin/full")
```

Listing folder contents

The `os.listdir()` function returns a list of files and folders. It takes a single argument which is a string containing the path of the folder whose contents you want to search. To get a list of the contents of the current working directory, use the string "." for the path:

```
for file_name in os.listdir("."):
    print("one file name is " + file_name)
```

To list the contents of a different folder, we just give the path as an argument:

```
for file_name in os.listdir("/home/martin"):
    print("one file name is " + file_name)
```

Running external programs

Another feature of Python that involves interaction with the operating system is the ability to run external programs. Just like file and folder manipulation, the ability to run other programs is very useful when using Python as part of a work flow. It allows us to use existing tools that would be very time consuming to recreate in Python, or that would run very slowly.

Running external programs from within your Python code can be a tricky business, and this feature wouldn't normally be covered in an introductory programming course. However, it's so useful for biology (and science in general) that we're going to cover it here, albeit in a simplified form.

As with the above section on file operations, the exact details of how external programs are run will vary with your operating system and the way your computer is set up. On UNIX-based systems, the program that you want to run might already be in your path, in which case you can simply use the name of the executable as the string to be executed. On Windows, you'll probably have to supply the full path to the program you want to run.

Running a program

The functions for running external program reside in the `subprocess()` module. The reasoning behind the name is slightly convoluted: when talking about operating systems, a running program is called a *process*, and a process that is started by another process is called a *subprocess*.

To run an external program, use the `subprocess.call()` function. This function takes a single string argument containing the path to the executable you want to run:

```
import subprocess
subprocess.call("/bin/date")
```

Any output that is produced by the external program is printed straight to the screen – in this case, the output from the Linux `date` program:

```
Fri Jul 26 15:15:26 BST 2013
```

If we want to supply command line options to the external program then we just include them in the string, and set the optional `shell` argument to `True`. Here we call the Linux `date` program with the options which cause it to just print the month:

```
subprocess.call("/bin/date +%B", shell=True)
```

```
July
```

Capturing program output

Often, we want to run some external program and then store the output in a variable so that we can do something useful with it. For this, we use `subprocess.check_output()`, which takes exactly the same arguments as `subprocess.call()`:

```
current_month = subprocess.check_output("/bin/date +%B", shell=True)
```

Just like when reading file contents, the output from an external program can run over multiples lines, so the string that's returned by `subprocess.check_output()` might contain multiple lines separated by newline characters.

User input makes our programs more flexible

The exercises and examples that we've seen so far in this book have used two different ways of getting data into a program. For small bits of data, like short DNA sequences, restriction enzyme motifs, and gene accession names, we've simply stored the data directly in a variable like this:

```
dna = "ATCGATCGTGACTAGCTACG"
```

When data is mixed in with the code in this manner, it is said to be *hard-coded*.

For larger pieces of data, like longer DNA sequences and tabular data, we've typically read the information from an external text file. For many purposes, this is a better solution than hard-coding the data, as it allows the separation of data and code, making our programs easier to read. However, in all the examples we've seen so far, the *names* of the files from which the data are read are still hard-coded.

Both of these approaches to getting data into our program have the same shortcomings – if we want to change the input data, we have to open the code and edit it. In the case of hard-coded variables, we have to edit the statement where the variables are created. In the case of files, we have two choices – we can either edit the contents of the file, or edit the hard-coded filename.

Real life useful programs don't generally work that way. Instead, they allow us to specify input files and options at the time when we run the program, rather than when we're writing it. This allows programs to be much more flexible and easier to use, especially for a person who didn't write the code in the first place.

In the next couple of sections we're going to see tools for getting user input, but more importantly we're going to talk about the transition from writing a program that's only useful to you, to writing one that can be used by other people. This involves starting to think about the experience of using a program from the perspective of a user.

There are many reasons why you might need your programs to be usable by somebody who's not familiar with the code. If you write a program that solves a problem for you, chances are that it could solve a problem for your colleagues and collaborators as well. If you write a program that forms a significant part of a piece of work which you later want to publish, you may have to make sure that whoever is peer reviewing your paper can get your program working as well. Of course, making your program easier to use for other people means that it will also be easier to use for you, a few months after you have written it when you have completely forgotten how the code works!

Interactive user input

To get interactive input from the user in our programs, we can use the `raw_input()` function[1]. `raw_input()` takes a single string argument, which is the prompt to be displayed to the user, and returns the value typed in as a string:

```
accession = input("Enter the accession name")
# do something with the accession variable
```

---

[1]   In Python 3, this function is called `input()`.

The `raw_input()` function behaves a little differently to other functions and methods we've seen, because it has to wait for something to happen before it can return a value – the user has to type in a string and press enter. The user input will be returned as a string (so if we need to use is as something else – e.g. a number – we'll have to do the conversion manually) and will end with a newline (so we might want to use `rstrip()` to remove it).

Capturing user input in this way requires us to think quite carefully about how our program behaves. Programs that we write to carry out analysis of large datasets will often take a considerable amount of time to run, so it's important that we minimize the chances of the user having to rerun them. When using the `raw_input()` function, there are two situations in particular that we want to avoid.

One is the situation where we have a long-running program that requires some user input, but doesn't make this fact clear to the user. What can happen in this scenario is that the user starts the program running and then switches their attention to something else, assuming that the program will continue to make progress in the background. If the user doesn't notice (or is not at their computer) when the program reaches the point where it requires input and halts, the program may be stuck waiting for input for a long time.

The other scenario to avoid is where a program runs for some time before asking the user for input, then fails to work due to an incorrect input or typo, requiring the user to restart the program from scratch.

A good way to avoid both of these problems is to design our programs so that they collect all necessary user input at the start, before any long-running tasks are carried out. We can also reduce the chances of incorrect input on the part of the user by offering clear instructions and documentation.

An important part of user input is *input validation* – checking that the input supplied by the user makes sense. For example, you might require that a particular input is a number between some minimum and maximum values, or that it's a DNA sequence without ambiguous bases, or that it's the name of a file that must exist. A good strategy for input validation is to check the input as soon as it's received, and give the user a second chance to enter their input if it's found to be invalid. Here's an example which uses a `while` loop to give the user multiple attemps to enter a number between 1 and 10:

```
answer = ""
while answer < 1 or answer > 10:
    answer = int(raw_input("enter a number between 1 and 10\n"))
print("final answer is " + str(answer))
```

validate_input.py

A better way to do this is to use Python exception system: see the chapter on exceptions in *Advanced Python for Biologists* for details.

One big drawback of getting user input interactively is that it makes it harder to run a program unsupervised as part of a work flow. For most biological analyses, specifying program options when it's run using command line arguments is a better approach.

Command line arguments

If you're used to using existing programs that have a command line user interface (as opposed to a graphical one) then you're probably familiar with command line arguments[1]. These are the strings that you type on the command line after the name of a program you want to run:

```
myprogram one two three
```

In the above command line, **one two** and **three** are the command line options. To use command line arguments in our Python scripts, we import the `sys` module. We can then access the command line arguments by using the special list `sys.argv`. Running the following code:

---

1   Not to be confused with the arguments that we pass to functions, although they do a similar job.

212

```
import sys
print(sys.argv)
```

with the command line:

```
python myprogram.py one two three
```

shows how the elements of `sys.argv` are made up of the arguments given on the command line:

```
['myprogram.py', 'one', 'two', 'three']
```

Note that the **first** element of `sys.argv` is always the name of the program itself, so the first command line argument is at index one, the second at index two, etc.

Just like with `raw_input()`, options and filenames given on the command line are stored as strings so if, for example, we want to use a command line argument as a number, we'll have to convert it with `int()`.

Command line arguments are a good way of getting input for your Python programs for a number of reasons. All the data your program needs will be present at the start of your program, so you can do any necessary input validation (like checking that files are present) before starting any processing. Also, your program will be able to be run as part of a shell script, and the options will appear in the user's shell history.

Recap

We started this chapter by examining two features of Python that allow your programs to interact with the operating system – file manipulation and external processes. We learned which functions to use for common file system operations, and which modules they belong to. We also saw two ways to call external programs from within your Python program.

When using these techniques to solve real life problems, or when working on the exercises, remember that you may encounter errors that are nothing to do with your program. For instance, when trying to rename a file you may get an error if a specified file doesn't exist or you don't have the necessary permissions to rename it. Similarly, if you get unexpected output when running an external program the problem may lie with the external program or with the way that you're calling it, rather than with your Python program. This is in contrast to the rest of the exercises in this book, which are mostly self contained. If you run into difficulties when using the tools in this chapter, check the external factors as well as checking your program code.

In the last portion of the chapter, we saw two different ways to get user input when your program runs. Using command line arguments is generally better for the type of programming that forms part of scientific research.

Exercises

In the *chapter_9* folder in the exercises download there is a collection of files with the extension *.dna* which contain DNA sequences of varying length, one per line. Use this set of files for both exercises.

Binning DNA sequences

Write a program which creates nine new folders – one for sequences between 100 and 199 bases long, one for sequences between 200 and 299 bases long, etc. Write out each DNA sequence in the input files to a separate file in the appropriate folder.

Your program will have to:

- iterate over the files in the folder
- iterate over the lines in each file
- figure out which bin each DNA sequence should go in based on its length
- write out each DNA sequence to a new file in the right folder

Kmer counting

Write a program that will calculate the number of all kmers of a given length across all DNA sequences in the input files and display just the ones that occur more than a given number of times. You program should take two command line arguments – the kmer length, and the cutoff number.

Solutions

Binning DNA sequences

The first job is to figure out how to read all the DNA sequences. We can get a list of all the files in the folder by using `os.listdir()`, but we'll have to be careful to only read DNA sequences from files that have the right filename extension. Here's a bit of code to start off with:

```python
import os

# for each file in the current folder...
for file_name in os.listdir("."):

    # ...only process it if contains DNA sequences
    if file_name.endswith(".dna"):
        print("reading sequences from " + file_name)
```

We can check the output to make sure that we're only going to process the correct files:

```
reading sequences from xag.dna
reading sequences from xaj.dna
reading sequences from xaa.dna
reading sequences from xab.dna
reading sequences from xai.dna
reading sequences from xae.dna
reading sequences from xah.dna
reading sequences from xaf.dna
reading sequences from xac.dna
reading sequences from xad.dna
```

The next step is to read the DNA sequences from each file. For each file that passes the name test, we'll open it❶, then process it one line at a time❷ and calculate the length of the DNA sequence❸:

```
# look at each file
for file_name in os.listdir("."):
    if file_name.endswith(".dna"):
        print("reading sequences from " + file_name)
        dna_file = open(file_name) ❶

        # look at each line
        for line in dna_file: ❷
            dna = line.rstrip("\n")
            length = len(dna) ❸
            print("found a dna sequence with length " + str(length))
```

Notice how we've used `rstrip()` to remove the newline character – we don't want to include it in the count of the sequence length, since it's not a base. With ten files, and ten DNA sequences per file, this program generates over a hundred lines of output – here are the first few:

```
reading sequences from xag.dna
found a dna sequence with length 432
found a dna sequence with length 818
found a dna sequence with length 604
found a dna sequence with length 879
found a dna sequence with length 619
found a dna sequence with length 500
found a dna sequence with length 119
found a dna sequence with length 341
found a dna sequence with length 303
found a dna sequence with length 469
reading sequences from xaj.dna
found a dna sequence with length 121
found a dna sequence with length 442
found a dna sequence with length 520
```

This looks good – we're getting a range of different sizes. Next we have to figure out which bin each of the sequences should go in. Because the limits of the bins follow a regular pattern, we can use the `range()` function to generate them. We can generate a list of the lower limits for each bin by taking a range of numbers from 100 to 1000 with a step size of 100, then adding 99 to get the upper limit of the bin. We'll go through this process for each sequence, checking if it belongs in each bin in turn:

```python
# go through each file in the folder
for file_name in os.listdir("."):❶

  # check if it ends with .dna
  if file_name.endswith(".dna"):❷
    print("reading sequences from " + file_name)

    # open the file and process each line
    dna_file = open(file_name)
    for line in dna_file:❸

      # calculate the sequence length
      dna = line.rstrip("\n")
      length = len(dna)
      print("sequence length is " + str(length))

      # go through each bin and check if the sequence belongs in it
      for bin_lower in range(100,1000,100):❹
        bin_upper = bin_lower + 99
        if length >= bin_lower and length <= bin_upper:❺
          print("bin is " + str(bin_lower) + " to " + str(bin_upper))
```

There are quite a few levels of indentation in the above code, so you might have to read it through a few times. We have

- the loop for each filename❶

- the **if** statement that checks the filename❷

- the loop for each sequence in a file❸

- the loop for each bin❹

- the if statement that checks if the sequence belongs in the bin❺

Having so many levels of indentation can make the code hard to read, so let's split it into a couple of functions. We'll take all the code that deals with a single sequence and turn it into a function:

```
def process_sequence(line):
    dna = line.rstrip("\n")
    length = len(dna)
    print("sequence length is " + str(length))
    for bin_lower in range(100,1000,100):
        bin_upper = bin_lower + 99
        if length >= bin_lower and length < bin_upper:
            print("bin is " + str(bin_lower) + " to " + str(bin_upper))

for file_name in os.listdir("."):
    if file_name.endswith(".dna"):
        print("reading sequences from " + file_name)
        dna_file = open(file_name)
        for line in dna_file:
            process_sequence(line)
```

and now the code is much easier to comprehend. The first few lines of the output
show that this approach works:

```
reading sequences from xag.dna
sequence length is 432
bin is 400 to 499
sequence length is 818
bin is 800 to 899
sequence length is 604
bin is 600 to 699
sequence length is 879
bin is 800 to 899
sequence length is 619
bin is 600 to 699
sequence length is 500
bin is 500 to 599
sequence length is 119
```

The final step is to create the new folders, and write each DNA sequence to the
appropriate one. We can reuse our `range()` idea to generate the folder names
and create them. The name of the folder for a given bin is the lower limit,
followed by an underscore, followed by the upper limit:

```
for bin_lower in range(100,1000,100):
    bin_upper = bin_lower + 99
    bin_folder_name = str(bin_lower) + "_" + str(bin_upper)
    os.mkdir(bin_folder_name)
```

When we want to write out DNA sequence to a file in a particular folder, we can use the same naming scheme to work out the name of the folder. Of course, we also have to figure out what to call the individual files. The exercise description didn't specify any kind of naming scheme, so we'll keep things simple and store the first DNA sequence in a file called *1.dna*, the second in a file called *2.dna*, etc. We'll need to create an extra variable to hold the number of DNA sequences we've seen, and to increment it after writing each DNA sequence. This sequence number will have to be passed to the `process_sequence()` function as an extra argument.

Here's the whole program – it's by far the largest one that we've written so far:

```
for bin_lower in range(100,1000,100):
    bin_upper = bin_lower + 99
    bin_folder_name = str(bin_lower) + "_" + str(bin_upper)
    os.mkdir(bin_folder_name)

def process_sequence(line, number):
    dna = line.rstrip("\n")
    length = len(dna)
    print("sequence length is " + str(length))
    for bin_lower in range(100,1000,100):
        bin_upper = bin_lower + 99
        if length >= bin_lower and length < bin_upper:
            print("bin is " + str(bin_lower) + " to " + str(bin_upper))
            bin_folder_name = str(bin_lower) + "_" + str(bin_upper)
            output_path = bin_folder_name + '/' + str(seq_number) + '.dna'
            output = open(output_path, "w")
            output.write(dna)
            output.close()

seq_number = 1
for file_name in os.listdir("."):
    if file_name.endswith(".dna"):
        print("reading sequences from " + file_name)
        dna_file = open(file_name)
        for line in dna_file:
            process_sequence(line, seq_number)
            seq_number = seq_number+1
```

binning_dna_sequences.py

Kmer counting

To come up with a plan of attack for this exercise, we must first think about the order in which we process the data. Can we simply read a single DNA sequence, count the kmers, and print the counts like we did for the dinucleotide example in chapter 8? Unfortunately the answer is no, because we only want to print the kmers which occur more than a given number of times across **all** sequences. In other words, we don't know which kmers we want to print the counts for until we have finished processing all the DNA sequences.

So, we will have to tackle this problem in two stages. First, we will go through each sequence one by one and gradually build up a list of kmer counts. Second, we will go through the list of counts and print only the ones whose count is above the cutoff.

How will we generate the kmer counts? A good first step would be to figure out how to split a DNA sequence into overlapping kmers of any given length. We can use a similar approach to the one in the DNA translation exercise in chapter 8: use the `range()` function to generate a list of the start positions of each kmer, then use substring notation to extract the kmer from the sequence. Here's a bit of code that prints all kmers of a given size. We'll use a short test DNA sequence for now:

```
test_dna = "ACTGTAGCTGTACGTAGC"
print(test_dna)
kmer_size = 4
for start in range(0,len(test_dna) - kmer_size + 1,1):
    kmer = test_dna[start:start+kmer_size]
    print(kmer)
```

The tricky bit is figuring out the arguments to the `range()` function. We know that we want to start at zero and increase by one each time. The finish position is the length of the sequence, minus the kmer size (to make sure there is one kmer's worth of bases after it) plus one (to allow for the fact that the finish position is exclusive). The `range()` function generates the start positions for each kmer, and to get the end positions we just add the kmer size. We can examine the output from this code and check that it agrees with our intuition:

```
ACTGTAGCTGTACGTAGC
ACTG
CTGT
TGTA
GTAG
TAGC
AGCT
GCTG
CTGT
TGTA
GTAC
TACG
ACGT
CGTA
GTAG
TAGC
```

To make it easier to test this bit of code, we'll turn it into a function. The function will take two arguments. The first argument will be the DNA sequence as a string, and the second argument will be the kmer size as a number. Instead of printing the list of kmers, it will return a list of them. Here's the code for the function and three statements to test it:

```
def split_dna(dna, kmer_size):
    kmers = []
    for start in range(0,len(dna) - kmer_size + 1,1):
        kmer = dna[start:start+kmer_size]
        kmers.append(kmer)
    return kmers


print(split_dna("AATGCTGCAT", 4))
print(split_dna("AATGCTGCAT", 5))
print(split_dna("AATGCTGCAT", 6))
```

As we can see from the output, running the function multiple times with the **same** DNA sequence but **different** kmer lengths gives different results, as expected:

```
['AATG', 'ATGC', 'TGCT', 'GCTG', 'CTGC', 'TGCA', 'GCAT']
['AATGC', 'ATGCT', 'TGCTG', 'GCTGC', 'CTGCA', 'TGCAT']
['AATGCT', 'ATGCTG', 'TGCTGC', 'GCTGCA', 'CTGCAT']
```

Now we can put this function together with the code we developed for looping through files from the previous exercise. To count the kmers, we will create an empty dict before we start processing the DNA sequences❶. Then for each kmer we find, we will look up the current count for it in the dict❷. If the k-mer is not found in the dictionary (i.e. this is the first time we've seen that particular k-mer) then we will say that the current count is zero. We'll then add one to the current count❸ and store the result back in the dictionary❹.

```python
import os
kmer_size = 6

def split_dna(dna, kmer_size):
    kmers = []
    for start in range(0,len(dna)-(kmer_size-1),1):
        kmer = dna[start:start+kmer_size]
        kmers.append(kmer)
    return kmers

kmer_counts = {}❶
for file_name in os.listdir("."):
    if file_name.endswith(".dna"):
        print("reading sequences from " + file_name)
        dna_file = open(file_name)
        for line in dna_file:
            dna = line.rstrip("\n")
            for kmer in split_dna(dna, kmer_size):
                current_count = kmer_counts.get(kmer, 0)❷
                new_count = current_count + 1❸
                kmer_counts[kmer] = new_count❹

print(kmer_counts)
```

This program generates a lot of output! Here are the first few lines so we can see that it's working:

```
{'gcagag': 11, 'aaataa': 13, 'ctttag': 11, 'gcagac': 14, 'ctttaa': 12,
'gcagaa': 15 ... }
```

As planned, we end up with a large dict where the keys are kmers and the values are their counts.

Next, we have to process the `kmer_counts` dictionary. We'll go through the items in a loop, and if the count is greater than some cutoff, we'll print the count. For testing, we'll fix the cutoff at 23 (later on we'll make this a command line option). Here's the code to process the dict:

```
count_cutoff = 23
for kmer, count in kmer_counts.items():
    if count > count_cutoff:
        print(kmer + " : " + str(count))
```

And here's the output we get:

```
agagat : 26
agcggg : 26
atcgga : 25
aaggag : 25
cccagc : 24
aggttc : 25
agatta : 24
tctagg : 24
gagtgg : 28
ccggtt : 26
gagcag : 24
ttctga : 26
agatgg : 24
tctgaa : 24
gcgggt : 25
ttcaaa : 25
gattaa : 25
ccagcg : 25
ggacgt : 27
atggct : 24
```

Nearly done. The final step is to replace the hard-coded values for the kmer size and the count cutoff with values read from the command line. We just have to import the `sys` module, and convert the arguments to numbers using the `int()` function. As specified in the exercise description, the first command line argument is the kmer size and the second is the cutoff. Here's the final code with comments:

```
import os
import sys

# convert command line arguments to variables
kmer_size = int(sys.argv[1])
count_cutoff = int(sys.argv[2])

# define the function to split dna
def split_dna(dna, kmer_size):
    kmers = []
    for start in range(0,len(dna)-(kmer_size-1),1):
        kmer = dna[start:start+kmer_size]
        kmers.append(kmer)
    return kmers

# create an empty dictionary to hold the counts
kmer_counts = {}

# process each file with the right name
for file_name in os.listdir("."):
    if file_name.endswith(".dna"):
        dna_file = open(file_name)

        # process each DNA sequence in a file
        for line in dna_file:
            dna = line.rstrip("\n")

            # increase the count for each k-mer that we find
            for kmer in split_dna(dna, kmer_size):
                current_count = kmer_counts.get(kmer, 0)
                new_count = current_count + 1
                kmer_counts[kmer] = new_count

# print k-mers whose counts are above the cutoff
for kmer, count in kmer_counts.items():
    if count > count_cutoff:
        print(kmer + " : " + str(count))
```

kmer_counting.py

Now we can specify the kmer length on the command line when we run the program, rather than having to edit the code each time we want to change it. With a kmer length of 6 and a cutoff of 25:

```
python kmer_counting.py 6 25
```

```
we get the output
agagat : 26
agcggg : 26
gagtgg : 28
ccggtt : 26
ttctga : 26
ggacgt : 27
```

With a kmer length of 3 and a cutoff of 900:

```
python kmer_counting.py 3 900
```

```
tct : 908
ttc : 924
gtt : 905
gat : 904
gga : 910
atc : 905
```

What have we learned?

These two exercises have been about something that's very common in scientific work: taking a collection of data spread over multiple files and processing it in some way. A lot of the logic that we have covered – including processing all files in a folder, and including or excluding files based on their name – will be useful when solving real world problems.

The second exercise has a particularly interesting twist: producing the output has to wait until all the input sequences have been processed. In your day to day research, you'll probably enounter both types of scenarios.

Afterword

This is the end of *Python for Biologists*; I hope you have enjoyed the book, and found it useful. If you've reached the end of the book without doing all the exercises, then I strongly recommend that you go back at some point and do them.

Where you go from here depends on your goals. If this book has been your first introduction to Python, then I have three suggestions for further study.

Firstly, explore the features of the Python **language** that we haven't had space to cover in this book. Python's object system, functional programming tools, comprehensions and exception handling are all interesting and useful topics, which are covered in detail in *Advanced Python for Biologists* by the same author.

Secondly, explore the **development tools**. Python has a wealth of tools designed to speed up and ease development of programs – things like automated testing, packaging, performance tuning, etc.

Thirdly, be aware of other people in your lab, institute, university or company who are using Python. Discussing programmin problems with other people is an excellent way to improve your skills.

Finally, remember that if you have any comments on the book – good or bad – I'd love to hear them; drop me an email at

martin@pythonforcompletebeginners.com

If you've found the book useful, please also consider leaving a Amazon **review**. You're now part of the Python community, and these reviews will help other people to find the book, and hopefully make learning Python a bit easier for everyone.

Index

Made in the USA
Lexington, KY
11 August 2016